EUROPEAN MINERALOGICAL UNION
NOTES IN MINERALOGY

Series editor: Giovanni Ferraris

Volume 11

LAYERED MINERAL STRUCTURES AND THEIR APPLICATION IN ADVANCED TECHNOLOGIES

UNIVERSITY TEXTBOOK

Edited by
M.F. Brigatti and A. Mottana

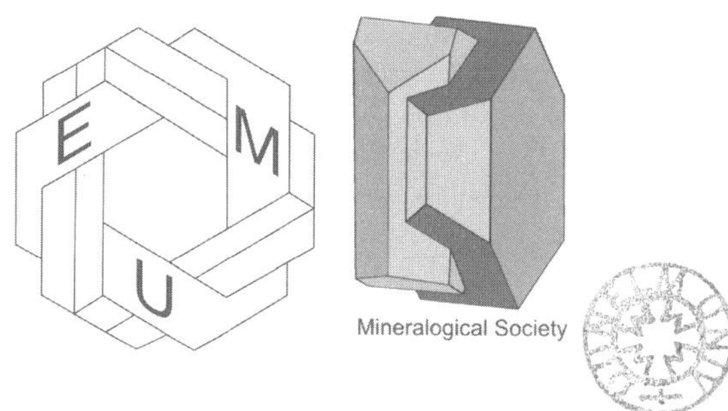

Published by the European Mineralogical Union and the
Mineralogical Society of Great Britain & Ireland
London, 2011

The publication of this textbook and the organization of the related school were sponsored by the European Mineralogical Union

EMU Notes in Mineralogy

A series published under the auspices of the European Mineralogical Union (EMU) in connection with the *EMU Schools meetings*.

Initiator of the EMU Schools and the EMU Notes in Mineralogy
Giovanni Ferraris, Torino, President of the EMU 1992–1996

Series Editor: Giovanni Ferraris, Torino (previous editors: Tamás G. Weiszburg and Gábor Papp, Budapest)

Editor of this Volume
M.F. Brigatti, Modena and Reggio Emilia University, and A. Mottana University of Rome TRE

Technical Editor and Indexer: Kevin Murphy, London

Front cover design: Michel H. Guay

On the front cover: (Background) AFM image of cleaved muscovite. (Foreground) Polyhedral structure of muscovite on (001); changes in the Si K-edge XANES spectrum of a muscovite crystal flake resulting from changing its orientation against the horizontally polarized synchrotron radiation beam from orthogonal to almost vertical; precession photograph of clintonite.

ISSN: 1417 2917
ISBN: 978-0-903056-29-8

Published by the European Mineralogical Union and the Mineralogical Society of Great Britain & Ireland (12, Baylis Mews, Amyand Park Road, Twickenham TW1 3HQ, UK)
Printed by Burlington, Cambridge, UK

The EMU Notes in Mineralogy Series

Published volumes

Volume	*Year*	*Editors*	*Title*
1	1997	S. Merlino	*Modular aspects of minerals*
2	2000	D.J. Vaughan R. Wogelius	*Environmental mineralogy*
3	2001	C.A. Geiger	*Solid solutions in silicate and oxide systems*
4	2002	C. Gramaccioli	*Energy modelling in minerals*
5	2003	D.A. Carswell R. Compagnoni	*Ultrahigh pressure metamorphism*
6	2004	A. Beran E. Libowitzky	*Spectroscopic methods in mineralogy*
7	2005	R. Miletich	*Mineral behaviour at extreme conditions*
8	2010	F.E. Brenker G. Jordan	*Nanoscopic approaches in Earth and Planetary Sciences*
9	2010	G.E. Christidis	*Advances in the characterization of Industrial minerals*
10	2010	M. Prieto	*Ion-partitioning in ambient-temperature aqueous systems*
11	2011	M.F. Brigatti A. Mottana	*Layered mineral structures and their application in advanced technologies*

Copies of the EMU Notes (volumes 1–7) are distributed in Europe by the larger member societies of the European Mineralogical Union:

Società Italiana di Mineralogia e Petrologia:
www.socminpet.it

Mineralogical Society of Great Britain & Ireland:
www.minersoc.org

Société Française de Minéralogie et de Cristallographie:
www.sfmc-fr.org

and by the Secretary of the EMU, Prof. Herta S. Effenberger:
www.univie.ac.at/Mineralogie/EMU

in America by the Mineralogical Society of America:
www.minsocam.org

Forthcoming volumes

12	2012	D.J. Vaughan R. Wogelius	*Environmental Mineralogy*
13	2012	J. Dubessy	*Raman spectroscopy and its applications to Earth Sciences and cultural heritage*
14	2013	F. Nieto	*Electron microscopy and nanoscale phenomena in minerals*

Institutional orders as well as individual requests from outside Europe and America should be sent also be sent to the the Secretary of the EMU, Prof. Herta S. Effenberger: www.univie.ac.at/Mineralogie/EMU

For volumes 8 onwards, the Mineralogical Society of Great Britain acts as co-publisher and copies may be ordered from www.minersoc.org

or from:
Mineralogical Society
12 Baylis Mews, Amyand Park Road
Twickenham TW1 3HQ
UK
E-mail: admin@minersoc.org
Tel. +44 (0)20 8891 6600
Fax: +44 (0)20 8891 6599

Contents

Preface XIII

Foreword XV

Chapter 1. Structure and mineralogy of layer silicates: recent perspectives and new trends
by Maria Franca **BRIGATTI**, Daniele **MALFERRARI**, Angela **LAURORA** and Chiara **ELMI** **1**

1. Introduction 1
2. Layer silicates: general structural overview 2
3. Mixed-layer structures and order-disorder in the layer stacking 8
   3.1. Mixed-layer structures 8
   3.2. Order-disorder of the layer stacking 9
4. The *cis*-vacant and the *trans*-vacant octahedral site 11
5. Layer charge 12
6. The 1:1 layer structure: kaolin-serpentine group 14
   6.1. Kaolin subgroup 14
      6.1.1. Kaolinite 15
      6.1.2. Dickite 19
      6.1.3. Nacrite 20
      6.1.4. Halloysite 21
      6.1.5. Hisingerite 23
   6.2. Odinite subgroup 23
   6.3. Serpentine subgroup 23
      6.3.1. Mg-rich species 24
      6.3.2. Fe-rich species 28
      6.3.3. Mn-rich species 29
      6.3.4. Ni-rich species 31
7. 2:1 layer structures 31
   7.1. Talc and pyrophyllite group 31
      7.1.1. Pyrophyllite and ferripyrophyllite 31
      7.1.2. Talc and talc-like minerals 32
   7.2. The mica group: some recent advances in crystal chemistry and structure of dioctahedral and trioctahedral micas 33
   7.3. The smectite group 35
   7.4. The vermiculite group 39
   7.5. The chlorite group 40
      7.5.1. Trioctahedral chlorites 43
      7.5.2. Di,trioctahedral and dioctahedral chlorites 44
   7.6. Some 2:1 layer silicates involving a discontinuous octahedral sheet and a modulated tetrahedral sheet 45
8. Imogolite and allophane 47

Acknowledgements 48

References 49

Chapter 2. An Overview of Order/Disorder in Hydrous Phyllosilicates
by STEPHEN GUGGENHEIM 73

1. Introduction 73
1.1. Order/disorder 74
1.2. Thermal disorder considerations 74
1.3. The idealized phyllosilicate model 75
1.4. Common structural distortions relating to the polyhedra and sheets 78
1.5. Common structural distortions relating to site substitutions 79
2. Layer charge, solid solution and exsolution 80
2.1. Layer charge and bulk structures 80
2.2. Solid solution and exsolution 81
2.2.1. Serpentine and kaolin 83
2.2.2. Talc and pyrophyllite 83
2.2.3. Mica 84
2.2.4. Chlorite 86
2.2.5. Smectite and vermiculite 86
2.3. Layer charge and surfaces 87
2.3.1. Cleaved surfaces: chemistry and reactivity of the interface 89
2.3.2. Confined surfaces: intercalation under controlled relative humidity in vermiculite 90
3. Stacking 91
3.1. Mica 91
3.2. Planar trioctahedral 1:1 layers 94
3.3. Planar 1:1 layers and kaolin: vacant octahedral sites and stacking 96
3.4. Chlorite 97
3.5. Vermiculite 98
3.6. Turbostratic (disorder) effects: smectite 99
3.7. Talc and pyrophyllite 99
3.8. Interstratifications, including intercalations 100
3.8.1. Reichweite 101
3.8.2. Exchange and solvation in swelling clays 102
4. Order and disorder effects on diffraction 102
4.1. Order/disorder in stacking 102
4.2. General diffraction characteristics 102
4.2.1. Turbostratic effects 104
4.2.2. Diffraction effects of interstratifications: rational and irrational effects 105
5. Modulated phyllosilicates, polysomatic structures, and cylindrical structures 107
5.1. Overview 107
5.2. Modulated 2:1 phyllosilicates involving a continuous octahedral sheet 107
5.3. Modulated 2:1 phyllosilicates involving a discontinuous octahedral sheet 109

5.4. Modulated 1:1 phyllosilicates 109
5.5. Rolled, tubular, cylindrical, curved 1:1 structures 112
5.6. Polysomes, order/disorder and diffraction effects 113
Acknowledgements 114
References 115

Chapter 3. Layered titanosilicates
by **ZHI LIN, FILIPE A. ALMEIDA PAZ and JOÃO ROCHA 123**
1. Introduction 123
2. Construction of layered titanosilicates 124
2.1. Building blocks 124
2.2. Layer construction 125
3. Representative structures 126
3.1. AM-1: $Na_4[Ti_2Si_8O_{22}]{\cdot}4H_2O$ 126
3.2. Lintisite-type structures 128
3.3. Natisite: $Na_2[TiSiO_5]$ 130
3.4. Jonesite: $Ba_2(K,Na)[Ti_2(Si_5Al)O_{18}(H_2O)]{\cdot}nH_2O$ 131
3.5. Lamprophyllite: $(Sr,Ba)_2[Na_3Ti_3O_2(Si_2O_7)_2(OH)_2]$ 132
3.6. Delindeite: $Ba_2(Na,K,\square)_3(Ti,Fe)[Ti_2(O,OH)_4Si_4O_{14}](H_2O,OH)_2$ 134
3.7. Heterophyllosilicates 136
3.7.1. Bafertisite-type 136
3.7.2. Astrophyllite- and Nafertisite-type 139
4. Preparation methods 141
5. Applications 142
6. Conclusions and outlook 144
Acknowledgements 145
References 145

Chapter 4. Modelling of X-ray diffraction profiles: Investigation of defective lamellar structure crystal chemistry
by **BRUNO LANSON 151**
1. Introduction 151
2. General background on structure defects and induced diffraction effects 153
2.1. Different structure defects 153
2.1.1. Local defects 153
2.1.2. Stacking defects 153
2.1.3. Interstratification 154
2.2. Interpretation of diffraction data from disordered solids 157
2.2.1. Crystals exhibiting local defects and/or random stacking faults 157
2.2.2. Well defined stacking defects/interstratification 158
2.3. XRD identification of defective structures 160
2.3.1. Calculation of XRD intensity from mixed layers 160

2.3.2. Usual identification methods ... 164
2.3.3. Multi-specimen method for XRD identification of mixed layers ... 166
3. Structural characterization of defective layered structures ... 168
3.1. Layered silicates: structural characterization of mixed layers the elementary layers of which differ from their basal spacings ... 168
3.1.1. Layered silicates: hydration and expansion heterogeneity, intercalation ... 168
3.1.2. Natural occurrence of multi-component mixed layers ... 171
3.1.3. Additional complexity of naturally occurring mixed layers ... 173
3.1.4. Recent developments and new insights into the actual structure of mixed layers: structure of elementary layers ... 173
3.1.5. Recent developments and new insights into the actual structure of mixed layers: intra-crystalline defects ... 176
3.1.6. Recent developments and new insights into the actual structure of mixed layers: outer surfaces of crystals ... 177
3.2. Natural layered silicates: interstratification of structural fragments differing in terms of their internal structure or their stacking mode ... 177
3.3. Interstratification of other lamellar structures: layered double hydroxides ... 180
3.4. Interstratification of other lamellar structures: layered oxides ... 183
3.4.1. Interstratification of commensurate polytype fragments ... 183
3.4.2. Interstratification of incommensurate fragments ... 185
3.5. Structural characterization of disordered layered structures ... 186
3.5.1. Layered silicates ... 187
3.5.2. Layered oxides ... 187
4. Conclusions ... 189
References ... 190

Chapter 5. Applications of computational atomistic methods to phyllosilicates
by C. IGNACIO SAINZ-DÍAZ ... 203
1. Introduction ... 203
2. Computational mineralogy ... 204
3. Methods of data management for analysis of experiments ... 205
4. Atomistic methods based on classic mechanics ... 206
5. Quantum mechanical methods ... 208
5.1. Theoretical approaches ... 209
5.2. Molecular cluster models ... 211
5.3. Periodical models for crystalline solids ... 212
6. Modelling tools ... 212
6.1. Minimization of geometry ... 212
6.2. Monte Carlo simulations ... 213

6.3. Molecular-dynamics simulations 213
6.4. Vibration spectra 214
7. Applications in phyllosilicates 215
7.1. Ordering in cation substitutions in phyllosilicates 215
7.2. *Cis*-vacant/*trans*-vacant polymorphism in dioctahedral phyllosilicates 223
7.3. Spectroscopic properties of clay minerals 225
7.4. Dehydroxylation-rehydroxylation of phyllosilicates 227
7.5. Adsorption phenomena on phyllosilicate surfaces 229
7.6. Effect of pressure 229
8. Future perspectives and conclusions 230
Acknowledgements 230
References 231

Chapter 6. The concept of layer charge of smectites and its implications for important smectite-water properties
by **George E. CHRISTIDIS 237**
1. Introduction 237
2. Structural features and layer charge of the smectites 239
3. Methods of determining layer charge 243
3.1. General characteristics 243
3.2. The structural formula method (SFM) 244
3.3. The alkylammonium method (AMM) 245
3.4. The K-saturation method (KSM) 246
4. Important smectite-water properties related to layer charge 247
4.1. Influence of layer charge on ion exchange 247
4.2. Influence of layer charge on swelling and viscosity 248
5. Fundamental particles, quasicrystals, aggregates and the concept of fundamental-particle charge 251
5.1. Fundamental particles, quasicrystals and aggregates 251
5.2. The concept of fundamental-particle charge 253
Acknowledgements 254
References 254

Chapter 7. Intercalation processes of layered minerals
by **Faïza BERGAYA and Gerhard LAGALY 259**
1. Introduction 259
1.1. Definitions 260
1.2. Chapter outline 260
2. Layered clay minerals: variable geometry, porosity and surfaces 261
3. Intercalation in non-swelling minerals 264
3.1. Kaolinite: a non-swelling natural clay mineral 264
3.1.1. Direct intercalation in kaolinite 265

3.1.2. Indirect intercalation in kaolinite by displacement reactions 266
3.2. Micas: natural non-swelling minerals 266
4. Intercalation in swelling clay minerals 267
4.1. Solvation by ion dipole and hydrogen-bonding interactions 267
4.2. Ion exchange 268
4.2.1. Exchange selectivity between inorganic cations 268
4.2.2. Organo-clay minerals (OC) 268
4.3. Protonation 269
4.4. Complexation 270
4.5. Electron transfer and redox reactions 270
4.6. Grafting 271
4.7. Co-intercalation of two compounds 271
5. Adsorption and intercalation 272
6. From intercalated clay minerals to pillared clay minerals (PILC) 273
7. Clay mineral polymer nanocomposites (CPN) 275
7.1. Intercalation from solvents 276
7.2. Melt intercalation 276
8. Intercalated *vs.* exfoliated structures 277
8.1. Intercalated structures 277
8.2. Exfoliated structures 277
9. Applications of intercalation/deintercalation processes 278
10. Conclusion 279
Acknowledgements 279
References 279

Chapter 8. Advanced techniques to define intercalation processes
by ANNIBALE **MOTTANA and** LUCA **ALDEGA 285**
1. Introduction 285
2. X-ray diffraction 287
2.1. Introduction 287
2.2. XRD measurement of crystal thickness 288
2.3. Grazing incidence X-ray diffraction 292
3. X-ray absorption spectroscopy 294
3.1. Preparing the sample 294
3.2. Choosing the strategy for recording the experimental spectrum 296
3.3. Recording and optimizing the experimental spectrum 296
3.4. Interpreting the normalized spectrum 297
3.5. Analysis and extraction of information from the spectral regions 298
3.5.1. Pre-edge 298
3.5.2. XANES 300
3.5.3 EXAFS 303
3.6. Conclusions 304

4. Practical examples 304
4.1. The structure of the interlayer of phyllosilicates (*e.g.* clays, illites, micas) 304
4.2. Graphene and its derivative structures 306
5. Conclusions..... 307
Acknowledgements..... 308
References..... 308

Chapter 9. Interaction of organic molecules with layer silicates, oxides and hydroxides and related surface-nano-characterization techniques
by GIOVANNI VALDRÈ, DANIELE MORO and GIANFRANCO ULIAN..... 313
1. Introduction 314
2. Nanoscale surface-imaging techniques of layer silicates, oxides and hydroxides 315
2.1. Nanotopography and cleavage patterns 315
2.2. Surface potential and related properties 318
2.3. Nanomechanical properties 321
3. Interaction of organic molecules with mineral and synthetic substrates 325
3.1. Layer silicates 325
3.2. Oxides, hydroxides and synthetic compounds 328
4. Comparison between experimental data and simulation..... 329
References..... 332

Chapter 10. The surface properties of clay minerals
by ROBERT A. SCHOONHEYDT and CLIFF T. JOHNSTON 335
1. Introduction 335
2. Clay mineral–water interactions 338
2.1 Origin of clay mineral–water interactions 338
2.2. Nanoconfined H_2O molecules in clay-mineral interlayers..... 342
2.3. Interlayer cations..... 343
2.4. Influence of clay mineral–water–cation interactions on clay-mineral structures 343
3. Transition metal ion complexes..... 345
3.1. Planar surfaces like planar complexes 345
3.2. Chiral clay minerals 347
4. Dyes on clay-mineral surfaces: from spectroscopy to optical materials 353
4.1 Spectroscopy of clay mineral–methylene blue systems..... 353
4.2 Organization of cationic dyes on the surface of clay minerals: towards optical materials 357
5. Amino acids and proteins on clay-mineral surfaces 360
6. Conclusions..... 361
Acknowledgments 362
References..... 362

Index 371

Preface

This volume covers the topics related to the 13th EMU School '*Layered Mineral Structures and their Application in Advanced Technologies*'. All of the selected topics, the school, and this volume are thus aimed at providing an in-depth knowledge of the complex field of layered materials, with an attempt to address several fundamental aspects, which range from crystal chemistry and structure to layer packing disorder, from surface properties to the description of the most advanced experimental techniques useful in the characterization of layered materials.

Layered materials, because of their particular atomic arrangement, are commonly characterized by physical and chemical properties of great interest in numerous technological and environmental processes and applications, as better detailed in the body of this volume. Most of these properties play a significant role in Earth sciences, environmental sciences, technology, biotechnology, material sciences and many other scientific areas. The surface properties of layered materials control important interaction processes, such as coagulation, aggregation, sedimentation, filtration, catalysis and ionic transport in porous media. Layered minerals also control many aspects of Earth's rheology, *i.e.* the movement of geological masses, at least as far down as the lower crust. Given this frameset, it should be no surprise that the extremely large field of investigation of these materials can, and in most of the cases must, be approached from several different viewpoints. However, providing full coverage of the immense literature devoted to all the topics above may be impractical, if not impossible. Nevertheless, providing our students, to whom this book is addressed, with fundamental knowledge of different disciplines and providing examples demonstrating the application of these foundations in their daily research, is feasible and certainly useful.

As previously suggested, this book was developed as training material for the 13th EMU school taking place in Rome. It is our duty and honour to acknowledge all the institutions and individuals that made the organization of the school and the production of this volume possible. Special thanks to Accademia Nazionale dei Lincei, which hosted the school, and to Associazione Amici dei Lincei, always keen to provide high-quality learning opportunities for students and researchers. Both of these institutions are always ready to sponsor and support cultural events, such as our school, in Rome, which contribute to the dissemination of knowledge and help to advance scientific research. The school was also supported by the European Community, which granted an IP Erasmus project to support attendance at the school by students and teachers. Further support was granted by Modena and Reggio Emilia University (Department of Earth Sciences) mainly in the areas of administration and management, and by the University of Rome TRE (Department of Geology), mainly by providing presentation tools and printing material. The Istituto Nazionale di Fisica Nucleare (INFN), and in particular Dr Augusto Marcelli, helped enormously in the organization of workshops. Active support was also provided by Bruker Optics S.r.l. (Milan) and CECOM S.r.l. (Rome).

The editors are grateful to all the referees who helped, with their reviews and editing, to make the chapters of this volume sound, and Prof. Giovanni Ferraris and Mr Kevin Murphy for their efforts in improving the volume. In particular, the editors are grateful to Dr Chiara Elmi and Dr Angela Laurora for their help in formatting the book chapters. They also thank the European Crystallographic Association and Associazione Italiana di Cristallografia for supporting those students attending the school from outside the European Community, as well as Fondazione Cassa di Risparmio di Modena, for helping with travel costs for the lecturers from outside the EU.

Last, but not least, the editors express their warmest thanks to all the contributors to this volume: their efforts and their clarity were essential in helping the students to gain a good understanding of the material being presented.

Maria Franca Brigatti
Annibale Mottana

Foreword

This new volume in the European Mineralogical Union (EMU) *Notes in Mineralogy* series is based on lectures delivered by scientists during the 2011 EMU school '*Layered Mineral Structures and their Application in Advanced Technologies*', organized by Maria Franca Brigatti (University of Modena and Reggio Emilia) and Annibale Mottana (University of Roma Tre). The volume was also edited by Professors Brigatti and Mottana.

There are three main reasons why this school holds particular appeal, and the EMU Executive Committee is very grateful to the organizing team for their efforts.

The first reason is the interdisciplinary nature of the topic, which uses modern approaches to fundamental mineralogy, such as the characterization of structure-function relations in mineral families, with its implications for and applications to technological issues.

The second reason is the support of the 2011 school by an Erasmus IP grant assigned to a consortium of 21 European Universities and coordinated by the University of Modena and Reggio Emilia. Support within the Erasmus IP framework had been a tradition in EMU schools but abandoned in the recent past. Such support provides a good opportunity to help students from many countries to meet colleagues and teachers, open their minds to research in a European context and to create collaborative links which may prove important in their later careers. Of course, the Erasmus IP grants also highlight the work of the EMU.

The third and perhaps most important reason is the collaboration with the Accademia Nazionale dei Lincei (Italy), the oldest scientific academy in the world, which has promoted excellence in scientific studies since 1603. The EMU is grateful for and honoured by the hospitality of the Accademia, and is proud to have held its 2011 school in such a prestigious location, benefiting from the high-quality services of the Accademia, including its guest house. Special thanks to the association Amici dei Lincei which helped enormously with the logistics for the school.

Starting from 2012, the EMU Notes in Mineralogy will welcome proposals for volumes which originate from workshops or other initiatives not necessarily related to EMU schools. The 'Notes' will be also distributed in electronic form from the website/online bookshop of the Mineralogical Society of Great Britain and Ireland (www.minersoc.org), our excellent new partner (in terms of the book series) since 2009. We hope that this exciting development will help to promote the methods and expertise of mineralogical research to a wider audience.

Giovanni Ferraris
Series Editor

Roberta Oberti
EMU President

EMU Notes in Mineralogy, Vol. 11 (2011), Chapter 1, 1–71

Structure and mineralogy of layer silicates: recent perspectives and new trends

MARIA FRANCA BRIGATTI, DANIELE MALFERRARI, ANGELA LAURORA and CHIARA ELMI

Dipartimento di Scienze della Terra, Università di Modena e Reggio Emilia, Largo S. Eufemia 19, I-41121 Modena, Italy
e-mail: mariafranca.brigatti@unimore.it; daniele.malferrari@unimore.it; angela.laurora@unimore.it; chiara.elmi@unimore.it

Because of their many novel and advanced applications, there is increasing interest in layer silicates from the scientific and technical communities. Appropriate application of these minerals requires deep understanding of their properties and of the processes where they are involved. This chapter, by providing fundamental definitions and crystal structural and chemical data pertaining to layer silicates, aims to introduce this field to new researchers and technicians, by describing the fundamental features leading to different behaviours of layer silicates in different natural or technical processes. The subject addressed is vast and so the reader is referred in some cases to work already published. The focus here is on layer silicates for which detailed crystal structures are given in the literature and which are likely to be used in an applied way in the future. Layer-silicate minerals fulfilling these requirements are: (1) kaolin-serpentine group (*e.g.* kaolinite, dickite, nacrite, halloysite, hisingerite, odinite, lizardite, berthierine, amesite, cronstedtite, nepouite, kellyite, fraipontite, brindleyite, guidottiite, bementite, greenalite, caryopilite; minerals of the pyrosmalite series); (2) talc and pyrophyllite groups (*e.g.* pyrophyllite, ferripyrophyllite, willemseite); (3) mica group (*i.e.* some recent advances in crystal chemistry and structure of dioctahedral and trioctahedral micas); (4) smectite group (*e.g.* montmorillonite, saponite, hectorite, sauconite, stevensite, swinefordite); (5) vermiculite group; (6) chlorite group (*e.g.* trioctahedral chlorite such as clinochlore, di,trioctahedral and dioctahedral chlorites such as cookeite and sudoite); (7) some 2:1 layer silicates involving a discontinuous octahedral sheet and a modulated tetrahedral sheet such as kalifersite, palygorskite and sepiolite; and (8) imogolite and allophane.

1. Introduction

There is great interest in the mineralogy, crystallography and potential novel technological applications of layer silicates, especially of those showing nano-sized dimensions, as recently reviewed by Bergaya *et al.* (2006). An increasing number of scientific publications devoted to such subjects is not limited to mineralogical, petrological and geochemical fields, but also includes other disciplines, such as Applied Chemistry, Environmental Sciences, Food and Soil Science, Ceramics, Cultural Heritage, Applied Physics, Engineering, Water Science and Biology. As a result, new possibilities and new priorities have arisen in research on layer silicates. Multi-competence and multidisciplinary approaches need to be supported and well integrated. Knowledge,

DOI: 10.1180/EMU-notes.11.1

down to an atomic level, of the mineral bulk and surface features and of the mineral interaction with its surrounding environment needs to be detailed to a level which allows the formulation of predictive models. The latter need to support better the existing and novel technological applications, which to date are largely based on phenomenological, empirical or trial-and-error approaches.

This chapter attempts to address both priorities: firstly, by introducing the fundamental aspects of layer silicates in terms of their structure and crystal chemistry and how these are related directly to mineral properties and to mineral interactions with the surrounding environment; secondly, by reviewing more recent findings and innovative experimental techniques, especially where they might contribute to novel or future applications, trying, at the same time, to overlap as little as possible, with recent, general purpose reviews on layer silicates (*e.g.* see Moore & Reynolds, 1997; Giese & van Oss, 2002; Mottana *et al.*, 2002a; Fleet, 2003; Meunier, 2005; Bergaya *et al.*, 2006; Deer *et al.*, 2009).

2. Layer silicates: general structural overview

The layer silicates considered in this chapter are commonly characterized by a continuous tetrahedral sheet (Fig. 1a). In the tetrahedral (T) sheet, individual TO_4 tetrahedra are interconnected, by sharing three corners each (*i.e.* the basal oxygen atoms, O_b), to form an infinite two-dimensional 'hexagonal' mesh pattern along the **a** and **b** crystallographic directions. Common tetrahedral cations are Si^{4+}, Al^{3+} and Fe^{3+}, but Be, B, Ga and Ge are also documented. The fourth oxygen atom (*i.e.* the apical oxygen atom, O_a) forms a corner of the octahedral coordination unit around the octahedral cations. In the octahedral sheet, each octahedron (M) connects to its neighbouring octahedra by sharing edges. Edge-shared octahedra form sheets of hexagonal or pseudo-hexagonal symmetry (Fig. 1b). Octahedral cations are usually Al^{3+}, Fe^{3+}, Mg^{2+}, Fe^{2+}, but other cations, such as Li^{+}, Mn^{2+}, Co^{2+}, Ni^{2+}, Cu^{2+}, Zn^{2+}, V^{3+}, Cr^{3+}, and Ti^{4+} are also observed. Octahedra show two different topologies, depending on the octahedral oxygen atom (O_o) position (*i.e.* the *cis*- and the *trans*-orientations, Fig. 1b). In the *trans*-orientation (M-*trans*), O_o lies across the octahedral diagonal, whereas in the *cis*-octahedra (M-*cis*) the O_o forms a shared edge between two octahedra. Common anions characterizing the O_o position, which lie near to the centre of each tetrahedral 6-fold ring, are O, OH, F, Cl (Brigatti & Guggenheim, 2002).

A continuous octahedral sheet is generated when the free tetrahedral corners (O_a) of all tetrahedra point to the same side of the sheet (Fig. 2a), thus connecting tetrahedral and octahedral sheets along a plane ideally containing O_o also. In some layer silicates, tetrahedral apices point in opposite directions and link octahedral 'ribbons'. These layer silicates are still characterized by a continuous two-dimensional tetrahedral sheet, but do not show any continuous octahedral sheet (Fig. 2b).

A layer is, by definition, the stacking of tetrahedral and octahedral sheets. Two basic stackings are observed: (1) the 1:1 (or TM), which consists of the repetition of one T and one M sheet along the crystallographic **c** direction; and (2) the 2:1, which consists of one M sheet sandwiched between two T sheets (Fig. 2a). Usually, in the 1:1 layer structure,

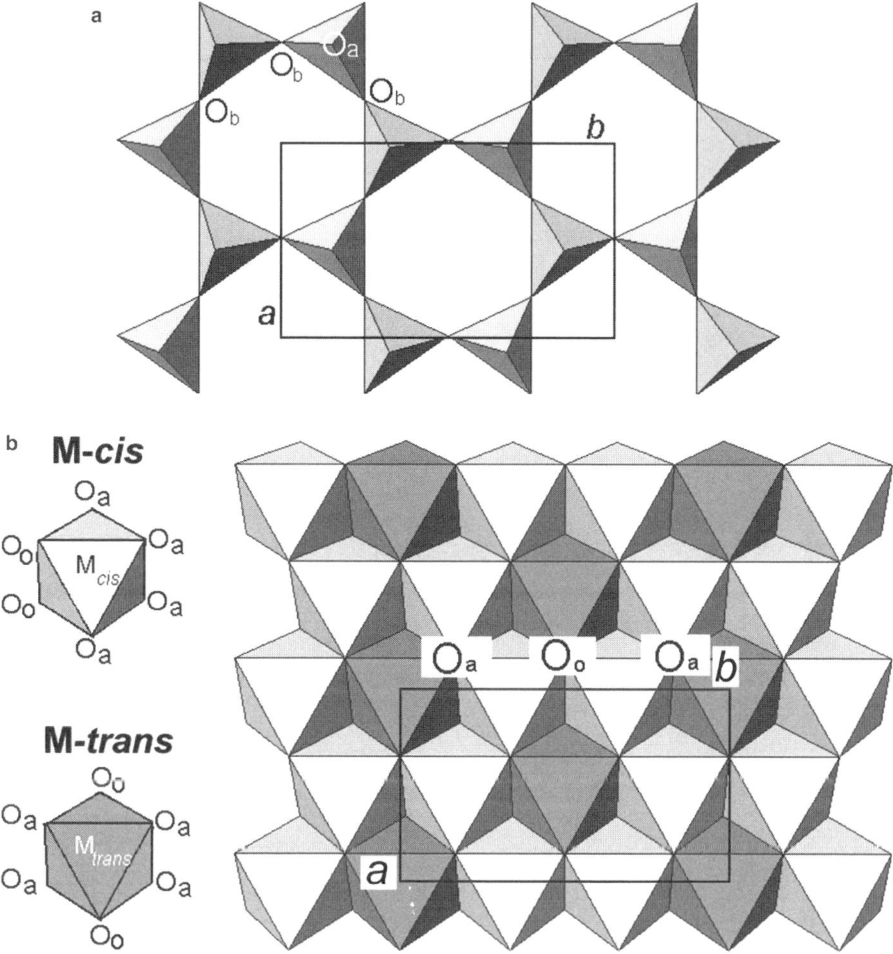

Fig. 1. (***a***) Tetrahedral sheet. O_a and O_b refer to apical and basal oxygen atoms, respectively. *a* and *b* are the unit-cell parameters; (***b***) The octahedral sheet and orientation of the *cis*-octahedron (M-*cis*) and *trans*-octahedron (M-*trans*). In M-*cis* oriented octahedra, two O_o (*i.e.* the octahedral anionic position) lie on an octahedral edge; in M-*trans* oriented octahedra two O_o are placed along the octahedron diagonal. O_a refers to apical oxygen atoms; *a* and *b* are the unit-cell parameters. Readers of the paper version of this chapter may wish to download a colour version of this figure from www.minersoc.org/emu-notes/emu-11/11-1-colour.pdf.

the unit cell includes six M positions (*i.e.* four M-*cis*- and two M-*trans*-oriented octahedra) and four T sites. In the 2:1 layer the unit cell contains six M sites (*i.e.* four M-*cis*- and two M-*trans*-oriented octahedra) and eight T sites. Structures with all six M sites occupied are defined as trioctahedral. If only four of the six M sites are occupied, the structure is dioctahedral. Sometimes terms such as 'brucite-like' and 'gibbsite-like' are, used instead of trioctahedral and dioctahedral sheets, respectively.

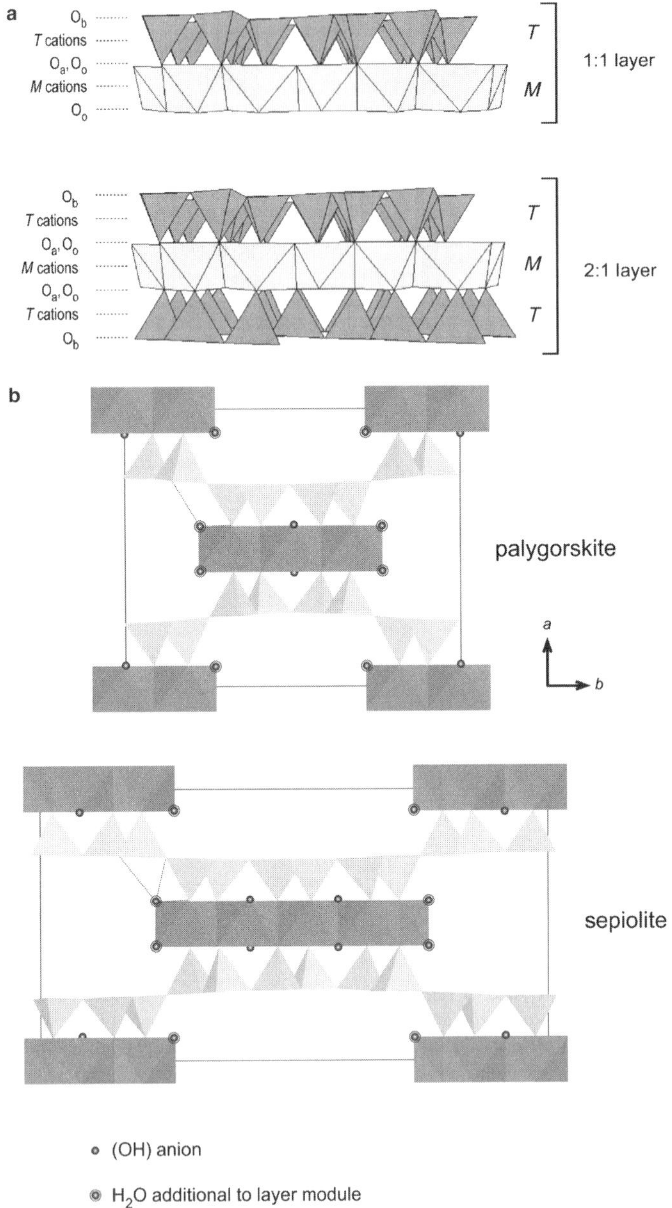

Fig. 2. (***a***) Models of a 1:1 and 2:1 layer structure. O_a, O_b and O_o refer to tetrahedral basal, tetrahedral apical and octahedral anionic positions, respectively. *M* and *T* indicate the octahedral and tetrahedral cations, respectively. (***b***) Modulated layer silicates characterized by a continuous two-dimensional tetrahedral sheet and by a discontinuous octahedral sheet: the schematic structure of palygorskite and sepiolite. Readers of the paper version of this chapter may wish to download a colour version of this figure from www.minersoc.org/emu-notes/emu-11/11-1-colour.pdf.

Two adjacent layers are separated by an interlayer space, which can be empty or occupied. The interlayer space can be occupied by cations, hydrated cations, organic material, hydroxide octahedra, and/or hydroxide octahedral sheets. The periodicity along the **c** axis of each 1:1 layer (usually evaluated by the d_{001} reflection) is $\sim$7 Å; in the 2:1 layers, it varies from 9.1 to 9.5 Å in talc and pyrophyllite, where the interlayer space is empty; it reaches $\sim$10 Å in micas, which present the interlayer occupied by anhydrous interlayer cations and 14 Å in chlorite, where the interlayer consists of octahedrally coordinated cations. In smectite and vermiculite the periodicity along the *c* crystallographic axis reflects the hydration of the interlayer cations. Generally, in smectite, the intercalation of 0, 1, 2 or 3 planes of H_2O molecules in the interlayer space corresponds to d_{001} = 10.0–10.2 Å (dehydrated layers), d_{001} = 11.6–12.9 Å (mono-hydrated layers), d_{001} = 14.9–15.7 Å (bi-hydrated layers), and d_{001} = 18.0–19.0 Å (tri-hydrated layers).

The periodicity along **c*** can vary, depending on polytypic arrangement because of the different number of layers involved in the stacking sequence, and because of different orientations. The theoretical principles of polytypism were reviewed in several works (*e.g.* see Baronnet, 1978; Bailey, 1988a,b; Takeda & Ross, 1995; Ďurovič, 1997, 1999; Nespolo *et al.*, 1997; Nespolo & Ďurovič, 2002) and will not be further discussed in this chapter.

A single layer, in 1:1 layer silicates, shows one negatively charged surface, as consisting entirely of oxygen atoms (O_b) belonging to the T sheet (Fig. 2a), and a positively charged one, when OH groups occupy the O_o position. In 2:1 layer silicates, two-thirds of the octahedral hydroxyl groups are thus replaced by tetrahedral apical oxygen atoms (Fig. 2a), also implying that both surfaces of such a layer are negatively charged, as mostly constituted by tetrahedral O_b.

Allophane and imogolite are commonly referred to in the literature as clay minerals, even though their structural arrangement deviates significantly from that of clays. These minerals are used widely in advanced industrial applications, and share with clay minerals nano-sized dimensions and chemical composition, thus attaining physical properties similar to, or even more pronounced than in clays.

Allophane consists of a 'gibbsite-like' sheet showing hollow spherical morphologies hosting SiO_4 tetrahedra attached to their inner surfaces. Imogolite shows a nano-tube structure. The tubes contain curved gibbsite sheets with silicate groups replacing hydroxyl groups on the inner surface, unlike the outer surfaces which contain $Al(OH)_3$ groups (Fig. 3).

Phyllosilicate layers may present numerous distortions and deviations from the ideal structural arrangement. These structural modifications, which are generally associated with chemical substitutions, could be observed and described in detail only for phyllosilicates with a crystal size suitable for single-crystal X-ray structural refinement. The distortion parameters more often discussed in the literature (mostly for dioctahedral and trioctahedral micas) are: (1) The tetrahedral flattening angle, τ, is derived from: $\tau = \sum_{i=1}^{3} (O_{apical} - T - O_{basal})_i/3$ (Fig. 4a). In an ideal tetrahedron, τ is equal to $\arccos(-1/3) \approx 109.47^\circ$. The τ parameter can deviate from its ideal value as a function of the relative position along **c** for the basal oxygen atoms with respect to the tetrahedral

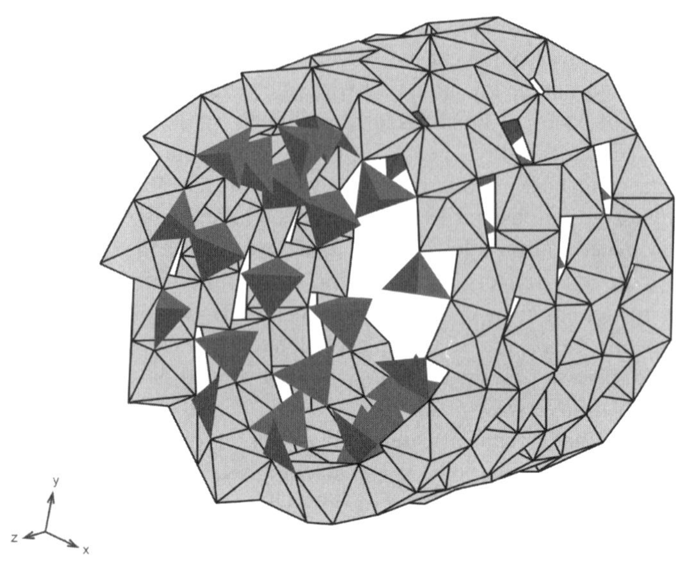

Fig. 3. Model structure of a nano-tube of imogolite. Readers of the paper version of this chapter may wish to download a colour version of this figure from www.minersoc.org/emu-notes/emu-11/11-1-colour.pdf.

cation and with respect to the mean basal-edge length and the mean tetrahedral-edge value (Brigatti & Guggenheim, 2002). (2) The tetrahedral rotation angle, α, is defined according to the following formula $\alpha = \frac{\sum_i^6 |120 - \varphi_i|}{12}$ where φ_i is a generic internal angle of the hexagon defined by basal oxygen atoms (Fig. 4b). Brigatti & Guggenheim (2002) demonstrated that α depends mostly on the ratio between the mean octahedral edges defined by O_a oxygen atoms and the distance defined by tetrahedral O_b: $\alpha = \cos^{-1}\left(\frac{\sqrt{3}}{2} \cdot \frac{\langle O_a - O_a \rangle}{\langle O_b - O_b \rangle}\right)$. Brigatti *et al.* (2003), for trioctahedral mica-1*M* with *C*2/*m* symmetry, demonstrated that the value of α depends mostly on tetrahedral distortions and that a precise estimation of the α value could thus be obtained from: $\cos(\alpha) = \frac{1}{\sqrt{1+3 \cdot [4 \cdot y(O1) - 1]^2}}$, where *y*(O1) represent the *y* atomic coordinate of one of two symmetry-independent O_b. (3) Basal oxygen-plane corrugation, Δz, is calculated from $\Delta z = (zO_{b(max)} - zO_{b(min)}) \times c \times \sin\beta$, where zO_b represents the *z* atomic coordinates of basal oxygen atoms. (4) The octahedral flattening angle, ψ, expresses the flattening of the octahedral sheet (Fig. 4c) and can be computed from: $\psi = \cos^{-1}\left(\frac{\text{octahedral thickness}}{2 \cdot \langle \text{M–O, OH, F, Cl, S} \rangle}\right)$, where the thickness of octahedral sheet is calculated from oxygen *z* coordinates of each M octahedron, including O_o. ⟨M-O, OH, F, Cl, S⟩ indicates the mean octahedral distance (Donnay *et al.*, 1964). (5) The ratio between shared and unshared edges, *eu*/*es*, is the ratio between the mean value of unshared and shared octahedral edges and expresses the distortion of each octahedron (Toraya, 1981). (6) The counter rotation angle, ω, defines the counter-rotation of upper and lower octahedral oxygen triads (Newnham, 1961; Appelo, 1978; Lin & Guggenheim,

1983). (7) The effective coordination number of the interlayer cation, ECoN, describes the effective coordination number of the octahedral cation. This parameter was derived for micas by Weiss *et al.* (1992) starting from the equation introduced by Hoppe (1979): $\mathrm{ECoN} = \sum_{j=1}^{j=12} C_j$, where $C_j = \exp[1.0 - (\mathrm{FIR_j/MEFIR})^6]$, $\mathrm{FIR_j}$ is calculated by dividing the interlayer cation (A) – basal oxygen ($O_{b,j}$) distances by the sum of anion and cation radii and then multiplying them by the cation radii; MEFIR is a weighted mean of FIR, *i.e.* $\mathrm{MEFIR} = \frac{\sum_{j=1}^{j=12} W_j \mathrm{FIR_j}}{\sum_{j=1}^{j=12} W_j}$, $w_j = \exp[1.0 - (\mathrm{FIR_j/FIR_{min}})^6]$, and $\mathrm{FIR_{min}}$ is the smallest $\mathrm{FIR_j}$ in the interlayer cation coordination. (8) Layer displacement is the displacement between two tetrahedral sheets across the interlayer regions. In 2:1 layer silicates, this parameter is defined as the sum of the intralayer displacement (*i.e.* the distance, ideally $a/3$, between two opposing tetrahedral cations, along the plane normal to **c***) and of interlayer displacement (*i.e.* the distance between the same tetrahedral cation positions along **c***). Both parameters are evaluated from tetrahedral atomic coordinates. (9) The thickness of tetrahedral and octahedral sheets is calculated from oxygen z coordinates of each polyhedron, including the OH group (or else any other anion). The interlayer separation is calculated by considering the tetrahedral basal oxygen z coordinates of adjacent 2:1 layers.

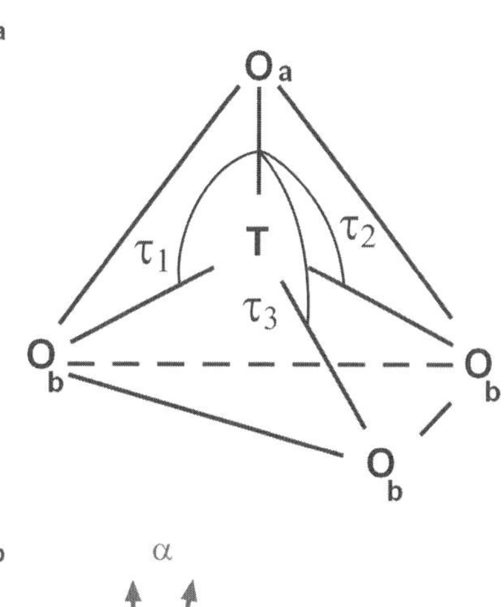

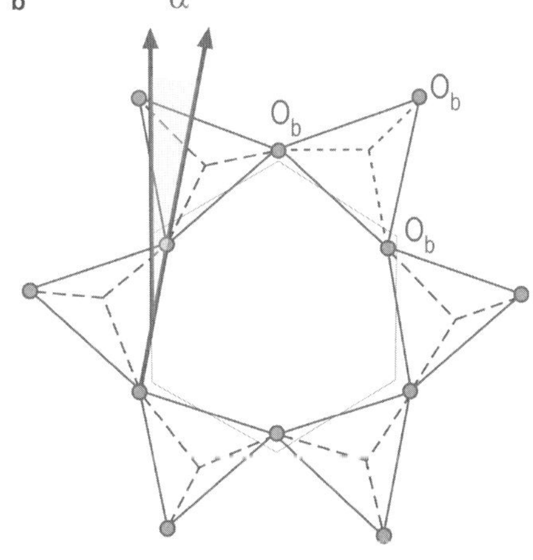

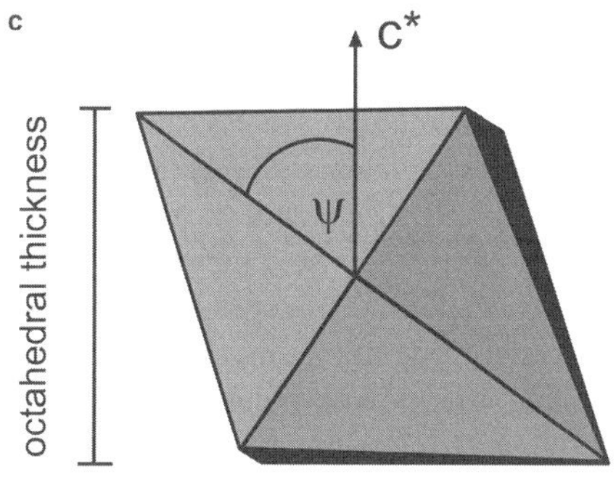

Fig. 4. (***a***) Tetrahedral flattening angle $\tau\left(\tau = \left(\frac{\tau_1+\tau_2+\tau_3}{3}\right)\right)$; (***b***) Definition of the tetrahedral rotation angle, α; (***c***) definition of the octahedral flattening angle ψ. Readers of the paper version of this chapter may wish to download a colour version of this figure from www.minersoc.org/emu-notes/emu-11/11-1-colour.pdf.

3. Mixed-layer structures and order-disorder in the layer stacking

3.1. Mixed-layer structures

Phyllosilicate crystals can also occur as randomly or regularly alternating layers of two or more phyllosilicate end-member minerals. Mixed-layer phyllosilicates (or interstratified phyllosilicates, or, more generally, interstratified clays) are usually considered the metastable intermediate products of a sequence of reactions involving end-member minerals (Merriman & Peacor, 1999).

Mixed-layer phyllosilicates are remarkable examples of order-disorder observed in natural and synthetic lamellar crystals. They consist of the alternation of layers revealing different structures and/or compositions in variable proportions (Lanson, 2011, this volume).

Regular interstratified minerals, often identified with specific mineral names, originate from a strictly regular periodic alternation of different layer types along the **c** axis (*e.g.* ABABAB). Interstratified minerals consist mostly of the alternation of 2:1 layers, with the exception of dozyite, which consists of the stacking of 1:1 serpentine layers and of 2:1 chlorite layers. The AIPEA Nomenclature Committee (Bailey, 1981) attributed special names to regular interstratifications of two layer types A and B, only if "such regularity of alternation is reached so that a well defined series of at least ten 00*l* summation spacings $d_{AB} = d_A + d_B$ is observed; suborders shall be integers and the even and odd suborders shall show very similar diffraction widths. Should any odd 00*l* suborders be missing, calculations shall be provided to demonstrate that their intensities are too small to be observed".

The coefficient of variation, CV, applied to at least ten 00*l* reflections values should be <0.75. The CV is calculated from: $\mathrm{CV} = \left(100 \times \left[\sum_{i=1}^{n} \frac{(X_i - \bar{X})^2}{(n-1)}\right]^{1/2} \frac{1}{\bar{X}}\right)$ where $X_i = l \times d_{001}$, and $\bar{X} = \sum_{i=1}^{n} \frac{X_i}{n}$) (Bailey, 1981; Reynolds, 1988; Guggenheim *et al.*, 2006).

Examples of regularly interstratified minerals are: (1) aliettite, a regular interstratification between trioctahedral talc and trioctahedral smectite in a 1:1 ratio (Veniale & van der Marel, 1969); (2) brinrobertsite, an ordered, mixed-layered, dioctahedral pyrophyllite-dioctahedral smectite (Dong *et al.*, 2002); (3) corrensite, a regular interstratification of either trioctahedral smectite or trioctahedral vermiculite in a 1:1 ratio with trioctahedral chlorite. Depending on whether trioctahedral smectite or trioctahedral vermiculite is involved, the mineral is termed 'low-charge corrensite' or 'high-charge corrensite', respectively (Lippman, 1954, 1960; Bailey, 1981; Guggenheim *et al.*, 2006); (4) dozyite, a 1:1 regular interstratification of serpentine (amesite) and chlorite (clinochlore) (Bailey *et al.*, 1995); (5) hydrobiotite, a regular 1:1 interstratification of biotite and vermiculite showing a large charge density within the vermiculite interlayer spaces (Bailey, 1989a); (6) kulkeite, a regular interstratification of trioctahedral talc and trioctahedral chlorite in a 1:1 ratio (Abraham *et al.*, 1980; Schreyer *et al.*, 1982); (7) rectorite, a regular interstratification of dioctahedral mica and dioctahedral smectite in a 1:1 ratio

(Caillère *et al.*, 1950; Brown & Weir, 1963; Bailey, 1981); (8) tosudite, a regular interstratification of chlorite and smectite in a 1:1 ratio where the smectite is mostly dioctahedral (Frank-Kamenetskii *et al.*, 1965).

In mixed-layer structures, the different layer types can either alternate randomly or else achieve some sort of ordering (avoiding the existence of pairs of the minor layer type) or segregation (clustering layers of a giving type). If a mineral consists of a random interstratification of two components, showing different *c* periodicity values, it is identified by using the name of the components separated by a hyphen, such as, for example, illite-smectite, illite-chlorite, illite-vermiculite and kaolinite-smectite. Some examples are represented by the irregular stacking of (1) illite ($c \approx 10$ Å) and smectite ($c \approx 14$ Å); (2) integral multiples of serpentine ($c \approx 7$ Å) and chlorite ($c \approx 14$ Å); or (3) layers with similar basal spacing, but with different local structure such as the random stacking of *trans*-vacant and *cis*-vacant illite layers (Drits, 1997, 2003).

Recognition of the interstratified character of a sequence and of the disorder in the layer stacking requires the modelling and the precise characterization of the different layer types (*i.e.* structure, composition, thickness), mode of distribution, and thickness distribution of coherent scattering domains.

3.2. Order-disorder of the layer stacking

Many studies were produced to establish the degree of ordering of a given sequence, mostly using models arising from the comparison of calculated and experimental *d* values of basal reflections and/or high-resolution electron microscopy observations.

The Kübler Index is defined as the full width at half-maximum height (FWHM) of the 10 Å X-ray diffraction (XRD) peak of illite-smectite interstratified minerals, measured on the <2 μm size fraction of an air-dried clay sample using Cu$K\alpha$ radiation (Kübler, 1964, 1967, 1984). Despite its widespread use, the Kübler Index remains controversial, mostly because of the numerous factors that affect the standardization and inter-laboratory calibration of its scale. Furthermore, the index is influenced by many parameters such as: (1) the mean size of crystal domains that scatter X-rays coherently (Weber *et al.*, 1976; Dunoyer de Segonzac & Bernoulli, 1976; Árkai & Tóth, 1983; Eberl & Velde, 1989; Drits *et al.*, 1997a); (2) crystallite size (*e.g.* see Eberl & Velde, 1989); (3) lattice strain (Árkai & Tóth, 1983; Árkai *et al.*, 1995, 1996).

The Árkai index (Árkai, 1991; Árkai *et al.*, 1996) is applied to chlorite and, like the Kübler index, involves the quantification of the width at the half-maximum height peaks of chlorite basal reflections. In this case, the index is derived from the 14 Å and 7 Å basal reflections.

The Reichweite index (R) is a simple method to obtain quantitative or semi-quantitative interpretation of the degree of ordering of an interstratified sequence from experimental *d* values of basal reflections. The Reichweite index expresses the probability, for a given layer of type A, that the next layer will be of type B (Drits *et al.*, 1994). This technique mostly applied to illite-smectite, uses graphical simulation of node positions in the reciprocal space along $\mathbf{c}^*$ and/or linear relationships (Guggenheim, 2011, this volume).

Several indices evaluate the kaolinite degree of ordering. These relations are based on changes observed in two specific groups of XRD reflections, namely: (1) the $02l$ and $11l$ sequences, which are sensitive to arbitrary and special interlayer displacements (such as $b/3$); and (2) the $13l$ and $20l$ sequences, which are affected by arbitrary displacements. Following this approach, several parameters could be defined, such as: (1) Hinckley index (HI) (Hinckley, 1963) and Range & Weiss index (QF) (Range & Weiss, 1969); (2) Stoch index (IK) (Stoch, 1974), which is similar to the previous two, but less sensitive to the presence of quartz; (3) Liètard index (R2) (Liètard, 1977), which, according to Cases *et al.* (1982) is sensitive to the presence of arbitrary defects only. Aparicio & Galán (1999), however, suggested that this parameter is also affected by the presence of associated phases (quartz, feldspar, iron and silica gels, illite, smectite and halloysite); (4) EXpert SYstem (EXSY), developed by Plançon & Zacharie (1990), which is based on parameters derived from the second basal reflection and from selected reflections belonging to (0 2 11) and (2 0 13) bands. The first one represents $02l$ and $11l$ reflections in the 2θ range from 19 to 24° and the second one $20l$ and $13l$ reflections in the 2θ range from 35 to 45° (Cu$K\alpha$ radiation); (5) Aparicio-Galán-Ferrell index (AGFI) $\left(\mathrm{AGFI} = \frac{(1\bar{1}0 + 11\bar{1})\ \text{peak heights}}{2 \times 020\ \text{peak height}}\right)$, which is determined after applying a peak-decomposition procedure by modelling the intensity of reflections in $02l$ and $11l$ sequences. This index seems to be less influenced by peak overlap and by the presence of accessory phases (Aparicio *et al.*, 1999).

Several programs were developed to simulate the diffraction patterns of layered materials showing typical long-tailed diffuse bands. *NEWMOD*™, developed by Reynolds (1985), calculates one-dimensional (oriented) XRD pattern profiles for pure phyllosilicates and for interstratifications of two phyllosilicates. In particular, once the diffraction patterns for each of the two constituent phyllosilicates are calculated, the program can simulate their combination consistently with a given mixture pattern. *NEWMOD*™ can describe effects associated with isomorphous substitutions, ordering of interstratifications, and with the particle size of the minerals present in the sample (for details of the statistical parameters defining the layer stacking of interstratified mineral see Reynolds, 1983; Drits & Tchoubar, 1990; Moore & Reynolds, 1997). Other codes that simulate defective layered structures are *MODXRSD* and *DIFFaX+* (Treacy *et al.*, 1991; Plançon, 2002; Leoni *et al.*, 2004). *MODXRSD* is a valuable tool for the simulation of powder patterns of disordered (both translational and rotational) sheet silicates or even mixed-layer crystals. It can also take into account variable crystal sizes and defects, such as cracks, inner-porosity, bent layers and edge dislocations. The *DIFFaX+* program allows only simulations of SAD (selected-area diffraction) and powder diffraction patterns (X-ray and neutrons). In general, *DIFFaX+* produces simulated powder patterns that can be compared visually to observed ones, thus providing confirmation of the reliability of a structure model.

Calculation of particle size on the diffraction peaks of phyllosilicates using peak broadening and shifts is also possible by means of the *MUDMASTER* and *GALOPER* programs. *MUDMASTER* (Eberl *et al.*, 1996) determines the crystallite size distribution and strain in the crystal direction corresponding to the peak by

Fourier analysis of XRD peak shape. *GALOPER* (Eberl *et al.*, 1998, 2000) calculates crystal-size distribution that results from crystal growth in open and closed systems using several growth mechanisms.

A good approach to solving the structure of regular mixed-layer structures is the multi-analytical approach, encompassing several experimental techniques, modelling of experimental XRD patterns and high-resolution electron microscopy (Środoń & Eberl, 1984; Drits & Tchoubar, 1990; Baronnet, 1992; Veblen, 1992; Banfield & Bailey, 1996; Drits, 1997; Moore & Reynolds, 1997; Lanson, 2005; Drits *et al.*, 2007; Lanson *et al.*, 2009). A good example of the application of this approach is the characterization of brinrobertsite (Dong *et al.*, 2002), using transmission electron microscopy (TEM), energy dispersive (ED) spectral analysis, thermal gravimetric analysis (TGA), and modelling of XRD results (Lanson, 2011, this volume).

The profile-fitting method calculates a complete XRD pattern from a structural model optimized for each clay species present (Drits & Tchoubar, 1990; Drits *et al.*, 1997b; Sakharov *et al.*, 1999). In the multi-specimen method, the optimized structural model should describe all XRD patterns obtained for a given sample following different treatments, such as saturation by different interlayer cations, ethylene-glycol solvation, heating, *etc.*, equally well. The multi-specimen method can be applied to mixed layers with more than two layer types whatever the layer stacking sequences are, and there is no *a priori* limitation to the nature of the species identified.

Structural details on irregular structures of mixed-layer crystals may be described in terms of a statistical probability-based model, where parameters define the proportion of structural fragments (in this case, layers) and the pattern of their distribution in the direction of periodicity loss. These parameters may be determined by modelling XRD patterns obtained by scanning in the above direction, seeking the best convergence of the experimental intensities with the intensities calculated by the theory of diffraction from irregular layered structures (Ivanova & Frank-Kamenetskaya, 2001).

Among mixed-layer minerals, the illite-smectite series was that most studied and, consequently, in greatest detail (*e.g.* see Altaner & Ylagan, 1997; Dong *et al.*, 1997; Dong, 2005; Lanson *et al.*, 2009). Several studies focused on the intercalation of these minerals with different molecules, mostly to support various applications and technological processes.

4. The *cis*-vacant and the *trans*-vacant octahedral site

Structural and crystal-chemical heterogeneity of layer silicates, especially in dioctahedral species, is affected significantly by the distribution of the octahedral cations and of vacancies in *trans*- and *cis*-oriented sites (Fig. 1b), as demonstrated by an extensive literature mostly concerning 2:1 layer silicates. In the beginning, only dioctahedral 2:1 layer silicates showing the *trans*-site vacant were described. The existence of 2:1 dioctahedral phyllosilicates with one of the *cis*-sites vacant was first reported by Méring & Oberlin (1971). Drits *et al.* (1984) derived unit-cell parameters and atomic coordinates for one-layer monoclinic *cis*-vacant illite; Tsipursky & Drits (1984) indicated that dioctahedral smectite can be both *cis*-vacant (mostly montmorillonite) and *trans*-vacant (mostly nontronite and beidellite). They also documented many cases showing interstratified *cis*- and *trans*-vacant layers. The major factor driving cation distribution on *cis*- and *trans*-vacant sites was recognized to be the ratio of Si-for-Al substitution. Reynolds (1983) demonstrated that illite in mixed-layer illite-smectite consists either of *trans*-vacant and *cis*-vacant layers or of an interstratification of both layer types. Drits *et al.* (2006) and Drits & Zviagina (2009), starting from powder XRD patterns calculated for different polytypes, consisting of either *trans*- or *cis*-vacant layers, demonstrated that: (1) Fe^{3+}- and Mg-rich dioctahedral species (celadonite, glauconite, Al-celadonite and the majority of phengites) usually present *trans*-vacant octahedra; (2) 1*M*-*cis*-vacant illite, as well as *cis*-vacant layers in the illite fraction of illite-smectite interstratified minerals, can be mostly associated with Fe- and Mg-poor varieties; (3) in illites and illite fundamental particles of illite-smectite consisting of *trans*-vacant and *cis*-vacant layers, *cis*-vacant layers prevail when the Al in octahedral and tetrahedral sites is > 3.10 and > 0.70 atoms per $O_{20}(OH)_4$, respectively; (4) Mg-rich *cis*-vacant smectite shows random distribution of isomorphous octahedral cations, whereas Mg-bearing *trans*-vacant smectite shows dispersed octahedral Mg cations to minimize, as much as possible, the formation of Mg-OH-Mg patterns. The structural characterization of *cis*-vacant and *trans*-vacant 2:1 layer silicates was obtained with different techniques, such as: (1) powder XRD patterns calculated for different polytypes consisting of either *trans*-vacant or *cis*-vacant layers (Zviagina *et al.*, 2007); (2) simulation of experimental XRD patterns corresponding to illite or illite fundamental particles in which *trans*-vacant and *cis*-vacant layers are interstratified (Drits & Zviagina, 2009); (3) semi-quantitative assessment of the relative content of the layer types in interstratified structures (McCarty *et al.*, 2009); (4) thermal analysis, conveniently exploiting the different dehydroxylation temperatures characterizing *trans*- and *cis*-vacant illite and smectite (Drits *et al.*, 1998; Wolters & Emmerich, 2007); (5) Mössbauer spectroscopy (Shabani *et al.*, 1998; Dainyak *et al.*, 2004, 2009; Dainyak & Drits, 2009). Using Mössbauer spectroscopy evidence, Shabani *et al.* (1998) discovered that $2M_1$ muscovite can present two types of spectra. The first type of muscovite crystals presents a well resolved Fe^{2+} quadrupole doublet, whereas a second type presents a single, broader Fe^{2+} quadrupole doublet. The model used for spectrum fitting, which assumes the occurrence of Fe^{2+} both in *cis*- and *trans*-sites, suggested that spectra of the second type can identify the presence of *cis*-vacant sites in muscovite, whereas spectra of the first type can allow an accurate calculation of the minimum number of *cis*-vacant sites. Dainyak & Drits (2009) considered the local cation arrangement around Fe^{2+} and Fe^{3+} and suggested that the main contribution to the first type muscovite is the 2(Fe^{2+}, Mg) Al local cation arrangement around Fe^{2+}, whereas, for the second type of muscovite, the 3Al and $2AlFe^{3+}$ local cation arrangements are more important.

5. Layer charge

When the tetrahedral and octahedral sheets are joined in a layer, the resulting layer can be either electrically neutral or negatively charged. Ideally, a layer is electrically neutral if: (1) the octahedral sheet contains trivalent cations (R^{3+}) at two octahedral sites (usually Al^{3+} or Fe^{3+}), with a vacancy (□) at the third octahedral position [R_2^{3+} □ $(OH)_6$], or else divalent cations (R^{2+}, usually Fe^{2+}, Mg^{2+}, Mn^{2+}) at all the octahedral sites [R_3^{2+} $(OH)_6$]; (2) the tetrahedral sheet contains Si^{4+} in all tetrahedra. However, substitutions within the layer can also produce neutrality (*e.g.* mutual tetrahedral and octahedral substitution producing layer neutrality $^{[iv]}(R_x^{3+}Si_{-x}^{4+})$ $^{[vi]}(R_x^{3+}R_{-x}^{2+})$). Negative layer charge arises from: (1) substitution of lower-charge cations at octahedral sites; (2) substitution of R^{3+} for Si^{4+} in tetrahedral sites; and (3) vacancies. The charge per formula unit, x, is the net negative charge per layer, expressed as a positive number. The net negative layer charge is balanced by the positively charged interlayer material.

Layer charge is a fundamental property of 2:1 phyllosilicates. In some minerals this charge is balanced by fixed cations (*e.g.* micas), whereas in others it is balanced by exchangeable cations (*e.g.* smectite and vermiculite), placed in the interlayer position.

Layer charge affects many properties of smectite, such as swelling (Laird, 2006), ion-exchange capacity, ion-exchange selectivity (Maes & Cremers, 1977; Shainberg *et al.*, 1987), and rheological properties (Christidis *et al.*, 2006). A number of independent studies using different analytical methods demonstrated that distribution of layer charge in smectite can vary considerably (Nadeau *et al.*, 1985; Lim & Jackson, 1986; Decarreau *et al.*, 1987; Christidis & Dunham, 1993, 1997; Lagaly, 1994; Mermut, 1994; Christidis, 2001, 2006; Christidis & Eberl, 2003). The layer charge of smectite can be estimated using a variety of analytical methods including: (1) the structural-formula method (Weaver & Pollard, 1973; Grim & Güven, 1978; Bain & Smith, 1987; Newman & Brown, 1987; Laird, 1994, 2006); (2) the alkylammonium method (Lagaly & Weiss, 1975; Lagaly, 1981, 1994); (3) the investigation of XRD traces of K-saturated, ethylene glycol solvated smectites (Christidis & Eberl, 2003). Additional methods, less frequently used, include: (1) NH^{4+} saturation and examination by infrared (IR) spectroscopy (Petit *et al.*, 2006); and (2) methylene blue absorption and examination by UV spectroscopy (Bujdák, 2006). However, all the methods discussed, aiming at providing an approximation of layer charge value, can be susceptible to deviations, sometimes significant, and can lead to erroneous interpretations if not used properly (Christidis, 2008). More quantitative details for this important topic are addressed extensively by Christidis (2011, this volume).

Layer charge can be located at different places in the layer, as affecting and being affected by isomorphous substitution involving either the tetrahedral or octahedral sheet. This aspect represents another important factor affecting both hydration and cation speciation in the interlayer of hydrated 2:1 layer silicates, such as smectites and vermiculites. In electrically neutral layers, the basal oxygen atoms act as a weak Lewis base (electron donor), forming weak hydrogen bonds with water molecules. When isomorphous substitution occurs, the basal oxygen atoms show an excess negative charge, and their electron-donating capacity increases. Sposito (1984) demonstrated that H-bonding between water molecules and basal oxygen atoms is enhanced by tetrahedral rather than by octahedral sheet substitutions. According to the HSAB (Hard and Soft Acid and Base) theory of Pearson (1963, 1968), the 2:1 silicate layers and the hydrated interlayer cations can be considered as Lewis bases and acids, respectively (Xu & Harsh, 1992). The location of the layer charge determines the strength of the Lewis base: when the layer charge derives from substitutions mostly at the octahedral sheet, the hydrated 2:1 layer silicate behaves as a soft base, as tetrahedral basal oxygen atoms are affected by a charge imbalance involving the whole layer rather than the tetrahedral sheet directly. On the contrary, when the layer charge derives from substitutions mostly involving the tetrahedral sheet, the hydrated 2:1 layer silicate behaves as a hard base because the layer charge related to tetrahedral substitutions affects the charge of coordinating basal oxygen atoms significantly. In this way, the 2:1 layer charge location affects the layer hydration as well as the cation-sorption process (Brigatti *et al.*, 2004), as hard and soft bases preferentially complex hard acids (cations) and soft acids, respectively. Clay minerals with charge located in the tetrahedral sheet, such as vermiculite, hydrate more strongly than those with charge located mainly in octahedral sheet, such as montmorillonite. Organic cations are thus adsorbed less strongly on the vermiculite surface because of the energy demand in displacing water from the adsorption site.

Residual negative charges can also develop along the edges of clay mineral particles, where Si–O–Si and Al–O–Al bonds are 'broken' and replaced by Si–OH and Al–OH groups (Güven, 1992).

6. The 1:1 layer structure: kaolin-serpentine group

6.1. Kaolin subgroup

Minerals in the kaolin subgroup consist of dioctahedral 1:1 layer structures (Fig. 2a), with general composition $Al_2Si_2O_5(OH)_4$. The layers are kept together by hydrogen bridges between surface hydroxyl groups on the octahedral sheet and basal oxygen atoms on the tetrahedral sheet. Documented polytypes for dehydrated kaolin minerals (Newnham, 1961) are: kaolinite, dickite, nacrite and halloysite-7 Å. These minerals are stacked with different positions of octahedral vacancies in successive layers (Fig. 5). Kaolinite is known to be the more abundant polytype, while dickite, nacrite and halloysite-7 Å are less common. Halloysite-10 Å, general stoichiometry $Al_2Si_2O_5(OH)_4 \cdot 2H_2O$, also belongs to kaolin subgroup and presents water molecules between two adjacent TO layers. The intercalated water is bonded weakly and can be removed readily and irreversibly.

Zvyagin (1954) investigated the polytypic arrangement of kaolinite, demonstrating 1-, 2- or 3-layer structures. Different polytypes of the kaolinite group could then be defined depending on the location of the vacant cavity (Bookin *et al.*, 1989; Adams, 1983; Thompson & Withers, 1987; Bish & Von Dreele, 1989; Smrčok *et al.*, 1990; Bish, 1993).

The composition of the kaolin group minerals is characterized by a predominance of Al^{3+} in octahedral sites. Isomorphous substitution of Mg^{2+}, Fe^{3+}, Ti^{4+}, and V^{3+} for Al^{3+} can also occur.

The kaolinite stacking sequence consists of identical layers with an interlayer shift of $2a/3$. Dickite and nacrite show a two-layer stacking sequence where the vacant site of the octahedral sheet alternates between two distinct sites (Brindley & Brown, 1980; Zheng & Bailey, 1994). Bailey (1963) demonstrated that both kaolinite and dickite are based on a 1*M* stacking sequence of layers. In the 1*M* structure there are three

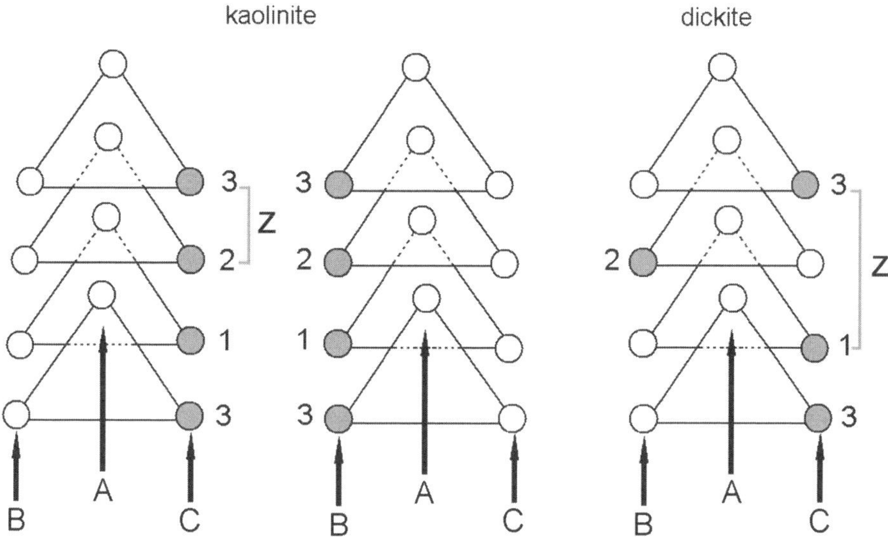

Fig. 5. Projection on the (001) plane of the octahedral sites in kaolinite and dickite showing the possible placement of the vacant octahedral site (filled circles) (modified after Bailey, 1963). Readers of the paper version of this chapter may wish to download a colour version of this figure from www.minersoc.org/emu-notes/emu-11/11-1-colour.pdf.

octahedral sites, denoted as A, B, or C (Fig. 5). In well crystallized kaolinite, each layer is identical and shows an octahedral site C (or B) vacant. In dickite the vacant site alternates between C and B in successive layers to create a two-layer structure. The sequence of layers in nacrite is in accord with the standard 6*R* polytype.

6.1.1. Kaolinite

Experiments aimed at investigating the kaolinite structure have reported conflicting results for space group and hydroxyl bond lengths and orientations. Explanations for such discrepancies have been discussed in the literature; the differences may be related to impurities in samples, temperature, preferential orientation of the crystallites and difficulties with structure refinement.

The first attempts to define the kaolinite structure date back to 1930, when Pauling, starting from models based on idealized polyhedra, provided a basic description. Several later studies were devoted to the definition of the space group for this mineral (Gruner, 1932a; Hendricks, 1938; Brindley & Robinson, 1945, 1946). Gruner (1932a) indicated, for kaolinite, a monoclinic *Cc* symmetry with $d_{001} = 14.3$ Å, corresponding to a two-layer structure. Brindley & Robinson (1945, 1946) recognized many reflections in the powder pattern that could not be indexed correctly on the basis of a monoclinic structure, and suggested a triclinic symmetry for the layer. This evidence was also confirmed by Zvyagin (1960), who suggested a rotation of SiO_4 tetrahedra, and by several other authors (*e.g.* see Brindley & Nakaira, 1958; Drits & Kashaev, 1960; Suitch & Young, 1983; Thompson & Withers, 1987; Bish & Von Dreele, 1989; Rocha & Pedrosa De

Jesus, 1994). In particular, Suitch & Young (1983) and Young & Hewat (1988) reported the appropriate space group for kaolinite as *P*1 and Bish (1993) reported unit-cell parameters (at 1.5 K) as: $a = 5.1535(3)$, $b = 8.9419(5)$, $c = 7.3906(4)$ Å, $\alpha = 91.926(2)$, $\beta = 105.046(2)$, $\gamma = 89.797(2)^\circ$ (space group *C*1). Furthermore, Rocha & Pedrosa De Jesus (1994) demonstrated the existence of two different octahedral coordinations for Al.

Ab initio energy-minimization method (Hobbs *et al.*, 1997; Castro & Martins, 2005), low-temperature neutron powder diffraction (Bish, 1993), refinement of the structure from single-crystal synchrotron data (Neder *et al.*, 1999) and molecular-dynamics simulations based on first-principles calculations within density functional theory (Sato *et al.*, 2004) further contributed to detailing and clarifying kaolinite structure, in particular by underlining that: (1) the average length of apical Si–O bonds (1.59 Å) is nearly equal to that characterizing basal Si–O bonds (1.60 Å) (Bish, 1993); (2) in the *C*-site polytype, Al–OH bonds (1.84 Å) are shorter than Al–O bonds observed elsewhere in kaolinite (1.90–2.04 Å) (Sato *et al.*, 2004); (3) low-temperature conditions mostly affect the interlayer separation, but not significantly, the tetrahedral and octahedral parameters.

The positions and orientations of OH groups have been investigated using different approaches. Raman and IR vibrational spectra (Farmer, 1974; Prost *et al.*, 1989; Johnston *et al.*, 1990; Frost, 1995; Frost & Van der Gaast, 1997; Shoval *et al.*, 1999, 2002; Farmer, 1998, 2000; Balan *et al.*, 2001) show four bands. The band at 3619 cm^{-1} is related to the stretching of the inner OH groups whereas the broad band observed at 3695 cm^{-1} is related to in-phase stretching mode of inner-surface OH groups. This band is shifted to a slightly lower frequency in the Raman spectrum (3679 cm^{-1}). The two other bands at 3651 and 3669 cm^{-1} are related to the two out-of-phase stretching modes of inner-surface OH groups.

Suitch & Young (1983) and Young & Hewat (1988) refined H-atom positions using neutron powder diffraction data, assuming a *P*1 space group. H-atom positions were determined also by Rietveld X-ray powder refinement (Adams, 1983), and by a Rietveld refinement of neutron powder diffraction data collected at low temperature in the space group *C*1 (Bish, 1993). This latter study revealed that the inner OH group is in the plane of the layers, and the inner-surface OH groups form angles in the range 60–73° with the (001) plane. Benco *et al.* (2001a,b) explained interlayer H bonding in kaolinite from *ab initio* molecular-dynamic simulations of a hypothetical isolated layer after identifying four distinct OH groups, in *P*1 symmetry. Of these four, two (OH3 and OH4) form weak H bonds with H...O distances of between 1.8 and 2.6 Å, and the other two (OH1 and OH2) are not involved in H bonding. Similar results were obtained by modelling IR spectra (Balan *et al.*, 2001, 2010) (Fig. 6).

White *et al.* (2009), using a density functional modelling approach, confirmed that kaolinite, at temperature values close to 0 K presents the inner OH groups nearly parallel (5.057°) and the inner-surface OH groups nearly perpendicular (78.272°, 84.585°, 68.582°) to the *a,b* plane. OH bond lengths vary between 0.970 and 0.974 Å.

The crystal-structure refinement of a deuterated kaolinite (Akiba *et al.*, 1997) suggested that the three inner OD vectors point towards the tetrahedral sheet, forming H bonding with basal oxygen atoms of the adjacent kaolinite layer. One of the three

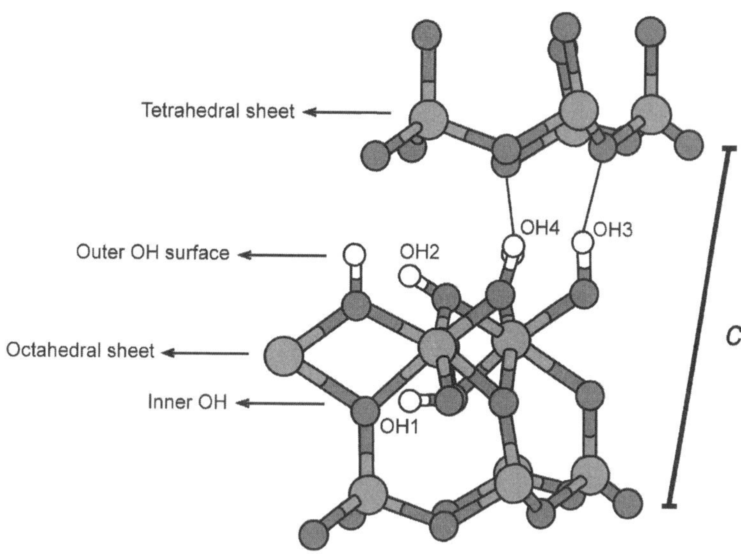

Fig. 6. Different OH orientations on the octahedral surface of kaolinite (modified after Benco *et al.*, 2001a). Readers of the paper version of this chapter may wish to download a colour version of this figure from www.minersoc.org/emu-notes/emu-11/11-1-colour.pdf.

vectors, however, differs from the other two in terms of bond angle, thus suggesting a different orientation of the bond.

Many other contributions have addressed the structure modifications of kaolinite following temperature/pressure variation (*e.g.* see Prost *et al.*, 1989; Mercier & Le Page, 2008, 2009; Mercier *et al.*, 2010; Welch & Crichton, 2010) or mechanical stress (*e.g.* Reynolds & Bish, 2002). Reynolds & Bish (2002) demonstrated that ordered kaolinite subjected to mechanical grinding presents an increase in disorder that could be modelled as a physical mixture of low- and high-defect material. Mercier & Le Page (2008, 2009) and Mercier *et al.* (2010) calculated enthalpies and cell volumes under pressure with models optimized for *ab initio* density functional theory calculations and identified new interlayer translations for kaolinite, (*i.e.* $-a/3$ and $(a+b)/3$), thus defining a new family of kaolin polytypes generating moderate pressure. Both translations place each silicon atom of each 1:1 layer on top of an OH group from the layer below, resulting in a triangular dipyramidal fivefold coordination for all silicon atoms. Those authors also indicated that kaolinite and dickite are the lowest-energy models at zero temperature and pressure (Mercier & Le Page, 2008). Welch & Crichton (2010) observed two-phase transitions in kaolinite. The ambient phase (kaolinite I) transforms reversibly into kaolinite II at 3.7 GPa, whereas kaolinite II transforms irreversibly into kaolinite III at 7.8 GPa.

Smirnov & Bougeard (1999) applied molecular dynamics to investigate the structure and short-time dynamics of interlayer water molecules of the 'hydrated kaolinite', *i.e.* halloysite with layer spacing of 8.5 and 10.0 Å. The structure of interlayer water is characterized by density profiles of the oxygen and hydrogen atoms of water molecules

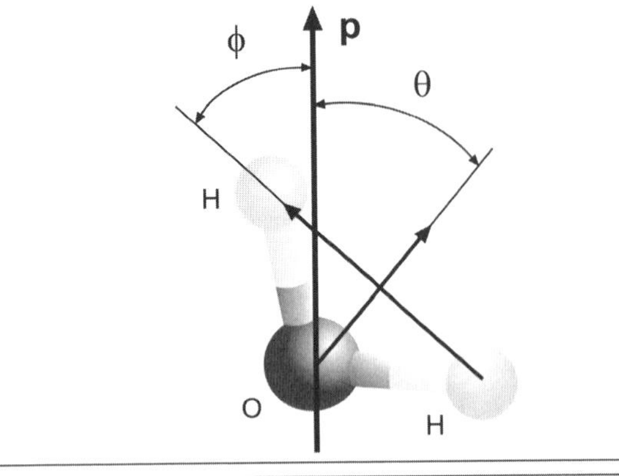

Fig. 7. Orientation of adsorbed water molecules with respect to the surface of kaolinite. **p** is the vector product of the **a** and **b** crystallographic axes vectors (**p** [**a** **b**]). The θ angle represents the orientations of the molecular dipole and the φ angle represents the **H–H** interatomic vector with respect to the **p** axis. The definition is after Smirnov & Bougeard (1999). Readers of the paper version of this chapter may wish to download a colour version of this figure from www.minersoc.org/emu-notes/emu-11/11-1-colour.pdf.

along the direction perpendicular to the surface of the clay layer. This direction (**p**) is the vector product of the **a** and **b** crystallographic axis vectors (**p** [**a** **b**]). Figure 7 also demonstrates the orientations of the water dipole (θ angle) and the **H–H** interatomic vector (φ angle) with respect to the **p** axis. These authors identified two types of adsorbed water molecules, which are characterized by different orientations with respect to the surface of the clay layer (Fig. 7). Adsorbed molecules of the first type are oriented with the **H–H** vector parallel to the surface and with water dipole inclined by 30° to the surface normal; molecules of the second type show **H–H** vectors and water dipole perpendicular to the surface and the surface normal, respectively. Such water molecules are located on surfaces of both tetrahedral and octahedral sheets. The increase in interlayer spacing and in the number of interlayer water molecules leads to the formation of a sheet consisting of associated water (Tarý *et al.*, 1999). This sheet is weakly bonded to OH groups in the octahedral sheet and more strongly to molecules adsorbed on the surface of the tetrahedral sheet.

Many theoretical studies have investigated the influence of kaolinite structural features and its ability to adsorb given molecules (*e.g.* see Michalková & Tunega, 2007; Vasconcelos *et al.*, 2007). For example, molecular-dynamics simulation demonstrated that the adsorption of cations and anions on kaolinite surfaces is controlled by the mineral surface charge, thus accounting for preferential adsorption of cations and anions on the basal tetrahedral surface and on the basal octahedral surface, respectively (Vasconcelos *et al.*, 2007).

Possible substitution of Al in octahedral sites and Si in tetrahedral sites was investigated by extended X-ray absorption fine structure (EXAFS), X-ray absorption near edge spectroscopy (XANES), electron paramagnetic resonance (EPR), magic angle spinning nuclear magnetic resonance (MAS-NMR) and Mössbauer spectroscopies (Bonnin *et al.*, 1982; Schroeder & Pruett, 1996; Gualtieri *et al.*, 2000; Balan *et al.*, 2001; He *et al.*, 2003). These studies have contributed to the following conclusions: (1) Al atoms cannot be substituted easily in both tetrahedral and octahedral sites, and (2) some cations such as Fe, often noted in chemical analyses, are related to associated phases (Bonnin *et al.*, 1982). Other contributions have suggested that well ordered kaolinite contains only [6]-fold coordinated Fe^{3+} (Balan *et al.*, 1999) whereas disordered kaolinite can also present Fe^{3+} substitutions in tetrahedral sites (Gualtieri *et al.*, 2000) (Mottana & Aldega, 2011, this volume).

The structural order in kaolin minerals is usually defined in terms of a series of either extended (*e.g.* stacking faults) or localized (*e.g.* due to impurities) faults (Bookin *et al.*, 1989; Plançon *et al.*, 1989; Zvyagin & Drits, 1996). Ordered and disordered kaolinites differ significantly in terms of diffraction patterns. The former shows sharp and narrow peaks, while the latter is characterized by less well defined, broad and asymmetrical peaks. A quantitative assessment of the degree of structural order in kaolinite can be derived from powder XRD (see section 3.2); IR spectroscopy (Schroeder, 2002) and thermal analysis (*e.g.* see Balek & Murat, 1996).

Stacking disorder in kaolinite was recently investigated using electron microscopy by Kogure *et al.* (2010), also demonstrating that the degree of stacking disorder is variable among individual grains. Stacking faults are mainly produced by disorder of alternating $-a/3$ and $-a/3 + b/3$ layer displacements. Furthermore, stacking faults can be isolated and can form different kinds of interstratification of two kinds of multilayer blocks, showing regular $a/3$ and $-a/3 + b/3$ layer displacements.

6.1.2. *Dickite*

(Unit-cell parameters at 12 K: $a = 5.1474(6)$, $b = 8.9386(10)$, $c = 14.390(2)$ Å and $\beta = 96.483(1)°$, space group: *Cc*, Bish & Johnston, 1993) The basic description of the dickite structure was provided by Gruner (1932b). The first refinement of the structure, except for the definition of hydrogen atom positions, was undertaken in the space group *Cc*, using single crystal X ray data, by Newnham & Brindley (1956, 1957). These studies and the subsequent results from Newnham (1961) showed that dickite exhibits significant distortions from the ideal kaolinite-type layer, including the rotation of SiO_4 tetrahedra and distortion of the octahedral sheet. The structural similarity of the dickite and kaolinite layers was first suggested by Brindley & Nakahira (1958) and confirmed later by Bailey (1963) in the space group *Cc*.

Structure refinements, achieved by single-crystal XRD measurements, were reported by Giese & Datta (1973), Adams & Hewat (1981), Suitch & Young (1983), Sen Gupta *et al.* (1984), Joswig & Drits (1986), and Dera *et al.* (2003). Bish & Johnston (1993) suggested that the inner OH group is approximately parallel to the (001) plane, inclined at just 1.3° to the tetrahedral sheet. The study of OH-group orientation was also addressed by IR (Farmer, 1974; Prost *et al.*, 1989; Johnston *et al.*, 1990; Bish & Johnston, 1993; Shoval *et al.*, 2001) and Raman spectroscopies (Johnston *et al.*, 1998; Shoval *et al.*, 2001), as well as by molecular-dynamics simulations. Balan *et al.* (2005) applied quantum mechanical calculations in the framework of density functional theory to explain experimental spectra of dickite. They concluded that: (1) the band observed at 3622 cm^{-1} may be related to the inner OH stretching; (2) the high-frequency bands, observed in IR spectra at 3711 cm^{-1} (3708 cm^{-1} in Raman spectra) and at

3655 cm^{-1} (3642 cm^{-1} in Raman spectra) correspond to transverse and longitudinal in-phase motion mode and are attributed to the two outer OH groups. Johnston *et al.* (2002) carried out a single-crystal Raman study of dickite at pressures up to 6.5 GPa using a diamond-anvil cell. They found a dramatic shift in the ν(OH) bands at ~2.2 GPa, indicating a significant change in the local environment of the interlayer OH groups. The lattice mode region showed only minor changes as a function of pressure, suggesting that the individual 1:1 layers did not change significantly with pressure. The spectra also showed that the phase transition is reversible.

Benco *et al.* (2001a) investigated the orientation of OH vectors by *ab initio* molecular dynamics and total-energy calculations and suggested that the inner OH and one inner-surface OH are oriented horizontally, while the other two OH are involved in interlayer bonding. On the contrary, Sato *et al.* (2004), starting from molecular-dynamics simulations based on first-principles calculations, concluded that the inner-surface OH groups are oriented perpendicular to (001) and form interlayer hydrogen bonding.

6.1.3. Nacrite

(Unit-cell parameters: $a = 8.910(3)$, $b = 5.144(2)$, $c = 14.593(3)$ Å, $\beta = 100.50(3)°$, space group: *Cc*; Zhukhlistov, 2008) The first crystal-structure refinement of nacrite was reported by Hendricks (1939) who suggested the space group *Cc* for this mineral. The structure is made up of six stacking layers, closely approaching rhombohedral symmetry with a pseudo-space group *R*3*c*. The refinement of the nacrite structure by XRD analysis (Blount *et al.*, 1969) and electron diffraction analysis (Zvyagin *et al.*, 1972) confirmed that the ideal structure of nacrite is based on a 6*R* stacking sequence of TM layers, simulating an *R*3*c* symmetry. However, the pattern of vacant octahedral sites reduces the symmetry to *Cc* symmetry. Successive refinement of the nacrite structure with the use of single-crystal precession (Zheng & Bailey, 1994) allowed the determination of the hydrogen atoms and of the OH groups' position in the difference electron density syntheses. It was shown that the O–H interatomic distance is ~0.8 Å and that the angles of inclination of the OH vectors with respect to the layer plane are ~60° for the outer OH groups, and ~−20° for the inner OH group.

Zhukhlistov (2008) refined the nacrite-2M_2 structure from the oblique-texture electron diffraction patterns in the space group *Cc*. Zhukhlistov (2008) suggested that: (1) the octahedra are oblate and their bases are rotated so that a ditrigonal pattern originates in the normal projection onto the *ab* plane; (2) octahedral shared edges are shorter with respect to other octahedral edges, and the OH–OH edge is the shortest among all the shared edges; (3) the Al cations are displaced towards the outer OH surface of the octahedral sheet; (4) the tetrahedra are slightly elongated and α is 7.8°; (5) the O–H interatomic distances and the angles of inclination of the O–H bond with respect to (001) are equal to 0.97 Å and −18.3° for the inner OH group and 0.92, 0.85, 0.93 Å, 60.8, 67.8, 58.4° for the outer OH groups; (6) the electrostatic potential distributions of the hydrogen atoms of the inner OH group and one of the outer OH groups located close to the pseudo-symmetry plane *m* of the layer are characterized by anisotropy, thus suggesting a statistical distribution of these hydrogen atoms.

Ben Haj Amara *et al.* (1997, 1998) described the structure of hydrated and dehydrated nacrite. The hydrated form is characterized by a basal distance of 8.42 Å, containing one water molecule per $Si_2Al_2O_5(OH)_4$ in the interlayer space. The interlayer water molecule is placed above the vacant octahedral site of the layer and is embedded in the ditrigonal cavity of the tetrahedral sheet of the upper layer.

6.1.4. Halloysite

(Unit-cell parameters: $a = 5.1$, $b = 8.9$, $c = 7.57$ Å, $\beta = 100°$, space group: *Cm* (dehydrated form); $a = 5.20$, $b = 8.92$ $c = 10.25$ Å, $\beta = 100°$, space group: *Cm* (hydrated form), Mehmel, 1935) Halloysite was first studied by Berthier (1826). Hofmann *et al.* (1934) proposed the presence of H_2O molecules in its interlayer space, giving the general formula $Si_2Al_2O_5(OH)_4{\cdot}2H_2O$ (Fig. 8). Hydrated halloysite presents layer periodicity close to 10 Å and is referred to as halloysite-10 Å (Brindley & Robinson, 1948). The interlayer water in halloysite can be removed easily, giving a dehydrated form with layer periodicity close to 7.2 Å (Bailey, 1989b). The symmetry of these crystals is either monoclinic ($C2/m$, $C2_1/c$ or Cc space groups were reported) or triclinic (Bates *et al.*, 1950; Honjo *et al.*, 1954; de Souza Santos *et al.*, 1965; Zvyagin, 1967). Kohyama *et al.* (1978) suggested a two-layer monoclinic structure with *Cc* space group for both halloysite-10 Å and halloysite-7 Å; unit-cell parameters are $a \approx 5.14$, $b \approx 8.90$, $c \approx 14.9$ Å, $\beta \approx 101.9°$ for the anhydrous form. The insertion of water molecules strongly influences the c parameter, which is increased to $c \approx 20.4$ Å for fully hydrated halloysite. The particles of halloysite can show different morphologies, such as spheres, tubes, plates (*e.g.* see Zvyagin *et al.*, 1966; Churchman & Theng, 1984; García *et al.*, 2009). The tubular halloysite is an attractive material from a technological

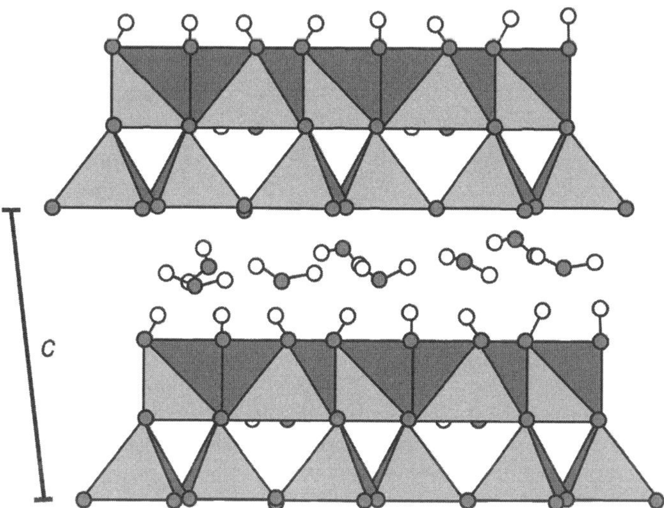

Fig. 8. Layer periodicity in halloysite-10 Å. Readers of the paper version of this chapter may wish to download a colour version of this figure from www.minersoc.org/emu-notes/emu-11/11-1-colour.pdf.

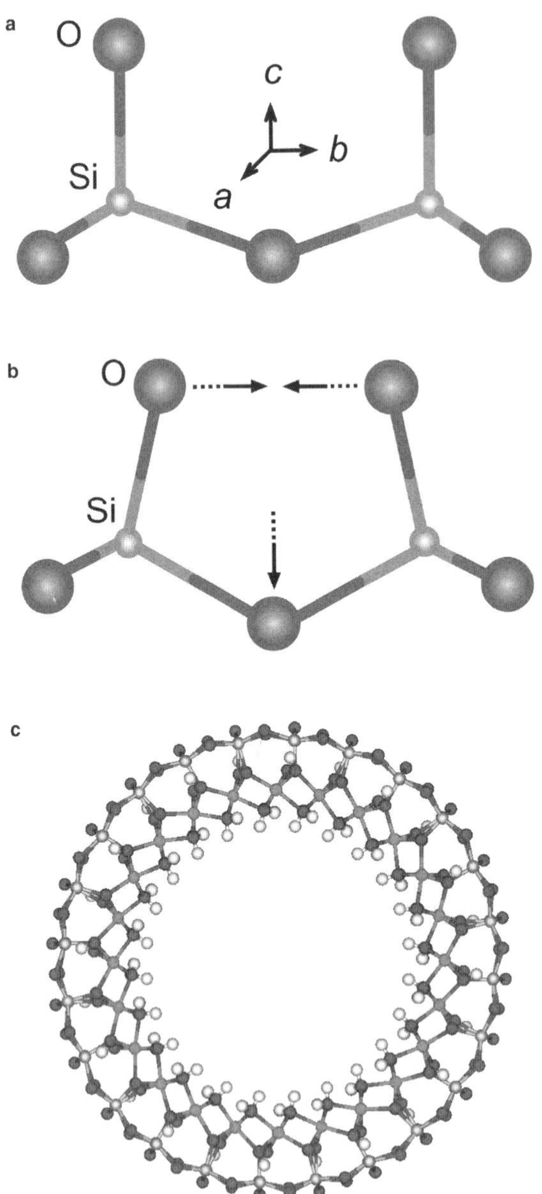

Fig. 9. The rolling mechanism in halloysite. (***a***) Halloysite planar tetrahedral sheet; (***b***) halloysite curved tetrahedral sheet. The distance between apical oxygen atoms in the curved sheet is reduced by decreasing Si–O–Si angle; (***c***) example of a halloysite nanotube (modified after Singh, 1996). Readers of the paper version of this chapter may wish to download a colour version of this figure from www.minersoc.org/emu-notes/emu-11/11-1-colour.pdf.

viewpoint, due to its availability and to the vast range of applications, including some showing remarkable biological relevance. Bailey (1989b) assumed that tubular halloysite is generated by a rolling mechanism promoted by Si–Si Coulomb repulsion in the Si plane. Singh (1996) suggested that the rolling mechanism is favoured, from a Si–Si repulsion perspective, in comparison to tetrahedral rotation at a given similar amount of misfit between tetrahedral and octahedral sheets (Fig. 9).

Tubular halloysite is a viable nanocage for biologically active molecules due to the empty space inside the nanotube, allowing entrance of molecules of specific sizes. Halloysite has been used as a support for immobilization of catalyst molecules such as metal complexes and for the controlled release of anticorrosion agents, herbicides and fungicides. It exhibits interesting features and offers a wide range of potential applications, such as entrapping hydrophilic and lipophilic active agents, acting as an enzymatic nanoscale reactor, closely controlling the release rate of drugs and other medical treatments in humans or animals, and improving mechanical performance of cements and polymers. A detailed review of the vast literature devoted to the application of this mineral is outside the scope of this chapter. Interested readers are directed to a recent review on this topic by Joussein *et al.* (2005).

6.1.5. Hisingerite

Hisingerite [$Fe_2^{3+}Si_2O_5(OH)_4{\cdot}2H_2O$] was originally described by Hisinger (1810) as an amorphous phase and, more recently, as an interstratified montmorillonite/chlorite (Lindqvist & Jansson, 1962) and as a poorly crystallized form of either Fe-rich saponite (*e.g.* see Whelan & Goldich, 1961; Brigatti, 1981) or nontronite (*e.g.* see Gruner, 1935; Sudo & Nakamura, 1952; Kohyama & Sudo, 1975; Mackenzie & Berezowski, 1980; Manceau *et al.*, 1995). Eggleton & Tilley (1998), based on TEM results, demonstrated for hisingerite a fabric of concentric spheres, with diameters of ~140 Å and with ~7 Å thick layers as walls. High-resolution images of these walls suggested a structure similar to that of spherical halloysite, thus leading to the conclusion that hisingerite is a ferric-iron member of the kaolin subgroup.

6.2. Odinite subgroup

The odinite structure, ideal composition $(Fe^{3+},Fe^{2+},Mg,Al,Ti,Mn^{2+})_{2.5}(Si,Al)_2O_5(OH)_4$, is based on a 1:1 layer, which is intermediate between dioctahedral and trioctahedral. The polytypic arrangement of this phase, which shows fine grain size and poor crystallinity, is mostly monoclinic (1*M*, space group *Cm*) and sometimes trigonal (or hexagonal) (1*T*) (Bailey, 1988b).

6.3. Serpentine subgroup

Minerals of the serpentine subgroup are commonly hydrous Mg-rich trioctahedral 1:1 phyllosilicates. Al and Fe^{3+} can substitute for Si in tetrahedral sites. Mg in octahedral sites can be substituted by Fe^{2+}, Fe^{3+}, Al, Cr, Ni and Mn^{2+}. They can present an extremely wide range of structural modifications, mainly to compensate for the geometrical misfit between the lateral dimensions of tetrahedral and octahedral sheets. Lizardite presents a flat structure, chrysotile is characterized by cylindrical or spiral tubes; antigorite shows a modulated wave-like shape (*e.g.* see Wicks & O'Hanley, 1988; Bailey, 1988a).

Like lizardite [$Mg_3Si_2O_5(OH)_4$], the trioctahedral 1:1 minerals berthierine [$(Fe^{2+},Al)_3(Si,Al)_2O_5(OH)_4$)], amesite [$(Mg,Al)_3(Si,Al)_2O_5(OH)_4$)], cronstedtite [$(Fe_2^{2+}Fe^{3+})(Si,Fe^{3+})O_5(OH)_4$], nepouite [$(Ni,Mg)_3Si_2O_5(OH)_4$)], kellyite [$(Mn^{2+},Mg,Al)_3(Si,Al)_2O_5(OH)_4$], fraipontite [$(Zn,Cu,Al)_3(Si,Al)_2O_5(OH)_4$], brindleyite [$(Ni_2Al)(SiAl)O_5(OH)_4$], and guidottiite [$(Mn_2^{2+}Fe^{3+})(SiFe^{3+})O_5(OH)_4$] present a planar structure. Like antigorite (general formula [$Mg_3Si_2O_5(OH)_4$]), bementite [$Mn_7^{2+}Si_6O_{15}(OH)_8$] shows a modulated layer structure, whereas greenalite [$(Fe^{2+},Fe^{3+})_{<3}Si_2O_5\,(OH)_4$], caryopilite [$(Mn^{2+},Mg)_3Si_2O_5(OH)_4$], and minerals of the pyrosmalite series (general formula [$(Mn^{2+},Fe)_8Si_6O_{15}(OH,Cl)_{10}$), such as friedelite [$Mn_8^{2+}Si_6O_{15}(OH,Cl)_{10}$], mcgillite [$(Mn^{2+},Fe^{2+})_8Si_6O_{15}(OH)_8\ Cl_2$], schallerite [$(Mn^{2+}Fe^{2+})_{16}Si_{12}As_3^{3+}O_{36}(OH)_{17}$], and nelenite [$(Mn^{2+},Fe^{2+})_{16}Si_{12}As_3^{3+}\ O_{36}\,(OH)_{17}$]), show islands of tetrahedral sheets (Mandarino & Back, 2004; Wahle *et al.*, 2010).

The serpentine subgroup also includes polygonal serpentine (*e.g.* see Middleton & Whittaker, 1976; Baronnet *et al.*, 1994; Baronnet & Devouard, 2005), and polyhedral serpentines (Baronnet *et al.*, 2007; Andreani *et al.*, 2008). Polygonal serpentines consist of fibres each showing 15 or 30 arranged sectors (Baronnet & Devouard, 2005); polyhedral serpentines comprise spheroids consisting of 92 or 176 lizardite crystals, each one defining a triangular facet of the spheroid (Baronnet *et al.*, 2007; Cressey *et al.*, 2010).

Because of small crystal dimensions and a low level of structural order, crystal-structure data are limited for most of these minerals. A short crystal-structure overview will thus be restricted to samples with 'applications' possibilities and a crystalline order enabling structural investigation.

6.3.1. Mg-rich species

Lizardite (Fig. 10). High quality three-dimensional structural studies for lizardite (Mellini, 1982; Mellini & Zanazzi, 1987, 1989; Mellini & Viti, 1994; Gregorkiewitz *et al.*, 1996; Zhukhlistov, 2007; Mellini *et al.*, 2010; Laurora *et al.*, 2011) were

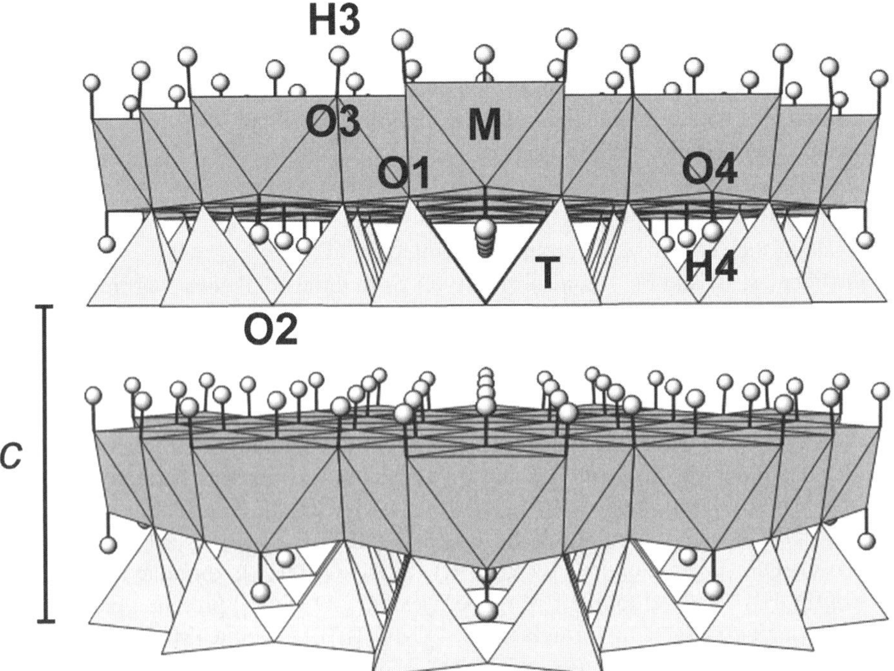

Fig. 10. Crystal structure of lizardite-1*T*. Tetrahedral sites (T), tetrahedral oxygen atoms (O1, O2), octahedral sites (M), (O1) oxygen atom is shared between tetrahedral and octahedral sheet; octahedral oxygen atoms (O3, O4), and hydrogen atoms (H3, H4) related to O3 and O4 oxygen atoms. Readers of the paper version of this chapter may wish to download a colour version of this figure from www.minersoc.org/emu-notes/emu-11/11-1-colour.pdf.

reported mostly for the $1T$ and $2H_1$ polytypes in the space groups $P31m$ (unit-cell parameters: $a = 5.332(3)$, $c = 7.233(4)$ Å, Mellini, 1982), and $P6_3cm$ (unit-cell parameters $a = 5.318(4)$, $c = 14.541(7)$ Å, Mellini & Zanazzi, 1987), respectively. Brigatti *et al.* (1997) examined the $2H_2$ form of Al-rich lizardite (intermediate between lizardite and amesite) in the space group $P6_3$. Previous studies have documented the effects of octahedral substitution on crystal structure, thus demonstrating that: (1) the substitution of trivalent cations in octahedral and tetrahedral sites promotes thermal stability (Caruso & Chernosky, 1979) and crystal order (Mellini & Viti, 1994); (2) homovalent Fe-for-Mg substitution produces a decrease in the octahedral site distortion and an increase in the c unit-cell parameter (Laurora *et al.*, 2011). Laurora *et al.* (2011) also suggested that the whole-rock composition affects lizardite crystal chemistry and structure, which are sensitive to the overprint of secondary, metasomatic events. Hydrogen positions were located on the *Fo* map and then refined (Mellini, 1982; Laurora *et al.*, 2011). Hydrogen bonding is not observed for the hydrogen atom placed at the centre of the tetrahedral hexagonal ring, unlike other hydrogen atoms which help to keep adjacent layers connected.

The effect of pressure on the structure of $1T$ lizardite was investigated by Mellini & Zanazzi (1989). Guggenheim & Zhan (1998) reported a high-temperature study of both $1T$ and $2H_1$ lizardite crystals. Geometrical changes induced by cation substitutions and details of the hydrogen bond were also studied by means of *ab initio* quantum chemistry calculations (Benco, 1997; Benco & Smrčok, 1998; Scholtzová *et al.*, 2000; Scholtzová & Smrčok, 2005; Auzuende *et al.*, 2006). All these studies suggested that: (1) cation substitutions account for geometrical changes in tetrahedral sheets, whereas octahedral sheets are almost unaffected; (2) substituted tetrahedra are tilted and their basal oxygen atoms are pushed below the plane defined from basal vertices of unsubstituted tetrahedra; (3) the hydrogen bond in lizardite consists of dipole to ion interaction with a total energy of 28.121 kJ mol^{-1}; (4) lizardite, as suggested by analysing the variation in cell dimensions following pressure increase, is stiffer in directions parallel to the layer than in the **c** direction, normal to the layer; (5) the cohesive energy between two successive layers along **c** is 0.33 eV (*i.e.* 0.11 eV per OH bond).

The internal dimensions of a 1:1 layer are remarkably less affected by chemical substitutions and by pressure or temperature increase, than interlayer thickness, which varies significantly with composition (Chernoski, 1975) or with pressure increase (Mellini & Zanazzi, 1989). As the interlayer thickness increases, the ditrigonalization of the tetrahedral sheet, which, depending on orientation, can be expressed either by a positive or negative angle, decreases. In the $1T$ polytype, the ditrigonal ring distortion changes from $-1.5°$ to $\sim 0°$ when the temperature changes from 20 to 480°C. In the $2H_1$ polytype, this structural parameter changes from 1.8°, at room temperature, to 1.3° at 300°C and remains unchanged up to 475°C. In the $2H_1$ polytype, the O–O distance, referring to the interlayer O–H...O bond, increases linearly from 3.08 to 3.15 Å as the temperature increases from 20 to 475°C. On the contrary, in the $1T$ polytype, this distance remains nearly constant up to 360°C. Above this temperature, the O–O distance increases slightly, thus apparently suggesting weaker hydrogen bonding for the $2H_1$ polytype than for the $1T$ polytype (Guggenheim & Zhan, 1998).

Antigorite (unit-cell parameters: $a = 43.505(6)$, $b = 9.251(1)$, $c = 7.263(1)$ Å, $\beta = 91.32(1)$, space group *Pm* for the $m = 17$ polysome, Capitani & Mellini, 2004; $a = 81.66(1)$, $b = 9.255(5)$, $c = 7.261(5)$ Å, $\beta = 91.409(5)$ space group $C2/m$ for the $m = 16$ polysome, Capitani & Mellini, 2006). Antigorite is a 1:1 layer silicate characterized by a modulated crystal structure. This mineral, because of its chemical composition and structural arrangement, belongs to the serpentine subgroup, without being considered a serpentine polymorph *sensu stricto*, due to the evident $Mg(OH)_2$ depletion, as indicated by the formula ${}^{vi}M_{3m-3}{}^{v}T_{2m}O_{5m}(OH)_{4m-6}$ (where M = Mg, Fe, Ni, Al; T = Si, Al; m = number of unique tetrahedra spanning a wavelength along the **a** axis) (Capitani & Mellini, 2004).

The antigorite polysomatic series (Spinnler, 1985; Ferraris *et al.*, 1986) was investigated by XRD (*e.g.* see Aruja, 1945; Kunze, 1956, 1958, 1961; Capitani & Mellini, 2004, 2006) and TEM (*e.g.* see Zussman *et al.*, 1957; Yada, 1979; Spinnler, 1985; Uehara & Shirozu, 1985; Mellini *et al.*, 1987; Wu *et al.*, 1989; Otten, 1993; Viti & Mellini, 1996; Uehara, 1998; Dódony *et al.*, 2002; Grobéty, 2003; Dódony & Buseck, 2004; Capitani & Mellini, 2005). The antigorite structure consists of positional modulations of the curved 1:1 silicate layer along the **a** direction (Fig. 11a). Although the structural continuity of both the tetrahedral and octahedral sheets is maintained, the modulation is due to the periodic reversal of the layer polarity (and of silicate tetrahedra), thus generating a wave-like effect, where each individual half wave is characterized by the same orientation of silicate tetrahedra. If m is even, two consecutive half waves are equal in dimension. Whe m is odd, one ('short half wave') is made from $(m/2)\ -0.5$ tetrahedra, whereas the following one ('long half wave') consists of $(m/2)\ +0.5$ tetrahedra. Most antigorite specimens are monoclinic, space groups $C2/m$ and *Pm*, for m even and odd, respectively (Capitani & Mellini, 2004, 2006). Two distinct reversals along the **a** direction are always observed across six-membered rings, where four contiguous tetrahedra point down and the two remaining tetrahedra point up, and *vice versa*. If the structural pattern for the first reversal is accepted, the second reversal is described with two different models. The first model assumes eight-membered rings (8-reversals) with four neighbouring tetrahedra pointing up and four neighbouring tetrahedra pointing down, which alternate along the **b** direction with four-membered rings (4-reversals) with two neighbouring tetrahedra pointing up and the other two tetrahedra pointing down, or *vice versa* (Wu *et al.*, 1989; Otten, 1993; Uehara & Kamata, 1994; Uehara, 1998; Grobéty, 2003, Capitani & Mellini, 2004, 2006; Capitani *et al.*, 2009). This pattern was first supported by HRTEM imaging (Wu *et al.*, 1989; Otten, 1993; Uehara & Kamata, 1994; Uehara, 1998; Grobéty, 2003), and, later confirmed by 3D refinement of X-ray data from the $m = 16$ (Capitani & Mellini, 2004) and $m = 17$ (Capitani & Mellini, 2006, 2008) polysomes.

Another structural model assumes 6-reversals (Dodòny *et al.*, 2002, 2006). Theoretical approaches based on density functional theory indicate that the structural model by Capitani & Mellini (2004) is energetically more favoured (Capitani *et al.*, 2009).

Chrysotile. A preliminary structure determination for chrysotile dates back to a study by Pauling (1930) who suggested for this commonly fibrous mineral, a curved hollow morphology, because of the need to fit tetrahedral and octahedral sheets. Further studies were promoted in the 1950s by Whittaker, starting from XRD data (Whittaker,

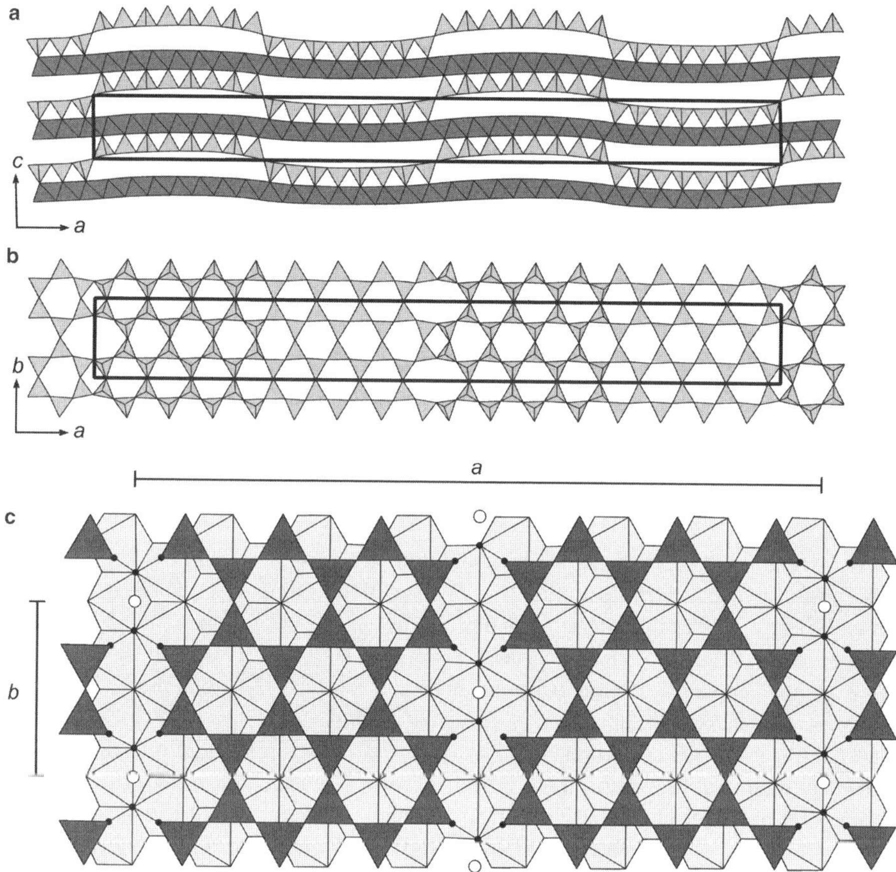

Fig. 11. (*a*, *b*) The antigorite polysome ($m = 16$). (*a*) [010] projection, (*b*) [001] projection of the modulated tetrahedral sheet (modified after Capitani & Mellini, 2006); (*c*) [001] projection of carlosturanite. Small filled circles represent OH groups, open circles refer to water molecules (modified after Mellini *et al.*, 1985). Readers of the paper version of this chapter may wish to download a colour version of this figure from www.minersoc.org/emu-notes/emu-11/11-1-colour.pdf.

1953, 1956a, 1956b, 1956c) and, later on, by Yada (1971), using high-resolution TEM. The latter author also defined three types of chrysotile: clinochrysotile, orthochrysotile and parachrysotile. Clinochrysotile, when its cylindrical structure is ideally developed into a plane, shows a two-layer monoclinic unit cell, where no rotation is observed among layers and where the x axis is parallel to the cylindrical axis. Unlike clinochrysotile, orthochrysotile shows an orthorhombic two-layer unit cell with 180 degrees rotation among the layers. Parachrysotile differs then from orthochrysotile as the y, instead of the x axis, is parallel to the cylindrical axis.

The technological application of chrysotile has decreased significantly over recent decades, substituted in traditional applications, by materials more compatible with human health and safety requirements. In recent times, however, the interest in the synthetic rather than the natural mineral has increased because of the interesting technological properties associated with the nanotube structure. This mineral shows various morphologies: hollow and full cylinders, tube-in-tube cylinders, conically wrapped fibres, cone-in-cone shaped concentric structures, spiral and multi-spiral (Yada, 1971).

Amesite (unit-cell parameters: $a = 5.319(2)$, $b = 9.208(3)$, $c = 14.060(5)$ Å, $\alpha = 90.01(3)$, $\beta = 90.27(3)$, $\gamma = 89.96(3)^\circ$, space group: $C1$; Hall & Bailey, 1979). Several studies have been devoted to the structure determination of amesite (*e.g.* see Brindley *et al.*, 1951; Steinfink & Brunton, 1956; Steadman & Nuttall, 1962; Hall & Bailey, 1979; Anderson & Bailey, 1981). All these studies agree in suggesting a two-layer polytype for this mineral and hexagonal or pseudohexagonal symmetry. Wiewióra *et al.* (1991) identified an uncommon amesite showing a triclinic symmetry and ordering of cations in both tetrahedral and octahedral sites [unit-cell parameters: $a = 5.31(1)$, $b = 9.212(2)$, $c = 14.401(7)$ Å, $\alpha = 102.11(3)$, $\beta = 90.2(1)$, $\gamma = 90.1(1)^\circ$]. Two stacking modes were recognized for amesite layers: the first shows a 180° rotation with no shift of adjacent layers, whereas the second shows a 180° rotation combined with a translation of $-b/3$.

Carlosturanite (Fig. 11c) shows a modulated structure (Compagnoni *et al.*, 1985), with the following general formula: $M_{21}[T_{12}O_{28}(OH)_4](OH)_{30}{\cdot}H_2O$ where M is predominantly Mg^{2+} with small amounts of Fe^{3+}, Mn^{2+}, Ti^{4+}, Cr^{3+}, and T is Si^{4+} or Al^{3+} and unit-cell parameters: $a = 36.70(3)$, $b = 9.41(2)$, $c = 7.291(5)$ Å, $\beta = 101.1(1)$, space group: Cm. Carlosturanite is poorer in Si and richer in H_2O than the common serpentine-group phases (Compagnoni *et al.*, 1985; Belluso & Ferraris, 1991). On the basis of its chemical properties and results of a detailed structural investigation, supported by high-resolution TEM, Mellini *et al.* (1985) proposed a structural model based on the interruption of the tetrahedral sheet (T_2O_5) and on the introduction of vacancies at tetrahedral sites along rows parallel to the direction of TO chains. In this model, the octahedral sheet of the serpentine structure is preserved, whereas $\frac{1}{7}$ of the $[Si_2O_7]^{6-}$ tetrahedral groups is replaced by $[(OH)_6H_2O]^{6-}$ groups, bonding to only one Si ion. The resulting layer of tetrahedra is thus formed from triple chains of tetrahedra bound to each other by means of H_2O molecules. Starting from a carlosturanite structural arrangement, Mellini *et al.* (1985) introduced the inophite family name to indicate a polysomatic series resulting from combination of serpentine S modules with composition $M_3T_2O_5(OH)_4$ (M = octahedral cations, T = tetrahedral cations) and X modules with composition $M_6T_2O_3(OH)_{14}{\cdot}H_2O$.

6.3.2. *Fe-rich species*

Cronstedtite. Kogure *et al.* (2001) discovered a great variety of polytypes in cronstedtite, *i.e.* all polytypes of groups A, C and D defined by Bailey (1969). Three-dimensional crystal-structure refinements for this mineral were provided by several authors and more recently determined or re-determined for polytype $1T$, space group $P31m$ [$a = 5.512(1)$, $c = 7.106(1)$ Å], for $3T$ polytype, space group: $P3_1$ [$a = 5.497(2)$, $c = 21.355(7)$ Å] and for $2H_2$ polytype, space group $P6_3$ [$a = 5.500(1)$, $c = 14.163(2)$ Å] (Smrčok *et al.*, 1994; Hybler *et al.*, 2000; Hybler *et al.*, 2002; Ďurovič *et al.*, 2004; Hybler,

2006). All polytypes show a full Fe octahedral occupancy, while differing for tetrahedral Si/Fe occupancy and/or ordering. In the $1T$ polytype, the Si/Fe ratio is ~3:1 (Hybler *et al.*, 2000). In the $2H_2$ polytype there is an Fe tetrahedral ordering, which is missing in the $3T$ polytype (Smrčok *et al.*, 1994).

Berthierine. The mineral, poorly crystallized, is commonly considered to be typical of marine sediments, probably because of its frequent occurrence in marine-oolitic ironstone formations (Brindley, 1982). However, non-marine occurrences have also been documented. The mineral is frequently associated (or interlayered) with Fe-rich chlorites, as demonstrated by electron diffraction studies (Jiang *et al.*, 1992). Berthierine is commonly a one-layer structure, showing semi-random stacking.

Greenalite. The structure of greenalite (Guggenheim & Eggleton, 1998) consists of octahedrally coordinated Fe and of continuous trioctahedral sheets. Six-member rings of tetrahedra link together to form triangular islands, made up of four or five tetrahedra and coordinating one octahedral sheet. The number of tetrahedra, constituting each island, varies constantly, according to a pattern assuring a limited, short-range order, but missing any long-range ordering, because of domain-boundary linkages and domain positioning. This structural modulation gives a chemical formula, deviating considerably from the stoichiometry characterizing serpentine minerals. In greenalite, Fe atoms are divalent and surrounded by six Fe atoms at 3.21–3.22 Å, consistently with edge-sharing Fe octahedra (Manceau *et al.*, 1995).

6.3.3. *Mn-rich species*

Kellyite is considered to be the Mn analogue of amesite. Peacor *et al.* (1974) identified two main polytypes: a six-layer rhombohedral phase and a two-layer hexagonal phase (space group $P6$).

Guidottiite is the Mn analogue of cronstedtite (Wahle *et al.*, 2010). This mineral, only recently described, presents the coexistence of $2H_1$ and $2H_2$ polytypes, together with areas dominated by disordered layer stacking. The single-crystal structure refinement suggests that the $2H_2$ polytype is hexagonal, space group $P6_3$ with $a = 5.5472(3)$, $c = 14.293(2)$ Å. Substitution of Fe^{3+} for Si occurs in tetrahedral sites.

Bementite. The description of bementite structure, a modulated phyllosilicate, was addressed by several authors (*e.g.* see Kato, 1963; Kato & Takeuchi, 1980), who suggested an orthorhombic symmetry (space group, $P222_1$). Heinrich *et al.* (1994) described the bementite structure in space group $P2_1/c$, with $a = 14.838(2)$, $b = 17.584(2)$, $c = 14.700(2)$ Å, $\beta = 95.54(2)^\circ$, as consisting of two hexagonal sheets of Mn octahedra, which are alternately rotated by 22° in the *ab* plane. These octahedral sheets are interlayered by a continuous tetrahedral sheet containing pairs of six-membered rings interconnected with five- and seven-membered rings (Fig. 12). Inverted tetrahedra form strips showing the same orientation (up or down) parallel to the ***a*** direction. Linked pairs of six-membered rings are rotated, with respect to pairs across the strip boundaries, by 22° to allow coordination with adjacent octahedral sheets. The bementite topology is closely related to that of armbrusterite $K_5Na_6Mn^{3+}Mn^{2+}_{14}[Si_9O_{22}]_4(OH)_{10} \cdot 4H_2O$, space group $C2/m$, with $a = 17.333(2)$, $b = 23.539(3)$, $c = 13.4895(17)$ Å, $\beta = 115.069(9)^\circ$. The structure of armbrusterite is

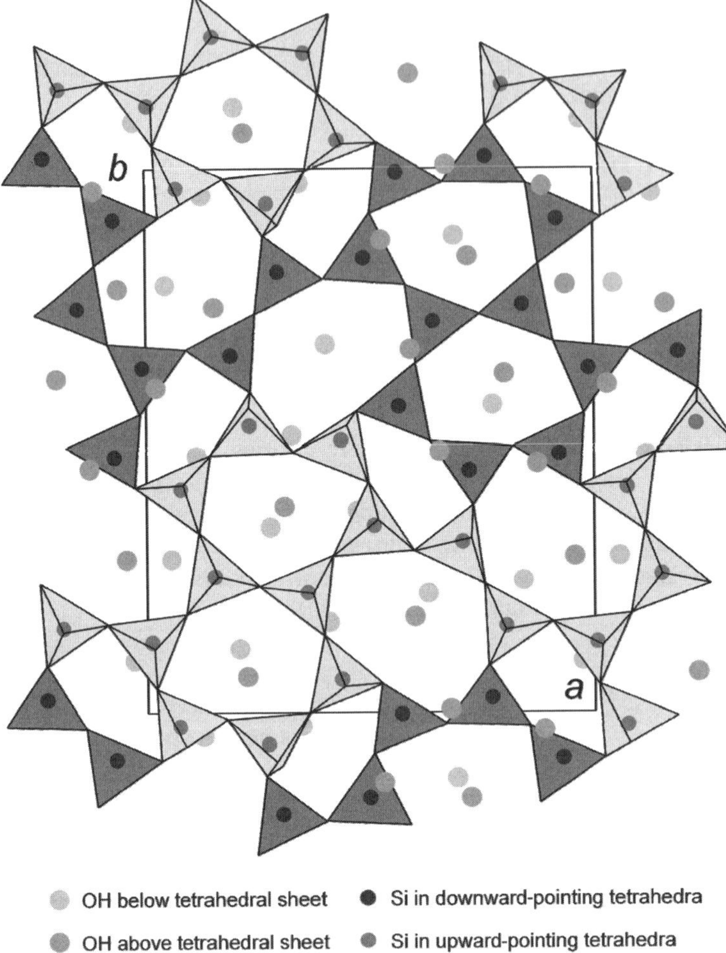

Fig. 12. Projection of bementite tetrahedral sheet on the *ab* plane. The tetrahedral sheet is sandwiched between two Mn octahedral sheets, alternately rotated by 22° in the *ab* plane (modified after Heinrich *et al.*, 1994). Readers of the paper version of this chapter may wish to download a colour version of this figure from www.minersoc.org/emu-notes/emu-11/11-1-colour.pdf.

based on double silicate sheets consisting of [5]-, [6]-, [7]- and [8]-membered tetrahedral rings. The sheets are linked by octahedral sheets formed by Na and Mn octahedral. K is located between two silicate sheets and links H_2O molecules (Yakovenchuk *et al.*, 2007).

Pyrosmalite group. Pyrosmalite is the name given to a relatively rare series of Fe-Mn silicates, usually containing some chlorine also. Mn-rich members of the series are more common. The name friedelite is reserved for the pure Mn end-member (Vaughan, 1986; Ozawa *et al.*, 1983). Friedelite, mcgillite and nelenite show 1*M* structure (space group *C*2/*m*). The basic X-ray reflections characterizing the monoclinic structure are sharp in mcgillite and diffuse in friedelite, which may thus be considered as the disordered equivalent of mcgillite.

Caryopilite structure is similar to that of greenalite, while differing from the latter because of a predominant Mn octahedral population.

6.3.4. Ni-rich species

Nepouite generally gives poorer XRD patterns than lizardite and is recognized by 7.2–7.3 Å basal spacing. The stacking of the layers is highly disordered. The large Ni content is indicated by the intense, dark green colour (Brindley & Wan, 1975). Manceau & Calas (1986) and O'Day *et al.* (1994), using EXAFS, suggested that Ni atoms, in Ni-rich serpentine species, cluster in octahedral domains with the Ni–O distance ranging from 2.03 to 2.08 Å and Ni–Ni distances ranging from 3.05 to 3.07 Å.

Fraipontite. Although several natural occurrences are reported, studies on fraipontite are mostly related to synthetic phases due to its considerable industrial and commercial interest (as a smoke suppressant, honeycomb forms for adsorbents and deodorants, in paint, in air-purification filters, as an odour adsorbent, in the treatment of waste gases, and as a blood coagulant for treatment of waste blood) (Kloprogge *et al.*, 2001). The mineral, poorly crystallized, presents a basal reflection at 7.44 Å.

Brindleyite is closely related in structure and stacking sequence to berthierine and occurs as mixtures of the groups A and C polytypes, as defined by Bailey (1969). The mineral is also characterized by a disordered cation arrangement (Maksimovic & Bish, 1978).

7. 2:1 layer structures

7.1. Talc and pyrophyllite group

Talc is a 2:1 layered, trioctahedral Mg silicate mineral; its ideal structural formula is $Mg_3Si_4O_{10}(OH)_2$. A small amount of water may also be present. Pyrophyllite is also a 2:1 layered hydrous silicate mineral. Its layer structure is dioctahedral and Al is located in the larger octahedron; its ideal structural formula is $Al_2Si_4O_{10}(OH)_2$ and, like talc, can include a small amount of water (Fig. 2a). The chemical composition of talc and pyrophyllite can vary depending on their geological history and parent-rock association. The ideal layer structure of talc and pyrophyllite is electrically neutral, and contiguous TOT layers are connected by van der Waals interactions, thus significantly impacting the physical (and mechanical) properties of both minerals. Both minerals are very soft: talc is the softest mineral on the Mohs hardness scale at 1, and pyrophyllite is at 1 to 2 on that scale of 1–10.

7.1.1. Pyrophyllite and ferripyrophyllite

The three-dimensional crystal structure refinement of pyrophyllite [space group $C\bar{1}$ with $a = 5.160(2)$, $b = 8.966(3)$, $c = 9.347(6)$ Å, $\alpha = 91.18(4)$, $\beta = 100.46(4)$, $\gamma = 89.64(3)$, Lee & Guggenheim, 1981] suggests that the tetrahedral $\langle T\text{–O}\rangle$ distance

(1.618 Å) is consistent with the lack of significant Al^{3+}-for-Si^{4+} substitutions and that the octahedral $\langle M\!-\!O\rangle$ distance (1.912 Å) matches well with complete Al occupancy. The pyrophyllite layer presents a tetrahedral rotation of $\sim 10^{\circ}$ and a strong corrugation of the basal oxygen-atom plane. Polytypism in pyrophyllite was addressed by several authors (see Evans & Guggenheim, 1988 for a review). Two dominant sequences were identified: a two-layer monoclinic, and a one-layer triclinic, sometime coexisting in the same sample. A significant number of theoretical studies has also been devoted to pyrophyllite, in particular suggesting that Mg^{2+}-for-Al^{3+} substitutions tend to be distributed in the octahedral sheet, whereas Fe^{3+}-for-Al^{3+} substitutions tend to be clustered (Sainz-Diaz *et al.*, 2002).

Ferripyrophyllite, $Fe^{3+}_2Si_4O_{10}(OH)_2$, is isostructural with the 2*M* modification of pyrophyllite and presents virtually no substitutions in either the tetrahedral or octahedral sheet, which are thus occupied exclusively by Si and Fe^{3+}, respectively. Ferripyrophyllite was first recognized by Chukhrov *et al.* (1979), who identified a ferric analogue of pyrophyllite in a precipitate from a low-temperature hydrothermal solution. After powder XRD measurements, these authors reported a monoclinic unit cell with $a = 5.26$, $b = 9.10$, $c = 19.1$ Å, and $\beta = 95.30^{\circ}$.

Ferriphyrophyllite can be considered as the Fe end-member of the series $(Fe,Al)_2Si_4O_{10}(OH)_2$. Coey *et al.* (1984) were the first to investigate the cation distribution in the layer by means of Mössbauer spectroscopy, in order to clear out divergent and conflicting interpretations of the spectra for ferric 2:1 layer minerals. Their experimental data demonstrated a Mössbauer spectrum at 4.2K, consisting of a single, resolved magnetic pattern with hyperfine field ($B_{hf} = 51.8$ T). The relatively high Néel temperature, compared to other dioctahedral ferric phyllosilicates, further demonstrated that Fe^{3+} cations tend to be ordered on *M*2 sites.

7.1.2. Talc and talc-like minerals

Pioneering studies described talc structure as monoclinic ($C2/c$ space group, Gruner, 1934a,b). More recently, the talc structure has been described as triclinic (space group $C\bar{1}$, Rayner & Brown, 1973), or as pseudomonoclinic (space group *Cc*), or else in the $C\bar{1}$ space group with $a = 5.290(3)$, $b = 9.173(5)$, $c = 9.460(3)$ Å, and $\alpha = 90.46(5)$, $\beta = 98.68(5)$, $\gamma = 90.09(5)^{\circ}$ (Perdikatsis & Burzlaff, 1981). The talc layer presents a nearly planar basal oxygen surface and the tetrahedral rotation at 3.6°. The cation distribution in the talc structure was described recently (Petit *et al.*, 2004; Martin *et al.*, 2006), following results obtained from Fourier-transform infrared (FTIR), Mössbauer and MAS-NMR spectroscopies. These studies quantified Fe^{2+}- (or Al^{3+}-) for-Mg substitution and the F content from the 2νOH region and suggested that Fe^{2+} (or Al^{3+}) substitutes randomly for Mg in octahedral sites. Detailed images of the talc basal oxygen plane at the molecular scale were obtained recently by atomic force microscopy (AFM) (Ferrage *et al.*, 2006), indicating for talc a flat surface and well defined tetrahedral rings, deviating little from hexagonal symmetry.

Different polytypic sequences of TOT layers were discussed by Weiss & Ďurovič (1984), who found 10 non-equivalent polytypes, where only seven of these can be distinguished by XRD.

Willemseite $(Ni,Mg)_3Si_4O_{10}(OH)_2$ is the Ni-bearing analogue of talc. Willemseite is a secondary mineral in a Ni-bearing layered igneous sill, and it is characterized by a fine-grained and massive structure, by a perfect {001} cleavage, and by a Mohs hardness scale equal to 2. De Waal (1970), who first suggested the name willemseite, indicated that willemseite is monoclinic (space group Cc or $C2/c$) with $a = 5.316(2)$, $b = 9.149(3)$, $c = 18\ 994(6)$ Å, and $\beta = 99.96(6)°$.

Ni should only be observed in octahedral coordination, as a substitute for Mg; nevertheless, Tejedor-Tejedor *et al.* (1983) demonstrated the presence of tetrahedral Ni by comparing the Ni/Si atomic ratio in extremely well crystallized willemseite, to its theoretical stoichiometry.

As demonstrated by Koshimizu *et al.* (1981), after measurements on various hydrothermally synthesized terms of the talc–willemseite series, the Ni contents could be derived both from the intensities of the absorption band in the OH-stretching region, and from the temperature value of the DTA endothermic effect related to dehydroxylation reaction. In particular, the greater the Ni^{2+}/Mg^{2+} ratio, the greater are both the intensity of the absorption band in the OH-stretching region and the temperature value at which the dehydroxylation reaction occurs.

Several investigations, based mainly on spectroscopic data (*e.g.* optical spectroscopy, IR spectroscopy, X-ray absorption spectroscopy (XAS), *etc.*), agree well, considering Ni as bivalent (Brindley *et al.*, 1979; Manceau *et al.*, 1984, 1985), without determining any appreciable distortion of the coordinating polyhedron when Ni substitutes for Mg (Manceau *et al.*, 1985). In addition, XAS measurements indicate that Ni atoms are segregated into domains inside the octahedral sheet, with an average size of at least 30 Å. However, experimental data could not exclude the possible existence of octahedral sheets completely occupied by Ni (Manceau & Calas, 1986).

7.2. The mica group: some recent advances in crystal chemistry and structure of dioctahedral and trioctahedral micas

The crystal chemistry and structure of dioctahedral and trioctahedral micas were reviewed recently by Brigatti & Guggenheim (2002). A complete description of the complex interrelationships between chemical composition and structure in micas is not given here, therefore. A short discussion of the main aspects of mica crystal chemistry and of some new trends and results is given, however.

Although the basic structure of a mica layer is well known, the accurate crystal-chemical description of micas is often a challenge because of: (1) the large number of cationic and anionic substitutions; (2) the presence of the same element in different structural sites; (3) different oxidation states for metals; (4) light elements not detected by electron microprobe analysis; (5) structural vacancies; and (6) cation ordering. Recent advances in mica crystal chemistry are mostly attributed to: (1) the determination of light elements, Fe speciation, hydrogen content (Scordari *et al.*, 2006; Mesto *et al.*, 2006; Scordari *et al.*, 2008; Di Vincenzo *et al.*, 2006; Gianfagna *et al.*, 2007; Matarrese *et al.*, 2008); (2) detailed structural evidence based on both X-ray and neutron diffractions at room and given temperature and pressure conditions

(Zanazzi & Pavese, 2002; Ferraris & Ivaldi, 2002; Pavese *et al.*, 2003; Comodi *et al.*, 2004; Welch *et al.*, 2004; Welch & Crichton, 2005; Curetti *et al.*, 2006; Dobson *et al.*, 2007; Zanazzi *et al.*, 2007a,b; Pavese *et al.*, 2007; Gemmi *et al.*, 2008; Ventruti *et al.*, 2009; Gatta *et al.*, 2009); (3) spectroscopic determinations on both powder and single crystals, *e.g.* IR (Scordari *et al.*, 2006; Beran, 2002; Chon *et al.*, 2006; Ferrage *et al.*, 2006; Boukili *et al.*, 2003; Busigny *et al.*, 2004; Asimow *et al.*, 2006; Mookherjee *et al.*, 2002; Piccinini *et al.*, 2006a,b), Raman (Rinaudo *et al.*, 2004; Comodi *et al.*, 2006), Mössbauer (Rancourt *et al.*, 2001; Dyar, 2002; Evans *et al.*, 2005; Redhammer *et al.*, 2005; Dyar *et al.*, 2008; Mercier *et al.*, 2006; Zazzi *et al.*, 2006), XAFS (Mottana *et al.*, 2002b; Cibin *et al.*, 2005, 2006, 2008; Dyar *et al.*, 2002a,b), XPS (Schingaro *et al.*, 2005), and NMR (Fechtelkord *et al.*, 2003a,b; Lee & Stebbins, 2003).

The study of polytypism, twinning, microstructures, intergrowths of different stacking sequences and zonation was recently further developed both experimentally (Dobson *et al.*, 2007; Ferraris *et al.*, 2000, 2005; Giorgetti *et al.*, 2000; Di Vincenzo *et al.*, 2003; Kogure, 2002; Kogure *et al.*, 2005, 2006; Kogure & Kameda, 2008; Viti *et al.*, 2004; Sugimori *et al.*, 2008) and theoretically (Nespolo & Ďurovič, 2002; Nespolo *et al.*, 2004). Theoretical models allow the description of cation ordering from structure determinations and spectroscopic data (Brigatti *et al.*, 2003, 2005, 2008a,b), from Monte-Carlo simulation (Palin & Dove, 2004; Palin *et al.*, 2004) and from MEM approaches (Merli *et al.*, 2009). The literature cited reveals widespread attention to micas, and the abundance of recent and relevant results means that they cannot be summarized here. Aspects which may bring together some features of mica crystal chemistry with those of 2:1 layer clay minerals, will be detailed only instead.

2:1 clay minerals sometimes show an octahedral occupancy that appears to be intermediate between dioctahedral and trioctahedral species. This crystal-chemical feature may either be attributed to an interlayering of dioctahedral- and trioctahedral-like layers or to a cation-exchange mechanism involving a vacant octahedral cavity. Usually the latter is placed in the *trans*-site for micas, whereas it can be in the *trans*- or in the *cis*-site for 2:1 illite, as suggested by Drits *et al.* (1984), who also predicted and outlined the powder XRD features of illites with *trans*-vacant and *cis*-vacant sites and structures where there is a statistical distribution of cations over all three sites (simulating a trioctahedral arrangement).

As mentioned, current experimental constraints prevent a final conclusion on most clay minerals, but in some cases for micas, a single-crystal structure solution is viable. The octahedral sheet of ideal muscovite consists of two Al-occupied octahedral *cis*-sites and one octahedral vacancy in the *trans*-site. Structural refinements on a great number of dioctahedral micas show a residual occupancy at the *trans*-site and Fe- or Mg-for-Al substitutions, connecting muscovite and celadonite end-members. Brigatti *et al.* (2005) demonstrated that the residual occupancy observed for the *trans*-octahedral site is related to octahedral Al substitution by assessing a direct relationship between the probability of finding a cation other than Al in octahedral coordination and the probability of finding a *trans*-octahedral site being occupied. This evidence seems to suggest that the residual occupancy at the *trans*-octahedral site should be interpreted as the substitution of a trioctahedral cell, consisting largely of divalent and monovalent cations, for a dioctahedral cell, rather than simply as the substitution of a monovalent or divalent cation for a vacancy. In micas, however, this substitution mechanism does not appear to be complete.

In other words, all the dioctahedral micas refined so far show an octahedral Al content greater than 1.4 atoms per formula unit (a.p.f.u.) calculated on $O_{10}(OH)_2$. Moreover, the maximum Al content in trioctahedral micas is <1 a.p.f.u., to reach ~1.1 a.p.f.u. in Li-rich micas. Therefore a composition gap in the range 1.1–1.4 a.p.f.u. seems to persist.

This behaviour may be attributed to crystal-chemical features and in particular to the requirements deriving from the matching of octahedral to tetrahedral sheets.

A discriminating feature between micas and most 2:1 clays is related to interlayer occupancy. The interlayer site in micas is usually fully occupied either by monovalent cations (mostly K) in true micas, or by divalent cations in brittle micas. On the contrary, the interlayer of most clay minerals is not fully occupied and is sometimes hydrated. Mica from some metamorphic environments may constitute an exception, as both chemical analyses and structure refinements seem to suggest that the interlayer site is not fully occupied (interlayer occupancy ≤ 0.90 a.p.f.u.), consequently presenting low interlayer charge. This feature is consistent either with $^{[XII]}(K,Na)^{+}_{-1}$ $^{[XII]}(H_3O)^{+}$ substitution or with $^{[XII]}(K, Na)^{+}_{-1}$ □ substitution. Reaching a final conclusion on this issue is extremely challenging, mostly because of the limited number of crystal-structure refinements for micas from metamorphic environments. Metamorphic micas, for which structural refinements are available, show that when the value of the interlayer occupancy decreases, the size of the interlayer site increases. This result is consistent with interlayer vacancies or with the substitution mechanism $^{[XII]}(K, Na)^{+}_{-1}$ $^{[XII]}(H_3O)^{+}$, or both. Furthermore, the tetrahedral flattening angle, τ, increases with interlayer occupancy and, unlike most micas described in the literature, it decreases with increasing Si content (Brigatti *et al.*, 2008b). These data are consistent with a reduced interlayer charge, thus suggesting the existence of interlayer vacancies in metamorphic micas.

7.3. The smectite group

In the smectite-group minerals, isomorphic substitutions in either tetrahedral or octahedral sites induce a permanent negative layer charge (between –0.1 and –0.6 per half unit cell), which is compensated by the presence of hydrated cations in the interlayer.

The octahedral sheet in smectite group minerals may either be mainly occupied by trivalent cations (dioctahedral smectites) or divalent cations (trioctahedral smectites).

The general formula is:

$$\left(M^{+}_{(x+y+2z+3w)}, M^{2+}_{\frac{(x+y+2z+3w)}{2}}\right) \times n\mathrm{H_2O}(R^{3+}_{2-(y+z+w)}R^{2+}_{y}R^{+}_{z}V_{w})(\mathrm{Si}^{4+}_{4-x}R^{3+}_{x})\mathrm{O}_{10}(\mathrm{OH})_2$$

for dioctahedral smectites and:

$$\left(M^{+}_{(x-y+z-2w)}, M^{2+}_{\frac{(x-y+z-2w)}{2}}\right) \times n\mathrm{H_2O}(R^{2+}_{3-(y+z+w)}R^{3+}_{y}R^{+}_{z}V_{w})(\mathrm{Si}^{4+}_{4-x}R^{3+}_{x})\mathrm{O}_{10}(\mathrm{OH})_2$$

for trioctahedral species, where *x*, *y* and *z* indicate the layer charge resulting from substitutions in tetrahedral and octahedral sites, respectively; R^+, R^{2+} and R^{3+} refer to a generic monovalent, divalent and trivalent octahedral cation; M^+ and M^{2+} refer to a generic monovalent and divalent interlayer cation; *V* indicates a vacancy.

A wide range of cations can occupy tetrahedral, octahedral and interlayer positions. Commonly Si^{4+}, Al^{3+} and Fe^{3+} are found in tetrahedral sites. Al^{3+}, Fe^{3+}, Fe^{2+}, Mg^{2+}, Ni^{2+}, Zn^{2+} and Li^+ generally occupy octahedral sites.

Most of the technological uses of smectite are related to reactions taking place in the interlayer space where cations such as Na^+, K^+, Ca^{2+} and Mg^{2+}, which balance the negative 2:1 layer charge, are commonly hydrated and exchangeable.

The ability of smectites to incorporate interlayer H_2O molecules and the subsequent change in basal spacing has been studied extensively for several decades (MacEwan & Wilson, 1980; Sposito & Prost, 1982; Newman, 1987; McBride, 1989; Güven, 1992; Brown *et al.*, 1995; Moore & Reynolds, 1997; Laird, 2006; Ferrage *et al.*, 2005a,b, 2007, 2010 should be consulted for more details). The hydration properties of smectites are seen to be controlled by several factors, such as the type of interlayer cation and the amount and location of layer charge. Crystalline swelling is controlled by the balance between the repulsive forces, following from 2:1 layer interactions, and the attractive forces between hydrated interlayer cations and the negatively charged surface of Si–O layers (Norrish, 1954; van Olphen, 1965; Kittrick, 1969a,b; Laird, 1996, 1999, 2006).

The observation of (00*l*) basal reflections on XRD patterns demonstrated that 2:1 layer periodicity along **c*** can swell from ~10.0 to 19.0 Å depending on the intercalation of layers of water molecules between individual 2:1 layers (*i.e.* in the interlayer positions). Pioneering XRD studies of smectite hydration analysed the position of (00*l*) basal reflections as a function of relative humidity, and revealed a stepwise expansion, where the different steps correspond to the intercalation of 0, 1, 2 or 3 planes of H_2O molecules in the interlayer (Moore & Reynolds, 1997). Different layer types were thus defined for smectite: dehydrated (d_{001}: 9.7–10.2 Å), monohydrated (d_{001}: 11.6–12.9 Å), bihydrated (d_{001}: 14.9–15.7 Å), and trihydrated (d_{001}: 18.0–19.0 Å). Structural heterogeneities in layer-charge location and/or in its value most often lead to the coexistence of different domains within smectite crystals (Ferrage *et al.*, 2005a,b). These heterogeneities are revealed by comparing XRD patterns recorded on the same smectite sample under contrasting relative-humidity conditions (Ferrage *et al.*, 2005a).

Laird (2006) suggested that the expandability of 2:1 phyllosilicates, saturated with alkali and alkaline earth cations in aqueous systems, is controlled by six different factors: (1) the crystalline swelling (deriving from the balance between electrostatic-attraction and hydration-repulsion forces); (2) the double-layer swelling (occurring when an electrostatic repulsion force develops between positively charged diffuse portions of double smectite layers overlapping in an aqueous suspension); (3) the formation and break-up of smectite layer-clusters; (4) cation demixing (depending by the nature of exchangeable cations); (5) the co-volume swelling (an entropy-driven process caused by restrictions on the rotational freedom of suspended smectite particles); and (6) Brownian swelling (resulting from random thermal motion of suspended smectite particles). The crystalline swelling occurs within each smectite layer and depends strongly on smectite layer charge: double-layer swelling, co-volume swelling, or Brownian swelling occur between two different smectite particles and are less affected by smectite layer charge. However, because layer charge influences the size and stability of smectite particles, layer charge may show an indirect effect also on double-layer and co-volume swelling. In mixed cation-smectite systems, layer charge significantly affects cation exchange selectivity and thus the relative proportions and distributions of the various cations in the interlayer and in exchange sites on external crystal surfaces. Demixing of cations, where, for example, Na preferentially concentrates in certain interlayers, significantly impacts on break-up and formation of smectite particles and thus on swelling.

In monohydrated smectite, interlayer cations are considered to be located in the mid-plane of the interlayer, together with H_2O molecules for one layer. For bi-hydrated layers, interlayer cations are also commonly assumed to be located in the mid-plane of the interlayer. In addition, it is usually assumed that two planes of H_2O molecules, each bearing 0.69 H_2O per $O_{20}(OH)_4$, are located at 0.35 and 1.06 Å from the cation along the **c*** axis, whereas a third plane (1.20 H_2O per $O_{20}(OH)_4$) is located further away from the central interlayer cation at 1.20 Å along the **c*** axis (Moore & Reynolds, 1997). For bi-hydrated smectites, Ferrage *et al.* (2005a,b, 2010) proposed an alternative configuration for H_2O molecules in the interlayer, gaining a better fit with the XRD spectrum in the high-angle region. This model considers a unique plane of H_2O molecules located, along the **c*** axis, on either side of the central interlayer cation. The layer hydration depends heavily on layer charge: as layer charge increases, the number of interlayer cations, and thus the total hydration, increases. Structural variations related to the transition from different hydration states were addressed by Ferrage *et al.* (2007). They suggested that the transition from bi- to mono-hydrated layers produced the maximum structural heterogeneity, with strong interlayer thickness fluctuation (in individual layers), and the presence of several elementary mixed-layer structures. In contrast, the transition from mono-hydrated to dehydrated layers occurs homogeneously within layers. The decrease in thickness of the bi-hydrated layer with dehydration is thus controlled by a mechanism of two-dimensional diffusion of water molecules through the interlayer space. Variation in the thickness of mono-hydrated layers is produced by localized layer collapse, following the restoration of the interlayer cation hydration shell (Ferrage *et al.*, 2007).

Saponite, ideally $(M^+_x \cdot nH_2O)Mg^{2+}_3(Si^{4+}_{4-x}Al^{3+}_x)O_{10}(OH)_2$, is a trioctahedral smectite, where the layer charge derives mostly from Al-for-Si tetrahedral substitutions. Octahedral Fe^{3+}- and Fe^{2+}-for-Mg substitutions are also possible, and several saponite with different Mg/Fe_{total} ratios are described in literature (*e.g.* see Güven, 1988). The interlayer is usually occupied by Ca, Na and K; basal spacing in air-dried species can vary from 13.5 up to 16.8 Å, where the larger limit applies to glycolated samples. The first crystallographic and structural modelling investigations were by Suquet *et al.* (1975, 1977) on a monoclinic, *C*-centred, Na-saturated saponite with a tetrahedral charge of 0.46 and unit-cell parameters $a = 5.333$, $b = 9.233$, $c = 15.42$ Å, $\beta = 96.66°$.

Nowadays, most of the applied and theoretical research on saponite refers to synthetic samples. Because of the potential applications of this mineral, much research has focused on the hydration properties and interlayer organization of water and cations, to better understand the reactivity of smectites towards water in natural media. Several experimental and computational investigations (*e.g.* see Coombes *et al.*, 2003) were carried out to model the disorder of interlayer H_2O molecules. Ferrage *et al.* (2010) integrated the positional disorder to model the fine interlayer structure of saponite and its evolution until dehydration, using low- and high-charge synthetic Na-saturated saponites. Modelling was achieved by collecting XRD patterns along a water-vapour-desorption isotherm and fitting with two randomly interstratified mixed-layer structures down to ~65% relative humidity. The XRD profile modelling can detail aspects of the structural evolution of synthetic saponites upon dehydration. Profile modelling can only be achieved, however, by assuming positional disorder of water molecules around one or two positions for mono- or bihydrated layers, respectively. Furthermore, the modelling procedure allows the discrimination of different types of H_2O molecules (*i.e.* crystalline water and pore-space network) present in saponite.

Hectorite, ideally $(M_y^+ \cdot nH_2O)(Mg_{3-y}^{2+}Li_y)Si_4^{4+}O_{10}(OH)_2$, is characterized by Li^+-for-Mg^+ and F^- for $(OH)^-$ substitutions in the octahedral sheet. Basal spacing, measured on an oriented Na-saturated clay film, gives a periodicity of 12.4 Å (McAtee & Lamkin, 1978). Only a few structure refinements are reported in the literature. Recently, Breu *et al.* (2003) identified a monoclinic cell (space group $C2/m$) with unit-cell parameters $a = 5.2401$, $b = 9.0942$, $c = 10.7971$ Å and $\beta = 99.21°$, for a Cs-saturated sample with formula $Cs_{0.601}(Mg_{2.398}Li_{0.602})Si_4O_{10}F_2$. Seidl & Breu (2005) provided a single-crystal structure refinement of a synthetic Cs-saturated hectorite intercalated with tetramethylammonium (TMA), determining a monoclinic cell (space group $C2/m$) and unit-cell parameters: $a = 5.2735(11)$, $b = 9.1165(14)$, $c = 13.5609(35)$, and $\beta = 97.693(3)°$. As in saponite, several investigations were focused on preparing and characterizing synthetic samples (*e.g.* see Sun *et al.*, 2008), and on modelling both water diffusion and stacking disorder in relation to the interlayer cation (Breu *et al.*, 2003; Malikova *et al.*, 2007). In particular, Malikova *et al.* (2007), by means of a quasi-elastic neutron scattering technique on a synthetic Na-hectorite, documented water-diffusion coefficients of $\sim 1.5 \times 10^{-10}$ m^2 s^{-1} and 4.5×10^{-10} m^2 s^{-1} for the monohydrated and bihydrated states, respectively.

Sauconite, ideally $(M_x^+ \cdot nH_2O)Zn_3^{2+}(Si_{4-x}^{4+}Al_x^{3+})O_{10}(OH)_2$, is a Zn-bearing smectite with octahedral Zn cations ranging from 1.48 to 2.89 a.p.f.u. and with the total octahedral occupancy varying from 2.70 to 3.06 a.p.f.u., thus suggesting that layer charge is dependent on tetrahedral substitution (Güven, 1988). Interlayer cations are commonly Ca, Na and K. Ross (1946) reported a monoclinic cell (space group $C2/m$) and unit-cell parameters $a = 5.34$, $b = 9.32$, $c = 15.8$ Å, and $\beta = 95°$ for sauconite. Another detailed investigation, also including several determinations at non-ambient temperature conditions, is that by Faust (1951). Thermal analyses reported a marked exothermic effect at ~810°C, closely related to the amount of Fe present in the mineral: the greater the Fe content, the lower the temperature of the exothermic reaction. This effect might be related to the formation of a Zn-spinel. Thermal analyses of saponite samples showed no higher-temperature exothermic reaction, which could be masked by a double endothermic reaction between 832 and 978°C (Faust, 1951).

Stevensite, ideally $(Na,Ca/2)_{0.3}Mg_3Si_4O_{10}(OH)_2$, is a trioctahedral smectite that derives its layer charge from both octahedral vacancies and Li-for-Mg substitution (Faust & Murata, 1953; Faust *et al.*, 1959; Ianovici *et al.*, 1990). It is a typical alteration product of sepiolite, and it is frequently observed to be interstratified with non-expanding talc-like minerals (*e.g.* see Martin de Vidales *et al.*, 1991).

Swinefordite, ideally $[(Ca,Na)_{0.3}(Li,Mg)_2(Si,Al)_4O_{10}(OH,F)_2 \cdot nH_2O]$, is a Li-bearing smectite. Initially it was assumed to show intermediate dioctahedral features (Tien *et al.*, 1975). Further investigations, however, taking the CEC into account also, in order to calculate the chemical formula, demonstrated octahedral occupancy similar to that characteristic of trioctahedral smectites. Only a few data about this mineral have been reported in literature, and a review was provided by Güven (1988).

7.4. The vermiculite group

Vermiculite presents hydrated interlayer cations that compensate the negative layer charge in a range from 0.6 to 0.9 a.p.f.u., *i.e.* greater than observed in smectite.

Vermiculite is generally trioctahedral and the layer charge is usually related to Al-for-Si substitution in tetrahedral sites. The structure of natural vermiculite is known from the pioneering investigations of Gruner (1934b) who proposed, using powder XRD data, a monoclinic *Cc* or *C*2/*c* cell. Later, this assumption was confirmed by Hendricks & Jefferson (1938), using single-crystal diffraction data. Water molecules and exchangeable cations were later demonstrated to occupy well defined sites within the interlayer space (Mathieson & Walker, 1954; Mathieson, 1958). Shirozu & Bailey (1966), after studying a 2-layer ordered vermiculite ($c = 28.89$ Å), suggested that the shift between successive 2:1 layers is always $-a/3$ along the x axis and alternates between $+b/3$ and $-b/3$ along the y axis. Vermiculites can present different layer-stacking sequences (de la Calle *et al.*, 1975a,b, 1978, 1985) depending on the nature of the interlayer cation and of the H_2O content. In fact the nature of the interlayer cation and the relative position of adjacent silicate layers influences the organization of the interlayer water molecules which generally show an ordered arrangement. Argüelles *et al.* (2010) obtained information on both the atomic positions and occupancies of exchangeable cations and water molecules in the interlayer space of vermiculite comparing experimental powder XRD patterns with those calculated by the program package *DIFFaX+*. Their results matched the conclusions of de la Calle *et al.* (1988) well and confirmed the suggestion that vermiculite is a semi-ordered crystalline material characterized by the existence of a large density of defects due to random $\pm b/3$ translations along the [010] direction.

The crystal structure of vermiculite modified after exchange with several different molecules has been investigated extensively. Some studies considered vermiculite intercalated with organic cations, such as tetramethylammonium, teramethylphosphonium (Vahedi-Faridi & Guggenheim, 1997, 1999) and polyvinyl acetate (Martynková *et al.*, 2007).

The arrangement and mobility of water molecules in one- and two-layer hydrates of vermiculite saturated with alkaline and alkali-earth cations was addressed by several authors (*e.g.* see Slade *et al.*, 1985; de la Calle *et al.*, 1984; Beyer & Von Reichenbach, 2001; Sanz *et al.*, 2006). Sanz *et al.* (2006) indicated two different orientations for water molecules, depending on the hydration state and on the sites occupied by interlayer cations. In one-layer-hydrated Na-vermiculite, two water molecules, coordinated to Na ions, are symmetrically disposed with respect to the vermiculite layers and the **H–H** vectors of water molecules are parallel to the $\mathbf{c}^*$ axis. In the case of one-layer hydrated Li- and Ba-vermiculites, water molecules, present at a rate of three or six per cation, respectively, are arranged in an asymmetric way, forming only one H bond with one of the adjacent layers. In this case, the angle between the **H–H** vector and the $\mathbf{c}^*$ axis is close to 38°. Furthermore, as the amount of water increases, hydrogen bond interactions between water molecules increase at the expense of water-silicate interactions. This effect favours water mobility.

Some simulation studies were devoted to the characterization of treated and untreated vermiculite. Monte Carlo simulation was applied to verify the interlayer molecular water structure in monolayer hydrated Na-vermiculite by Skipper *et al.* (1995). They found that water molecules in the interlayer are affected significantly by the magnitude and distribution of layer charge. As layer charge increases, water molecules tend to increase their occupancy of the interlayer midplane, interacting with tetrahedral basal oxygen atoms and adopting an orientation with their dipole moment vectors parallel to the tetrahedral sheet surface. Tunega & Lishka (2003) employed density functional theory to study the effect of the Si/Al distribution in the tetrahedral sheets on the vermiculite layer. They concluded that the dominant factor affecting the layer stacking is the formation of very strong hydrogen bonds between water molecules coordinating the interlayer Mg^{2+} cations and the basal oxygen atoms of the 2:1 layers. Furthermore the most stable vermiculite layer is observed when only Si occupies tetrahedral sites. Arab *et al.* (2004), using molecular-dynamics simulation techniques, suggested that in a water-free structure, the Zn^{2+} ions are adsorbed on the surface of the clay layers. On the contrary, when H_2O molecules are present in the interlayer space, $Zn(H_2O)_6^{2+}$ complexes are built, after migration of the ions to the midplane of the interlayer space. The complexes are oriented in the interlayer space so that at least four water molecules interact by their H atoms with the O atoms of the clay surfaces.

7.5. The chlorite group

Chlorites are 2:1 layer silicates characterized by an octahedral interlayer sheet. They are usually trioctahedral with Mg^{2+}, Al^{3+}, Fe^{2+} and Fe^{3+} in both octahedral sheets. More rarely, octahedra are occupied by Cr^{3+}, Mn^{2+}, Ni^{2+}, V^{3+}, Cu^{2+}, Zn^{2+} and Li^{+}. Tetrahedral cations are Si^{4+} and Al^{3+}. Si^{4+} can occasionally be substituted by Fe^{3+}, Zn^{2+}, Be^{2+} or B^{3+} (Bailey, 1988a).

Several proposals for the classification of chlorites were suggested over the years, following advances related to the definition of their composition, structure and properties. However, the difficulty in designing a fully satisfactory classification system is related to the complexity of these minerals, to the complexity of their structure and to the many and various solid solutions being documented. The AIPEA Nomenclature Committee (Bailey, 1980) proposed that the group of chlorites be divided into four distinct subgroups according to the octahedral occupancy of both octahedral sheets (*i.e.* the octahedral sheet of the 2:1 layer and the octahedral sheet in the interlayer): (1) trioctahedral chlorites, where both octahedral sheets are trioctahedral (*e.g.* clinochlore, chamosite, pennantite, nimite, baileychlore); (2) dioctahedral chlorites, where both octahedral sheets are dioctahedral (*e.g.* dombassite); (3) di,trioctahedral chlorites characterized by a dioctahedral 2:1 octahedral sheet and by a trioctahedral interlayer (*e.g.* cookeite, sudoite); (4) tri,dioctahedral, characterized by a trioctahedral 2:1 octahedral sheet and by a dioctahedral interlayer.

Bayliss (1975) introduced a nomenclature for trioctahedral chlorites based on five end-members and discrediting all other varieties: (1) clinochlore $Mg_5^{2+}Al^{3+}(Si_3^{4+}Al^{3+})O_{10}(OH)_8$; (2) chamosite $Fe_5^{2+}Al^{3+}(Si_3^{4+}Al^{3+})O_{10}(OH)_8$; (3) pennantite $Mn_5^{2+}Al^{3+}(Si_3^{4+}Al^{3+})O_{10}(OH)_8$; (4) nimite $Ni_5^{2+}Al^{3+}(Si_3^{4+}Al^{3+})O_{10}(OH)_8$; and (5) baileychlore $Zn_5^{2+}Al^{3+}(Si_3^{4+}Al^{3+})O_{10}(OH)_8$.

More recently, classification schemes for chlorites were suggested by Wiewióra & Weiss (1990), using a unified system of composition projection, and by Zane & Weiss (1996), using electron microprobe data evaluation.

The chlorite-model structure consists of negatively charged 2:1 layers and positively charged interlayer sheets, connected by long hydrogen bonds. However, different 2:1 layer and O sheet sequences are ~14 Å thick and produce a large number of polytypes (Bailey, 1988a). These different sequences are generated by the displacement of two adjacent 2:1 layers either by $+a/3$ or by $-a/3$. Furthermore, the O sheet can be oriented in two different ways with respect to the 2:1 layer, thus being referred to as type I and type II: type I shows the octahedra in both the interlayer and the 2:1 layer oriented in the same way, whereas type II is characterized by an opposed octahedral orientation (Fig. 13, after Brown & Bailey, 1963). The interlayer sheet, in either type I or type II orientation, needs to match the 2:1 layer to form H-bonds between basal oxygen atoms and OH groups of the O interlayer. This requirement is satisfied by six different geometrical arrangements which can be divided into two sets (A and B). In set A, one of the three interlayer cations, if projected on the basal plane, overlaps with the H placed at the centre of the hexagonal ring. In set B, the interlayer sheet is displaced by $a/3$ and thus the projection of the octahedral cation on the basal plane matches the H position in the adjacent 2:1 layer in the tetrahedral sheet below. The triclinic IIb-4 polytype, with symmetry $C\bar{1}$, and the monoclinic IIb-2 polytype, with symmetry $C2/m$, are the most abundant regular stacking one-layer chlorites occurring in nature. Brown & Bailey (1963) suggested that the natural abundances of chlorite polytypes is related to two main structural controls: (1) cation–cation repulsive forces between the interlayer and the tetrahedral sheet, and (2) hydrogen-bonded configurations that lead to short hydrogen bonds. The dominance of the IIb polytype results from the absence of control and from the presence of a structurally favourable hydrogen-bonding configuration. Starting from different combinations of 14 Å units, Lister & Bailey (1967) and Drits & Karavan (1969) considered the geometric relationships between the layers for a regular 28 Å stacking.

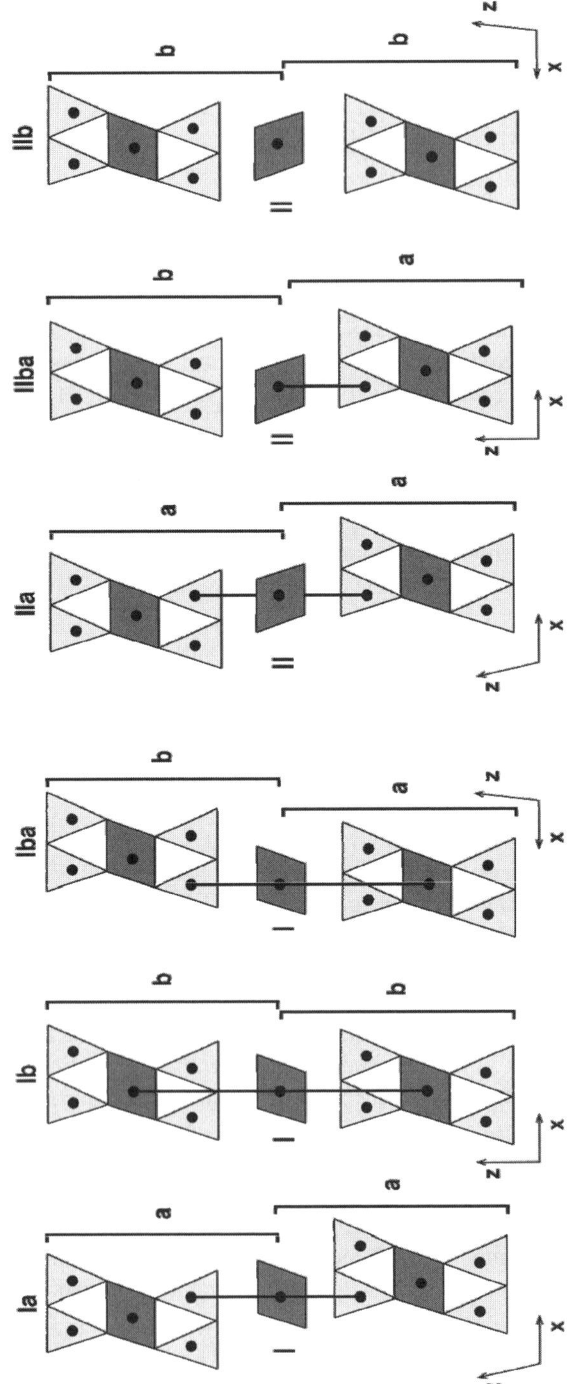

Fig. 13. Relationships between the 2:1 layer and the octahedral sheet in the I chlorite polytype (***a***) and in the II chlorite polytype (***b***). Readers of the paper version of this chapter may wish to download a colour version of this figure from www.minersoc.org/emu-notes/emu-11/11-1-colour.pdf.

7.5.1. Trioctahedral chlorites

The trioctahedral chlorite structure can be described as formed by 2:1 layers, negatively charged, with ideal composition $(R^{2+},R^{3+})_3(Si_{4-x}Al_x)O_{10}(OH)_2$, separated by an interlayer brucite-like octahedral sheet (O), positively charged, with composition $(R^{2+},R^{3+})_3(OH)_6$ (Bailey, 1988a). The interlayer sheet shows a positive charge, due to R^{3+}-for-R^{2+} substitution in order to balance the total negative charge of the 2:1 layer.

Large cations, such as Na^+, K^+ and Ca^{2+}, usually cannot be accommodated in the interlayer octahedral sheet. Ca^{2+} was indeed observed in franklinfurnaceite, a zinc-silicate intermediate between chlorite and mica (Peacor *et al.*, 1988) and Na^+ in glagolevite where Na atoms, sevenfold coordinated, are located between the 2:1 layers and the interlayer octahedral sheets (Krivovichev *et al.*, 2004). The main exchange vectors in trioctahedral chlorites are: (1) isovalent substitution (*e.g.* $^{[VI]}Fe^{2+}\ ^{[VI]}Mg^{2+}_{-1}$); (2) heterovalent substitutions, also known as Tschermak substitution (*e.g.* $^{[IV]}Si^{4+}_{-1}\ ^{[VI]}R^{2+}_{-1}\ ^{[IV]}Al^{3+}\ ^{[VI]}Al^{3+}$); (3) dioctahedral substitution $^{[VI]}(Mg^{2+}, Fe^{2+})_{-3}\ ^{[VI]}Al^{3+}_2\ ^{[VI]}(\square)$ leading to the generation of octahedral vacancies (Laird, 1988). Most of the structural data on trioctahedral chlorites refers to clinochlore.

The triclinic structure (space group $C\bar{1}$, unit-cell parameters $a = 5.3301(4)$; $b = 9.2511(6)$; $c = 14.348(1)$ Å; $\alpha = 90.420(3)$; $\beta = 97.509(3)$; $\gamma = 89.996(4)°$) was refined and described by several authors in IIb-4 polytype (Steinfink, 1958b; Phillips *et al.*, 1980; Zheng & Bailey, 1989; Joswig & Fuess, 1990; Nelson & Guggenheim, 1993; Smyth *et al.*, 1997; Joswig *et al.*, 1980; Zanazzi *et al.*, 2006; Valdré *et al.*, 2009). The monoclinic structure was described by Steinfink (1958a), Zheng & Bailey (1989), Rule & Bailey (1987), Joswig & Fuess (1989) and Zanazzi *et al.* (2007a) mostly in IIb-2 polytype (space group $C2/m$, $a = 5.327(2)$, $b = 9.227(2)$, $c = 14.327(5)$ Å, $\beta - 96.81(3)°$).

For both polytypes, these studies generally suggested significant ordering in the interlayer octahedral sheet. The two octahedra in the interlayer differ considerably in terms of volume, distortion and mean cation–oxygen bond distance. Trivalent cations (Al^{3+}, Cr^{3+} and Fe^{3+}) concentrate in one of the two independent octahedral sites in the interlayer (usually called M4) creating a net positive charge to balance the net negative charge on the 2:1 layer. Bish & Giese (1981), starting from energy considerations, showed that the ordering of trivalent cations in the M4 site significantly increases the energy of the interlayer bonding and consequently the stability of the structure. A disordered Si/Al distribution was suggested for IIb-4 polytype following single-crystal structure refinements (Rule & Bailey, 1987). However, Welch *et al.* (1995) on the basis of MAS-NMR spectroscopy results, indicated the existence of considerable short-range order.

Joswig *et al.* (1980) suggested that **O–H** vectors of the interlayer OH in IIb-4 polytype are tilted slightly away from the vertical towards their basal oxygen acceptors to form bent hydrogen bonds. The O–OH distances range from 2.87 to 2.91 Å. In the monoclinic polytype IIb-2, the two OH dipoles are roughly perpendicular to the interlayer sheet, forming weak to medium hydrogen bonds with O...O distances of 2.859 and 2.881 Å. The OH-dipole of the 2:1 layer is perpendicular to the (001) plane (Joswig & Fuess, 1989; Welch & Marshall, 2001; Welch *et al.*, 1995).

Valdré *et al.* (2009) studied the relationships between clinochlore cleavage characteristics, in terms of nano-morphology, and surface potential, as a function of average crystal chemistry and topology. In IIb-4 clinochlore, octahedral sites of the silicate layer are equal and equally occupied by Mg, whereas the octahedral sites in the interlayer show different sizes and are mostly completely occupied by divalent (Mg^{2+} and Fe^{2+}) or trivalent (Al^{3+}) cations. The clinochlore cleaved surface is present in two forms: (1) the stripe type, which is characterized by bands along the [100] crystal direction (4.0 Å in height, up to several μm long and ranging from a few nm to a few μm wide); (2) the triangular type (4.0 Å in height), which is characterized by triangular areas where the surface shows missing interlayer regions, with average lateral sizes ranging from a few to more than hundreds of nanometers. Both features may result either from interlayer sheets, where cleavage directions are induced by different octahedral site occupancy in the interlayer, or by weak interlayer bonding along specific directions of the connected 2:1 layer. The cleaved surface, particularly at the cleaved edges, presents high DNA affinity, which is directly related to an average positive surface and ledge potential.

Structural results on chlorites, other than clinochlore, are limited in number and could provide only a partial description. Some interesting results were presented by Rule & Radke (1988) and Walker & Bish (1992).

7.5.2. Di,trioctahedral and dioctahedral chlorites

The known di,trioctahedral chlorites are cookeite and sudoite. They present a dioctahedral 2:1 layer and a trioctahedral interlayer. *Dombassite* is a dioctahedral chlorite and presents 2:1 layer and interlayer both dioctahedral.

Cookeite, ideal composition $LiAl_4(Si_3Al)O_{10}(OH)_8$, is a di,trioctahedral chlorite in which the 2:1 layer is dioctahedral and the hydroxide interlayer is trioctahedral. Cookeite differs from other Al-, Li-rich chlorite species (*i.e.* sudoite and dombassite) due to the greater Li content (~3–4 wt.% Li_2O). Although several cookeite polytypes were reported (*e.g.* Vrublevskaja *et al.*, 1975), the structure of this mineral was detailed by Zheng & Bailey (1997) only in space group *Cc* (two-layer 'r' structure of *Iaa* polytype, $a = 5.158(1)$, $b = 8.940(2)$, $c = 28.498(6)$ Å, $\beta = 96.60(3)°$). Mean *T*–O bond lengths indicate a partly ordered but asymmetric distribution of tetrahedral Si and Al. The two tetrahedral sheets within the 2:1 layer show different compositions and charges. The Al-rich, higher-charge tetrahedral sheet is thicker and more closely approaches the interlayer sheet than the Si-rich, lower-charged sheet. Two Al cations occupy the *cis*-octahedra in the dioctahedral 2:1 layer. Mean bond lengths in the trioctahedral interlayer sheet indicate a partly ordered distribution of octahedral Al and Li. The Li-rich, lower-charge octahedron in the interlayer is located on a vertical straight line between an Al-rich tetrahedron and a Si-rich tetrahedron. The two higher-charge interlayer Al cations are located vertically between a Si-rich tetrahedron and the centre of a six-membered ring. This pattern of ordering minimizes the cation-cation repulsion and provides the best local charge balance. The protons of the six surface OH groups tilt away from the two Al-rich interlayer sites towards the lower-charge Li site. Recently borocookeite was reported, where $^{[IV]}Al$ is replaced by B (Zagorsky *et al.*, 2003).

Sudoite is a di,trioctahedral Mg-rich chlorite with ideal composition: $(Al_3Mg_2)(Si_3Al)O_{10}(OH)_8$. Natural sudoite samples are usually poorly crystalline and fine-grained, thus rendering a complete understanding of the structure of this mineral extremely difficult. Eggleston & Bailey (1967) partially refined the structure of sudoite (space group $C2/m$, $a = 5.237$, $b = 9.070$, $c = 14.285$ Å, $\beta = 97.03°$) and confirmed that the mineral is composed of dioctahedral 2:1 layers separated by interlayer trioctahedral sheets with a IIb-4 stacking sequence. High-resolution TEM observations indicate that the stacking sequence is characterized by a largely uniform intralayer shift of $a/3$ in the $-\mathbf{X_1}$ direction ($\mathbf{X_1}$ represent the directions along the pseudohexagonal axes) and by an interlayer displacement of similar magnitude in either the $-\mathbf{X_2}$ or $-\mathbf{X_3}$ direction. Stacking disorder is primarily caused by the mixing of interlayer displacements in the two directions (Kameda *et al.*, 2007).

7.6. Some 2:1 layer silicates involving a discontinuous octahedral sheet and a modulated tetrahedral sheet

Structure determinations of modulated 2:1 layer silicates are limited due to a lack of suitable crystals for single-crystal XRD. As a result, numerous models were produced regarding cell dimensions and indexing of powder patterns (Christ *et al.*, 1969; Jones and Galán, 1988) and polymorphs and structure models (Bradley, 1940; Preisinger, 1963; Gard & Follet, 1968; Drits and Sokolova, 1971; Chisholm, 1992).

Sepiolite (Fig. 2b). The first structural pattern for sepiolite was introduced by Nagy & Bradley (1955), who suggested the $C2/m$ ($A2/m$) space groups as being the most appropriate. Later, Brauner & Preisinger (1956) and Preisinger (1959) proposed another structural model for sepiolite under space group *Pnan*. The difference between monoclinic and orthorhombic models is generated by the tetrahedral inversion at the edge of the ribbons, either occurring along the middle of the zigzag Si O Si chains (Nagy & Bradley, 1955) or along their edges (Brauner & Preisinger, 1956). In the Brauner & Preisinger (1956) model, adjacent inverted ribbons are joined by a single basal oxygen (instead of two as in the Nagy-Bradley model), and there are eight octahedral sites in a ribbon (instead of nine), four OH (instead of six), and eight water molecules (instead of six). The model of Brauner & Preisinger (1956) contains three hydrous species: (1) OH anions; (2) structural H_2O at the edges of the octahedral strips; (3) three zeolitic H_2O positions inside the channels. Electron density patterns from single sepiolite fibres obtained by Brindley (1959), Zvyagin (1967) and Gard & Follet (1968) confirmed that the systematic absences are in agreement with the space group *Pnan*. The Brauner-Preisinger model for sepiolite was also confirmed and refined by Rautureau *et al.* (1972), Rautureau & Tchoubar (1974), Yucel *et al.* (1981) and Post *et al.* (2007). Post *et al.* (2007), based on the results obtained from a Rietveld refinement, using synchrotron powder XRD data, confirmed the general structural model determined by Brauner & Preisinger (1956), suggested an additional zeolitic H_2O site inside the channels and demonstrated that two zeolitic H_2O sites are fully occupied, one half occupied and the last only one-third occupied. Sepiolite unit-cell parameters are $a = 5.2750(1)$, $b = 27.016(1)$, and $c = 13.405(1)$ Å (space group *Pnan*; Post *et al.*, 2007).

Palygorskite (Fig. 2b). Bradley (1940) suggested a model for palygorskite structure with a probable $A2/m$ space group, differing from the sepiolite model, because of a shorter b dimension following from the presence of only two tetrahedral ribbons. Like sepiolite, the palygorskite structural model (Bradley, 1940) includes three types of H_2O molecules: (1) (OH) groups bonded to some of the Mg and Al atoms; (2) H_2O molecules that complete the coordination of the Mg atoms at the edges of the octahedral strips; and (3) zeolitic H_2O molecules in the tunnels. Drits & Sokolova (1971) confirmed the Bradley model and measured a β angle of 107°. Preisinger (1963) reported an orthorhombic model for palygorskite, similarly to that previously discussed for sepiolite, except for the ribbon width. Christ *et al.* (1969) analysed powder XRD data for five palygorskite samples and concluded that palygorskite exists in structurally related orthorhombic and monoclinic forms. Chisholm (1992) compared observed and theoretical XRD powder patterns and noted that most palygorskite samples are mixtures of monoclinic and orthorhombic polymorphs. Recently, Rietveld refinements, using powder X-ray and neutron diffraction data, confirmed the basic monoclinic and orthorhombic palygorskite structure models (Artioli *et al.*, 1994; Giustetto & Chiari, 2004). Post & Heaney (2008) refined the structure of pure monoclinic palygorskite samples by Rietveld refinements on synchrotron powder XRD data. Palygorskite unit-cell parameters (monoclinic $A2/m$ space group) are $a = 5.2419(2)$, $b = 17.8476(7)$, $c = 13.2858(8)$ Å, $\beta = 107.560(5)°$ (Post & Heaney, 2008).

Galán & Carretero (1999) reviewed chemical analyses from the literature and concluded that: (1) sepiolite is a true trioctahedral mineral with eight octahedral positions filled with Mg^{2+}, and showing a formula close to $Mg_8Si_{12}O_{30}(OH)_4(OH_2)_4(H_2O)_8$; (2) palygorskite is an intermediate dioctahedral-trioctahedral mineral, showing a formula close to $(Mg_2R^{3+}_2\square_1)(Si_{8-x}Al_x)O_{20}(OH)_2\cdot R^{2+}_{x/2}\cdot(H_2O)_4$, where $\square$ stands for vacancy, R stands primarily for Al^{3+}, Fe^{3+}, Fe^{2+} and Mn^{2+} and x ranges from 0 to 0.5. Furthermore, Post & Heaney (2008) discovered Al and Mg octahedral ordering, with Mg cations placed in octahedral sites at the edge of the channels, for monoclinic palygorskite.

Sepiolite and palygorskite are commonly used in technical and mostly in sorption applications (Alvarez, 1984; Galán, 1996). The sorption capacity is controlled primarily by the fibrous mineral surface and by zeolite-like channels. Mineral-fibre surfaces include silanol groups, surface oxygen atoms and structural H_2O along with octahedral broken bonds. The zeolite-like channels allow the exchange of metal cations with the surrounding environment. Krekeler & Guggenheim (2008) observed that the different adsorption behaviour of several sepiolite and palygorskite samples can also be associated with defects, which alternatively enhance or reduce the sorption ability. A fascinating use of palygorskite is related to the manufacturing of a dye, Maya blue, long used by the ancient Maya people. The crystal structure of Maya Blue was investigated by Chiari *et al.* (2003) using the Rietveld method and synchrotron powder XRD data.

Kalifersite, with ideal composition $(K,Na)_5(Fe^{3+})_7[Si_{20}O_{50}](OH)_6\cdot 12H_2O$ (space group $P\bar{1}$, $a = 14.86$, $b = 20.54$, $c = 5.29$ Å, $\alpha = 95.6$, $\beta = 92.3$, $\gamma = 94.4°$), was described by Ferraris *et al.* (1998) after noting a modular relationship between this mineral, sepiolite and palygorskite. Kalifersite shows an alternation of one module of sepiolite (S) and palygorskite (P) along [010]. This mineral can thus be considered as the S_1P_1 term of the palysepiole polysomatic series (P_pS_S).

Tuperssuatsiaite shows a diffraction pattern similar to that of palygorskite, thus suggesting a similar crystal structure (Camara *et al.*, 2002). Tuperssuatsiaite is monoclinic, space group $C2/m$, with $a = 14.034(7)$, $b = 17.841(7)$, $c = 5.265(2)$ Å, and $\beta = 103.67(4)^\circ$. The chemical composition is $Na_{1.87}Fe_{2.14}Mn_{0.48}Ti_{0.14}Al_{0.03}Mg_{0.02}[Si_8O_{20}](OH)_2{\cdot}n(H2O)$. Like palygorskite, the tuperssuatsiaite structure consists of an octahedral sheet sandwiched between two opposing tetrahedral sheets. The octahedral sheet forms strips which are three octahedra wide and defines channels which could be occupied by H_2O. This mineral is also characterized by Na in octahedral coordination and by notable Fe^{2+} content.

8. Imogolite and allophane

Imogolite (Cradwick *et al.*, 1972) and allophane (Wada, 1967) are both nano-sized aluminosilicate minerals with a predominantly hollow structure and different Si/Al ratio. Imogolite consists of single-walled nanotubes (Fig. 3), unlike allophane, which presents a hollow spherical morphology.

Natural imogolite, mostly observed in soils originating from volcanic ashes and weathered pumice sand spodosols, is characterized by the ideal chemical formula $(OH)_3Al_2O_3SiOH$, thus meaning a Si/Al ratio of 0.5. Imogolite is commonly characterized by poor crystallinity, with short-range order structures defined by hollow tubes of curved gibbsite sheets with SiO_4 tetrahedra replacing OH groups at the inner surface.

A first approach to imogolite synthesis was defined by Farmer & Fraser (1979) starting from diluted solutions of $Al(ClO_4)_3$ and silanol $(Si)(OH)_4$. Other methods include the hydrolysis of fused sodium silicates, leading to the initial formation of amorphous protoimogolite or allophane and then followed by a subsequent growth of crystalline imogolite tubes. Other methods for imogolite synthesis include tetraethoxysilane hydrolysis and hydrolysis of commercial sodium silicate.

The imogolite tube shows an inner diameter of 0.5–0.9 nm, as measured by nitrogen adsorption or by transmission electron microscopy (TEM), and an outer diameter of 2.0–2.5 nm. Tube length ranges from several hundreds of nm to 1 μm.

Solid-state NMR, powder XRD and TEM allowed the definition of a structural imogolite atomic model, *i.e.* the wall of the tube consists of a single curved sheet of gibbsite, wherein the OH on one face are replaced by orthosilicate groups, with Al only occurring in octahedral coordination and Si in tetrahedral coordination. Electron diffraction measurements performed by Cradwick *et al.* (1972) suggested that the most likely structure model of natural imogolite contains ten gibbsite units around the circumference, corresponding to a tubular diameter of 21 ± 5 Å. On the other hand, Farmer & Fraser (1979) concluded that natural imogolite consists of 12 units. Si can be partly or completely substituted by Ge. The synthesized Ge-rich products are similar to natural imogolite in their tubular morphology. However the external diameter of the tube increases with increasing Ge substitution up to ~33 Å. This evidence suggests that Ge-for-Si substitution causes a decrease in the curvature of the gibbsite-like sheet bonding to Si or Ge polyhedra. The number of gibbsite unit cells forming the circumference of the tube is 10–12 in natural imogolite and 18 in the Ge-substituted phase (Wada & Wada, 1982). The repeat distance along the tube axis is ~8.4 Å in both Si-rich and Ge-rich imogolite. Theoretical studies (Alvarez-Ramirez, 2009) involving imogolite-like single-wall nanotubes confirmed that, by increasing the Ge/(Si + Ge) ratio, the gibbsite-like units increase from 9 to 13. Maillet *et al.* (2010) suggested that the structure of Ge-substituted imogolite is a double-walled nanotube which consists of two concentric tubes of equal length and identical wall structure.

Allophane is a weathering or hydrothermal alteration product of feldspars and other primary minerals. Its ideal chemical formula is $Al_2O_3{\cdot}(SiO_2)_{1.3\text{-}2}{\cdot}2.5\text{–}3(H_2O)$. The first model describing the structural arrangement of allophane was proposed by Wada (1967). This model describes the mineral structure as consisting of a silica tetrahedral chain sharing corners with one or two chains of AlO_6 octahedra, thus giving rise to allophane with the Si/Al ratios of 1 and 0.5, respectively. Models based on kaolinite structure were introduced later (Milestone, 1971; Henmi & Wada, 1976; Okada & Ossaka, 1983; Van der Gaast *et al.*, 1985). Okada *et al.* (1975) proposed a scheme based on a two-dimensional kaolin-like structure with Si atoms in the tetrahedral sheet partially substituted by Al. The kaolin-based model was further supported by Wada *et al.* (1979), who described the synthesis process of allophane. Starting from electronic microscopy data, Kitagawa (1971) and Henmi & Wada (1976) suggested that an individual allophane is a sphere- or polyhedron-like hollow particle with external diameter ranging from 35 to 50 Å. These authors also suggested that the particles may show openings to admit water or other guest species inside the structure. Further experimental measurements, detailing the density of allophane, confirmed their hypothesis, and suggested a wall thickness ranging from 7 to 10 Å (Wada & Wada, 1977; Wada, 1995). Another model for allophane structure was proposed by Parfitt & Henmi (1980). The strong similarity between allophane and imogolite, as suggested by IR spectra, led the authors to conclude that allophane consists of imogolite structural units. Parfitt *et al.* (1980) documented a 'protoimogolite' allophane sphere with diameter of 40 Å. This latter is made up of 125 unit cells of imogolite and should present at least six openings with ~4 Å diameter. In the following paper, Parfitt *et al.* (1980) discriminated two types of allophane. The first type is characterized by Si/Al ratios of ~0.5 and is built from imogolite units. The second type of allophane shows a Si/Al ratio close to 1, contains condensed silicate units, and presents a halloysite-like structure. The two structural models of allophane were further examined in a number of studies, some of which proposed a model combining kaolin-like and imogolite-like units (Mackenzie *et al.*, 1991). This latter model was supported by the ESCA data from He *et al.* (1995).

Acknowledgements

The authors acknowledge the Italian Ministry for Education and Research for PRIN 2008 ('*Phyllosilicates of particular petrologic relevance: chemical and physical features, and their variation in different environments either natural or simulated*') and

Fondazione Cassa di Risparmio di Modena for the International grant '*The relationships of bulk structure, surface structure, chemistry and physical properties of mineral phases with six-membered silicate rings*'. A grant-in-aid by LLP-ERASMUS (2010-1-IT2-ERA-10-16274) made the organization of the school possible at which this short presentation was used as didactic support. Stephen Guggenheim kindly reviewed the chapter and suggested helpful improvements.

References

Abraham, K., Schreyer, W., Medenbach, O. & Gebert, W. (1980) Kulkeit, ein geordnetes 1:1 Mixed-Layer-Mineral zwischen Klinochlor und Talk. In: *Fortschritte der Mineralogie, Beiheft 1, Abstracts*, **5E**, 4–5.

Adams, J.M. (1983) Hydrogen atom positions in kaolinite by neutron profile refinement. *Clays and Clay Minerals*, **31**, 352–356.

Adams, J.M. & Hewat, A.W. (1981) Hydrogen atom positions in dickite. *Clays and Clay Minerals*, **29**, 316–319.

Akiba, E., Hayakawa, H., Hayashi, S., Miyawaki, R., Tomura, S., Shibasaki, Y., Izumi, F., Asano, H. & Kamiyama, T. (1997) Structure refinement of synthetic deuterated kaolinite by Rietveld analysis using time-of-flight neutron powder diffraction data. *Clays and Clay Minerals*, **45**, 781–788.

Altaner, S.P. & Ylagan, R.F. (1997) Comparison of structural models of mixed-layer illite/smectite and reaction mechanisms of smectite illitization. *Clays and Clay Minerals*, **45**, 517–533.

Alvarez, A. (1984) Sepiolite: properties and uses. In: *Palygorskite–Sepiolite: Occurrences, Genesis, and Uses* (A. Singer & E. Galán, editors). Elsevier, New York, pp. 253–287.

Alvarez-Ramirez, F. (2009) Theoretical study of $(OH)_3N_2O_3MOH$, M = C, Si, Ge, Sn and N = Al, Ga, In, with imogolite-like structure. *Journal of Computational and Theoretical Nanoscience*, **6**, 1120–1124.

Anderson, C.S. & Bailey, S.W. (1981) A new cation ordering pattern in amesite-$2H_2$. *American Mineralogist*, **66**, 185–195.

Andreani, M., Grauby, O., Baronnet, A. & Munoz, M. (2008) Occurrence, composition, and growth of polyhedral serpentine. *European Journal of Mineralogy*, **20**, 159–171.

Aparicio, P. & Galán, E. (1999) Mineralogical interference on kaolinite crystallinity index measurements. *Clays and Clay Minerals*, **47**, 12–27.

Aparicio, P., Ferrell, R.E. & Galán, E. (1999) A new kaolinite crystallinity index from mathematical modelling of XRD data. In *Abstracts volume of the 9th EUROCLAY Conference, Kraków, Poland*, 57.

Appelo, C.A.J. (1978) Layer deformation and crystal energy of micas and related minerals. I. Structural models for $1M$ and $2M_1$ polytypes. *American Mineralogist*, **63**, 782–792.

Arab, M., Bougeard, D. & Smirnov, K.S. (2004) Structure and dynamics of interlayer species in a hydrated Zn-vermiculite. A molecular dynamics study. *Physical Chemistry Chemical Physics*, **6**, 2446–2453.

Argüelles, A., Leoni, M., Blanco, J.A. & Marcos, C. (2010) Semi-ordered crystalline structure of the Santa Olalla vermiculite inferred from X-ray powder diffraction. *American Mineralogist*, **95**, 126–134.

Árkai, P. (1991) Chlorite crystallinity: an empirical approach and correlation with illite crystallinity, coal rank and mineral facies as exemplified by Palaeozoic and Mesozoic rocks of northeast Hungary. *Journal of Metamorphic Geology*, **9**, 723–734.

Árkai, P. & Tóth, N.M. (1983) Illite crystallinity: combined effects of domain size and lattice distortion. *Acta Geologica Hungarica*, **26**, 341–358.

Árkai, P., Sassi, F.P. & Sassi, R. (1995) Simultaneous measurements of chlorite and illite crystallinity: a more reliable tool for monitoring low- to very low-grade metamorphisms in metapelites. A case study from the Southern Alps (NE Italy). *European Journal of Mineralogy*, **7**, 1115–1128.

Árkai, P., Merriman, R.J., Roberts, B., Peacor, D.R. & Tóth, M. (1996) Crystallinity, crystallite size and lattice strain of illite-muscovite and chlorite: comparison of XRD and TEM data for diagenetic and epizonal pelites. *European Journal of Mineralogy*, **8**, 1119–1137.

Artioli, G., Galli, E., Burattini, E., Cappuccio, G. & Simeoni, S. (1994) Palygorskite from Bolca, Italy: A characterization by high-resolution synchrotron radiation powder diffraction and computer modeling. *Neues Jahrbüch für Mineralogie Monatshefte*, **1994**, 271–229.

Aruja, E. (1945) An X-ray study of the crystal structure of antigorite. *Mineralogical Magazine*, **27**, 65–74.

Asimow, P.D., Stein, L.C., Mosenfelder, J.L. & Rossman, G.R. (2006) Quantitative polarized infrared analysis of trace OH in populations of randomly oriented mineral grains. *American Mineralogist*, **91**, 278–284.

Auzende, A.L., Pellenq, R.J.-M., Devouard, B., Baronnet, A. & Grauby, O. (2006) Atomistic calculations of structural and elastic properties of serpentine minerals: the case of lizardite. *Physics and Chemistry of Minerals*, **33**, 266–275.

Bailey, S.W. (1963) Polymorphism of the kaolin minerals. *American Mineralogist*, **48**, 1196–1209.

Bailey, S.W. (1969) Polytypism of trioctahedral 1:1 layer silicates. *Clays and Clay Minerals*, **17**, 355–371.

Bailey, S.W. (1980) Summary of recommendations of AIPEA nomenclature committee on clay minerals. *American Mineralogist*, **65**, 1–7.

Bailey, S.W. (1981) A system of nomenclature for regular interstratifications. *The Canadian Mineralogist*, **19**, 651–655 (reprinted in: Bailey *et al.* (1982) Report of the Clay Minerals Society Nomenclature Committee for 1980–1981: Nomenclature for regular interstratifications. *Clays and Clay Minerals*, **30**, 76–78; and in: Bailey *et al.* (1982) Report of the AIPEA Nomenclature Committee. *AIPEA Newsletter* No. 18, Supplement of February, 1982).

Bailey, S.W. (1988a) Structure and composition of other trioctahedral 1:1 phyllosilicates. In: *Hydrous Phyllosilicates (Exclusive of Micas)* (S.W. Bailey, editor). Reviews in Mineralogy, **19**. Mineralogical Society of America, Washington, D.C., 169–186.

Bailey, S.W. (1988b) Odinite, a new dioctahedral-trioctahedral Fe^{3+}-rich 1:1 clay mineral. *Clay Minerals*, **23**, 237–247.

Bailey, S.W. (1989a) Report of the AIPEA Nomenclature Committee. *AIPEA Newsletter*, **26**, 17–18.

Bailey, SW. (1989b) Halloysite–A critical assessment. In: *Proceedings of 9th International Clay Conference, Strasbourg, France* (V.C. Farmer & Y. Tardy, editors). Sciences Géologiques Mémoire, **86**, 89–98.

Bailey, S.W., Banfield, J.F., Barker, W.W. & Katchan, G. (1995) Dozyite, a 1:1 regular interstratification of serpentine and chlorite. *American Mineralogist*, **80**, 65–77.

Bain, D.C. & Smith, B.L.F. (1987) Chemical analysis. In: *A Handbook of Determinative Methods in Clay Mineralogy* (M.J. Wilson, editor). Blackie, Glasgow, London, pp. 248–274.

Balan, E., Allard, T., Boizot, B., Morin, G. & Muller, J.-P. (1999) Structural Fe^{3+} in natural kaolinites: new insights from electron paramagnetic resonance spectra fitting at X and Q-band frequencies. *Clays and Clay Minerals*, **47**, 605–616.

Balan, E., Saitta, A.M., Mauri, F. & Calas, G. (2001) First-principles modeling of the infrared spectrum of kaolinite. *American Mineralogist*, **86**, 1321–1330.

Balan, E., Lazzeri, M., Saitta, A.M., Allard, T., Fuchs, Y. & Mauri, F. (2005) First-principles study of OH-stretching modes in kaolinite, dickite, and nacrite. *American Mineralogist*, **90**, 50–60.

Balan, E., Delattre, S., Guillaumet, M. & Salje, E.K.H. (2010) Low-temperature infrared spectroscopic study of OH-stretching modes in kaolinite and dickite. *American Mineralogist*, **95**, 1257–1266.

Balek, V. & Murat, M. (1996) The emanation thermal analysis of kaolinite clay minerals. *Thermochimica Acta*, **282/283**, 385–397.

Banfield, J.F. & Bailey, S.W. (1996) Formation of regularly interstratified serpentine-chlorite minerals by inversion in long-period serpentine polytypes. *American Mineralogist*, **81**, 79–91.

Baronnet, A. (1978) Some aspects of polytypism in crystals. *Progress in Crystal Growth Characterization*, **1**, 151–211.

Baronnet, A. (1992) Polytypism and stacking disorder. In: *Minerals and Reactions at the Atomic Scale: Transmission Electron Microscopy* (P.R. Buseck, editor). Reviews in Mineralogy, **27**, 231–282. Mineralogical Society of America, Washington, D.C.

Baronnet, A. & Devouard, B. (2005) Microstructures of common polygonal serpentines from axial HRTEM imaging, electron diffraction, and lattice-simulation data. *The Canadian Mineralogist*, **43**, 513–542.

Baronnet, A., Mellini, M. & Devouard, B. (1994) Sectors of polygonal serpentine. A model based on dislocations. *Physics and Chemistry of Minerals*, **21**, 330–343.

Baronnet, A., Andreani, M., Grauby, O., Devouard, B., Nitsche, S. & Chaudanson, D. (2007) Onion morphology and microstructure of polyhedral serpentine. *American Mineralogist*, **92**, 687–690.

Bates, T.F., Hildebrand, F.A. & Swineford, A. (1950) Morphology and structure of endellite and halloysite. *American Mineralogist*, **35**, 463–484.

Bayliss, P. (1975) Nomenclature of the trioctahedral chlorites. *The Canadian Mineralogist*, **13**, 178–180.

Belluso, E. & Ferraris, G. (1991) New data on balangeroite and carlosturanite from alpine serpentinites. *European Journal of Mineralogy*, **3**, 559–566.

Ben Haj Amara, A., Ben Brahim, J., Plançon, A., Ben Rhaiem, H. & Besson, G. (1997) Etude structurale d'une nacrite Tunisienne. *Journal of Applied Crystallography*, **30**, 338–344.

Ben Haj Amara, A., Ben Brahim, J., Plançon, A. & Ben Rhaiem, H. (1998) The structure of 8.4 Å hydrated and dehydrated nacrite determined by X-ray diffraction. *Journal of Applied Crystallography*, **31**, 654–662.

Benco, L. (1997) Electron densities in hydrogen bonds: Lizardite-1*T*. *European Journal of Mineralogy*, **9**, 811–819.

Benco, L. & Smrčok, L. (1998) Hartree–Fock study of pressure-induced strengthening of hydrogen bonding in lizardite-1*T*. *European Journal of Mineralogy*, **10**, 483–490.

Benco, L., Tunega, D., Hafner, J. & Lischka, H. (2001a) Orientation of OH groups in kaolinite and dickite: *ab initio* molecular dynamics study. *American Mineralogist*, **86**, 1057–1065.

Benco, L., Tunega, D., Hafner, J. & Lischka, H. (2001b) Upper limit of the O–H–O hydrogen bond. *Ab initio* study of the kaolinite structure. *Journal of Physical Chemistry*, **B105**, 10812–10817.

Beran, A. (2002) Infrared Spectroscopy of Micas. In: *Micas: Crystal Chemistry and Metamorphic Petrology* (A. Mottana, F.P. Sassi, J.B. Thompson Jr. & S. Guggenheim, editors). Reviews in Mineralogy and Geochemistry, **46**. Mineralogical Society of America, Chantilly, Virginia, USA, pp. 351–370.

Bergaya, F., Theng, B.K.G. & Lagaly, G. (editors) (2006) *Handbook of Clay Science*. Elsevier, Amsterdam, 1224 pp.

Berthier, P. (1826) Analyse de l'halloysite. *Annales de Chimie et de Physique*, **32**, 332.

Beyer, J. & Von Reichenbach, H.G. (2001) An extended revision of the interlayer structures of one- and two-layer hydrates of Na-vermiculite. *Clay Minerals*, **37**, 157–168.

Bish, D.L. (1993) Rietveld refinement of the kaolinite structure at 1.5 K. *Clays and Clay Minerals*, **41**, 738–744.

Bish, D.L. & Giese, R.F.Jr. (1981) Interlayer bonding in [Group] IIb chlorite. *American Mineralogist*, **66(11-12)**, 1216–1220.

Bish, D.L. & Johnston, C.T. (1993) Rietveld refinement and Fourier-transform infrared spectroscopy study of the dickite structure at low temperature. *Clays and Clay Minerals*, **41**, 297–304.

Bish, D.L. & Von Dreele, R.B. (1989) Rietveld refinement of non-hydrogen atom position in kaolinite. *Clays and Clay Minerals*, **37**, 289–296.

Blount, A.M., Threadgold, I.M. & Bailey, S.W. (1969) Refinement of the crystal structure of nacrite. *Clays and Clay Minerals*, **17**, 185–194.

Bonnin, D., Muller, S. & Calas, G. (1982) Iron in kaolins: EPR, Mössbauer, EXAFS spectrometric studies. *Bulletin de Mineralogie*, **105**, 467–475.

Bookin, A.S., Drits, V.A., Plançon, A. & Tchoubar, C. (1989) Stacking faults in kaolin-group minerals in the light of real structural features. *Clays and Clay Minerals*, **37**, 297–307.

Boukili, B., Holtz, F., Robert, J.L., Joriou, M., Beny, J.M. & Naji, M. (2003) Infrared spectra of annite in the OH stretching vibrational range. *Schweizerische Mineralogische und Petrographische Mitteilungen*, **83**, 331–340.

Bradley, W.F. (1940) The structural scheme of attapulgite. *American Mineralogist*, **25**, 405–410.

Brauner, K. & Preisinger, A. (1956) Struktur und Entstehung des Sepioliths. *Tschermaks Mineralogische und Petrographische Mitteilungen*, **6**, 120–140 (in German).

Breu, J., Seidl, W. & Stoll, A. (2003) Hectorite. *Zeitschrift für anorganische und allgemeine Chemie*, **629**, 503–515.

Brigatti, M.F. (1981) Hisingerite: A review of its crystal chemistry. *Proceedings of 7th International Clay Conference. Developments in Sedimentology*, **35**, 97–110.

Brigatti, M.F. & Guggenheim, S. (2002) Mica crystal chemistry and the influence of pressure, temperature, and solid solution on atomistic models. In: *Micas: Crystal Chemistry and Metamorphic Petrology*

(A. Mottana, F.P. Sassi, J.B. Thompson Jr. & S. Guggenheim, editors). Reviews in Mineralogy and Geochemistry, **46**. Mineralogical Society of America, Chantilly, Virginia, USA, 1–97.

Brigatti, M.F., Galli, E., Medici, L. & Poppi, L. (1997) Crystal structure refinement of aluminian lizardite-2*H*2. *American Mineralogist*, **82**, 931–935.

Brigatti, M.F., Guggenheim, S. & Poppi, M. (2003) Crystal chemistry of the 1*M* mica polytype: The octahedral sheet. *American Mineralogist*, **88**, 667–675.

Brigatti, M.F., Colonna, S., Malferrari, D. & Medici, L. (2004) Characterization of Cu-complexes in smectite with different layer charge location: chemical, thermal and EXAFS study. *Geochimica et Cosmochimica Acta*, **68**, 781–788.

Brigatti, M.F., Malferrari, D., Poppi, M. & Poppi, L. (2005) The $2M_1$ dioctahedral mica polytype: A crystal chemical study. *Clays and Clay Minerals*, **53**, 190–197.

Brigatti, M.F., Malferrari, D., Poppi, M., Mottana, A., Cibin, G., Marcelli, A. & Cinque, G. (2008a) Interlayer potassium and its surrounding in micas: Crystal chemical modeling and XANES spectroscopy. *American Mineralogist*, **93**, 821–830.

Brigatti, M.F., Guidotti, C.V., Malferrari, D. & Sassi, F.P. (2008b) Single-crystal X-ray studies of trioctahedral micas coexisting with dioctahedral micas in metamorphic sequences from Western Maine. *American Mineralogist*, **93**, 396–408.

Brindley, G.W. (1959) X-ray and electron diffraction data for sepiolite. *American Mineralogist*, **44**, 495–500.

Brindley, G.W. (1982) Chemical compositions of berthierines – a review. *Clays and Clay Minerals*, **30**, 153–155.

Brindley, G.W. & Brown, G. (editors) (1980) *Crystal Structures of Clay Minerals and their X-ray Identification.* Mineralogical Society, Monograph **5**, London, 495 pp.

Brindley, G.W. & Nakahira, M. (1958) Further consideration of the crystal structure of kaolinite. *Mineralogical Magagazine*, **31**, 781–786.

Brindley, G.W. & Robinson, K. (1945) Structure of kaolinite. *Nature*, **156**, 661–663.

Brindley, G.W. & Robinson, K. (1946) The structure of kaolinite. *Mineralogical Magazine*, **27**, 242–253.

Brindley, G.W. & Robinson, K. (1948) X-ray studies of halloysite and metahalloysite. I. The structure of metahalloysite, an example of random layer lattice. *Mineralogical Magazine*, **28**, 393–406.

Brindley, G.W. & Wan, H.-M. (1975) Compositions, structures, and thermal behavior of Nickel-containing minerals in the lizardite-nepouite series. *American Mineralogist*, **60**, 863–871.

Brindley, G.W., Oughton, B.M. & Youell, R.F. (1951) The crystal structure of amesite and its thermal decomposition. *Acta Crystallographica*, **4**, 552–557.

Brindley, G.W., Bish, D.L., & Hsien-Ming, W. (1979) Compositions, structures, and properties of nickel-containing minerals in the kerolite-pimelite series. *American Mineralogist*, **64**, 615–625.

Brown, B.E. & Bailey, S.W. (1963) Chlorite polytypism: II. Crystal structure of a one-layer Cr chlorite. *American Mineralogist*, **48**, 42–61.

Brown, G. & Weir, A.H. (1963) The identity of rectorite and allevardite. In: *Proceedings of the International Clay Conference, Stockholm*, 27–35 (1) and 87–90 (2).

Brown, G.E., Parks, G.A. & O'Day, P.A. (1995) Sorption at mineral-water interfaces: macroscopic and microscopic perspectives. In: *Mineral Surfaces* (D.J. Vaughan & R.A.D. Pattrick, editors). Mineralogical Society Series, **5**. Chapman & Hall, London, pp. 129–183.

Bujdák, J. (2006) Effect of the layer charge of clay minerals on optical properties of organic dyes. A review. *Applied Clay Science*, **34**, 58–73.

Busigny, V., Cartigny, P., Philippot, P. & Javoy, M. (2004) Quantitative analysis of ammonium in biotite using infrared spectroscopy. *American Mineralogist*, **89**, 1625–1630.

Caillère, S., Mathieu-Sicaud, A. & Hénin, S. (1950) Nouvel essai d'identification du minéral de la Table près Allevard, l'allevardite. *Bulletin de la Société Française de Minéralogie et de Cristallographie*, **73**, 193–201 (in French).

Cámara, F, Garvie, L.A.J., Devouard, B, Groy, T.L. & Buseck, P.R. (2002) The structure of Mn-rich tuperssuatsiaite: A palygorskite-related mineral. *American Mineralogist*, **87**, 1458–1463.

Capitani, G.C. & Mellini, M. (2004) The crystal structure of antigorite: the $m = 17$ polysome. *American Mineralogist*, **89**, 147–158.

Capitani, G.C. & Mellini, M. (2005) HRTEM evidence for 8-reversals in the $m = 17$ antigorite polysome. *American Mineralogist*, **90**, 991–998.

Capitani, G.C. & Mellini, M. (2006) The crystal structure of a second antigorite polysome ($m = 16$), by single-crystal synchrotron diffraction. *American Mineralogist*, **91**, 394–399.

Capitani, G.C. & Mellini, M. (2008) Rationale for the existence of four- and eight-reversals in antigorite. *American Mineralogist*, **93**, 796–799.

Capitani, G.C., Stixrude, L. & Mellini, M. (2009) First-principles energetics and structural relaxation of antigorite. *American Mineralogist*, **94**, 1271–1278.

Caruso, L.J. & Chernosky, J.V. (1979) The stability of lizardite. *The Canadian Mineralogist*, **17**, 757–769.

Cases, J.M., Liètard, O., Yvon, J. & Delon, J.F. (1982) Étude des propriétés cristallochimiques, morphologiques, superficielles de kaolinites désordonnées. *Bulletin de Minéralogie*, **105**, 439–455 (in French).

Castro, E.A.S. & Martins, J.B.L. (2005) Theoretical study of kaolinite. *International Journal of Quantum Chemistry*, **103**, 550–556.

Chernosky, J.V. (1975) Aggregate refractive indices and unit cell parameters of synthetic serpentine in the system MgO-Al_2O_3-SiO_2-H_2O. *American Mineralogist*, **60**, 2000–2008.

Chiari, G., Giustetto, R. & Ricchiardi, G. (2003) Crystal structure refinements of palygorskite and Maya Blue from molecular modeling and powder synchrotron diffraction. *European Journal of Mineralogy*, **15**, 21–33.

Chisholm, J.E. (1992) Powder diffraction patterns and structural models for palygorskite. *The Canadian Mineralogist*, **30**, 61–73.

Chon, C.M., Lee, C.K., Song, Y. & Kim, S.A. (2006) Structural changes and oxidation of ferroan phlogopite with increasing temperature: in situ neutron powder diffraction and Fourier transform infrared spectroscopy. *Physics and Chemistry of Minerals*, **33**, 289–299.

Christ, C.L., Hathaway, J.C., Hostetler, P.B. & Shepard, A.O. (1969) Palygorskite: New X-ray data. *American Mineralogist*, **54**, 198–205.

Christidis, G.E. (2001) Formation and growth of smectites in bentonites: a case study from Kimolos Island, Aegean, Greece. *Clays and Clay Minerals*, **49**, 204–215.

Christidis, G.E. (2006) Genesis and compositional heterogeneity of smectites. Part III: alteration of basic pyroclastic rocks. A case study from the Troodos ophiolite complex, Cyprus. *American Mineralogist*, **91**, 685–701.

Christidis, G.E. (2008) Validity of the structural formula method for layer charge determination of smectites: A re-evaluation of published data. *Applied Clay Science*, **42**, 1–7.

Christidis, G.E. (2011) The concept of layer charge of smectites and its implications on important smectite-water properties. In: *Layered Mineral Structures and their Application in Advanced Technologies* (M.F. Brigatti & A. Mottana, editors). EMU Notes in Mineralogy, **11**, pp. 237–258.

Christidis, G. & Dunham, A.C. (1993) Compositional variations in smectites derived from intermediate volcanic rocks. A case study from Milos Island, Greece. *Clay Minerals*, **28**, 255–273.

Christidis, G. & Dunham, A.C. (1997) Compositional variations in smectites. Part II: alteration of acidic precursors. A case study from Milos Island, Greece. *Clay Minerals*, **32**, 255–273.

Christidis, G.E. & Eberl, D.D. (2003) Determination of layer charge characteristics of smectites, *Clays and Clay Minerals*, **51**, 644–655.

Christidis, G.E., Blum, A.E. & Eberl, D.D. (2006) Influence of layer charge and charge distribution of smectites on the flow behaviour and swelling of bentonites. *Applied Clay Science*, **34**, 125–138.

Chukhrov, F.V., Zvyagin, B.B., Drits, V.A., Gorshkov, A.I., Ermilova, L.P., Goilo, E.A. & Rudnitskaya, E.S. (1979) The ferric analogue of pyrophyllite and related phases. In: *Proceedings of the VI International Clay Conference, Oxford* (M.M. Mortland & V.C. Farmer, editors), 55–64. Elsevier, Amsterdam.

Churchman, G.J. & Theng, B.K.G. (1984) Interactions of halloysites with amides: Mineralogical factors affecting complex formation. *Clay Minerals*, **19**, 161–175.

Cibin, G., Mottana, A., Marcelli, A. & Brigatti, M.F. (2005) Potassium coordination in trioctahedral micas investigated by K-edge XANES spectroscopy. *Mineralogy and Petrology*, **85**, 67–87.

Cibin, G., Mottana, A., Marcelli, A. & Brigatti, M.F. (2006) Angular dependence of potassium K-edge XANES spectra of trioctahedral micas: Significance for the determination of the local structure and electronic behavior of the interlayer site. *American Mineralogist*, **91**, 1150–1162.

Cibin, G., Cinque, G., Marcelli, A., Mottana, A. & Sassi, R. (2008) The octahedral sheet of metamorphic $2M_1$-phengites: A combined EMPA and AXANES study. *American Mineralogist*, **93**, 414–425.

Coey, J.M.D., Chukhrov, F.V. & Zvyagin, B.B. (1984) Cation distribution, Mössbauer spectra, and magnetic properties of ferripyrophyllite. *Clays and Clay Minerals*, **32**, 198–204.

Comodi, P., Fumagalli, P., Montagnoli, M. & Zanazzi, P.F. (2004) A single crystal study on the pressure behavior of phlogopite and petrological implications. *American Mineralogist*, **89**, 647–653.

Comodi, P., Cera, F., Dubrovinsky, L. & Nazzareni, S. (2006) The high-pressure behavior of the 10 Å phase: A spectroscopic and diffractometric study up to 42 GPa. *Earth and Planetary Science Letters*, **246**, 444–457.

Compagnoni, R., Ferraris, G. & Mellini, M. (1985) Carlosturanite, a new asbestiform rock-forming silicate from Val Varaita, Italy. *American Mineralogist*, **70**, 767–772.

Coombes, D.S., Catlow, C., Richard, A. & Garces, J.M. (2003) Computational studies of layered silicates. *Modelling and Simulation in Materials Science and Engineering*, **11**, 301–306.

Cradwick, P.D.G., Farmer, V.C., Russell, J.D., Masson, C.R., Wada, K. & Yoshinaga, N. (1972) Imogolite, a hydrated aluminum silicate of tubular structure. *Nature Physical Science*, **240**, 187–189.

Cressey, G., Cressey, B.A., Wicks, F.J. & Yada, K. (2010) A disc with five-fold symmetry: the proposed fundamental seed structure for the formation of chrysotile asbestos fibres, polygonal serpentine fibres and polyhedral lizardite spheres. *Mineralogical Magazine*, **74**, 29–37.

Curetti, N., Levy, D., Pavese, A. & Ivaldi, G. (2006) Elastic properties and stability of coexisting $3T$ and $2M_1$ phengite polytypes. *Physics and Chemistry of Minerals*, **32**, 670–678.

Dainyak, L.G. & Drits, V.A. (2009) A model for the interpretation of Mössbauer spectra of muscovite. *European Journal of Mineralogy*, **21**, 99–106.

Dainyak, L.G., Drits, V.A. & Lindgreen, H. (2004) Computer simulation of octahedral cation distribution and interpretation of the Mössbauer Fe^{2+} components in dioctahedral trans-vacant micas. *European Journal of Mineralogy*, **16**, 451–468.

Dainyak, L.G., Rusakov, V.S., Sukhorukov, I.A., Zviagina, B.B. & Drits, V.A. (2009) An improved model for the interpretation of Mössbauer spectra of dioctahedral 2:1 *trans*-vacant Fe-rich micas: refinement of parameters. *European Journal of Mineralogy*, **21**, 995–1008.

de la Calle, C., Dubernat, J., Suquet, H., Pezerat, H., Gaulthier, J. & Mamy, J. (1975a) Crystal structure of two-layer Mg-vermiculites and Na-, Ca-vermiculites. In: *Proceedings of the International Clay Conference, Mexico City* (S.W. Bailey, editor). Applied Publishing, Wilmette, Illinois, USA, pp. 201–209.

de la Calle, C., Suquet, H. & Pezerat, H. (1975b) Glissement de feuillets accompagnant certains echanges cationiques dans les vermiculites. *Bulletin du Groupe Français des Argiles*, **27**, 31–49 (in French).

de la Calle, C., Suquet, H., Dubernat, J. & Pezerat, H. (1978) Mode d'Empilement des feuillets dans les vermiculites hydratees a 'deux couches. *Clay Minerals*, **13**, 275–297 (in French).

de la Calle, C., Plançon, A., Pons, C.H., Dubernat, J., Suquet, H. & Pezerat, H. (1984) Mode d'empilement des feuillets dans la vermiculite sodique hydraté à une couche (Phase à 11 85 Å). *Clay Minerals*, **19**, 563–578 (in French).

de la Calle, C., Suquet, H. & Pezerat, H. (1985) Vermiculites hydratees a une couche. *Clay Minerals*, **20**, 221–230.

de la Calle, C., Suquet, H. & Pons, C.H. (1988) Stacking order in a 14.30 Å Mg-vermiculite. *Clays and Clay Minerals*, **36**, 481–490.

de Souza Santos, P., Brindley, G.W. & de Souza Santos, H. (1965) Mineralogical studies of kaolinite-halloysite clays. III. A fibrous kaolin mineral from Piedade, Sao Paulo, Brazil. *American Mineralogist*, **50**, 619–28.

De Waal, S.A. (1970) Nickel minerals from Barberton, South Africa. III. Willemseite, a nickel-rich talc. *American Mineralogist*, **55**, 31–42.

Decarreau, A., Colin, F., Herbillon, A., Manceau, A., Nahon, D., Paquet, H., Trauth-Badeaud, D. & Trescases, J.J. (1987) Domain segregation in Ni–Fe–Mg–smectites, *Clays and Clay Minerals*, **35**, 1–10.

Deer, W.A., Howie, R.A. & Zussman, J. (2009) *Rock-Forming Minerals, vol. 3B, Layered Silicates Excluding Micas and Clay Minerals.* 2nd edition. Geological Society, London, 314 pp.

Dera, P., Prewitt, C.T., Japel, S., Bish, D.L. & Johnston, C.T. (2003) Pressure-controlled polytypism in hydrous layered materials. *American Mineralogist*, **88**, 1429–1425.

Di Vincenzo, G., Viti, C. & Rocchi, S. (2003) The effect of chlorite interlayering on ^{40}Ar-^{39}Ar biotite dating: an ^{40}Ar-^{39}Ar laser-probe and TEM investigations of variably chloritised biotites. *Contributions to Mineralogy and Petrology*, **145**, 643–658.

Di Vincenzo, G., Tonarini, S., Lombardo, B., Castelli, D. & Ottolini, L. (2006) Comparison of ^{40}Ar-^{39}Ar and Rb-Sr data on phengites from the UHP Brossasco-Isasca Unit (Dora Maira Massif, Italy): Implications for dating white mica. *Journal of Petrology*, **47**, 1439–1465.

Dobson, D.P., De Ronde, A.A., Welch, M.D. & Meredith, P.G. (2007) The acoustic emissions signature of a pressure-induced polytypic transformation in chlorite. *American Mineralogist*, **92**, 437–440.

Dódony, I. & Buseck, P.R. (2004) Serpentines close-up and intimate: An HRTEM view. *International Geology Review*, **46**, 507–527.

Dódony, I., Pósfai, M. & Buseck, P.R. (2002) Revised structure models for antigorite: An HRTEM study. *American Mineralogist*, **87**, 1443–1457.

Dódony, I., Posfai, M. & Buseck, P.R. (2006) Does antigorite really contain 4- and 8-membered rings of tetrahedra? *American Mineralogist*, **91**, 1831–1838.

Dong, H. (2005) Interstratified illite-smectite: a review of contributions of TEM data to crystal chemical relations and reaction mechanisms. *Clay Science*, **12**, 6–12.

Dong, H., Peacor, D.R. & Freed, R.L. (1997) Phase relations among smectite, R1 illite/smectite and illite. *American Mineralogist*, **82**, 379–391.

Dong, H., Peacor, D.R., Merriman, R.J. & Kemp, S.J. (2002) Brinrobertsite: a new R1 interstratified pyrophyllite/smectite-like clay mineral: characterization and geological origin. *Mineralogical Magazine*, **66**, 605–617.

Donnay, G., Morimoto, N., Takeda, H. & Donnay, D.H. (1964) Trioctahedral one-layer micas: 1. Crystal structure of a synthetic iron mica. *Acta Crystallographica*, **17**, 1369–1373.

Drits, V.A. (1997) Mixed layer minerals. In: *Modular Aspects of Minerals* (S. Merlino, editor). EMU Notes in Mineralogy, **1**, Eötvös University Press, Budapest, pp. 153–190.

Drits, V.A. (2003) Structural and chemical heterogeneity of layer silicates and clay minerals. *Clay Minerals*, **38**, 403–432.

Drits, V.A. & Karavan, Y.V. (1969) Polytypes of the two-packet chlorites. *Acta Crystallographica, Section B: Structural Crystallography and Crystal Chemistry*, **25**, 632–639.

Drits, V.A. & Kashaev, A.A. (1960) The structure of kaolinite. *Materialy po geologii i poleznym iskopaemym vostochnoĭ Sibiri, Irkutsk*, 264–266.

Drits, V.A. & Sokolova, G.V. (1971) Structure of palygorskite. *Soviet Physics, Crystallography*, **16**, 183–185.

Drits, V.A. & Tchoubar, C. (1990) *X-ray Diffraction by Disordered Lamellar Structures: Theory and Applications to Microdivided Silicates and Carbons.* Springer-Verlag, Berlin, 371 pp.

Drits, V.A. & Zviagina, B.B. (2009) *Trans*-vacant and *cis*-vacant 2:1 layer silicates: structural features, identification, and occurrence. *Clays and Clay Minerals*, **57**, 405–415.

Drits, V.A., Plançon, A., Sakharov, B.A., Besson, G., Tsipursky, S.I. & Tchoubar, C. (1984) Diffraction effects calculated for structural models of K-saturated montmorillonite containing different types of defects. *Clay Minerals*, **19**, 541–562.

Drits, V.A. Varaxina, T.V., Sakharov, B.A. & Plançon, A. (1994) A simple technique for identification of one-dimensional powder X-ray diffraction patterns for mixed-layer illite-smectites and other interstratified minerals. *Clays and Clay Minerals*, **42**, 382–90.

Drits, V.A., Środoń, J. & Eberl, D.D. (1997a) XRD measurements of mean crystallite thickness of illite and illite/smectite: reappraisal of the Kübler index and the Scherrer equation. *Clays and Clay Minerals*, **45**, 461–475.

Drits, V.A., Lindgreen, H., Sakharov, B.A. & Salyn, A.L. (1997b) Sequential structure transformation of illite-smectite-vermiculite during diagenesis of Upper Jurassic shales, North Sea. *Clay Minerals*, **33**, 351–371.

Drits, V.A., Lindgreen, H., Salyn, A.L., Ylagan R. & McCarty, D.K. (1998) Semiquantitative determination of *trans*-vacant and *cis*-vacant 2:1 layers in illites and illite-smectites by thermal analysis and X-ray diffraction. *American Mineralogist*, **83**, 1188–1198.

Drits, V.A., McCarty, D.K. & Zviagina, B.B. (2006) Crystal-chemical factors responsible for the distribution of octahedral cations over *trans*- and *cis*-sites in dioctahedral 2:1 layer silicates. *Clays and Clay Minerals*, **54**, 131–152.

Drits, V.A., Lindgreen, H., Sakharov, B.A., Jakobsen, H.J., Fallick, A.E., Salyn, A.L., Dainyak, L.G., Zviagina, B.B. & Barfod, D.N. (2007) Formation and transformation of mixed-layer minerals by Tertiary intrusives in Cretaceous mudstones, West Greenland. *Clays and Clay Minerals*, **55**, 260–283.

Dunoyer de Segonzac, G. & Bernoulli, D. (1976) Diagenese et métamorphisme des argiles dans le Rhétien Sud-alpin et Austro-alpin (Lombardie et Grisons). *Bulletin Societé Géologique de la France*, **18**, 1283–1293 (in French).

Ďurovič, S. (1997) Fundamentals of OD theory. In: *Modular Aspects of Minerals* (S. Merlino, editor). EMU Notes in Mineralogy, **1**. Eötvös University Press, Budapest, pp. 1–28.

Ďurovič, S. (1999) Layer stacking in general polytypic structures. In: *International Tables for Crystallography, Volume C* (A.C.J. Wilson & E. Prince, editors). Kluwer Academic Publishers, Dordrecht, The Netherlands, pp. 752–765.

Ďurovič, S., Hybler, J. & Kogure, T. (2004) Parallel intergrowths in cronstedtite-1*T*: Implications for structure refinement. *Clays and Clay Minerals*, **52**, 613–622.

Dyar, M.D. (2002) Optical and Mössbauer spectroscopy of iron in micas. In: *Micas: Crystal Chemistry and Metamorphic Petrology* (A. Mottana, F.P. Sassi, J.B. Thompson Jr. & S. Guggenheim, editors). Reviews in Mineralogy and Geochemistry, **46**. Mineralogical Society of America, Chantilly, Virginia, USA, pp. 313–349.

Dyar, M.D., Gunter, M.E., Delaney, J.S., Lanzarotti, A. & Sutton, S.R. (2002a) Systematics in the structure and XANES spectra of pyroxenes, amphiboles, and micas as derived from oriented single crystals. *The Canadian Mineralogist*, **40**, 1375–1393.

Dyar, M.D., Lowe, E.W., Guidotti, C.V. & Delaney, J.S. (2002b) Fe^{3+} and Fe^{2+} partitioning among silicates in metapelites: A synchrotron micro-XANES study. *American Mineralogist*, **87**, 514–522.

Dyar, M.D., Schaefer, M.W., Sklute, E.C. & Bishop, J.L. (2008) Mössbauer spectroscopy of phyllosilicates: effects of fitting models on recoil-free fractions and redox ratios. *Clay Minerals*, **43**, 3–33.

Eberl, D.D. & Velde, B. (1989) Beyond the Kübler index. *Clay Minerals*, **24**, 571–577.

Eberl, D.D., Drits, V., Środoń, J. & Nüesch, R. (1996) *MudMaster: A program for calculating crystallite size distributions and strain from the shapes of X-ray diffraction peaks.* U.S. Geological Survey Open-File Report 96-171, 46 pp.

Eberl, D.D., Drits, V.A. & Środoń, J. (1998) Deducing growth mechanisms for minerals from the shapes of crystal size distributions. *American Journal of Science*, **298**, 499–533.

Eberl, D.D., Drits, V.A. & Środoń, J. (2000) *User's guide to Galoper—a program for simulating the shapes of crystal size distributions—and associated programs.* U.S. Geological Survey Open-File Report 00-505, 44 pp.

Eggleston, R.A. & Bailey, S.W. (1967) Structural aspects of dioctahedral chlorite. *American Mineralogist*, **52**, 673–689.

Eggleton, R.A. & Tilley, D.B. (1998) Hisingerite: a ferric kaolin mineral with curved morphology. *Clays and Clay Minerals*, **46**, 400–413.

Evans, B.W. & Guggenheim, S. (1988) Talc, pyrophyllite, and related minerals. In: *Hydrous Phyllosilicates (Exclusive of Micas)* (S.W. Bailey, editor). Reviews in Mineralogy, **19**. Mineralogical Society of America, Washington, D.C., 225–294.

Evans, R., James, R., Denis, G. & Grodzicki, M. (2005) Hyperfine electric field gradient tensors at Fe^{2+} sites in octahedral layers: Toward understanding oriented single crystal Mössbauer spectroscopy measurements of micas. *American Mineralogist*, **90**, 1540–1555.

Farmer, V.C. (1974) *The Infrared Spectra of Minerals.* Mineralogical Society, London, 539 pp.

Farmer, V.C. (1998) Differing effect of particle size and shape in the infrared and Raman spectra of kaolinite. *Clay Minerals*, **33**, 601–604.

Farmer, V.C. (2000) Transverse and longitudinal crystal modes associated with OH stretching vibrations in single crystals of kaolinite and dickite. *Spectrochimica Acta Part A*, **56**, 927–930.

Farmer, V.C. & Fraser, A.R. (1979) Synthetic imogolite, a tubular hydroxyaluminium silicate. In: *International Clay Conference, Oxford* (M.M. Mortland & V.C. Farmer, editors). Elsevier, Amsterdam, pp. 547–553.

Faust, G.T. (1951) Thermal analysis and X-ray studies of sauconite and of some zinc minerals of the same paragenetic association. *American Mineralogist*, **36**, 795–822.

Faust, G.T. & Murata, K.J. (1953) Stevensite, redefined as a member of the montmorillonite group. *American Mineralogist*, **38**, 973–987.

Faust, G.T., Hathway, J.C. & Millot, G. (1959) A restudy of stevensite and allied minerals. *American Mineralogist*, **44**, 343–370.

Fechtelkord, M., Behrens, H., Holtz, F., Bretherton, J.L., Fyfe, C.A., Groat, L.A. & Raudsepp, M. (2003a) Influence of F content on the composition of Al-rich synthetic phlogopite: Part II. Probing the structural arrangement of aluminum in tetrahedral and octahedral layers by ^{27}Al MQMAS and ^{1}H/^{19}F-^{27}Al HETCOR and REDOR experiments. *American Mineralogist*, **88**, 1046–1054.

Fechtelkord, M., Behrens, H., Holtz, F., Fyfe, C.A., Groat, L.A. & Raudsepp, M. (2003b) Influence of F content on the composition of Al-rich synthetic phlogopite: Part I. New information on structure and phase-formation from ^{29}Si, ^{1}H, and ^{19}F MAS NMR spectroscopies. *American Mineralogist*, **88**, 47–53.

Ferrage, E., Lanson, B., Malikova, N., Plançon, A., Sakharov, B.A. & Drits, V.A. (2005a) New insights on the distribution of interlayer water in bi-hydrated smectite from X-ray diffraction profile modeling of 00l reflections. *Chemistry of Materials*, **17**, 3499–3512.

Ferrage, E., Lanson, B., Sakharov, B.A. & Drits, V.A. (2005b) Investigation of smectite hydration properties by modeling of X-ray diffraction profiles. Part 1. Montmorillonite hydration properties. *American Mineralogist*, **90**, 1358–1374.

Ferrage, E., Seine, G., Gaillot, A.C., Petit, S., de Parseval, P., Boudet, A., Lanson, B., Ferret, J. & Martin, F. (2006) Structure of the {001} talc surface as seen by atomic force microscopy. Comparison with X-ray and electron diffraction results. *European Journal of Mineralogy*, **18**, 483–491.

Ferrage, E., Lanson, B., Sakharov, B.A., Jacquot, E., Geoffroy, N. & Drits, V.A. (2007) Investigation of smectite hydration properties by modeling of X-ray diffraction profiles. Part 2. Influence of layer charge and charge location. *American Mineralogist*, **92**, 1731–1743.

Ferrage, E., Lanson, B., Michot, L.J. & Robert, J-L. (2010) Hydration properties and interlayer organization of water and ions in synthetic Na-smectite with tetrahedral layer charge. Part 1. Results from X-ray diffraction profile modeling. *Journal of Physical Chemistry*, **114**, 4515–4526.

Ferraris, G. & Ivaldi, G. (2002) Structural features of micas. In: *Micas: Crystal Chemistry and Metamorphic Petrology* (A. Mottana, F.P. Sassi, J.B. Thompson Jr. & S. Guggenheim, editors). Reviews in Mineralogy and Geochemistry, **46**. Mineralogical Society of America, Chantilly, Virginia, USA, pp. 118–153.

Ferraris, G., Mellini, M. & Merlino, S. (1986) Polysomatism and the classification of minerals. *Rendiconti della Società Italiana di Mineralogia e Petrografia*, **41**, 181–192.

Ferraris, G., Khomyakov, A.P., Belluso, E. & Soboleva, S.V. (1998) Kalifersite, a new alkaline silicate from Kola Peninsula (Russia) based on a palygorskite-sepiolite polysomatic series. *European Journal of Mineralogy*, **10**, 865–874.

Ferraris, C., Chopin, C. & Wessicken, R. (2000) Nano to micro scale decompression products in ultrahigh pressure phengite: HRTEM and AEM study, and some petrological implications. *American Mineralogist*, **85**, 1195–1201.

Ferraris, C., Castelli, D. & Lombardo, B. (2005) SEM/TEM-AEM characterization of micro- and nano-scale zonation in phengite from a UHP Dora-Maira marble: Petrologic significance of armoured Si-rich domains. *European Journal of Mineralogy*, **17**, 453–464.

Fleet, M.E. (2003) *Rock-Forming Minerals, Volume 3A. Sheet Silicates: Micas.* 2nd edition, Geological Society, London, 758 pp.

Frank-Kamenetskii, V.A., Logvinenko, N.V. & Drits, V.A. (1965) Tosudite – a new mineral, forming the mixed-layer phase in alushtite. In: *Proceedings of the International Clay Conference, Stockholm*, pp. 181–186 (2).

Frost, R.L. (1995) Fourier transform Raman spectroscopy of kaolinite, dickite and halloysite. *Clays and Clay Minerals*, **43**, 191–195.

Frost, R.L. & Van der Gaast, S.J. (1997) Kaolinite hydroxyls; a Raman microscopy study. *Clay Minerals*, **32**, 471–484.

Galán, E. (1996) Properties and applications of palygorskite-sepiolite clays. *Clay Minerals*, **31**, 443–453.

Galán, E. & Carretero, M.I. (1999) A new approach to compositional limits for sepiolite and palygorskite. *Clays and Clay Minerals*, **47**, 399–409.

García, F.J., García Rodríguez, S., Kalytta, A. & Reller, A. (2009) Study of natural halloysite from the Dragon Mine, Utah (USA). *Zeitschrift für Anorganische und Allgemeine*, **635**, 790–795.

Gard, J.A. & Follet, E.A. (1968) A structural scheme for palygorskite. *Clay Minerals*, **7**, 367–369.

Gatta, D., Rotiroti, N., Pavese, A., Lotti, P. & Curetti, N. (2009) Structural evolution of a 3*T* phengite up to 10 GPa: an *in situ* single-crystal X-ray diffraction study. *Zeitschrift für Kristallographie*, **224**, 302–310.

Gemmi, M., Merlini, M., Pavese, A. & Curetti N. (2008) Thermal expansion and dehydroxylation of phengite micas. *Physics and Chemistry of Minerals*, **35**, 367–379.

Gianfagna, A., Scordari, F., Mazziotti-Tagliani, S., Ventruti, G. & Ottolini, L. (2007) Fluorophlogopite from Biancavilla (Mt. Etna, Sicily, Italy): crystal structure and crystal chemistry of a new F-dominant analog of phlogopite. *American Mineralogist*, **92**, 1601–1609.

Giese, R.F. & Datta, P. (1973) Hydroxyl orientation in kaolinite, dickite, and nacrite. *American Mineralogist*, **58**, 471–479.

Giese, R.F. & van Oss, C.J. (2002) *Colloid and Surface Properties of Clays and Related Minerals*. Marcel Dekker, New York, 295 pp.

Giorgetti, G., Tropper, P., Essene, E.J. & Peacor, D.R. (2000) Characterization of non equilibrium and equilibrium occurrences of paragonite/muscovite intergrowths in an eclogite from the Sesia Lanzo Zone (Western Alps, Italy). *Contributions to Mineralogy and Petrology*, **138**, 326–336.

Giustetto, R. & Chiari, G. (2004) Crystal structure refinement of palygorskite from neutron powder diffraction. *European Journal of Mineralogy*, **16**, 521–532.

Gregorkiewitz, M., Lebech, B., Mellini, M. & Viti, C. (1996) Hydrogen positions and thermal expansion in lizardite-1*T* from Elba: a low-temperature study using Rietveld refinement of neutron diffraction data. *American Mineralogist*, **81**, 1111–1116.

Grim, R.E. & Güven, N. (1978) *Bentonites*. Elsevier, New York, 256 pp.

Grobéty, B. (2003) Polytypes and higher-order structures of antigorite: A TEM study. *American Mineralogist*, **88**, 27–36.

Gruner, J.W. (1932a) The crystal structure of kaolinite. *Zeitschrift für Kristallographie, Kristallgeometrie, Kristallphysik, Kristallchemie*, **83**, 75–80.

Gruner, J.W. (1932b) The crystal structure of dickite. *Zeitschrift für Kristallographie, Kristallgeometrie, Kristallphysik, Kristallchemie*, **83**, 394–404.

Gruner, J.W. (1934a) The crystal structure of talc and pyrophyllite. *Zeitschrift für Kristallographie*, **88**, 412–419.

Gruner, J.W. (1934b) The structures of vermiculites and their collapse by dehydration. *American Mineralogist*, **19**, 557–575.

Gruner, J.W. (1935) The structural relationships of nontronite and montmorillonite. *American Mineralogist*, **20**, 475–483.

Gualtieri, A.F., Moen, A. & Nicholson, D.G. (2000) XANES study of the local environment of iron in natural kaolinites. *European Journal of Mineralogy*, **12**, 17–23.

Guggenheim, S. (2011) An overview of order/disorder in hydrous phyllosilicates. In: *Layered Mineral Structures and their Application in Advanced Technologies* (M.F. Brigatti & A. Mottana, editors), EMU Notes in Mineralogy, **11**, pp. 73–121.

Guggenheim, S. & Eggleton, R.A. (1998) Modulated crystal structures of greenalite and caryopilite: a system with long-range, in-plane structural disorder in the tetrahedra sheet. *The Canadian Mineralogist*, **36**,163–179.

Guggenheim, S. & Zhan, W. (1998) Effect of temperature on the structures of lizardite-1*T* and lizardite-2H_1. *The Canadian Mineralogist*, **36**, 1587–1594.

Guggenheim, S., Adams, J.M., Bain, D.C., Bergaya, F., Brigatti, M.F., Drits, V.A., Formoso, M.L.L., Galán, E., Kogure, T. & Stanjek, H. (2006) Summary of recommendations of nomenclature committees relevant to clay mineralogy: report of the Association Internationale Pour l'Etude des Argiles (AIPEA) nomenclature committee for 2006. *Clays and Clay* Minerals, **54**, 761–772.

Güven, N. (1988) Smectites. In: *Hydrous Phyllosilicates (Exclusive of Micas)* (S.W Bailey, editor). Reviews in Mineralogy, **19**. Mineralogical Society of America, Washington, D.C., pp. 497–559.

Güven, N. (1992) Molecular aspects of clay-water interactions. In: *Clay-Water Interface and its Rheological Implications* (N. Güven & R.M. Pollastro, editors). CMS Workshop Lectures, **4**. The Clay Minerals Society, Boulder, Colorado, USA, pp. 1–80.

Hall, S.H. & Bailey, S.W. (1979) Cation ordering pattern in amesite. *Clays and Clay Minerals*, **27**, 241–247.

He, H., Barr, T.L. & Klinowski, J. (1995) ESCA and solid-state NMR studies of allophane. *Clay Minerals*, **30**, 201–209.

He, H.P., Guo, J.G., Zhu, J.X. & Hu, C. (2003) ^{29}Si and ^{27}Al MAS NMR study of the thermal transformations of kaolinite from North China. *Clay Minerals*, **38**, 551–559.

Heinrich, A.R., Eggleton, R.A. & Guggenheim, S. (1994) Structure and polytypism of bementite, a modulated layer silicate. *American Mineralogist*, **79**, 91–106.

Hendricks, S.B. (1938) The crystal structure of the clay minerals: dickite, halloysite and hydrated halloysite. *American Mineralogist*, **23**, 295–301.

Hendricks, S.B. (1939) The crystal structure of nacrite, $Al_2O_3{\cdot}2SiO_2{\cdot}2H_2O$, and the polymorphism of the kaolin minerals. *Zeitschrift für Kristallographie*, **100**, 509–518.

Hendricks, S.B. & Jefferson, M.E. (1938) Crystal structure of vermiculites and mixed vermiculite-chlorites. *American Mineralogist*, **23**, 851–862.

Henmi, T. & Wada, K. (1976) Morphology and composition of allophane. *American Mineralogist*, **61**, 379–390.

Hinckley, D.N. (1963) Variability in "crystallinity" values among the kaolin deposits of the coastal plain of Georgia and South Carolina. *Clays and Clay Minerals*, **11**, 229–235.

Hisinger, W. (1810) Undersökning af Svenska Mineralier. VI. Svart Stenart från Gillinge Jern-Grufva i Södermanland. *Afhandlingar i Fysik, Kemi, och Mineralogi*, **3**, 304–306 (in Swedish).

Hobbs, J.D., Cygan, R.T., Nagy, K.L., Schultz, P.A. & Sears, M.P. (1997) All-atom *ab initio* energy minimization of the kaolinite crystal structure. *American Mineralogist*, **82**, 657–662.

Hofmann, U., Endell, K. & Wilm, D. (1934) X-ray and colloid-chemical studies of clay. *Angewandte Chemie*, **47**, 539–547.

Honjo, G., Kitamura, N. & Mihama, K. (1954) A study of clay minerals by means of single-crystal electron-diffraction diagrams – the structure of tubular kaolin. *Clay Minerals Bulletin*, **2**, 133–141.

Hoppe, R. (1979) Effective coordination numbers (ECON) and mean fictive ionic radii (MEFIR). *Zeitschrift für Kristallographie*, **150**, 23–52.

Hybler, J. (2006) Parallel intergrowths in cronstedtite-1*T*: determination of the degree of disorder. *European Journal of Mineralogy*, **18**, 197–205.

Hybler, J., Petricek, V., Ďurovič, S. & Smrčok, L. (2000) Refinement of the crystal structure of cronstedtite-1*T*. *Clays and Clay Minerals*, **48**, 331–338.

Hybler, J., Petricek, V., Fabry, J. & Ďurovič, S. (2002) Refinement of the crystal structure of cronstedtite-$2H_2$. *Clays and Clay Minerals*, **50**, 601–613.

Ianovici, V., Neacsu, G., & Neacsu, V. (1990) Li-bearing stevensite from Moldova Noua, Romania. *Clays and Clay Minerals*, **38**, 171–178.

Ivanova, T.I. & Frank-Kamenetskaya, O.V. (2001) Using the statistical probability-based model of an irregular mixed-layer structure for describing the real structures of chemically nonhomogeneous single crystals. *Journal of Structural Chemistry*, **42**, 126–143.

Jiang, W.T., Peacor, D.R. & Slack, J.F. (1992) Microstructures, mixed layering, and polymorphism of chlorite and retrograde berthierine in the Kidd Creek massive sulfide deposit, Ontario. *Clays and Clay Minerals*, **40**, 501–514.

Johnston, C.T., Agnew, S.F. & Bish, D.L. (1990) Polarized single-crystal Fourier-transform infrared microscopy of Ouray dickite and Keokuk kaolinite. *Clays and Clay Minerals*, **38**, 573–583.

Johnston, C.T., Helsen, J., Schoonheydt, R.A., Bish, D.L. & Agnew, S.F. (1998) Single-crystal Raman spectroscopic study of dickite. *American Mineralogist*, **83**, 75–84.

Johnston, C.T., Wang, S.-L., Bish, D.L., Dera, P., Agnew, S.F. & Kenney III, J.W. (2002) Novel pressure-induced phase transformations in hydrous layered materials. *Geophysical Research Letters*, **29**, 1770.

Jones, B.F. & Galán, E. (1988) Sepiolite and palygorskite. In: *Hydrous Phyllosilicates* (S.W. Bailey, editor). Reviews in Mineralogy, **19**. Mineralogical Society of America, Chantilly, Virginia, USA, pp. 631–674.

Joswig, W. & Drits, V.A. (1986) The orientation of the hydroxyl groups in dickite by X-ray diffraction. *Neues Jahrbuch für Mineralogie Monatshefte*, **1**, 19–22.

Joswig, W. & Fuess, H. (1989) Neutron diffraction study of a one-layer monoclinic chlorite. *Clays and Clay Minerals*, **37**, 511–514.

Joswig, W. & Fuess, H. (1990) Refinement of a one-layer triclinic chlorite. *Clays and Clay Minerals*, **38**, 216–218.

Joswig, W., Fuess, H., Rothbauer, R., Takeuchi, Y. & Mason, S.A. (1980) A neutron diffraction study of a one-layer triclinic chlorite (penninite). *American Mineralogist*, **65**, 349–352.

Joussein, E., Petit, S., Churchman, J., Theng, B., Righi, D. & Delvaux, B. (2005) Halloysite clay minerals – a review. *Clay Minerals*, **40**, 383–426.

Kameda, J., Miyawaki, R., Kitagawa, R. & Kogure, T. (2007) XRD and HRTEM analyses of stacking structures in sudoite, di-trioctahedral chlorite. *American Mineralogist*, **92**, 1586–1592.

Kato, T. (1963) New data on the so called bementite. *Mineralogical Journal of Japan*, **6**, 93–103.

Kato, T. & Takeuchi, Y. (1980) Crystal structures and submicroscopic textures of layered manganese silicates. *Mineralogical Journal of Japan*, **14**, 165–178.

Kitagawa, Y. (1971) The unit particle of allophane. *American Mineralogist*, **56**, 465–475.

Kittrick, J.A. (1969a) Interlayer forces in montmorillonite and vermiculite. *Soil Science Society of America Journal*, **33**, 217–222.

Kittrick, J.A. (1969b) Quantitative evaluation of the strong-force model for expansion and contraction of vermiculite. *Soil Science Society of America Journal*, **33**, 222–225.

Kloprogge, J.T., Hammond, M., Hickey, L. & Frost, R.L. (2001) A new low temperature synthesis route of fraipontite $(Zn,Al)_3(Si,Al)_2O_5(OH)_4$. *Materials Research Bulletin*, **36**, 1091–1098.

Kogure, T. (2002) Investigations of micas using advanced transmission electron microscopy. In: *Micas: Crystal Chemistry and Metamorphic Petrology* (A. Mottana, F.P. Sassi, J.B. Thompson Jr. & S. Guggenheim, editors). Reviews in Mineralogy and Geochemistry, **46**. Mineralogical Society of America, Chantilly, Virginia, USA, pp. 300–380.

Kogure, T. & Kameda, J. (2008) High-resolution TEM and XRD simulation of stacking disorder in 2:1 phyllosilicates. *Zeitschrift für Kristallographie*, **223**, 69–75.

Kogure, T., Hybler, J. & Ďurovič, S. (2001) A HRTEM study of cronstedtite: determination of polytypes and layer polarity in trioctahedral 1:1 phyllosilicates. *Clays and Clay Minerals*, **49**, 310–317.

Kogure, T., Miyawaki, R. & Banno, Y. (2005) The true structure of wonesite, an interlayer deficient trioctahedral sodium mica. *American Mineralogist*, **90**, 725–731.

Kogure, T., Jige, M., Kameda, J., Yamagishi, A., Miyawaki, R. & Kitagawa, R. (2006) Stacking structures in pyrophyllite revealed by high-resolution transmission electron microscopy (HRTEM). *American Mineralogist*, **91**, 1293–1299.

Kogure, T., Elzea-Kogel, J., Johnston, C. T. & Bish, D.L. (2010) Stacking disorder in a sedimentary kaolinite. *Clays and Clay Minerals*, **58**, 62–71.

Kohyama, N. & Sudo, T. (1975) Hisingerite occurring as a weathering product of iron-rich saponite. *Clays and Clay Minerals*, **23**, 215–218.

Kohyama, N., Fukushima, K. & Fukami, A. (1978) Observation of the hydrated form of tabular halloysite by an electron microscope equipped with an environmental cell. *Clays and Clay Minerals*, **26**, 25–40.

Koshimizu, H., Higuchi, S. & Otsuka, R. (1981) Hydrothermal synthesis and some properties of talc-willemseite solid solutions. *Nendo Kagaku*, **21**, 61–71.

Krekeler, M.P.S. & Guggenheim, S. (2008) Defects in microstructure in palygorskite–sepiolite minerals: A transmission electron microscopy (TEM) study. *Applied Clay Science*, **39**, 98–195.

Krivovichev, S.V., Armbruster, T., Organova, N.I., Burns, P.C., Seredkin, M.V. & Chukanov, N.V. (2004) Incorporation of sodium into the chlorite structure: the crystal structure of glagolevite, $Na(Mg,Al)_6[Si_3AlO_{10}](OH,O)_8$. *American Mineralogist*, **89,** 1138–1141.

Kübler, B. (1964) Les argiles, indicateurs de métamorphisme. *Revue de l'Institut Francaise du Petrole*, **19**, 1093–1112 (in French).

Kübler, B. (1967) La cristallinité de l'illite et les zones tout á fait supécieures du métamorphisme. In Étages Tectoniques. *Collogue de Neuchâtel*, **1996**, 105–121 (in French).

Kübler, B. (1984) Les indicateurs des transformations physiques et chimiques dans la diagenèse, température et calorimétrie. In: *Thermométrie et Barométrie Géologiques* (M. Lagache, editor). Société Française Minèralogie et de Cristallographie, Paris (in French), pp. 489–596.

Kunze, V.G. (1956) Die gewellte Struktur des Antigorits, I. *Zeitschrift fiir Kristallographie*, **108**, 82–107 (in German).

Kunze, V.G. (1958) Die gewellte Struktur des Antigorits, II. *Zeitschrift für Kristallographie*, **110**, 282–320 (in German).

Kunze, V.G. (1961) Antigorit: Strucktur theoretische Grundlagen und ihre praktische Bedeutung für die weiters Serpentinforschung. *Fortschritte der Mineralogie*, **39**, 206–324 (in German).

Lagaly, G. (1981) Characterization of clays by organic compounds. *Clay Minerals*, **16**, 1–21.

Lagaly, G. (1994) Layer charge determination by alkylammonium ions. In: *Layer Charge Characteristics of 2:1 Silicate Clay Minerals* (A.R. Mermut, editor). CMS Workshop Lectures, **6**. The Clay Minerals Society, Boulder, Colorado, USA, pp. 2–46.

Lagaly, G. & Weiss, A. (1975) The layer charge of smectitic layer silicates. In: *Proceedings of the International Clay Conference, 1975, Mexico City, Mexico*, pp. 157–172.

Laird, D.A. (1994) Evaluation of the structural formula and alkylammonium methods of determining layer charge. In: *Layer Charge Characteristics of 2:1 Silicate Clay Minerals* (A.R. Mermut, editor). CMS Workshop Lectures, **6**. The Clay Minerals Society, Boulder, Colorado, USA, pp. 80–103.

Laird, D.A. (1996) Model for crystalline swelling of 2:1 phyllosilicates. *Clays and Clay Minerals*, **44**, 553 559.

Laird, D.A. (1999) Layer charge influences on the hydration of expandable 2:1 phyllosilicates. *Clays and Clay Minerals*, **47**, 630–636.

Laird, D.A. (2006) Influence of layer charge on swelling of smectites. *Applied Clay Science*, **34**, 74–87.

Laird, J. (1988) Chlorites: Metamorphic Petrology. In: *Hydrous Phyllosilicates (exclusive of Micas)* (S. W. Bailey, editor). Reviews in Mineralogy, **19**. Mineralogical Society of America, Chantilly, Virginia, USA, pp. 405–453.

Lanson, B. (2005) Crystal structure of mixed-layer minerals and their X-ray identification: new insights from X-ray diffraction profile modeling. *Clay Science*, **12**, 1–5.

Lanson, B. (2011) Modelling of X-ray diffraction profiles: Investigation of defective lamellar structure crystal chemistry. In: *Layered Mineral Structures and their Application in Advanced Technologies* (M.F. Brigatti & A. Mottana, editors). EMU Notes in Mineralogy, **11**, 151–202.

Lanson, B., Sakharov, B.A., Claret, F. & Drits V.A. (2009) Diagenetic smectite to illite transition in clay-rich sediments: a reappraisal of X-ray diffraction results using the multi-specimen method. *American Journal of Science*, **300**, 476–516.

Laurora, A., Brigatti, M.F., Malferrari, D., Galli, E. & Rossi, A. (2011) Crystal chemistry of lizardite-1*T* from Northern Apennines ophiolites, near Modena, Italy. *The Canadian Mineralogist* (in press).

Lee, J.H. & Guggenheim, S. (1981) Single crystal x-ray refinement of pyrophyllite 1*Tc*. *American Mineralogist*, **66**, 350–357.

Lee, S.K. & Stebbins, J.F. (2003) O atom sites in natural kaolinite and muscovite: O-17 MAS and 3QMAS NMR study. *American Mineralogist*, **88**, 493–500.

Leoni, M., Gualtieri, A.F. & Roveri, N. (2004) Simultaneous refinement of structure and microstructure of layered materials. *Journal of Applied Crystallography*, **37**, 166–173.

Liètard, O. (1977) *Contribution à l'étude des propiétés phisicochimiques, cristallographiques et morphologiques des kaolins*. Thèse Doctoral Science Physique, Nancy, France, 345 pp. (in French).

Lim, C.H. & Jackson, M.L. (1986) Expandable phyllosilicate reactions with lithium on heating, *Clays and Clay Minerals*, **34**, 346–352.

Lin, J.-C. & Guggenheim, S. (1983) The crystal structure of a Li, Be-rich brittle mica: a dioctahedral-trioctahedral intermediate. *American Mineralogist*, **68**, 130–142.

Lindqvist, B. & Jansson, S. (1962) On the crystal chemistry of hisingerite. *American Mineralogist*, **47**, 1356–1362.

Lippmann, F. (1954) Uber einen Keuperton von Zaiserweiher bei Maulbronn. *Heidelberger Beitrdge zur Mineralogie und Petrographie*, **4**, 130–134 (in German).

Lippmann, F. (1960) Corrensit. In: *Handbuch der Mineralogie* (C. Hintze, editor). Ergänzungsband II (F. Chudoba, editor). Neue Mineralien und Neue Mineralnamen, Teil III (in German), pp. 688–691.

Lister, J.S. & Bailey, S.W. (1967) Chlorite polytypism. IV. Regular two-layer structures. *American Mineralogist*, **52**, 1614–1631.

MacEwan, D.M.C. & Wilson, M.J. (1980) Interlayer and intercalation complexes of clay minerals. In: *Crystal Structures of Clay Minerals and Their X-ray Identification* (G.W. Brindley & G. Brown, editors). Monograph **5**, Mineralogical Society, London, pp. 197–248.

Mackenzie, K.J.D. & Berezowski, R.M. (1980) Thermal and Mössbauer studies of iron-containing hydrous silicates. II. Hisingerite. *Thermochimica Acta*, **41(3)**, 335–355.

MacKenzie, K.J.D., Bowden, M.E. & Meinhold, R.H. (1991) The structure and thermal transformations of allophanes studied by ^{29}Si and ^{27}Al high resolution solid-state NMR. *Clays and Clay Minerals*, **39**, 337–346.

Maes, A. & Cremers, A. (1977) Charge density effects in ion exchange. Part 1. Heterovalent exchange equilibria. *Faraday Transactions Royal Society*, **73**, 1807–1814.

Maillet, P., Levard, C., Larquet, E., Mariet, C., Spalla, O., Menguy, N., Masion, A., Doelsch, A., Rose, J. & Thill, A. (2010) Evidence of double-walled Al-Ge imogolite-like nanotubes. A Cryo-TEM and SAXS investigation. *Journal of the American Chemical Society*, **132**, 1208–1209.

Maksimovic, Z. & Bish, D.L. (1978) Brindleyite, a nickel-rich aluminous serpentine mineral analogous to berthierine. *American Mineralogist*, **63**, 484–489.

Malikova, N., Cadene, A., Dubois, E., Marry, V., Durand-Vidal, S., Turq, P., Breu, J., Longeville, S. & Zanotti, J.-M. (2007) Water diffusion in a synthetic hectorite clay studied by quasi-elastic neutron scattering. *Journal of Physical Chemistry C*, **111**, 17603–17611.

Manceau, A. & Calas, G. (1986) Nickel-bearing clay minerals: II. Intracrystalline distribution of nickel: An X-ray absorption study. *Clay Minerals*, **21**, 341–360.

Manceau, A., Calas, G. & Petiau, J. (1984) Cation ordering in nickel-magnesium phyllosilicates of geological interest. In: *Springer Proceedings in Physics, 2 (EXAFS Near Edge Struct. 3)*, 358–361.

Manceau, A., Calas, G. & Decarreau, A. (1985) Nickel-bearing clay minerals: I. Optical spectroscopic study of nickel crystal chemistry. *Clay Minerals*, **20**, 367–387.

Manceau, A., Ildefonse, Ph., Hazemann, J.L., Flank, A.-M. & Gallup, D. (1995) Crystal chemistry of hydrous iron silicate scale deposits at the Salton Sea geothermal field. *Clays and Clay Minerals*, **43**, 304–317.

Mandarino, J.A. & Back, E.B. (2004) *Fleischer's Glossary of Mineral Species 2004*. The Mineralogical Record Inc., Tuscon, Arizona, USA, 310 pp.

Martin de Vidales, J.L., Pozo, M., Alia, J.M., Garcia-Navarro, F. & Rull, F. (1991) Kerolite–stevensite mixed-layers from the Madrid Basin, central Spain. *Clays and Clay Minerals*, **26**, 329–342.

Martin, F., Ferrage, E., Petit, S., de Parseval, P., Delmotte, L., Ferret, J., Arseguel, D. & Salvi, S. (2006) Fine-probing the crystal-chemistry of talc by MAS-NMR spectroscopy. *European Journal of Mineralogy*, **18**, 641–651.

Martynková, G.S., Valášková, M. & Šupová, M. (2007) Organo-vermiculite structure ordering after PVAc introduction. *Physica Status Solidi*, **204**, 1870–1875.

Matarrese, S., Schingaro, E., Scordari, F., Stoppa, F., Rosatelli, G., Pedrazzi, G. & Ottolini L. (2008) Crystal chemistry of phlogopite from Vulture–S. Michele Subsynthem volcanic rocks (Mt. Vulture, Italy) and volcanological implications. *American Mineralogist*, **93**, 426–437.

Mathieson, A.M. (1958) Mg-vermiculite: A refinement of the crystal structure of the 14.36 Å phase. *American Mineralogist*, **43**, 300–304.

Mathieson, A.M. & Walker, G.F. (1954) Crystal structure of magnesium-vermiculite. *American Mineralogist*, **39**, 231–255.

McAtee, J.L. & Lamkin, G. (1978) A modified freeze-drying procedure for the electron microscopy examination of hectorite. *Clays and Clay Minerals*, **27**, 293–296.

McBride, M.B. (1989) Surface chemistry of soil minerals. In: *Minerals in Soil Environments, Second edition* (J.B. Dixon & S.B. Weed, editors). Soil Science Society of America, Madison, Wisconsin, USA, pp. 35–88.

McCarty, D.K., Sakharov, B.A. & Drits, V.A. (2009) New insights into smectite illitization: a zoned K-bentonite revisited. *American Mineralogist*, **94**, 1653–1671.

Mehmel, M. (1935) Uber die Struktur von Halloysit und Metahalloysit. *Zeitschrift fur Kristallographie*, **90**, 35–43.

Mellini, M. (1982) The crystal structure of lizardite 1*T*: Hydrogen bonds and polytypism. *American Mineralogist*, **67**, 587–598.

Mellini, M. & Viti, C. (1994) Crystal structure of lizardite-1*T* from Elba, Italy. *American Mineralogist*, **79**, 1194–1198.

Mellini, M. & Zanazzi, P.F. (1987) Crystal structures of lizardite-1*T* and lizardite-$2H_1$ from Coli, Italy. *American Mineralogist*, **72**, 943–948.

Mellini, M. & Zanazzi, P.F. (1989) Effects of pressure on the structure of lizardite-1*T*. *European Journal of Mineralogy*, **1**, 13–19.

Mellini, M., Ferraris, G. & Compagnoni, R. (1985) Carlosturanite: HRTEM evidence of a polysomatic series including serpentine. *American Mineralogist*, **70**, 773–781.

Mellini, M., Trommsdorff, V. & Compagnoni, R. (1987) Antigorite polysomatism: behaviour during progressive metamorphism. *Contribution to Mineralogy and Petrology*, **97**, 147–155.

Mellini, M., Cressey, G., Wicks, F.J. & Cressey, B.A. (2010) The crystal structure of Mg end-member lizardite-I*T* forming polyhedral spheres from the Lizard, Cornwall. *Mineralogical Magazine*, **74**, 277–284.

Mercier, P.H.J. & Le Page, Y. (2008) Kaolin polytypes revisited *ab initio*. *Acta Crystallographica*, **B64**, 131 143.

Mercier, P.H.J. & Le Page, Y. (2009) *Ab initio* exploration of layer slipping transformations in kaolinite up to 60 GPa. *Materials Science and Technology*, **25**, 437–442.

Mercier, P.H.J., Rancourt, D.G., Redhammer, G.J., Lalonde, A.E., Robert, J.L., Berman, R.G. & Kodama, H. (2006) Upper limit of the tetrahedral rotation angle and factors affecting octahedral flattening in synthetic and natural 1*M* polytype $C2/m$ space group micas. *American Mineralogist*, **91**, 831–849.

Mercier, P.H.J., Le Page, Y. & Desgreniers, S. (2010) Kaolin polytypes revisited *ab initio* at 10 GPa. *American Mineralogist*, **95**, 1117–1120.

Méring, J. & Oberlin, A. (1971) Smectite. In: *The Electron-Optical Investigation of Clays* (J.A. Gard, editor). Monograph **2**. Mineralogical Society, London, pp. 193–229.

Merli, M., Pavese, A. & Curetti, N. (2009) Maximum entropy method: an unconventional approach to explore observables related to the electron density in phengites. *Physics and Chemistry of Minerals*, **36**, 19–28.

Mermut, A.R. (editor) (1994) *Layer Charge Characteristics of 2:1 Silicate Clay Minerals*. CMS Workshop Lectures, **6**, 134 pp.

Merriman, R.J. & Peacor, D.R. (1999) Very low-grade metapelites; mineralogy, microfabrics and measuring reaction progress. In: *Lowgrade Metamorphism* (M. Frey & D. Robinson, editors), Blackwell Sciences Ltd., Oxford, UK, pp. 10–60.

Mesto, E., Schingaro, E., Scordari, F. & Ottolini, L. (2006) An electron microprobe analysis, secondary ion mass spectrometry, and single-crystal X-ray diffraction study of phlogopites from Mt. Vulture, Potenza, Italy: Consideration of cation partitioning. *American Mineralogist*, **91**, 182–190.

Meunier, A. (2005) *Clays*. Spinger-Verlag, Berlin, 472 pp.

Michalkova, A. & Tunega, D. (2007) Kaolinite: dimethylsulfoxide intercalate – A theoretical study. *Journal of Physical Chemistry C*, **111(30)**, 11259–11266.

Middleton, A.P. & Whittaker, E.J.W. (1976) The structure of Povlen-type chrysotile. *The Canadian Mineralogist*, **14**, 301–306.

Milestone, N.B. (1971) Allophane. Its structure and possible uses. *Chemistry in New Zealand*, **35**, 191–197.

Mookherjee, M., Redfern, S.A.T., Zhang, M. & Harlov, D.E. (2002) Orientational order-disorder of ND_4^+/ NH_4^+ in synthetic ND_4/NH_4-phlogopite: a low-temperature infrared study. *European Journal of Mineralogist*, **14**, 1033–1039.

Moore, D.M. & Reynolds, R.C.Jr. (1997) *X-Ray Diffraction and the Identification and Analysis of Clay Minerals*. Oxford University Press, New York, 378 pp.

Mottana, A & Aldega L. (2011) Advanced techniques to define intercalation processes. In: *Layered Mineral Structures and their Application in Advanced Technologies* (M.F. Brigatti & A. Mottana, editors). EMU Notes in Mineralogy, **11**, 285–312.

Mottana, A., Sassi, F.P., Thompson, J.B., Jr. & Guggenheim S. (editors) (2002a) *Micas: Crystal Chemistry and Metamorphic Petrology*. Reviews in Mineralogy and Geochemistry, **46**. Mineralogical Society of America, Chantilly, Virginia, USA.

Mottana, A., Marcelli, A., Cibin, G. & Dyar, M.D. (2002b) X-ray absorption spectroscopy of the micas. In: *Micas: Crystal Chemistry and Metamorphic Petrology* (A. Mottana, F.P. Sassi, J.B. Thompson Jr. & S. Guggenheim, editors). Reviews in Mineralogy and Geochemistry, **46**, 371–411. Mineralogical Society of America, Chantilly, Virginia, USA.

Nadeau, P.H., Wilson, M.J., McHardy, W.J. & Tait, M.J. (1985) The conversion of smectite to illite during diagenesis: Evidence from some illitic clays from bentonites and sandstones. *Mineralogical Magazine*, **49**, 393–400.

Nagy, B. & Bradley, W.F. (1955) Structure of sepiolite. *American Mineralogist*, **40**, 885–892.

Neder, R.B., Burghammer, M., Grasl, Th., Schulz, H., Bram, A. & Fiedler, S. (1999) Refinement of the kaolinite structure from single-crystal synchrotron data. *Clays and Clay Minerals*, **47**, 487–494.

Nelson, D.O. & Guggenheim, S. (1993) Inferred limitations to the oxidation of Fe in chlorite: a high-temperature single-crystal X-ray study. *American Mineralogist*, **78**, 1197–1207.

Nespolo, M. & Ďurovič, S. (2002) Crystallographic basis of polytypism and twinning in micas. In: *Micas: Crystal Chemistry and Metamorphic Petrology* (A. Mottana, F.P. Sassi, J.B. Thompson Jr. & S. Guggenheim, editors). Reviews in Mineralogy and Geochemistry, **46**. Mineralogical Society of America, Chantilly, Virginia, USA, pp. 155–272.

Nespolo, M., Takeda, H. & Ferraris, G. (1997) Crystallography of mica polytypes. In: *Modular Aspects of Minerals* (S. Merlino, editor). EMU Notes in Mineralogy, **1**. Eötvös University Press, Budapest, pp. 81–118.

Nespolo, M., Ferraris, G., Ďurovič, S. & Takeuchi, Y. (2004) Twins *vs.* modular crystal structures. *Zeitschrift für Kristallographie*, **219**, 773–778.

Newman, A.C.D. (1987) The interaction of water with clay mineral surfaces. In: *Chemistry of Clays and Clay Minerals* (A.C.D. Newman, editor). Monograph **6**, Mineralogical Society, London, pp. 237–274.

Newman, A.C.D. & Brown, G. (1987) The chemical constitution of clays. In: *Chemistry of Clays and Clay Minerals* (A.C.D. Newman, editor). Monograph **6**, Mineralogical Society, London, pp. 1–128.

Newnham, R.E. (1961) A refinement of the dickite structure and some remarks on polymorphism in kaolin minerals. *Mineralogical Magazine*, **32**, 683.

Newnham, R.E. & Brindley, G.W. (1956) The crystal structure of dickite. *Acta Crystallographica*, **9**, 759–764.

Newnham, R.E. & Brindley, G.W. (1957) The structure of dickite: correction. *Acta Crystallographica*, **10**, 88.

Norrish, K. (1954) The swelling of montmorillonite. *Discussions of the Faraday Society*, **18**, 120–133.

O'Day, P.A., Rehr, J.J., Zabinsky, S.I. & Brown, G.E. Jr. (1994) Extended X-ray absorption fine structure (EXAFS) analysis of disorder and multiple-scattering in complex crystalline solids. *Journal of American Chemical Society*, **116**, 2938–2949.

Okada, K. & Ossaka, J. (1983) Dehydration reaction of interlayer water of halloysite by heating. *Nendo Kagaku*, **23**, 27–31.

Okada, K., Morikawa, H., Iwai, S., Ohira, Y. & Ossaka, J. (1975) Structure model of allophane. *Clay Science*, **4**, 291–303.

Otten, M.T. (1993) High-resolution electron microscopy of polysomatism and stacking defects in antigorite. *American Mineralogist*, **78**, 75–84.

Ozawa, T., Takeuchi, Y., Takahata, T., Donnay, G. & Donnay, J.D.H. (1983) The pyrosmalite group of minerals. II. The layer structure of mcgillite and friedelite. *The Canadian Mineralogist*, **21**, 7–17.

Palin, E.J. & Dove, M.T. (2004) Investigation of Al/Si ordering in tetrahedral phyllosilicate sheets by Monte Carlo simulation. *American Mineralogist*, **89**, 176–184.

Palin, E.J., Dove, M.T., Hernandez Laguna, A. & Sainz Diaz, C.I. (2004) A computational investigation of the Al/Fe/Mg order disorder behavior in the dioctahedral sheet of phyllosilicates. *American Mineralogist*, **89**, 164–175.

Parfitt, R.L. & Henmi, T. (1980) Structure of some allophanes from New Zealand. *Clays and Clay Minerals*, **28**, 285–294.

Parfitt, R.L., Furkert, R.J. & Hemmi, T. (1980) Identification and structure of two types of allophane from volcanic ash soils and tephra. *Clays and Clay Minerals*, **28**, 328–334.

Pauling, L. (1930) The structures of the micas and related minerals. *Proceedings of the National Academy of Sciences of the United States of America*, **16**, 123–129.

Pavese, A., Levy, D., Curetti, N., Diella, V., Fumagalli, P. & Sani, A. (2003) Equation of state and compressibility of phlogopite by *in situ* high-pressure X-ray powder diffraction. *European Journal of Mineralogy*, **15**, 455–463.

Pavese, A., Curetti, N., Diella, V., Levy, D., Dapiaggi, M. & Russo, U. (2007) P-V and T-V equations of state of natural biotite: an *in situ* high-pressure and high-temperature powder diffraction study, combined with Mössbauer spectroscopy. *American Mineralogist*, **92**, 1158–1164.

Peacor, D.R., Essene, E.J., Simmons, W.B. Jr. & Bigelow, W.C. (1974) Kellyite, a new manganese-aluminum member of the serpentine group from Bald Knob, North Carolina, and new data on grovesite. *American Mineralogist*, **59**, 1153–1156.

Peacor, D.R., Rouse, R.C. & Bailey, S.W. (1988) Crystal structure of franklinfurnaceite: A tri-dioctahedral zincosilicate intermediate between chlorite and mica. *American Mineralogist*, **73**, 876–887.

Pearson, R.G. (1963) Hard and soft acids and bases. *Journal of the American Chemical Society*, **85**, 3533–3539.

Pearson, R.G. (1968) Hard and soft acids and bases HSAB, part 1: Fundamental principles. *Journal of Chemical Education*, **45**, 581–587.

Perdikatsis, B. & Burzlaff, H. (1981) Refinement of talc's ($Mg_3[(OH)_2Si_4O_{10}]$) structure. *Zeitschrift für Kristallographie*, **156**, 177–186.

Petit, S., Martin, F., Wiewióra, A., de Parseval, P. & Decarreau, A. (2004) Crystal-chemistry of talcs: a near infrared (NIR) spectroscopy study. *American Mineralogist*, **89** 319–326.

Petit, S., Righi, D. & Madejová, J. (2006) Infrared spectroscopy of NH_4^+-bearing and saturated clay minerals: a review of the study of layer charge. *Applied Clay Science*, **34**, 22–30.

Phillips, T.L., Loveless, J.K. & Bailey, S.W. (1980) Cr^{3+} coordination in chlorites: a structural study of ten chromian chlorites. *American Mineralogist*, **65**, 112–122.

Piccinini, M., Cibin, G., Marcelli, A., Della Ventura, G., Bellatreccia, F. & Mottana, A. (2006a) Synchrotron radiation FT-IR micro-spectroscopy of fluorophlogopite in the O–H stretching region. *Vibrational Spectroscopy*, **42**, 59–62.

Piccinini, M., Cibin, G., Marcelli, A., Mottana, A., Della Ventura, G. & Bellatreccia, F. (2006b) Synchrotron radiation FTIR micro-spectroscopy of natural layer-silicates in the O–H stretching region. *Infrared Physics and Technology*, **49**, 64–68.

Plançon, A. (2002) New modeling of X-ray diffraction by disordered lamellar structures, such as phyllosilicates. *American Mineralogist*, **87**, 1672–1677.

Plançon, A. & Zacharie, C. (1990) An expert system for the structural characterization of kaolinites. *Clay Minerals*, **25**, 249–260.

Plançon, A., Giese, R.F. Jr., Snyder, R., Drits, V.A. & Bukin, A.S. (1989) Stacking faults in the kaolin-group minerals: Defect structures of kaolinite. *Clays and Clay Minerals*, **37**, 203–210.

Post, J.E. & Heaney, P.J. (2008) Synchrotron powder X-ray diffraction study of the structure and dehydration behavior of palygorskite. *American Mineralogist*, **93**, 667–675.

Post, J.E., Bish, D.L. & Heaney, P.J. (2007) Synchrotron powder X-ray diffraction study of the structure and dehydration behaviour of sepiolite. *American Mineralogist*, **92**, 91–97.

Preisinger, A. (1959) X-ray study of the structure of sepiolite. *Clays and Clay Minerals*, **6**, 61–67.

Preisinger, A. (1963) Sepiolite and related compounds: Its stability and application. *Clays and Clay Minerals*, **10**, 365–371.

Prost, R., Dameme, A., Huard, E., Driard, J. & Leydecker, J.P. (1989) Infrared study of structural hydroxyl in kaolinite, dickite, nacrite, and poorly crystalline kaolinite at 5 to 600 K. *Clays and Clay Minerals*, **37**, 464–468.

Rancourt, D.G., Mercier, P.H.J., Cherniak, D.J., Desgreniers, S., Kodama, H., Robert, J.L. & Murad, E. (2001) Mechanisms and crystal chemistry of oxidation in annite: resolving the hydrogen loss and vacancy reactions. *Clays and Clay Minerals*, **49**, 455–491.

Range, K.J. & Weiss, A. (1969) Uber das Verhalten von kaolinitit bei hohen Drücken. *Berichte Deutsche Keramische Gesellschaft*, **46**, 231–238 (in German).

Rautureau, M. & Tchoubar, C. (1974) Precisions concernant l'analyse structurale de la sepiolite par microdiffraction electronique. *Comptes Rendues de l'Académie des Sciences, Paris*, **278**, 25–28 (in French).

Rautureau, M., Tchoubar, C. & Méring, J. (1972) Analyse structurale de la sepiolite par microdiffraction electronique. *Comptes Rendues de l'Académie des Sciences, Paris*, **274**, 269–271 (in French).

Rayner, J.H. & Brown, G. (1973) Crystal structure of talc. *Clays and Clay Minerals, Proceedings of the Conference 1973*, **21**, 103–314.

Redhammer, G.J., Amthauer, G., Lottermoser, W., Bernroider, M., Tippelt, G. & Roth, G. (2005) X-ray powder diffraction and ^{57}Fe- Mössbauer spectroscopy of synthetic trioctahedral micas $\{K\}[Me_3]O_{10}(OH)_2$, Me = Ni^{2+}, Mg^{2+}, Co^{2+}, Fe^{2+}; T = Al^{3+}, Fe^{3+}. *Mineralogy and Petrology*, **85**, 89–115.

Reynolds, R.C. Jr. (1983) Three-dimensional X-ray diffraction from disordered illite: simulation and interpretation of the diffraction pattern. In: *Computer Application in X-ray Diffraction Methods* (R.C. Reynolds & I. Walker, editors). CMS Workshop Lectures, **5**. The Clay Minerals Society, Bloomington, Indiana, USA, pp. 44–78.

Reynolds, R.C. Jr. (1985) *NEWMOD a computer program for the calculation of one-dimensional diffraction pattern of mixed layer clays*. R.C. Reynolds, 9 Brook Rd., Hanover, New Hampshire 03755, USA.

Reynolds, R.C. Jr. (1988) Mixed layer chlorite minerals. In: *Hydrous Phyllosilicates (Exclusive of Micas)* (S.W. Bailey, editor). Reviews in Mineralogy, **19**. Mineralogical Society of America, Washington, D.C., pp. 601–629.

Reynolds, R.C. Jr. & Bish, D.L. (2002) The effects of grinding on the structure of a low-defect kaolinite. *American Mineralogist*, **87**, 1626–1630.

Rinaudo, C., Roz, M., Boero, M. & Franchini, A.M. (2004) FT-Raman spectroscopy on several di- and tri-octahedral T-O-T phyllosilicates. *Neues Jahrbuch Fur Mineralogie Monatshefte*, **12**, 537–554.

Rocha, J. & Pedrosa De Jesus, J.D. (1994) ^{27}Al satellite transition MAS-NMR spectroscopy of kaolinite. *Clay Minerals*, **29**, 287–291.

Ross, C.S. (1946) Sauconite – a clay mineral of the montmorillonite group. *American Mineralogist*, **31**, 411–24.

Rule, A.C. & Bailey, S.W. (1987) Refinement of the crystal structure of a monoclinic ferroan clinochlore. *Clays and Clay Minerals*, **35**, 129–138.

Rule, A.C. & Radke, F. (1988) Baileychlore, the zinc end member of the trioctahedral chlorite series. *American Mineralogist*, **73**, 135–139.

Sainz-Diaz, C.I., Timon, V., Botella, V., Artacho, E. & Hernandez-Laguna, A. (2002) Quantum mechanical calculations of dioctahedral 2:1 phyllosilicates: Effect of octahedral cation distributions in pyrophyllite, illite, and smectite. *American Mineralogist*, **87**, 958–965.

Sakharov, B.A., Lindgreen, H., Salyn, A.L. & Drits, V.A. (1999) Determination of illite-smectite structures using multispecimen XRD profile fitting. *Clays and Clay Minerals*, **47**, 555–566.

Sanz, J., Herrero C.P. & Serratosa J.M. (2006) Arrangement and mobility of water in vermiculite hydrates followed by 1H NMR spectroscopy. *Journal of Physical Chemistry B*, **110**, 7813–7819.

Sato, H., Ono, K., Johnston, C.T. & Yamagishi, A. (2004) First-principle study of polytype structures of 1:1 dioctahedral phyllosilicates. *American Mineralogist*, **89**, 1581–1585.

Schingaro, E., Scordari, F., Mesto, E., Brigatti, M.F. & Pedrazzi, G. (2005) Cation-site partitioning in Ti-rich micas from Black Hill (Australia): A multi-technical approach. *Clays and Clay Minerals*, **53**, 179–189.

Scholtzová, E. & Smrčok, L. (2005) On local structural changes in lizardite-1*T*: {Si^{4+}/Al^{3+}}, {Si^{4+}/Fe^{3+}}, [Mg^{2+}/Al^{3+}], [Mg^{2+}/Fe^{3+}] substitutions. *Physics and Chemistry of Minerals*, **32**, 362–373.

Scholtzová, E., Smrčok, L., Tunega, D. & Turi Nagy, L. (2000) *Ab initio* 2-D periodic Hartree-Fock study of Fe-substituted lizardite 1*T* – a simplified cronstedtite model. *Physics and Chemistry of Minerals*, **27**, 741–746.

Schreyer, W., Medenbach, O., Abraham, K., Gebert, W. & Mueller, W.F. (1982) Kulkeite, a new metamorphic phyllosilicate mineral: ordered 1:1 chlorite/talc mixed-layer. *Contributions to Mineralogy and Petrology*, **80**, 103–109.

Schroeder, P.A. (2002) Infrared Spectroscopy in clay science. In: *Teaching Clay Science* (A. Rule & S. Guggenheim, editors). CMS Workshop Lectures, **11**. The Clay Mineral Society, Aurora, Colorado, USA, pp. 181–206.

Schroeder, P.A. & Pruett, R.J. (1996) Fe ordering in kaolinite: insights from ^{29}Si and ^{27}Al MAS NMR spectroscopy. *American Mineralogist*, **81**, 26–38.

Scordari, F., Ventruti, G., Sabato, A., Bellatreccia, F., Della Ventura, G. & Pedrazzi G. (2006) Ti-rich phlogopite from Mt. Vulture (Potenza, Italy) investigated by a multianalytical approach: substitutional mechanisms and orientation of the OH dipoles. *European Journal of Mineralogy*, **18**, 379–391.

Scordari, F., Schingaro, E., Ventruti, G., Lacalamita, M. & Ottolini, L. (2008) Red micas from basal ignimbrites of Mt. Vulture (Italy): interlayer content appraisal by a multi-methodic approach. *Physics and Chemistry of Minerals*, **35**, 163–174.

Seidl, W. & Breu, J. (2005) Single crystal structure refinement of tetramethylammonium-hectorite. *Zeitschrift für Kristallographie*, **220**, 169–176.

Sen Gupta, P.K., Schlemper, E.O., Johns, W.D. & Ross, F. (1984) Hydrogen positions in dickite. *Clays and Clay Minerals*, **32**, 483–485.

Shabani, A.A.T., Rancourt, D.G. & Lalonde, A.E. (1998) Determination of *cis* and *trans* Fe^{2+} populations in $2M_1$ muscovite by Mössbauer spectroscopy. *Hyperfine Interactions*, **117**, 117–129.

Shainberg, I., Alperovitch, N.I. & Keren, R. (1987) Charge density and Na–K–Ca exchange on smectites. *Clays and Clay Minerals*, **35**, 68–73.

Shirozu, H. & Bailey, S.W. (1966) Crystal structure of a two-layer Mg-vermiculite. *American Mineralogist*, **51**, 1124–1143.

Shoval, S., Yariv, S., Michaelian, K.H., Lapides, I., Boudeulle, M. & Panczer, G. (1999) A fifth OH-stretching band in IR spectra of kaolinites. *Journal of Colloid and Interface Science*, **212**, 523–529.

Shoval, S., Yariv, S., Michaelian, K.H., Boudeulle, M. & Panczer, G. (2001) Hydroxyl-stretching bands in curve-fitted micro-Raman, photoacoustic, and transmission infrared spectra of dickite from St. Claire, Pennsylvania. *Clays and Clay Minerals*, **49**, 347–354.

Shoval, S., Yariv, S., Michaelian, K.H., Boudeulle, M. & Panczer, G. (2002) Hydroxyl-stretching bands in polarized micro-Raman spectra of oriented single-crystal Keokuk kaolinite. *Clays and Clay Minerals*, **50**, 56–62.

Singh, B. (1996) Why does halloysite roll?–A new model. *Clays and Clay Minerals*, **44**, 191–196.

Skipper, N.T., Sposito, L.G. & Chou Chang, F.-R. (1995) Monte Carlo simulation of interlayer molecular structure in swelling clay minerals. 2. monolayer hydrates. *Clays and Clay Minerals*, **43**, 294–303.

Slade, P.G., Stone, P.A. & Radoslovich, E.W. (1985) Interlayer structures of the two-layer hydrates of Naand Ca-vermiculites. *Clays and Clay Minerals*, **33**, 51–61.

Smirnov, K.S. & Bougeard, D. (1999) A molecular dynamics study of structure and short-time dynamics of water in kaolinite. *Journal of Physical Chemistry B*, **103**, 5266–5273.

Smrčok, L., Gyepesová, D. & Chmielová, M. (1990) New X-ray Rietveld refinement of kaolinite from Keokuk, Iowa. *Crystal Research and Technology*, **25**, 105–110.

Smrčok, L., Ďurovič, S., Petricek, V. & Weiss, Z. (1994) Refinement of the crystal structure of cronstedtite-3*T*. *Clays and Clay Minerals*, **42**, 544–551.

Smyth, J.R., Dyar, M.D., May, H.M., Bricker, O.P. & Acker, J.G. (1997) Crystal structure refinement and Mössbauer spectroscopy of an ordered, triclinic clinochlore. *Clays and Clay Minerals*, **45**, 544–550.

Spinnler, G.E. (1985) *HRTEM study of antigorite, pyroxene-serpentine reactions and chlorite*. Ph.D. thesis, Arizona State University, Tempe, Arizona, USA, 248 pp.

Sposito, G. (1984) *The Surface Chemistry of Soils*. Oxford University Press, New York, 234 pp.

Sposito, G. & Prost, R. (1982) Structure of water adsorbed on smectites. *Chemical Reviews*, **82**, 553–573.

Środoń, J. & Eberl, D.D. (1984) Illite. In: *Micas* (S.W. Bailey, editor). Review in Mineralogy, **13**. Mineralogical Society of America, Washington D.C., pp. 495–544.

Steadman, R. & Nuttall, P.M. (1962) The crystal structure of amesite. *Acta Crystallographica*, **15**, 510–511.

Steinfink, H. (1958a) Crystal structure of chlorite. I. A monoclinic polymorph. *Acta Crystallographica*, **11**, 191–195.

Steinfink, H. (1958b) Crystal structure of chlorite. II. A triclinic polymorph. *Acta Crystallographica*, **11**, 195–198.

Steinfink, H. & Brunton, G. (1956) The crystal structure of amesite. *Acta Crystallographica*, **9**, 487–492.

Stoch, L. (1974) *Mineraly Ilaste ('Clay Minerals')*. Geological Publishers, Warsaw, pp. 186–193.

Sudo, T. & Nakamura, T. (1952) Hisingerite from Japan. *American Mineralogist*, **37**, 618–621.

Sugimori, H., Iwatsuki, T. & Murakami, T. (2008) Chlorite and biotite weathering, Fe^{2+}-rich corrensite formation, and Fe behaviour under low P_{O2} conditions and their implication for Precambrian weathering. *American Mineralogist*, **93**, 1080–1089.

Suitch, P.R. & Young, R.A. (1983) Atom position in highly ordered kaolinite. *Clays and Clay Minerals*, **31**, 357–366.

Sun, H., Liu, Y., Peng, T. & Yue, M. (2008) Hydrothermal synthesis and characterization of hectorite. *Rengong Jingti Xuebao*, **37**, 844-848.

Suquet, H., de la Calle, C. & Pezerat, H. (1975) Swelling and structural organization of saponite. *Clays and Clay Minerals*, **23**, 1–9.

Suquet, H., Iiyama, J.T., Kodama, H. & Pezerat, H. (1977) Synthesis and swelling properties of saponite with increasing layer charge. *Clays and Clay Minerals*, **25**, 231–242.

Takeda, H. & Ross, M. (1995) Mica polytypism: identification and origin. *American Mineralogist*, **80**, 715–724.

Tarý, G., Bobos, I., Gomes, C.S.F. & Ferriera, J.M.F. (1999) Modification of surface charge properties during kaolinite to halloysite-7 Å transformation. *Journal of Colloid and Interface Science*, **210**, 360–366.

Tejedor-Tejedor, M.I., Anderson, M.A. & Herbillon, A.J. (1983) An investigation of the coordination number of Ni^{2+} in nickel bearing phyllosilicates using diffuse reflectance spectroscopy. *Journal of Solid State Chemistry*, **50**, 153–162.

Thompson, J.G. & Withers, R.L. (1987) A transmission electron microscopy contribution to the structure of kaolinite. *Clays and Clay Minerals*, **35**, 237–239.

Tien, P.L., Leavens, P.B. & Nelen, J.A. (1975) Swinefordite, a dioctahedral-trioctahedral Li-rich member of the smectite group from Kings Mountain, North Carolina. *American Mineralogist*, **60**, 540–547.

Toraya, H. (1981) Distortions of octahedra and octahedral sheets in 1*M* micas and the relation to their stability. *Zeitschrift für Kristallographie*, **157**, 173–190.

Treacy, M.M.J., Newsam, J.M. & Deem, M.W. (1991) A general recursion method for calculating diffracted intensities from crystals containing planar faults. *Proceedings: Mathematical and Physical Sciences*, **433**, 499–520.

Tsipursky, S.I. & Drits, V.A. (1984) The distribution of octahedral cations in the 2:1 layers of dioctahedral smectites studied by oblique-texture electron diffraction. *Clay Minerals*, **19**, 177–193.

Tunega, D. & Lischka, H. (2003) Effect of the Si/Al ordering on structural parameters and the energetic stabilization of vermiculites – a theoretical study. *Physics and Chemistry of Minerals*, **30**, 517–522.

Uehara, S. (1998) TEM and XRD study of antigorite superstructures. *The Canadian Mineralogist*, **36**, 1595–1606.

Uehara, S. & Kamata, K. (1994) Antigorite with a large supercell from Saganoseki, Oita Prefecture, Japan. *The Canadian Mineralogist*, **32**, 93–103.

Uehara, S. & Shirozu, H. (1985) Variations in chemical composition and structural properties of antigorite. *Mineralogical Journal*, **12**, 299–318.

Vahedi-Faridi, A. & Guggenheim, S. (1997) Crystal structure of TMA-exchanged vermiculite. *Clays and Clay Minerals*, **45**, 859–866.

Vahedi-Faridi, A. & Guggenheim, S. (1999) Structural study of tetramethylphosphonium-exchanged vermiculite. *Clays and Clay Minerals*, **47**, 219–225.

Valdrè, G., Malferrari, D. & Brigatti, M.F. (2009) Crystallographic features and cleavage nanomorphology of chlinochlore: specific applications. *Clays and Clay Minerals*, **57**, 183–193.

Van der Gaast, S.J., Wada, K., Wada, S.I. & Kakuto, Y. (1985) Small-angle X-ray powder diffraction, morphology, and structure of allophane and imogolite. *Clays and Clay Minerals*, **33**, 237–243.

Van Olphen, H. (1965) Thermodynamics of interlayer adsorption of water in clays. I Sodium vermiculite. *Journal of Colloid Science*, **20**, 822–837.

Vasconcelos, I.F., Bunker, B.A. & Cygan, R.T. (2007) Molecular dynamics modelling of ion adsorption to the basal surfaces of kaolinite. *Journal of Physical Chemistry C*, **111**, 6753–6762.

Vaughan, J.P. (1986) The iron end-member of the pyrosmalite series from the Pegmont lead-zinc deposit, Queensland. *Mineralogical Magazine*, **50**, 527–531.

Veblen, D.R. (1992) Electron microscopy applied to nonstoichiometry, polysomatism, and replacement reactions in minerals. In: *Minerals and Reactions at the Atomic Scale: Transmission Electron Microscopy* (P.R. Buseck, editor). Reviews in Mineralogy, **27**. Mineralogical Society of America, Washington, D. C., pp. 181–223.

Veniale, F. & van der Marel, H.W. (1969) Identification of some l:l regular interstratified trioctahedral clay minerals. In: *Proceedings of the International Clay Conference, Tokyo*, 233–244.

Ventruti, G., Levy, D., Pavese, A., Scordari, F. & Suard, E. (2009) High-temperature treatment, hydrogen behaviour and cation partitioning of a Fe-Ti bearing volcanic phlogopite by in-situ neutron powder diffraction and FTIR spectroscopy. *European Journal of Mineralogy*, **21**, 385–396.

Viti, C. & Mellini, M. (1996) Vein antigorite from Elba Island, Italy. *European Journal of Mineralogy*, **8**, 423–434.

Viti, C., Di Vincenzo, G. & Mellini M. (2004) Thermal transformations in laser-heated chloritized annite. *Physics and Chemistry of Minerals*, **31**, 92–101.

Vrublevskaya, Z.V., Delitsin, I.S., Zvyagin, B.B. & Soboleva, S.V. (1975) Cookeite with a perfect regular structure, formed by bauxite alteration. *American Mineralogist*, **60**, 1041–1046.

Wada, K. (1967) A structural scheme of soil allophane. *American Mineralogist*, **52**, 690–708.

Wada, K. (1995) Structure and formation of non- and para-crystalline aluminosilicate clay minerals: a review. In: *Clays Controlling the Environment, Proceedings of the 10th International Clay Conference, Adelaide* (G.J. Churchman, R.W. Fitzpatrick & R.A. Eggleton, editors), pp. 443–448.

Wada, S. & Wada, K. (1977) Density and structure of allophane. *Clay Minerals*, **12**, 289–298.

Wada, S. & Wada, K. (1982) Effects of substitution of germanium for silicon in imogolite. *Clays and Clay Minerals*, **30**, 123–128.

Wada, S.I., Eto, A. & Wada, K. (1979) Synthetic allophane and imogolite. *Journal of Soil Science*, **30**, 347–355.

Wahle, M.W., Bujnowski, T.J., Guggenheim, S. & Kogure, T. (2010) Guidottiite, the Mn-analogue of cronstedtite: a new serpentine-group mineral from South Africa. *Clays and Clay Minerals*, **58**, 364–376.

Walker, J.R. & Bish, D.L. (1992) Application of Rietveld refinement techniques to a disordered IIb Mg-chamosite. *Clays and Clay Minerals*, **40**, 319–322.

Weaver, C.E. & Pollard, L.D. (editors) (1973) *The Chemistry of Clay Minerals. Developments in Sedimentology*. Elsevier, Amsterdam, 221 pp.

Weber, F., Dunoyer de Segonzac, G. & Economou, C. (1976) Une nouvelle expression de la 'crystallinité'de l'illite et des micas. Notion d'epaisseur des cristallites. *Compte Rendu Sommaire de Seances de la Societé Géologique de France*, **5**, 225–227 (in French).

Weiss, Z. & Ďurovič, S. (1984) Polytypism of pyrophyllite and talc. Part II. Classification and X-ray identification of MDO polytypes. *Silikaty*, **28**, 289–309.

Weiss, Z., Rieder, M. & Chmielová, M. (1992) Deformation of coordination polyhedra and their sheets in phyllosilicates. *European Journal of Mineralogy*, **4**, 665–682.

Welch, M.D. & Crichton, W.A. (2005) A high-pressure polytypic transformation in type-I chlorite. *American Mineralogist*, **90**, 1139–1145.

Welch, M.D. & Crichton, W.A. (2010) Pressure-induced transformations in kaolinite. *American Mineralogist*, **95**, 651–654.

Welch, M.D. & Marshall, W.G. (2001) High-pressure behavior of clinochlore. *American Mineralogist*, **86**, 1380–1386.

Welch, M.D., Barras, J. & Klinowski, J. (1995) A multinuclear NMR study of clinochlore. *American Mineralogist*, **80**, 441–447.

Welch, M.D., Kleppe, A.K. & Jephcoat, A.P. (2004) Novel high pressure behaviour in chlorite: A synchrotron XRD study of clinochlore to 27 GPa. *American Mineralogist*, **89**, 1337–1340.

Whelan, J.A. & Goldich, S.S. (1961) New data for hisingerite and neotocite. *American Mineralogist*, **46**, 1412–1423.

White, C.E., Provis, J.L., Riley, D.P., Kearley, G.J. & van Deventer, J.S.J. (2009) What Is the Structure of Kaolinite? Reconciling Theory and Experiment. *Journal of Physical Chemistry B*, **113**, 6756–6765.

Whittaker, E.J.W. (1953) The structure of chrysotile. *Acta Crystallographica*, **6**, 747–748.

Whittaker, E.J.W. (1956a) The structure of chrysotile. II. Clinochrysotile. *Acta Crystallographica*, **9**, 855–862.

Whittaker, E.J.W. (1956b) The structure of chrysotile. III. Orthochrysotile. *Acta Crystallographica*, **9**, 862–864.

Whittaker, E.J.W. (1956c) The structure of chrysotile. IV. Parachrysotile. *Acta Crystallographica*, **9**, 865–867.

Wicks, F.S. & O'Hanley, D.S. (1988) Serpentine minerals; structures and petrology. In: *Hydrous Phyllosilicates (Exclusive of Micas)* (S.W. Bailey, editor). Reviews in Mineralogy, **19**. Mineralogical Society of America, Washington, D.C., pp. 91–167.

Wiewióra, A. & Weiss, Z. (1990) Crystallochemical classifications of phyllosilicates based on the unified system of projection of chemical composition: II. The chlorite group. *Clay Minerals*, **25**, 83–92.

Wiewióra, A., Rausell-Colom, J.A. & Garcia-Gonzalez, T. (1991) The crystal structure of amesite from Mount Sobotka: a nonstandard polytype. *American Mineralogist*, **76**, 647–652.

Wolters, F. & Emmerich, K. (2007) Thermal reactions of smectites – Relation of dehydroxylation temperature to octahedral structure. *Thermochimica Acta*, **462**, 80–88.

Wu, X.J., Li, F.H. & Hashimoto, H. (1989) High-resolution transmission electron microscopy study of the superstructure of Xiuyan Jade and Matterhorn serpentine. *Acta Crystallographica*, **B45**, 129–136.

Xu, S. & Harsh, J.B. (1992) Alkali cation selectivity and surface charge of 2:1 clay minerals. *Clays and Clay Minerals*, **40**, 567–574.

Yada, K. (1971) Study of the microstructure of chrysotile asbestos by high resolution electron microscopy. *Acta Crystallographica*, **27**, 659–664.

Yada, K. (1979) Microstructures of chrysotile and antigorite by high-resolution electron microscopy. *The Canadian Mineralogist*, **13**, 227–243.

Yakovenchuk, V.N., Krivovichev, S.V., Pakhomovsky, Y.A, Ivanyuk, G.Y, Selivanova, E.A., Men'shikov, Y.P. & Britvin, S.N. (2007) Armbrusterite, $K_5Na_6Mn^{3+}Mn^{2+}_{14}[Si_9O_{22}]_4(OH)_{10} \cdot 4H_2O$, a new Mn hydrous heterophyllosilicate from the Khibiny alkaline massif, Kola Peninsula, Russia. *American Mineralogist*, **92**, 416–423.

Young, R.A. & Hewat, A.W. (1988) Verification of the triclinic crystal structure of kaolinite. *Clays and Clay Minerals*, **36**, 225–232.

Yucel, A.M., Rautureau, M., Tchoubar, D. & Tchoubar, C. (1981) Calculation of the X-ray powder reflection profiles of very small needle-like crystal. II. Quantitative results on Eskischir sepiolite fibers. *Journal of Applied Crystallography*, **14**, 431–454.

Zagorsky, V.Y., Peretyazhko, I.S., Sapozhnikov, A.N., Zhukhlistov, A.P. & Zvyagin, B.B. (2003) Borocookeite, a new member of the chlorite group from the Malkhan gem tormaline deposit, Central Transbaikalia, Russia. *American Mineralogist*, **88**, 830–836.

Zanazzi, P.F. & Pavese, A. (2002) Behaviour of micas at high pressure and high temperature conditions. In: *Micas: Crystal Chemistry and Metamorphic Petrology* (A. Mottana, F.P. Sassi, J.B. Jr. Thompson &

S. Guggenheim, editors). Reviews in Mineralogy and Geochemistry, **46**, Mineralogical Society of America, Chantilly, Virginia, USA, pp. 99–116.

Zanazzi, P.F., Montagnoli, M., Nazzareni, S. & Comodi, P. (2006) Structural effects of pressure on triclinic chlorite: a single-crystal study. *American Mineralogist*, **91(11-12)**, 1871–1878.

Zanazzi, P.F., Montagnoli, M., Nazzareni, S. & Comodi, P. (2007a) Structural effects of pressure on monoclinic chlorite: A single-crystal study. *American Mineralogist*, **92**, 655–661.

Zanazzi, P.F., Comodi, P., Nazzareni, S., Rotiroti, N. & Smaalen, S. (2007b) Behavior of 10-Å phase at low temperatures. *Physics and Chemistry of Minerals*, **34**, 23–29.

Zane, A. & Weiss, Z. (1996) A procedure for classifying rock-forming chlorites based on microprobe data. *Rendiconti Lincei*, **9**, 51–56.

Zazzi, A., Hirsch, T.K., Leonova, E., Kaikkonen, A., Grins, J., Annersten, H. & Eden, M. (2006) Structural investigations of natural and synthetic chlorite minerals by X-ray diffraction, Mössbauer spectroscopy and solid-state nuclear magnetic resonance. *Clays and Clay Minerals*, **54**, 252–265.

Zheng, H. & Bailey, S.W. (1989) Structures of intergrown triclinic and monoclinic IIb chlorites from Kenya. *Clays and Clay Minerals*, **37**, 308–316.

Zheng, H. & Bailey, S.W. (1994) Refinement of the nacrite structure. *Clays and Clay Minerals*, **42**, 46–52.

Zheng, H. & Bailey, S.W. (1997) Refinement of the cookeite "r" structure. *American Mineralogist*, **82**, 1007–1013.

Zhukhlistov, A.P. (2007) Electron Diffraction Study of Lizardite 1*T* with Stacking Faults. *Crystallography Reports*, **52**, 208–214.

Zhukhlistov, A.P. (2008) Crystal Structure of Nacrite from the Electron Diffraction Data. *Crystallography Reports*, **53**, 76–82.

Zussman, J., Brindley, G.W. & Comer, J.J. (1957) Electron diffraction studies of serpentine minerals. *American Mineralogist*, **42**, 133–153.

Zvyagin, B.B. (1954) Electron diffraction of the minerals of the kaolinite group. *Doklady Akademii Nauk SSSR*, **96**, 809–812.

Zvyagin, B.B. (1960) Determination of the structure of kaolinite by electron diffraction. *Doklady Akademii Nauk SSSR*, **130**, 1023–1026.

Zvyagin, B.B. (1967) *Electron Diffraction Analysis of Clay Mineral Structures*. Plenum Press, New York, 364 pp.

Zvyagin, B.B. & Drits, V.A. (1996) Interrelated features of structure and stacking of kaolin mineral layers. *Clays and Clay Minerals*, **44**, 297–303.

Zvyagin, B.B., Berkhin, S.I. & Gorshkov, A.I. (1966) Structural characteristics of halloysite based on X-ray and electron-diffraction patterns. *Akademii Nauk SSSR*, 69–93 (in Russian, English abstract).

Zvyagin, B.B., Soboleva, S.V. & Fedotov, A.F. (1972) Refinement of nacrite structure by a high-voltage electron-diffraction method. *Kristallografiya*, **17**, 514–520 (in Russian, abstract in English).

Zviagina, B.B., Sakharov, B.A. & Drits, V.A. (2007) X-ray diffraction criteria for the identification of trans- and cis-vacant varieties of dioctahedral micas. *Clays and Clay Minerals*, **55**, 467–480.

EMU Notes in Mineralogy, Vol. 11 (2011), Chapter 2, 73–121

An Overview of Order/Disorder in Hydrous Phyllosilicates

STEPHEN GUGGENHEIM

Department of Earth and Environmental Sciences, University of Illinois at Chicago, 845 W. Taylor St., Chicago, Illinois 60607 USA
e-mail: xtal@uic.edu

The purpose of the chapter is to provide a condensed introduction for later chapters. Major features of atomic order and disorder in hydrous phyllosilicates are presented. Emphasis is on layer charge and the effects of cation distributions, layer stacking (polytypism), interstratified systems, modulated phyllosilicates and non-planar structures. Idealized phyllosilicate models are presented along with adjustments that structures require to compensate for lateral misfit of the semi-rigid component tetrahedral and octahedral sheets. Standard polytypes and interstratified structures are described along with the X-ray diffraction effects of both. Details of modulated systems and polysome relationships are given. In addition, the effects of the phyllosilicate interface with aqueous fluids and biomolecules are discussed.

1. Introduction

This chapter attempts to provide the background needed to understand and appreciate how the variability in crystal structure and chemical composition of the phyllosilicate group minerals greatly influences the identification and physical properties of these minerals. To achieve such an overview, the chapter describes how structure and chemistry interact by examining short-range and long-range atom interactions in phyllosilicates in general, with an emphasis here on natural materials and X-ray techniques, primarily diffraction. Therefore, atomic order and disorder among individual atoms and groups of atoms are considered. Structural distortions that can potentially affect stability are discussed also. In contrast to this 'broad-brush' approach, more detailed approaches primarily by describing individual mineral species are given by Brigatti *et al.* (2011, this volume), Brigatti *et al.* (2006), and by Brigatti & Guggenheim (2002) for mica minerals. In addition, much can be learned of structures by computational methods using first principles (*i.e.* classic mechanics, quantum mechanics) by simulating atomic arrangements, and this topic is discussed by Sainz-Diaz (2011, this volume).

For the purposes of this chapter, order-disorder phenomena in phyllosilicates include thermal disorder, cation distributions and layer charge, polytypism (layer stacking), interstratified (mixed-layer) systems, modulated phyllosilicates and non-planar layers. Provided that layers have similar chemical compositions, stacking layers on top of adjacent layers can produce different one-dimensional structures, known as 'polytypes' (*e.g.* Guinier *et al.*, 1984). In phyllosilicates, the origin of these structures may be related to

DOI: 10.1180/EMU-notes.11.2

how sheets within a layer mesh or how layers stack on adjacent layers, or both. Polytype derivations are discussed briefly below for the major phyllosilicate subgroups. Polysomatic series (Baronnet, 1997) are discussed briefly also, where a polysome is derived from two or more structural units (or modules). Polysomatic series may include interstratification (mixed-layer) structures of, for example, chlorite and lizardite (see below), or 'modulated' phyllosilicate structures (Guggenheim & Eggleton, 1987, 1988), where there is periodic perturbation to the structure that may involve tetrahedral reversals or incomplete octahedral sheets.

1.1. Order/disorder

'A crystalline solid is defined as a solid consisting of atoms, ions or molecules packed together in a periodic arrangement', whereas an 'amorphous solid is one in which the constituent components are arranged randomly' (Guggenheim *et al.*, 2002). As noted by Guggenheim *et al.* (2002) variations occur between the two extremes and even within an extreme case. For example, Si can be coordinated by four oxygen atoms, joined as a silicate chain by sharing O corners, but each chain displaced randomly so that there is 'limited' order (four O atoms about the Si) but without periodicity between chains. 'Short-range order' exists where nearest neighbours about a site are comparable in distribution by translation or rotation to all other equivalent sites (site equivalence requires the same chemical species in the site, same geometry, same topology, *etc.*). 'Long-range order' involves a degree of translational periodicity that incorporates short range order, but extends the periodicity throughout large portions of the crystal or the entire crystal. The terms 'order' and 'disorder' are thus used in a qualitative way to describe the degree of perfection (order) or disturbance (disorder) of periodicity. 'Crystallinity', a common term used to describe the degree of perfection of translational periodicity, is often described by diffraction techniques.

1.2 Thermal disorder considerations

'Perfect crystallinity' implies that atomic regularity extends throughout the crystal with each unit cell containing distributions of atoms that are identical in symmetry, geometry, and chemical content. Departures from these conditions imply disorder. Disorder induced by thermal motion of atoms is present in all substances, and although it is convenient to ignore correlated vibrational effects between bonded atoms in diffraction experiments, these effects are nonetheless present. X-ray crystal structure analyses involve techniques where each (X-ray) photon that impinges the crystal has an interaction time that is much shorter than atom thermal vibrations (see *e.g.* Massa, 2004). Therefore, modelling thermal vibrations in diffraction experiments involves averaging the atom position over time. However, disorder of the atomic position from unit cell to unit cell produces the same effect, and it is for this reason that structure-analysis methods avoid the use of the term 'thermal parameters', but instead use the more proper description 'displacement parameters' to account for both vibration effects and positional disorder. Displacement of atomic positions is especially notable where chemical substitutions occur between atoms of different sizes.

1.3. The idealized phyllosilicate model

The basis of a phyllosilicate structure involves a series of 'planes' of atoms, each of which is defined as a single set of atoms stretching in two dimensions. Such a plane may be a set of like atoms, such as oxygen atoms, or dissimilar atoms, such as Si and Al atoms. Three planes, taken together: an *O*-atom (anion) plane, a *T*-atom (cation) plane, and an O + OH plane, form a 'tetrahedral sheet'. Many workers do not include the OH as part of the tetrahedral sheet because it is not directly linked to a tetrahedral (*T*) cation, although they do include the OH when making calculations involving sheet thicknesses. The oxygen atom plane, which forms the base of the model, is often referred to as the 'basal plane'. The *T* cations are commonly Si and Al atoms, which are considerably smaller in size than oxygen atoms, and they sit nestled in the dimple formed from three adjacent basal oxygen atoms. To complete four-fold coordination about the Si, a fourth oxygen atom, the apical oxygen atom, sits adjacent to the Si and off the basal plane. Thus, the tetrahedral sheet consists ideally of corner-sharing tetrahedra which form hexagonal-like rings extending in two dimensions as shown in Figure 1. The OH group is in the centre of the hexagonal ring and planar with the apical-oxygen plane.

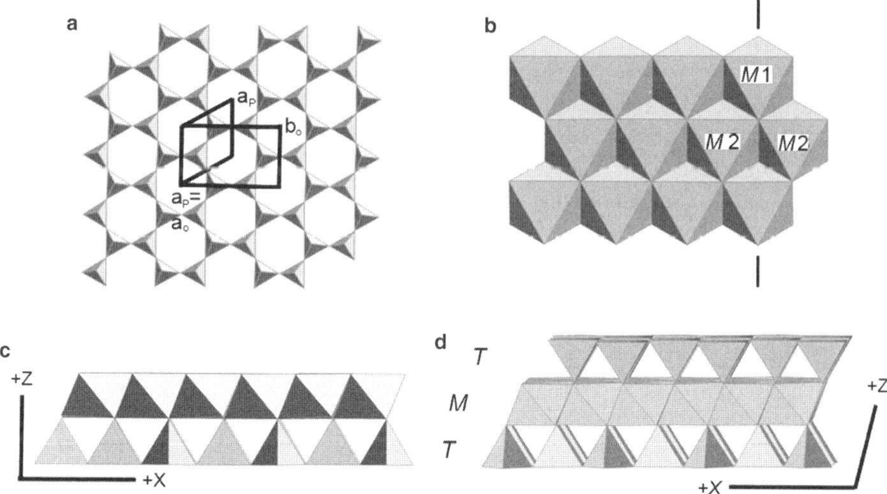

Fig. 1. The idealized phyllosilicate structure consists of hexagonal silicate rings forming a tetrahedral sheet (***a***) and an octahedral sheet (***b***) with *M* cation sites. All cation sites are shown by polyhedral linkages of anions. These sheets form either a 1:1 layer (one tetrahedral sheet + one octahedral sheet) or a 2:1 layer (two tetrahedral sheets + one octahedral sheet), shown in (***c***) and (***d***), respectively. Two unit cells are depicted in (*a*), a (primitive) hexagonal-based cell, a_P, and an orthohexagonal-based cell, a_o and b_o. The orthohexagonal cell is *C* centred, and is conveniently used for comparison between phyllosilicate structures. In (***b***), the *M* cation sites are labelled; *M*1 is located on a mirror plane (designated as a pair of lines above and below the octahedral sheet), whereas the two *M*2 sites are related by this *m* plane. In (***c***) and (***d***), the tetrahedral and octahedral sheets, shown in side view, are designated *T* and *M*, respectively. Note that the apical oxygen atoms of the tetrahedra form a common plane with the octahedra (after Bailey, 1980).

The top part of the tetrahedral sheet, the O + OH plane, is also the bottom part of the 'octahedral sheet', and thus this plane is a common junction between the two sheets. The ideal octahedral sheet differs for different types of phyllosilicates. For example, for an ideal model of the serpentine and kaolin mineral groups, the octahedral sheet consists of the O + OH plane, an *M*-atom (cation) plane, and a terminating plane of OH groups. The *M* cations (commonly Mg, Fe, Al and many others) sit in the dimple formed by two apical oxygen atoms and the adjacent OH group on the top part of the tetrahedral sheet, and this arrangement forms half of the six-fold coordination of the *M* cations. The remaining portion of the coordination unit involves the plane of OH groups, with the *M* cation closest packed with OH groups by residing in the dimples formed from three adjacent OH groups. Each *M* cation is coordinated to two O atoms and four OH groups. This layer is referred to as a '1:1 layer' because of the ratio of tetrahedral sheets to octahedral sheets. For the remaining types of phyllosilicate minerals, a '2:1 layer' is the basis of the structure, which is formed where a second tetrahedral sheet is inverted relative to the first, such that O + OH planes occur on both sides of the *M* atom plane (Fig. 1c). In this case, each *M* cation is coordinated to four O atoms and two OH groups, and the OH groups are well within the layer (instead of on the top surface as in 1:1 layers). The construction of the model defines the different portions of the phyllosilicate structure precisely with increasing thicknesses: planes, sheets and layers, and these terms should not be used interchangeably.

A completed model (Fig. 2) involves the placement of material between the layers (in the 'interlayer') for 2:1 layer structures. For both 1:1 and 2:1 structures, the placement of another layer above the first establishes a layer sequence, and this is referred to as 'stacking'. Most natural interlayer materials are cations, cation-hydrate complexes, hydroxide sheets, or no interlayer material. Where interlayer material is present, this material has a net positive charge. Control of what enters the interlayer depends on bulk chemistry and the charge on the layer ('layer charge'). The net charge on the layer is either zero or negative. For 2:1 phyllosilicates, the talc/pyrophyllite group has no net layer charge and no interlayer material, and the layers are bonded by van der Waals interactions. For micas, an interlayer cation resides in the silicate ring, with a univalent cation in the true micas (and the layer charge of -1) and a divalent cation in the brittle micas (and the layer charge of -2). The vermiculites and smectites, sometimes considered as separate groups and sometimes not, have a layer charge of approximately -0.2 to -0.6 (smectite) and -0.6 to -0.85 (vermiculite). Both have cations + H_2O molecules in the interlayer, with the cations offsetting the net negative charge on the layer and the H_2O variable in content, depending on the cation and the activity of H_2O in the environment around the particles. Because of the ability of vermiculite and smectite to expand owing to the variable interlayer H_2O, these clay minerals are referred to as 'swelling clays'. Finally, the chlorites consist of the 2:1 layer and a positively charged interlayer of ideal composition of $(R^{2+}, R^{3+})_3(OH)_6$, where *R* is equal to a cation. A combination of hydrogen bonds and electrostatic interactions bond the interlayer to the 2:1 layer in chlorites. In general, 1:1 phyllosilicates also have both hydrogen bonding and electrostatic interactions linking the layers, but unlike the 2:1 layer structures, the layers do not have a significant net layer charge.

In the ideal case, a 'trioctahedral sheet' forms where divalent cations only reside in all possible *M* site positions. The smallest structural unit contains three octahedral cations and a trioctahedral arrangement indicates a charge of $3 \times 2^+ = 6^+$ charges. A 'dioctahedral sheet' forms where one of the three sites is vacant, and to maintain charge balance, the remaining *M* sites of the smallest structural unit must be filled with trivalent cations, hence $2 \times 3^+ = 6^+$ charges. In general, a group (Fig. 2) is defined on the basis of both the type of layer and the layer charge, subgroups are based on the presence of trioctahedral or dioctahedral varieties, and chemical composition determines differences among species. The octahedral sites are usually designated as *M*1, and reside on the mirror plane in ideal symmetry (in micas, $C2/m$: $a \approx 5.4$ Å, $b \approx 9.5$ Å, $c \approx 10.0$ Å,

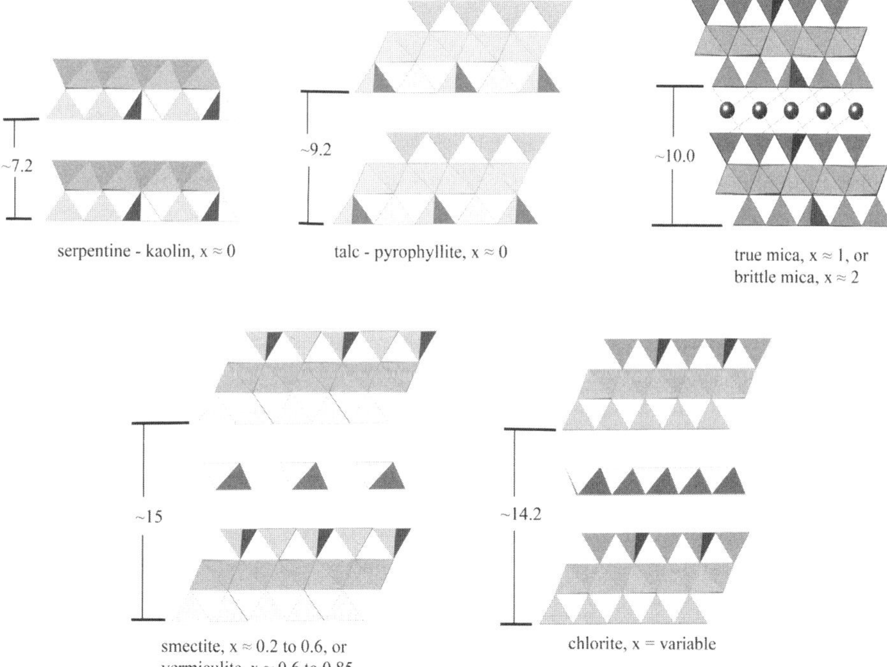

Fig. 2. The five basic phyllosilicate groups. The groups are separated by layer type and interlayer charge, x, per formula unit. The layer charge is numerically equivalent to the interlayer charge, but is negative in value. The names of the serpentine-kaolin and the talc-pyrophyllite groups are based on combining trioctahedral-dioctahedral subgroup names. Dioctahedral and trioctahedral species of the other groups exist, but are not used in group names. The micas and smectite-vermiculite are further divided based on interlayer charge. Depending on the phyllosilicate group, the material between the layers may be cations (*e.g.* mica), hydrate (cation + H_2O) complexes (*e.g.* smectite), hydroxide sheets (chlorite), or no interlayer material (*e.g.* talc-pyrophyllite). Smectite and vermiculite have swelling properties, and the isolated interlayer octahedra are approximations of the interlayer environment under controlled (humidity) conditions. The layer-to-layer spacing (d_{001} values, in Å) to the left of each group is useful for identification based on XRD where samples are prepared as oriented aggregates on a glass or other plate (after Bailey, 1980).

$\beta \approx 100^\circ$) for a one-layer structure, and two *M*2 sites, which are related by reflection across the mirror plane. The *M*1 is often referred to as the '*trans*' site because the OH groups are located in opposing corners of the octahedron, whereas the *M*2 sites are '*cis*' sites because these groups are at adjacent corners.

The 1:1 layer structure is non-centric because the layer is composed of a single tetrahedral sheet residing on one side of the octahedral sheet. The serpentine minerals are trioctahedral and, for the ideal one-layer platy structure (*e.g.* lizardite), there is only one symmetry unique octahedral site and one tetrahedral site. In contrast, the dioctahedral kaolin minerals (kaolinite, nacrite, dickite) are more complex primarily because of the arrangement of the octahedral vacant site.

In chlorite, there are two independent interlayer sites; the *M*3 site and the *M*4 site. The *M*3 sites surround the *M*4 sites. Nelson & Guggenheim (1993) showed that Mg-rich chlorite thermally expands nearly isotropically, which is unlike other phyllosilicates where thermal expansion along the *c* axis dominates. This result suggests that the H bond network in Mg-rich chlorite is relatively strong.

1.4. Common structural distortions relating to the polyhedra and sheets

Major structural distortions in phyllosilicates (Fig. 3) generally compensate for a lateral misfit between the tetrahedral sheet and the octahedral sheet. In ideal planar structures, a relatively larger tetrahedral sheet can reduce its lateral dimensions *via* tetrahedral rotation, the in-plane rotation of adjacent tetrahedra in opposite directions around the silicate ring, which is measured by the tetrahedral rotation angle, α. A fully extended tetrahedral sheet produces a tetrahedral rotation angle of zero, and this results in a silicate ring that is hexagonal and an interlayer cation site with a coordination number of 12 nearest neighbours. In contrast, if the misfit is such that the tetrahedral sheet is larger than the octahedral sheet, the tetrahedral rotation angle deviates from zero, the silicate ring distorts to ditrigonal (a polyhedron where pairs of faces are related by three-fold symmetry), and the interlayer cation site becomes six-coordinated. Lateral size adjustments of the component sheets are commonly related to Al content, where Al-for-Si substitution in the tetrahedral sheet enlarges this sheet, whereas Al-for-R^{2+} substitutions in the octahedral sheet reduce the lateral dimensions of the octahedral sheet. Observed α values range from 0 to 24.9° (maximum value possible is 30°), but values of $<12^\circ$ are common. At $\alpha = 10^\circ$, the reduction in the tetrahedral sheet is near 1.5% and at $\alpha = 20^\circ$, the reduction is near 6%. Other adjustments (*e.g.* sheet thinning, sheet thickening) are possible also to reduce misfit, but these adjustments are of less importance and it is not clear if they are only related to the compensation of misfit. The parameter, tau (τ), or the 'tetrahedral flattening angle' of Brigatti & Guggenheim (2002), is defined as the angle $O_{apical}-T-O_{basal}$ and is ideally 109.47°. Usually, the average of the three $O_{apical}-T-O_{basal}$ values for a tetrahedron is given as the τ value. The tetrahedral sheet becomes thinner as the observed τ value(s) decreases from the ideal and thicker as the value(s) increases. The parameter, psi (ψ), or the 'octahedral flattening angle', measures the thickness of an octahedron and is defined as the angle between the body diagonal and the vertical; the ideal value is 54.73°, which decreases as the octahedron thickens. Derived ψ values are conveniently calculated from

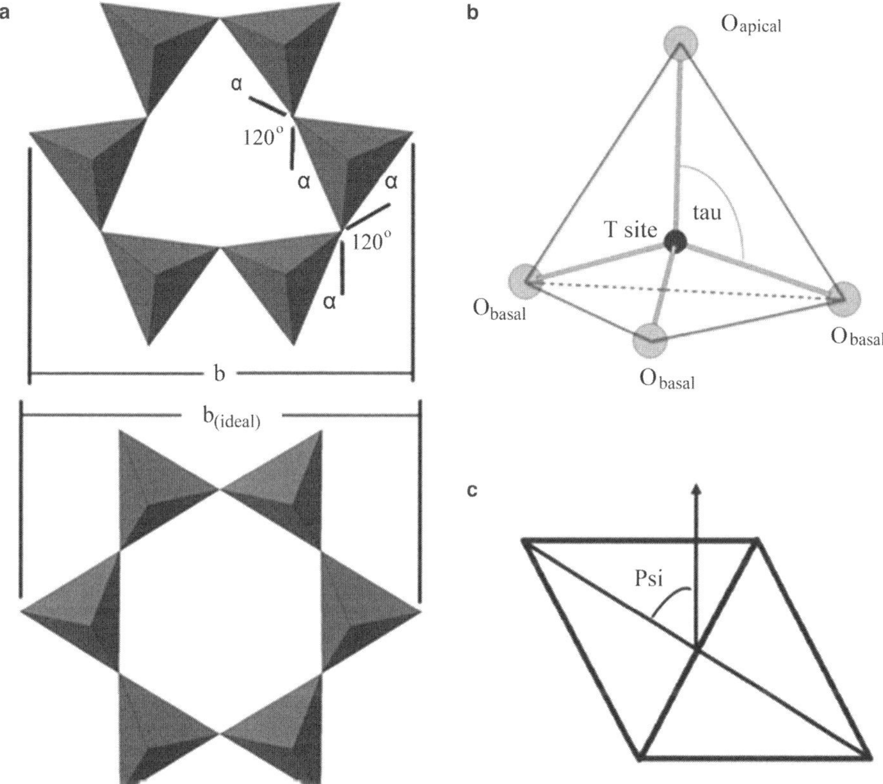

Fig. 3. Common structural distortions. The tetrahedral rotation angle, α, is defined as half the angular deviation from 120° of adjacent tetrahedra (***a***). Note that the shape of the silicate ring distorts to ditrigonal from ideally hexagonal and that the lateral dimensions, illustrated along the *b* axis, decreases. The parameter, τ, is related to the thickness of a tetrahedron, the ideal value is 109.47°, which increases in value by thickening the tetrahedron (***b***). The psi (ψ) parameter is defined as the angle between the body diagonal and the vertical (***c***) (modified from Guggenheim and Krekeler, 2011; with permission).

$\cos\psi$ = oct. thickness/2(*M*-O,OH,F) where *M*-O,OH,F is the mean *M* site size (see Bailey, 1984; Guggenheim & Eggleton, 1987, 1988; Brigatti & Guggenheim, 2002 for discussions of these parameters). For cases where the lateral dimensions of the tetrahedral sheet are smaller than the lateral dimensions of the octahedral sheet, the mechanisms to compensate for misfit are considerably more complex and may involve perturbations that include tetrahedral inversions, incomplete tetrahedral sheets, *etc.* These mechanisms are discussed, in part, in section 5 on modulated phyllosilicates.

1.5. Common structural distortions relating to site substitutions

In general, dioctahedral phyllosilicates have a larger, vacant *M*1 site and a smaller *M*2 site occupied by R^{3+} cations. Thus, for most common dioctahedral phyllosilicates, there are six $M2$ sites distributed as second-nearest neighbours around a single $M1$ site, which requires an adjustment of apical oxygen atoms to accommodate the different size cations and their distributions (*e.g.* Bailey, 1980). These adjustments to accommodate the larger $M1$ include a distortion of the near hexagonal rings of apical oxygen atoms, a counter-rotation of apical oxygen + OH triads (*i.e.* the upper triangular octahedral face relative to the lower face), and an out-of-plane tetrahedral tilt that results from the rigid tetrahedra tilting in response to the changes in the positions of the apical oxygen atoms. The out-of-plane tetrahedral tilt therefore affects the basal oxygen plane to produce what has been described as a 'corrugation' effect (*i.e.* the resulting off-plane displacement of the basal oxygen atoms), measured as Δz (the difference of the maximum deviations of the z coordinates of the basal oxygen atoms times c sinβ). Bailey (1984) and Brigatti & Guggenheim (2002) discuss the crystal chemical causes of these distortions. In contrast to dioctahedral phyllosilicates, the trioctahedral varieties generally have $M1$ and $M2$ sites of nearly equal size and thus, fewer distortions. Offsets between layers and within layers refer to displacements between layers or between tetrahedral sheets within a layer, respectively, and are not considered here except as they are required in the development of stacking; see Bailey (1984) and Brigatti & Guggenheim (2002) for details.

However, there are cases where cation ordering affects site sizes in ways that differ from the above general examples, but such cases appear to be relatively rare. For cases where the symmetry may be lower than ideal and no mirror plane exists, then the two $M2$ sites are designated as $M2$ and $M3$. For well crystallized phyllosilicates (and those that have become equilibrated with their surroundings), the $M1$ site is the vacant site (or nearly so) in dioctahedral 2:1 layers, and this site may be referred to as '*trans* vacant'. In some poorly crystalline clays (*e.g.* smectite, illite), the $M2$ or $M3$ site may be the vacant site, and this site may be referred to as the '*cis*-vacant' site. Ďurovič (1994) introduced terms to identify the ions present in an octahedral sheet: 'homo-octahedra' have all the M sites occupied by like ions; 'meso-octahedra' describe two M sites that are ordered with like ions, which differ from the third; and 'hetero-octahedra' involve three different octahedra, each with ordering of a different ion. Different octahedral sheets can lead to different 2:1 layer types. Other parameters useful to describe ordering and distortions are summarized by Brigatti & Guggenheim (2002).

2. Layer charge, solid solution and exsolution

Phyllosilicate minerals are chemically complex, and this is a testament to their ability to accommodate different sized cations with differing charges. Variable composition with resulting structure modifications leads to many of the physical properties unique to phyllosilicate (clay) minerals.

2.1. Layer charge and bulk structures

Layer charge is used to distinguish the different groups of phyllosilicates (Fig. 2), based on the net layer charge or average composition of the layer. In addition to general classification, the layer-charge concept is, in part, fundamental to our understanding of how phyllosilicates respond in general and swelling clays (smectite, vermiculite) in particular respond to the environment, especially with regard to rheological properties and sorption of inorganic and organic complexes. In the latter cases, the specific origin (*e.g.* distribution of charge between the tetrahedral and octahedral sheets, size of charge) of layer charge becomes important.

The major ideal phyllosilicate groups are distinguished by comparing layer-to-layer distances, see Table 1, and these distances are related to the layer structure and interlayer charge. Where the interlayer charge is high such as for the micas, an interlayer cation balances the charge on the layer and the layer-to-layer distance is fixed. Rieder *et al.* (1998) have suggested from published mica analyses that the lower-limit charge threshold for dioctahedral micas is 0.85, but that trioctahedral micas may be different (although probably similar). There is a class of micas referred to as 'interlayer deficient micas' (more properly, 'interlayer-cation-deficient micas'), with an interlayer charge from $\geq$0.6 to $<$0.85 and, by definition, these micas must not swell. Where the interlayer charge is low ($\sim$0.25 to $\sim$0.6 for smectite; $\sim$0.6 to $\sim$0.85 for vermiculite) and the phyllosilicate has swelling capabilities, the importance of the origin of layer charge (tetrahedral, octahedral, or both) increases as the charge becomes more localized or diffuse on the basal plane, and the charge distribution becomes especially important for sorption and wetting properties (*e.g.* Giese & van Oss, 1993), reactivity (*e.g.* Yariv, 1992) and weathering, colloid behaviour (rheology), *etc.* The interlayer charge of $\sim$0.6 to $\sim$0.85 associated with (non-swelling) interlayer-deficient micas and with (swelling) vermiculites is similar, suggesting that charge distributions within the layer may account for the differences in swelling behaviour, but this observation requires further study. The nature of layer charge in smectite is provided in detail by Christidis (2011, this volume).

The differentiation between vermiculite and smectite is based on how these materials respond to Mg exchange and glycerol solvation (see section 3.8.2 for a formal definition of exchange and solvation). With Mg saturation, vermiculite will have a resultant $d_{001} \approx 14.5$ Å value, whereas smectite has a 17.7 Å value after solvation. Samples with intermediate d_{001} values are most probably a mixture of the two phases as oriented intergrowths or, less likely, phases with characteristics between the two. The smectite group has been separated into low-charge *vs.* high-charge varieties (*e.g.* at –0.46 equivalents per half cell) by various workers (*e.g.* Christidis & Eberl, 2003), although such a classification is not widely accepted. Both vermiculite and smectite are greatly affected by relative humidity and thus have dynamic interlayer configurations. Details about the possible interlayer configuration in a 'confined' interlayer (fixed by a specified relative humidity) for vermiculite are given in section 2.3.2.

2.2. Solid solution and exsolution

Solid solution involves one or more cation (or anion or vacancy) substitutions at an atomic site. Ideally, substitutional solid solutions involve a random substitution at a site over many unit cells. When referred to a single unit cell, all substituting cations

Table 1. Classification of planar hydrous phyllosilicates (from Guggenheim *et al.*, 2006, 2007).

Layer type	Interlayer material[1]	Group	Octahedral character	Species[2]
1:1	None or H_2O only ($x \approx 0$)	Serpentine-kaolin	Trioctahedral	lizardite, berthierine, amesite, cronstedtite,
			Dioctahedral	kaolinite, dickite, nacrite, halloysite (planar)
			Di,trioctahedral	odinite
2:1	None ($x \approx 0$)	Talc-pyrophyllite	Trioctahedral	talc, willemseite, kerolite, pimelite
			Dioctahedral	pyrophyllite, ferripyrophyllite
	Hydrated exchangeable cations ($x \approx 0.2–0.6$)	Smectite	Trioctahedral	saponite, hectorite, sauconite, stevensite, swinefordite
			Dioctahedral	montmorillonite, beidellite, nontronite, volkonskoite
	Hydrated exchangeable cations ($x \approx 0.6–0.9$)	Vermiculite	Trioctahedral	trioctahedral vermiculite
			Dioctahedral	dioctahedral vermiculite
	Non-hydrated mono- or divalent cations ($x \approx 0.6–0.85$)	Interlayer-deficient mica	Trioctahedral	wonesite[3,4]
			Dioctahedral	none[4]
	Non-hydrated monovalent cations ($\geq$50% monovalent, $x \approx 0.85–1.0$ for dioctahedral)	True (flexible) mica	Trioctahedral	phlogopite, siderophyllite, aspidolite
			Dioctahedral	muscovite, celadonite, paragonite
	Non-hydrated divalent cations ($\geq$50% divalent, $x \approx 1.8–2.0$)	Brittle mica	Trioctahedral	clintonite, kinoshitalite, bityite, anandite
			Dioctahedral	margarite, chernykhite
	Hydroxide sheet (x = variable)	Chlorite	Trioctahedral	clinochlore, chamosite, pennantite, nimite, baileychlore
			Dioctahedral	donbassite
			Di,trioctahedral	cookeite, sudoite
			Tri,dioctahedral	none
2:1	Regularly interstratified (x = variable)	Variable	Trioctahedral	corrensite, aliettite, hydrobiotite, kulkeite
			Dioctahedral	rectorite, tosudite, brinrobertsite
1:1, 2:1			Trioctahedral	dozyite

[1] x is net layer charge per formula unit, given as a positive number
[2] not an exhaustive list of species; in general, listed in order of abundance
[3] net layer charge may be <0.6, but this is an exception
[4] 'series' names are given in Rieder *et al.* (1998) as a convenient way to describe incompletely investigated micas. For example, biotite is a trioctahedral true-mica series name for certain dark micas that may be used as a field term, and illite is a dioctahedral interlayer-deficient series name to describe certain micas after only optical microscopic data become available. Other dioctahedral interlayer-deficient micas of a series type are glauconite and brammallite.

average to a resultant size and a resultant electron density, and this average atom is called a 'hybrid' atom. Where several atomic sites may be involved and substituting cations have different charges, which must balance, these substitutions are known as 'coupled substitutional solid solutions'. Both substitutional solid solutions and coupled substitutional solid solutions involve the disordered state. In contrast, where cations either partially or completely enter sites preferentially, either over large numbers of unit cells of the crystal (long range order) or groups of unit cells (short-range order within domains), then an ordered state is described. Unmixing or 'exsolution' is a special type of intergrowth which suggests an equilibrium texture has been achieved. Although relatively rare in phyllosilicates, there are examples for some micas; see below.

2.2.1. *Serpentine and kaolin*

To maintain a platy morphology, the lateral dimensions (*a*, *b*) of the octahedral sheet must equal the lateral dimensions of the tetrahedral sheet for 1:1 layers, or a structural mechanism, such as tetrahedral rotation, must reduce an overly large tetrahedral sheet to allow articulation to a smaller octahedral sheet. The planar serpentine structures generally have R^{3+} cations, such as Al and/or Fe^{3+}, substituting in the T and M sites, and the platy, wave-like antigorite (Mg) structure may have lesser amounts of these substitutions. Octahedral substitutions include Mg (lizardite), Mg,Fe^{2+}(amesite), Mg,Mn (kelleyite), Fe^{2+} (cronstedtite), Mn^{2+} (guidottiite), and Fe,Mn,Mg (berthierine); *e.g.* crystal structure analyses of Mellini and co-workers, including Mellini (1982), Mellini *et al.* (2010), and Anderson & Bailey (1981), Hybler *et al.* (2002), and Wahle *et al.* (2010). In this way, a smaller R^{3+} cation, by substituting for a larger octahedral cation, reduces the lateral size of an octahedral sheet. In contrast, the R^{3+} cation increases the lateral size of the tetrahedral sheet by substituting for a smaller Si. Some overlap in composition between rolled morphologies (*e.g.* chrysotile) and platy structures occurs (Bailey, 1980; Chernosky, 1975), and the difference in morphology is believed to be related to the strength of interlayer H bonding; see section 5.5 on modulated/rolled structures. Other combinations of substitutions exist that do not resolve misfit between the component sheets, such as those with large R^{2+} octahedral cations (*e.g.* greenalite: Fe,Mn,Mg; caryopilite: Mn, Zn and small amounts of Al), but small, nearly pure Si tetrahedral sheets. Therefore, greenalite and caryopilite form modulated structures (Guggenheim & Eggleton, 1998). It is unclear how limited the octahedral substitutions (of Fe^{2+} or Mn^{2+}) may be in planar structures before a structure requires a modulated configuration. The kaolins, *i.e.* kaolinite, nacrite, dickite and halloysite, do not show any significant solid solutions (<1% Al for Si), and are essentially pure aluminosilicates, $Al_2Si_2O_5(OH)_4$.

2.2.2. *Talc and pyrophyllite*

Talc is generally very Mg-rich, although Fe and F may be present and trace amounts of Mn, Ti, Cr, Ni, Ca, Na and K have been reported (Evans & Guggenheim, 1988). Reports of Fe^{3+} in tetrahedral coordination also may be a result of analytical inaccuracies or contamination, as may be the Ca, Na and K. A very carefully analysed talc (Martin *et al.*,

1999) gave the following formula: $(Mg_{2.978}Fe^{2+}_{0.019}Mn^{2+}_{0.001}Fe^{3+}_{0.005}Al_{0.007})(Si_{3.984}Fe^{3+}_{0.004}Al_{0.007})O_{10}(OH)_{1.952}F_{0.048}$. Kager & Oen (1983) reported an iron talc of near $Mg_{1.80}Fe_{1.17}Si_4O_{10}(OH)_2$ associated with minnesotaite, and Robinson & Chamberlain (1984) reported a hydrothermal talc of $(Mg_{1.90}Fe^{2+}_{0.98}Fe^{3+}_{0.06}Al_{0.02})$ $Si_{4.04}O_{10}(OH)_2$. More commonly, however, are talcs that have only a small amount of iron (<2.5 wt.%; *e.g.* Petit *et al.*, 2004). Very rich iron 'talcs' are probably minnesotaite (see section 5.2). Willemseite is the Ni-rich end member of talc. Pyrophyllites are also of near end-member compositions (Evans & Guggenheim, 1988). Small amounts (<0.3 a.p.f.u.) of Al may substitute for Si, with substitutions near 0.1 typical. Small amounts of Fe^{2+}, Fe^{3+}, Mg and Ti may substitute in the M sites, with typical octahedral sums ranging from 1.95 to 2.05 per two octahedral sites. F substitution (0.35 wt.%) for OH has been documented by Sykes & Moody (1978), but most workers do not analyse for F. Ferripyrophyllite is the ferric iron analogue of pyrophyllite.

2.2.3. *Mica*

Within the 2:1 layer, common solid-solution substitutions for natural samples (modified from Bailey, 1984) include: (1) R^{3+} or R^{2+} (relatively small size cations) substituting for Si^{4+} in T sites; (2) R^{1+} or R^{2+} (medium size cations) substituting for R^{2+} or R^{3+} in M sites; (3) anion substitutions at the OH site; and (4) vacancy substitutions (omission of a cation) in M sites. An excess negative charge develops where a cation of lesser positive charge substitutes for a cation of greater charge. For example, the substitution of Al^{3+} for Si^{4+} creates a net negative charge in the tetrahedral sheet relative to the ideal. The excess negative charge within the 2:1 layer (offset by interlayer material, see above) may occur by substitutional solid solution in the tetrahedral sheet, the octahedral sheet, or within both sheets. Bailey (1984) considered the micas and found that where the excess tetrahedral charge was > -1.0 (true micas) or -2.0 (brittle micas), a substitution of commonly R^{3+} for R^{2+} in the octahedral sheet occurs to compensate, in addition to the presence of the positive interlayer cation (layer charge is expressed as a negative value, interlayer charge is equal in value, but opposite in sign). The most common tetrahedral substitutions involve the replacement of Si by Al^{3+} and Fe^{3+}, although Be^{2+} is also known to occur in (bityite) mica and chlorite and B^{3+} and Zn^{2+} in chlorite. Medium sized cations generally substitute in M sites, and commonly include Mg, Al, Fe^{2+} and Fe^{3+}, with less common Cr, Mn, Ni, V, Cu, Zn and Li. The rare K-rich mica, shirokshinite (Pekov *et al.*, 2003), contains 0.92 Na per three octahedral sites, indicating that the relatively large Na cation can enter the M sites also in considerable number. However, Na is not a common substituting cation at this site in natural samples, and the Na substitutions may result because of the very low Al and Li concentrations in the hyperalkaline environment where this mica is found.

The site normally occupied by OH in micas is often replaced either entirely or in part by F or, very rarely, entirely by S (in anandite mica). The negative charge and spherical shape of F^-, as compared to the $O\text{-}H^+$ vector with the positive charge pointing towards the interlayer cation, increases the stability of a mica structure to higher temperatures (Bassett, 1960) in trioctahedral micas. Chlorine is known to replace OH, and synthesis studies (*e.g.* Chevychelov *et al.*, 2008; Nazzareni *et al.*, 2008) indicate that solid solution of Cl is limited in biotite (possibly at <0.39 a.p.f.u., the maximum found by Nazzareni *et al.*, 2008). The substitution of O for OH ('oxy-component') is also possible and occurs to completion in norrishite mica (Eggleton & Ashley, 1989; Tyrna & Guggenheim, 1991). For 2:1 layers in general, the O for OH substitution is often related to the oxidation/reduction of iron, $Fe^{2+} + (OH)^- = Fe^{3+} + O^{2-}$, in the octahedral sites and can be reversible or partly reversible (*e.g.* Takeda & Ross, 1975). Dehydroxylation events, where 2(OH) are lost as H_2O and a residual O remains in the structure also affect this site, but the residual oxygen atom is part of five-fold coordination around the *M* site and a dehydroxylate structure results (*e.g.* Guggenheim *et al.*, 1987).

For trioctahedral micas from high metamorphic grade rocks with large Al content, and small Fe and Mn contents, vacancy content is <0.057 per three octahedral sites (Bujnowski *et al.*, 2010). Dioctahedral micas, by definition, have a vacancy at *M*1. Some micas (*e.g.* Toraya *et al.*, 1976, 1978; Lin & Guggenheim, 1983) have octahedral vacancies midway between trioctahedral and dioctahedral. These micas generally either simulate a trioctahedral mica where the *M*1 and *M*2 sites are equal or nearly equal in size (and contain cations of similar mean electron counts), or a dioctahedral mica where the *M*1 site is much larger than *M*2 (and contains a cation of low charge or a vacancy). It is important, however, to rule out the possibility of interlayering of dioctahedral and trioctahedral forms (*e.g.* Levinson, 1953). Brigatti & Guggenheim (2002) discuss these trends and exceptions to these more common ordering schemes.

Ferraris *et al.* (2001a) found phlogopite-muscovite exsolution textures suggesting crystallization temperatures near 500–600°C. This unmixing primarily involves the octahedral sites. Ruiz Cruz & Sanz de Galdeano (2009) found exsolution textures between annite and NH_4-bearing muscovite, with exsolution occurring near 450°C and ~3 kbar. They believe that NH_4 occurred only in the muscovite at early metamorphic conditions, but later became redistributed between the muscovite and annite, with the annite containing greater amounts of NH_4. Although a temperature where unmixing occurred could not be determined, Ferrow *et al.* (1990) presented evidence that celadonite (Fe- and Mg-rich and Al-poor muscovite) and muscovite components exsolved in very fine (~100 Å) lamellae. Presumably, both the octahedral and the tetrahedral sites are involved in unmixing.

Ordering of interlayer cations in micas has not been reported, and thus substitutional disorder is the rule. One argument as to why interlayer cations are disordered may be related to the relatively high symmetry of the site, especially if the coordination of the interlayer site is twelve. In addition, however, the interlayer-site size is partially controlled by the misfit between the tetrahedral sheet and the octahedral sheet where the tetrahedral sheet has a reduction in size via tetrahedral rotation, and this multi-sheet interaction may be a primary reason why interlayer cations remain disordered. This effect distorts the tetrahedral ring from an ideally hexagonal shape to ditrigonal, which produces a change in the interlayer cation coordination from twelve to six. The size of the interlayer-cation site, because the cation resides in the cavity formed by opposing rings from adjacent 2:1 layers, is thus affected by the compositions of the attached sheets (as well as the field strength of the interlayer cation and other effects as discussed in Brigatti & Guggenheim, 2002).

Unmixing (exsolution) involving the interlayer site occurs also. Wonesite (Spear *et al.*, 1981) is an interlayer-deficient, sodium-dominated biotite-like mica with a layer charge (~ -0.5) very unlike traditional micas. Under ordinary circumstances, this phase should swell. However, lamellae of talc and Na-rich biotite oriented at an inclination of near 37° from the (001) exist, which lock the layers and prevent expansion (Veblen, 1983). Kogure *et al.* (2005) showed that the 2:1 layers in wonesite are shifted, like talc, across the interlayer (layer offset) by 1.25 Å along the $[1\bar{1}0]$ direction with Na atoms partially occupying the interlayer site.

2.2.4. Chlorite

The common constituents of the octahedral sheets in chlorite are Fe^{2+}, Mg, Al and Fe^{3+} with lesser amounts of Cr, Ni, Mn, V, Cu, Zn and Li. Bailey (1988a), in a review of the compositional data for chlorites, suggested that the entire 2:1 layer has a net -1.0 charge that is compensated by a $+1.0$ net charge of the interlayer for the stable arrangement. A similar conclusion was reached by Lee *et al.* (2007) with respect to layer charge, and where Al saturation was found to occur in metamorphic chlorites at $Al_T/(Al_T + Mg + Fe) \approx 0.4$ (where Al_T = total Al content). Foster (1962) examined chlorite compositions, but rarely considered the importance of bulk composition and petrologic conditions. Nonetheless, she found that Al is generally not balanced between the octahedral and tetrahedral sites and octahedral Al is generally low, with other R^{3+} cations compensating in the tetrahedral sheets. Also, where ${}^{oct}R^{3+} \approx {}^{tet}R^{3+}$, octahedral occupancy is vacancy-free, but if ${}^{oct}R^{3+} > {}^{tet}R^{3+}$, then the excess ${}^{oct}R^{3+}$ replace R^{2+} by a ratio of 2:3 with octahedral vacancies.

For chlorite, because of the nearly same size and charge of Fe^{2+} and Mg, there is a continuous solid solution series between the two end members, often with appreciable Al present. A complete coupled substitution exists between Al for Si and Mg following $Si^{4+} + Mg^{2+} = {}^{tet}Al^{3+} + {}^{oct}Al^{3+}$ (Foster, 1962). Bailey (1988a) discussed individual structure refinements and (limited) trends. Partial solid solutions exist in the Al-rich Mg−Mn−Fe trioctahedral chlorites (Schreyer *et al.*, 1986), with a miscibility gap near $Mn/(Mg + Mn)$ from 0.25 to 0.50. Further study by Abad-Ortega & Nieto (1995) showed that the gap closes near $Fe/(Mg + Mn) \approx 0.25$. Nelson & Guggenheim (1993) showed that R^{3+} cations are more stable in the *M*4 site based on charge considerations, but this site is larger at higher temperatures than the surrounding *M*3 sites, which limits Fe oxidation. Low-temperature chlorites of diagenetic or low-metamorphic grade studied by Inoue *et al.* (2009) generally have greater Si contents, smaller Fe + Mg contents, and greater numbers of octahedral vacancies than higher-grade metamorphic chlorites, and these samples contained significant Fe^{3+} at <14 wt.%.

2.2.5. Smectite and vermiculite

Dioctahedral smectites show primarily Mg, Fe^{2+} and Fe^{3+} substitutions in the *M* sites and Si, Al and Fe^{3+} in the *T* sites, and species are distinguished based on where the dominant negative charge on the 2:1 layer exists. For example, montmorillonite, with a dominant net negative (less positive) charge originating in the octahedral sites usually involves Mg^{2+} substitution for Al^{3+} [commonly at $Al_{1.58}Mg_{0.42}$ per $Si_4O_{10}(OH)_2$], whereas beidellite involves primarily tetrahedral substitutions of Al^{3+}-for-Si^{4+} [near $Al_2(Si_{3.58}Al_{0.42})O_{10}(OH)_2$]. The excess negative charge is compensated by hydrated interlayer cations. Where the octahedral content is Fe^{2+} and limited, but significant amounts of tetrahedral substitutions occur for Si, the resulting species is nontronite, $Fe^{3+2}(Si_{4-x}Al_x)O_{10}(OH)_2 \cdot R^{+x} \cdot nH_2O$. Trioctahedral smectite, saponite, has an excess negative layer charge that originates tetrahedrally by the substitution of Al for Si, but with a partial compensation by Al^{3+} for Mg^{2+} in the octahedra. Vacancies may occur also in the octahedra. More unusual substitutions occur with Fe^{2+} or Fe^{3+} forming Fe-rich saponite. Significant amounts of Li, Cr, Ni, V, Zn and Cu are known to occur in smectites, as well as F^- (Newman & Brown, 1987). Vermiculites, which commonly evolve from the alteration of micas with compositions near the phlogopite–annite join, have a composition similar to the parent. Iron in octahedral sites is oxidized relative to the mica parent, which reduces the net negative charge on the 2:1 layer and permits swelling by the exchange and incorporation of, most commonly, $Mg + H_2O$ in the interlayer. Other compositional differences may occur also (*e.g.* see Newman & Brown, 1987), such that tetrahedral Al:Si ratios usually exceed 1:3 and octahedral substitutions of R^{3+} occur for R^{2+} cations relative to the mica parent. Thus, the source of negative charge often originates in the tetrahedral sheet and not the octahedral sheet.

2.3. Layer charge and surfaces

Random substitutional solid solution over many unit cells produces a net negative charge on the layer (see above). Figure 4 shows the effect of individual substitutions on the basal plane for an ideal 2:1 layer. As discussed by Johnston (1996; see also Farmer, 1978; Sposito, 1984; Johnston, 2010), for substitutions in the octahedra where R^{2+} replaces R^{3+} or R^{1+} replaces R^{2+} (*i.e.* a cation of lesser positive charge replaces a cation of greater positive charge), a net negative charge results, and this charge difference affects the coordinating anions about the cation. To compensate, the tetrahedral cations readjust their positions, which affects the charge distribution on the coordinating tetrahedral anions, including the anions of the basal plane. For each octahedral substitution, the charge distribution affecting the basal plane occurs over a group of oxygen anions (~5 atoms per each side of the layer) and the charge on the basal plane is relatively diffuse. In contrast, because a tetrahedral site substitution of Al^{3+} for Si^{4+} affects the three basal oxygen atoms of the tetrahedron more directly, the charge distribution is localized. Taken separately, these are the two end-member situations, but tetrahedral and octahedral substitutions can both occur in the structure. Therefore, the resulting site energies at the interlayer can be diverse and the net negative charge on the basal plane can be compensated in many ways, by exchangeable cations, organic cations, polar complexes, *etc.* This discussion is greatly simplified, but useful as a teaching tool: Bleam (1990) provides a more formal and detailed account. The diversity of origin of the layer charge suggests that clay minerals, and especially smectites, are ideal candidates to 'sample' the Martian surface for possible organic cations, in addition to the fact that these clays are common weathering products.

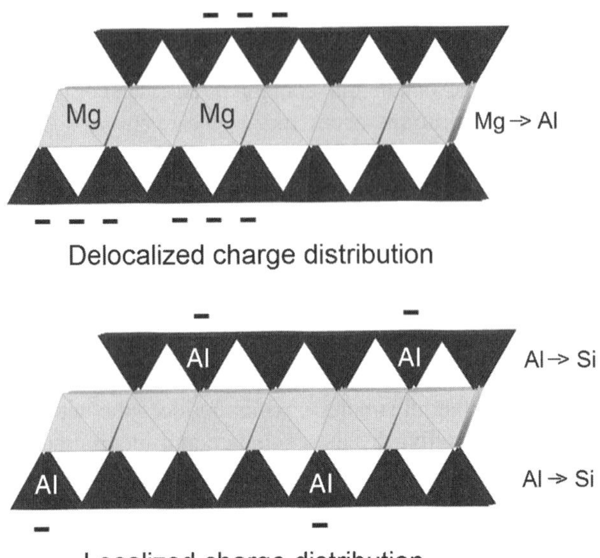

Fig. 4. Schematic showing the distribution of layer charge in a montmorillonite that occurs by substitutional solid solution in the octahedral sheet *vs.* in the tetrahedral sheets. In both cases, the substitutions involve cations with lesser positive charge for a cation of greater positive charge (and hence an increase in the net negative charge of the layer), and these substitutions are random. Substitutions well within the 2:1 layer produce a delocalized charge on the interlayer surface whereas substitutions near the edge of the layer create a localized charge. The effect of these different distributions is interlayer sites with different site energies. Random substitutions imply that the spacings differ between substitutions (either within tetrahedra, octahedra or both) which produce many unique site energies (after Johnston, 1996; with permission of The Clay Minerals Society).

Organic molecules can interact with layer silicate surfaces in several ways; see Schoonheydt & Johnston (2011, this volume). In addition to the direct layer-charge effect, other considerations include: (1) neutral sites on the basal oxygen surface forming Si tetrahedra (siloxane if H or hydrocarbons are bonded to the Si, although 'siloxane surface' is often used to describe the basal oxygen surface linked to the Si; *e.g.* Güven, 1992) of, for example, talc, pyrophyllite and kaolins, (2) interlayer cation (alkali, alkaline earth, transition metals, *etc.*) to organic solute interactions, (3) H_2O molecules that are polarized by the interlayer cation, (4) hydrophobic sites where organic molecules replace inorganic interlayer cations and where weak but significant van der Waals interactions between the organic molecules occur, and (5) broken-edge sites associated with the OH groups. Some of these sites are not as readily available because of either van der Waals interactions (*e.g.* pyrophyllite, talc) or hydrogen bonding (*e.g.* kaolins) between layers, and thus layers do not easily separate to accommodate organic molecules. Other sites, such as the broken edge sites, are very abundant and reactive. Because of this reactivity, many natural surfaces are coated, which influences sorption behaviour in complex ways. One way to examine the behaviour of ideal surfaces is to use freshly exposed surfaces, and this commonly involves freshly cleaved surfaces for phyllosilicates.

2.3.1. Cleaved surfaces: chemistry and reactivity of the interface

There are two important overall aspects to examining cleaved surfaces: (1) the atomic/electronic structure of the surface itself and (2) the effect of the surface on the other component of the (gas, liquid or solid) interface. Simplifying experiments tend to be the norm, so that contaminated surfaces are avoided and the surface itself is studied. Of particular importance is the organization of the interfacial water in a bulk-water environment, because of the ubiquitous nature of aqueous fluids in the natural environment and the importance of understanding rheological/colloidal properties, such as flocculation and dispersion. In addition, the biogeochemistry of these surfaces is commonly studied both because of the possible linkage of clay surfaces to the origin of life and because of the use of clays as catalysts to synthesize organic compounds.

Colloid chemists (*e.g.* see Lyklema, 1991) use various diffuse double layer theories to describe the interaction of aqueous fluids with a clay surface, where the clay surface is balanced by counter ions of opposite charge in the aqueous fluid. A simplified picture is that the counter ions balance the surface charge in a bound (although perhaps transient) compact layer (inner Helmholtz plane), whereas a somewhat more realistic picture involves counter ions attached to the surface along with counter ions diffusing into the liquid, with the concentration diminishing with distance, to form a diffuse double layer (Gouy-Chapman double layer). Additional refinements in theory, for purposes of quantification, involve using ionic sizes rather than point charges to the counter ions. Ionic sizes limit the approach of the counter ions to a charged surface (to form a Stern plane) and the possibility of adsorption. Cation adsorption complexes may form part of the Stern plane as either inner-sphere surface complexes with no H_2O molecules between the cation and the surface or as outer-sphere surface complexes with one or more H_2O molecules (Sposito *et al.*, 1999). Cheng *et al.* (2001), using high-resolution X-ray reflectivity, showed that adsorbed H_2O molecules (plus some H_3O^+, if the counter ion is removed) occur at 1.3 Å above the average (001) plane at the surface of muscovite near the ditrigonal hole (adsorbate layer). Another H_2O plane forms at near 2.5 Å (first hydration layer) from the (001) plane. Finally, a more disordered H_2O layer (second hydration layer at 3.7 Å) marks the beginning of a liquid-like (more disordered) structure outward from the (001), showing a diminishing oscillatory H_2O distribution. The adsorbate layer has about one H_2O molecule associated with each ditrigonal hole. This result is an effect of the layer charge caused by Al for Si substitutions and the effect of the (weak) solvation of K^+ that allows H_2O molecules to move between the counter ion and the oxygen plane. Park & Sposito (2002) showed that Monte Carlo calculations were consistent with this model. Based on these calculations, they suggested that the first hydration layer had the H pointing to the (001) surface, consistent with H bonding. Also, K^+ counter ions, which were removed by washing in the experiments of Cheng *et al.* (2001) are surrounded by H_2O molecules and the K^+ forms two distinct populations (described as 'planes') at 2.1 and 2.5 Å from the (001). In a high-resolution X-ray reflectivity study, Schlegel *et al.* (2006) showed that where K^+ is present in chloride solutions (0.01, 0.5 M) on the muscovite surface, the adsorbate layer and the first hydration layer merge (first solution layer, 1.7–1.8 Å), but the second hydration layer (= second solution layer) remains about the same (3.6–4.2 Å). Counter ion charge plays an important role in the nature of the double layer (*i.e.* diffuse, 'condensed'); see Park *et al.* (2008) and the references therein. Schlegel *et al.* (2006) noted also that relaxation effects of basal oxygen atoms were small, at $\sim\leq 0.07$ Å, but occurred over long distances, 30–40 Å away (1–2 unit cells) from the (001) outer-surface basal plane. In contrast, when considering the geometry of (110)- and (010)-type edge-surface structures for dioctahedral structures, Bickmore *et al.* (2003) calculated that surface relaxation effects are significant for various edge protonization schemes. Skipper *et al.* (1995) examined Li-H_2O complexes in confined spaces (*i.e.* an interlayer) at a fixed relative humidity and referred to the $Li(H_2O)_6$ complex as an electric double layer, which is described in more detail below under 'confined surfaces'.

The adhesion of biomolecules on the cleaved surface of phyllosilicates has recently been studied by Valdré and co-workers (Valdré *et al.*, 2004; Antognozzi *et al.*, 2006; Valdré, 2007; see Valdré *et al.*, 2011, this volume). Using scanning probe microscopy techniques, Antognozzi *et al.* (2006) showed that talc and brucite have a much greater affinity for DNA molecules than muscovite, and biotite is between these two extremes. In contrast, clinochlore cleavages (Valdré, 2007), where portions of the brucite-like and the 2:1 layer are alternatively exposed on the surface, show filaments of DNA molecules that are stretched across the negatively charged 2:1 layer and pinned to the brucite-like positively charged sections, indicating the effects of differing surface potentials. Thus, there are interesting possible applications of manipulating DNA assemblages on a non-conducting substrate in industry (*i.e.* biomolecule self-assembly, fluidity, microelectronics) and possible relevance to the origin of life.

2.3.2. *Confined surfaces: intercalation under controlled relative humidity in vermiculite*

Skipper and co-workers (*e.g.* Skipper *et al.*, 1991, 1994, 1995) used ($00l$, $l \leq 27$–30) reflections from neutron diffraction experiments to obtain information about the interlayer cation coordination for fixed relative humidities (RH) for Na-, Ni-, Ca- and Li-exchanged vermiculite. At a RH = 88% (layer-to-layer spacing of 14.96 Å), the Na cation is approximately coordinated by an average of 4.9 H_2O molecules with half of the molecules oriented to form a H bond to the basal oxygen surface of the 2:1 layer; clearly, the coordination units are poorly organized. Non-coordinated H_2O resides in the silicate ring at the same height as the basal oxygen plane. The interlayer Na ions could not be located definitively, but Skipper *et al.* (1991) suggested that they may not lie midway between the layers. In contrast, the best model for Ca-exchanged vermiculite (RH = 100%, $c\sin\beta = 15.05$ Å, Skipper *et al.*, 1994) is that the Ca is in slightly distorted octahedral coordination. This coordination involves two H_2O molecules in the central plane with Ca, and two H_2O molecules each on opposing sides to form hydrogen bonds to the basal plane of the adjacent layer. Approximately 0.6 H_2O fills sites in the centre of the silicate ring at the basal oxygen-plane height. The Li-exchanged vermiculite (Skipper *et al.*, 1995), which is analogous to the Ni sample, forms a distorted $Li(H_2O)_6$ complex oriented with Li in the central plane and three H_2O molecules on each of two sides ('electric double layer'). Four H_2O molecules form H bonds to the basal oxygen plane of adjacent 2:1 layers. Those authors believe that ~0.25 Li migrated to fill vacancies in the octahedral sheet of the 2:1 layer. Details of the structures of natural Mg-rich vermiculite are given in section 3.5.

3. Stacking

The order of presentation of descriptions for polytypism and stacking in the phyllosilicates is given in a pedagogic sequence, rather than following a sequence where the layers increase in thickness or interlayer charge increases, as was done above for solid solutions.

3.1. Mica

Smith & Yoder (1956) described layer stacking in micas as a result of layer rotations of an idealized, trioctahedral 2:1 layer. An 'interlayer stacking angle' requires definition, and this angle involves a vector from the centre of the lower tetrahedral ring to the upper tetrahedral ring within a layer as projected on the (001), and the interlayer stacking angle is the angle between vectors from an adjacent layer based on the clockwise direction as +. Possible interlayer stacking angles are 0°, 60° (= −300°), 120° (= −240°), and 180°. If only one stacking angle in either + or +/− directions is considered, then six 'standard' polytypes may form with regular stacking sequences of 1 to 6 layers. The polytypes are conveniently specified using Ramsdell (1947) notation, where the notation specifies the number of repeating layers, the overall symmetry, and a subscript to separate duplicate notations. Of the six standard polytypes, the *6H* polytype has not been found, and *2O* (as determined by the layer sequence in a fluorophlogopite, Ferraris *et al.*, 2001b; and as determined from octahedral ordering in an anandite, Filut *et al.*, 1985) is very rare. A seventh form exists that shows long-range stacking disorder (d) and this is referred to as $1M_d$ because the ideal base-layer is monoclinic (*M*). Bailey (1984) noted that there is compositional control of layer sequences with, for example, trioctahedral micas often crystallizing as *1M* and dioctahedral micas as $2M_1$, although there are many examples where this is not the case. Conditions of increasing temperature or time have been shown to produce regularity of layer sequence or increasing long-range stacking order, such as $1M_d$ to *1M* to $2M_1$ in illite/muscovite (*e.g.* Levinson, 1955).

The approach of using interlayer stacking angles between 2:1 layers has proven to be particularly useful in the development of computer code to calculate atomic positions of different stacking sequences for the modelling of diffraction patterns. However, from an atomistic view, the process of polytype formation is unlikely to involve crystallization where large groups of atoms must rotate, as is implied in the Smith & Yoder derivation. Furthermore, the procedure cannot be used to derive polytypes of 1:1 layers and chlorites. Bailey (1969, 1980, 1988a) developed an alternative method to derive layer stacking which involves a more atomistic approach of building layers from planes of atoms,

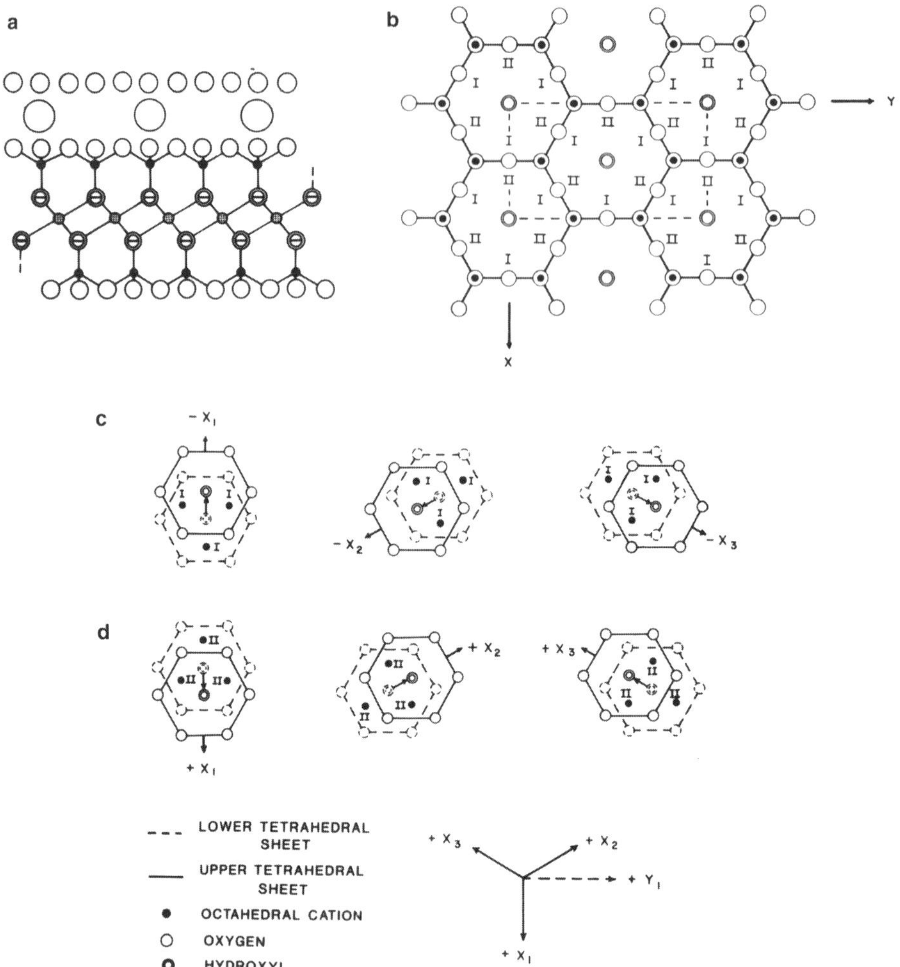

Fig. 5. Six standard mica polytypes. Part (***a***) shows the basic mica layer; and (***b***) the octahedral cation site position, labelled either as I or II, relative to the tetrahedral sheet and orthohexagonal unit cell. Octahedral cations can occupy either set I or set II, but not both sets, within a layer. Part (***c***) shows the direction of shift of the upper tetrahedral sheet relative to the lower tetrahedral sheet within a layer if set I is occupied, (***d***) or if set II is occupied, based on hexagonal axes (lower right of part ***d***). Note that where set I occurs, the shift is in a negative direction along an axis, whereas set II occupancy is always along a positive axial direction. The direction of shift is a requirement for closest packing to occur based on the set occupied. Part (***e***), continued on the next page, shows the derivation of the six 'standard' mica polytypes assuming that shifts can occur along one *a* axis, two axes, or three axes (after Bailey, 1984; with permission of the Mineralogical Society of America).

and this approach can be used to unify stacking for all phyllosilicates. Both approaches, that of Smith & Yoder and that of Bailey, produce identical results, and it is sometimes convenient to refer to either one or the other in explaining stacking.

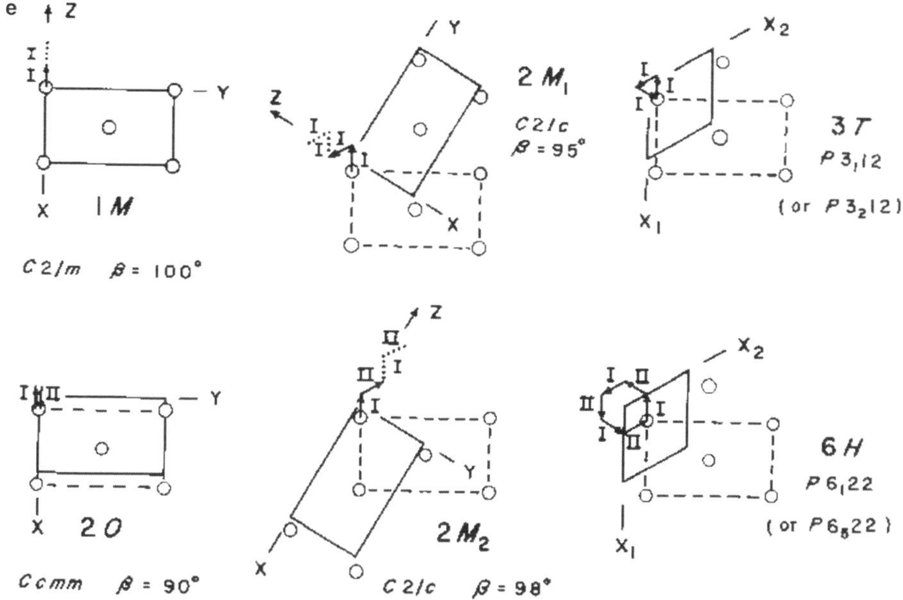

Fig. 5. Continued.

Like Smith & Yoder (1956), Bailey (1988b) assumes an ideal trioctahedral 2:1 layer, with polymerization of Si tetrahedra forming an ideally hexagonal sheet. However, the tetrahedral sheet interlayer cation-tetrahedral sheet junction does not produce the effect of a rotation, as it does in the Smith & Yoder approach where the layer is the basic unit, because any $n60°$ (*i.e.* 0, 60, 120, 180°) rotation of the hexagonal tetrahedral sheet is equivalent. The stacking variation is related to how the octahedral cations reside on the O_{apical}/OH plane. To maintain closest packing, octahedral cations can sit in the dimples formed from an OH and two adjacent O atoms in one of two ways, labelled set I or set II positions (Fig. 5). The occupancy of either set I or set II becomes significant only if adjacent layers show an alternation (otherwise structures with only set I or only set II are equivalent by being enantiomorphic). Defects with a mixture of set I and set II within a single octahedral plane are unknown (cation-to-cation distances are too short to be stable) and therefore do not need to be considered. Because the O_{apical}/OH atoms of the upper tetrahedral sheet obtain closest packing, they fit within the dimples formed by the octahedral cation plane, and thus there is a relationship between how closest packing is obtained and the set occupied. If only set I is occupied, the upper tetrahedral sheet has a stagger of $-1/3a_1$ (= $1M$) if one pseudohexagonal axis is considered, $-1/3a_1$ followed by $-1/3a_2$ (= $2M_1$) if two axes are considered, or $-1/3a_1$ followed by $-1/3a_2$ and then $-1/3a_3$ (= $3T$) if all pseudohexagonal axes are considered. Note that the set I shifts are always in the $-1/3a$. In contrast, the stagger for set II occupancy is positive $a/3$. The direction of shift for set I and set II is a result of maintaining closest packing within the layer. Therefore, if set I alternates with set II positions in adjacent layers, then shifts are $-1/3a_1$ followed by $+1/3a_1$ (= 2*O*) for one pseudohexagonal axis and $-1/3a_1$ followed by $+1/3a_2$ (= $2M_2$) for two pseudohexagonal axes. A six-unit (= 6*H*) structure involves alternating − and + directions along three pseudohexagonal axes. The reason why the Bailey approach of building layers is equivalent to the rotation of layers of Smith & Yoder is that set I octahedral positions can be symmetry related to set II positions by $\pm 60°$ or by 180° rotations.

The standard mica polytypes are the common polytypes observed, but structures are not limited to just these polytypes. Non-standard polytypes also exist, although these polytypes are not common, *e.g.* see Nespolo *et al.* (1997) and Nespolo & Ďurovič (2002). In keeping with an overview, the non-standard polytype derivations are not discussed here. The assumptions of the ideal trioctahedral configuration, including the idealized hexagonal tetrahedral sheet symmetry, and overall layer symmetry are useful in the derivation, but these assumptions are not always necessary to describe real structures. Most well crystalline, dioctahedral micas have the vacant *trans* site (*M*1) on the mirror plane (in $C2/m$ symmetry, one-layer structure) and the overall symmetry remains the same as ideal trioctahedral micas. In addition, most micas have major distortions that are involved with adjustments that allow the tetrahedral sheet to link to the octahedral sheet and around octahedra that are unequal in size to other octahedra.

3.2. Planar trioctahedral 1:1 layers

First, consider how stacking for 1:1 layers is similar to or different from the micas. For the micas, set I is defined as the first of two possible positions of the octahedral cations, with the important feature being either that only one set is occupied throughout the structure or set I alternates with set II. In 1:1 layer stacking, set II is defined as the first set when describing a sequence, but again the important aspect is that set II occurs throughout the structure or set II alternates with set I. Also, in contrast to mica stacking as described by Bailey (1988b), the variable in stacking does not occur within the layer as in micas, but between 1:1 layers. To maintain closest packing, the OH surface terminating the octahedral sheet shows another important feature of the effect of set I or set II occupancy, and that is the opposite slant or offset that occurs.

Bailey (1969) identified three ways (Fig. 6) to obtain hydrogen bonding between adjacent 1:1 layers and these include: (1) no layer shift involved with the stacking of the second layer on top of the first. In this case, two polytypes are formed, one for set II occupancy throughout the structure (1*T*) and one where set II alternates with set I ($2H_1$). (2) A layer shift of $1/3a_1$, $1/3a_2$ or $1/3a_3$ combined with either set I or set II occupancy. To maintain closest packing, the direction of shift along the *a* axis involved is linked to the octahedral set occupied, with a $-1/3a$ if set II is occupied and a $+1/3a$ if set I is occupied. Six polytypes are derived by displacement of layers only along a_1 and either having set II always occupied (1*M*) or alternating set II and set I (2*O*), by alternating the displacement of layers along a_1 and a_2 and either having set II always occupied ($2M_1$) or alternating between set II and set I ($2M_2$), and by alternating the displacement of layers among all three *a* axes and either having set II always occupied

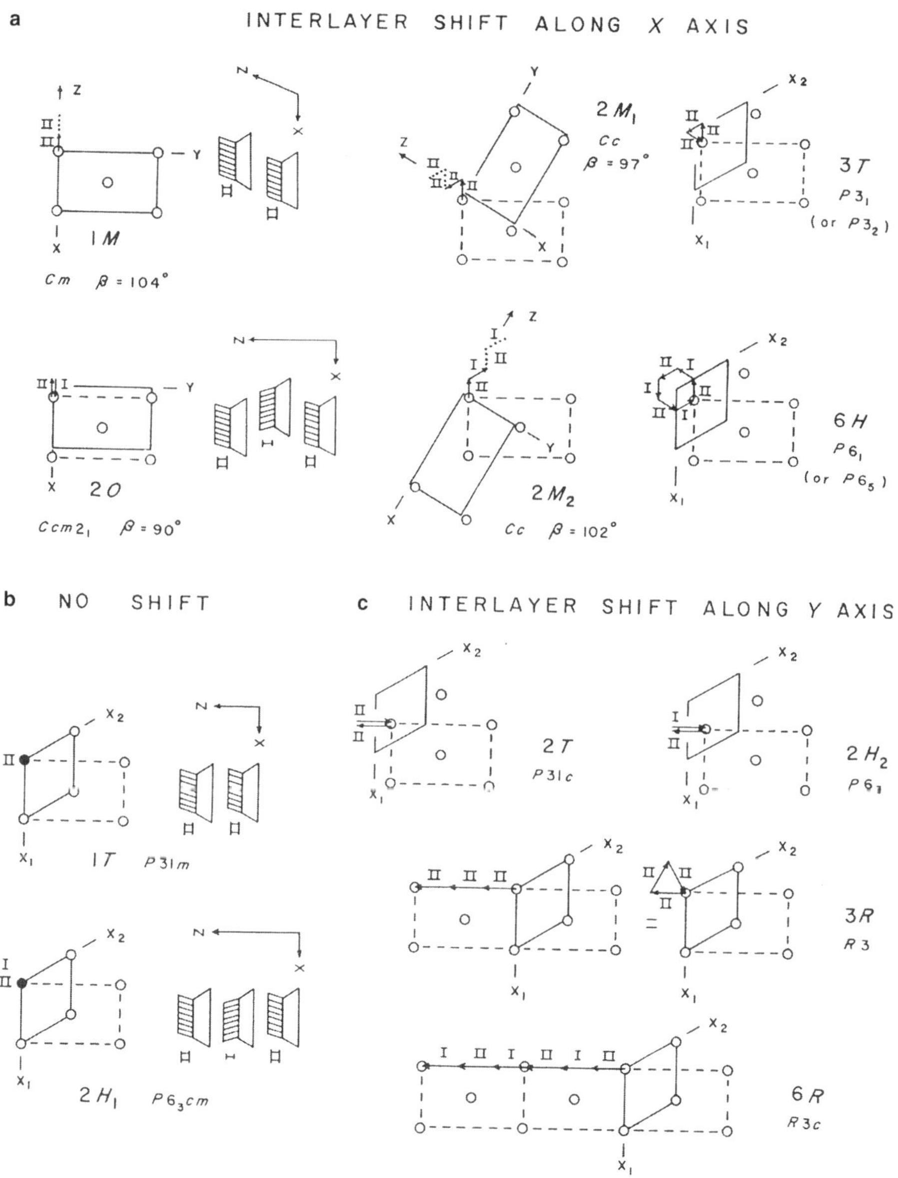

Fig. 6. The 1:1 layer and twelve standard polytypes. Stacking in the serpentine minerals involves shifts between 1:1 layers along one of the three *a* (hexagonal) axes, two of the three axes, or all three axes (part ***a***), or no shifts between 1:1 layers (***b***), or shifts along one *b* axis (***c***). Set II occupancy of the first layer is assumed. Dashed lines represent the unit cell of the first layer, solid lines are the resultant cell. The 3*R* polytype may be derived in two ways, as shown (slightly modified from Figure 1.5 of Bailey, 1980; with kind permission of the Mineralogical Society of Great Britain & Ireland).

($3T$) or alternating between set II and set I ($6H$). Finally, four polytypes form where (3) interlayer shifts of the second layer occur along $\pm 1/3b$. In these cases, idealized layers produce identical results regardless of the b axis chosen (b is defined as the unit direction perpendicular to a), and the direction of the b axis shift is unrelated to the octahedral set occupied. Two unique polytypes form by combining a b-axis shift first in a + and then a − direction with set II always occupied ($2T$), and by alternating set II and set I ($2H_2$). Finally, two additional polytypes, for a total of twelve structures, are derived by combining only a $+b$ axis shift with set II always occupied ($3R$), and by alternating set II and set I ($6R$).

3.3. Planar 1:1 layers and kaolin: vacant octahedral sites and stacking

Bailey (1963) defined the three octahedral sites as A, B and C (Fig. 7). The vacant site in kaolinite is the B site to form a triclinic structure. Alternatively, the vacant site in kaolinite may be in position C, which would produce a mirror image of the structure with the vacant B site. The vacant site in dickite alternates between the B and C sites from layer to layer, thus producing a monoclinic two-layer structure. In nacrite, the vacant

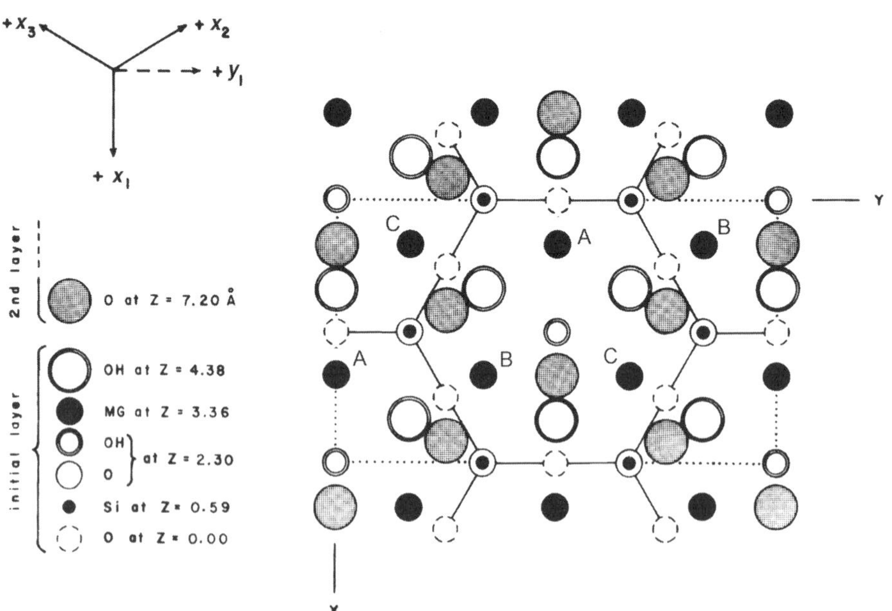

Fig. 7. Octahedral sites (labelled A, B and C) defined for kaolin minerals. One of these sites is vacant depending on whether the dioctahedral species is kaolinite, dickite or nacrite (see text). In addition, the hydrogen bond between the H of the OH group and the basal oxygen atom of the next adjacent 1:1 layer is optimized, where pairing occurs of these two atoms (OH, O) as illustrated in this projection (slightly modified from Fig. 1.3 of Bailey, 1980; with kind permission of the Mineralogical Society of Great Britain & Ireland).

octahedral site also alternates between B and C sites in successive layers, but the stacking differs from either kaolinite or dickite. In nacrite the stacking is based on an ideal trioctahedral six-layer (6*R*) structure, which reduces to a monoclinic structure, and an alternate choice of axes allows *a* for a two-layer, monoclinic unit cell. Thus, the kaolin minerals have essentially the same compositions and different layer structures (*i.e.* polymorphs) and, if nomenclature rules are followed, they should be identified by a single species name with a suffix symbol identifying the layer stacking sequence. However, the different vacancy location makes a stacking symbol insufficient to define the phase precisely, and the names are now well established (Guggenheim *et al.*, 2006). From transmission electron microscopy (TEM), Kogure & Inoue (2005) showed that stacking sequences are possible for the kaolin minerals in addition to those described by X-ray diffraction (Bailey, 1963), which further suggests the value of having separate names to describe these aluminosilicate species.

3.4. Chlorite

The potential number of chlorite structures based on different stacking arrangements is well over a thousand polytypes, but the vast majority of chlorites form either a one-layer structure (the 2:1 layer plus interlayer) or one that is disordered. Chlorite structure variations involving a single layer indicate that a model mica-like 2:1 layer can be used as the starting layer (*e.g.* set I occupied, $-1/3a_1$ shift). The interlayer octahedral sheet also may be described as either set I (same as the 2:1 layer) or II (different from the 2:1 layer). This sheet may be stacked in one of six ways to obtain hydrogen bonds between the OH of the interlayer and the basal oxygen atom plane of the 2:1 layer (Fig. 8). Because of the high (trigonal) symmetry of the interlayer sheet, three of these positions are equivalent and can be labelled *a*, and the other three are equivalent and can be labelled *b*. In position *a*, the interlayer-sheet cations project to the centre of the hexagonal ring or on top of the tetrahedral cations of the upper tetrahedral sheet of the 2:1 layer below. In position *b*, the projection of the interlayer-sheet cations is midway between, symmetrically arranged as triads (Fig. 8). Four interlayer + 2:1 layer assemblages are thus defined as: I*a*, I*b*, II*a* and II*b*. For regular one-layer structures, the next 2:1 layer can be placed on top of the interlayer in six ways to maintain hydrogen bonding, much like the interlayer was positioned above the initial 2:1 layer. In literature prior to 1988, Bailey used polytype symbols of the form of I*a* − 1 ... I*a* − 6 (along with the β angle in some cases to give additional information about directions of shift) with a total of 24 structures, although this number reduces to 12 unique structures when duplications are removed. For semi-random stacking, the projected position of the second 2:1 layer can be described also as position *a* or position *b* (instead of number 1 through 6) and six polytypes result: I*aa*, I*bb*, II*aa*, II*bb*, I*ba* = I*ab*, and II*ab* = II*ba* and this symbolism is now used. For regular one-layer polytypes, Bailey (1988b) used the form I*aa* − 1 ... I*aa* − 6 to designate the polytypes, and the added information can be used to determine if the polytype is monoclinic or triclinic (see the 1988b paper for additional details). Multi-layer chlorite polytypes are known also (*e.g.* Lister & Bailey, 1967), but these structures are rare and not considered further here.

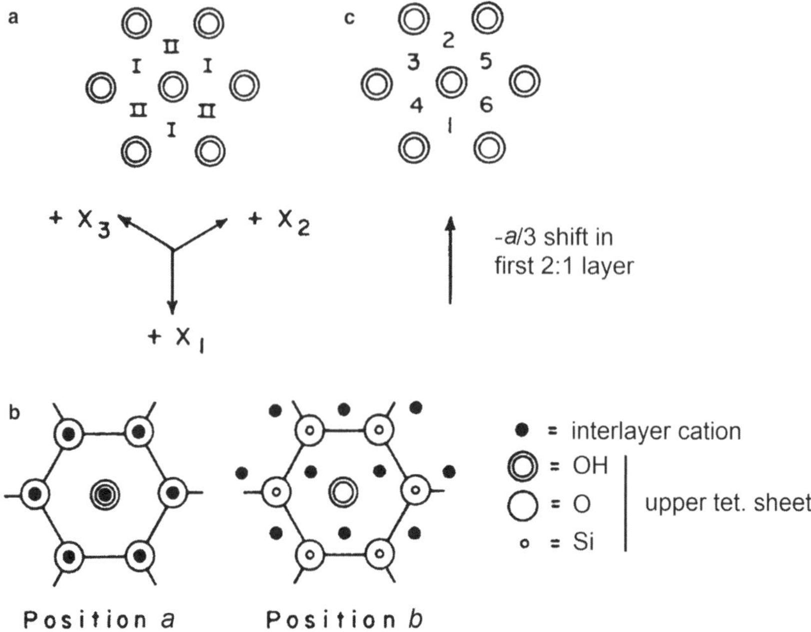

Fig. 8. Chlorite polytypes. Part (***a***) defines the pseudohexagonal set of axes and shows the positions, in projection, of the octahedral cations of the interlayer sheet for set I and set II. Set I is the same orientation as the octahedral occupancy in the underlying 2:1 layer ($-a_1/3$ shift in the 2:1 layer). The interlayer sheet may be positioned over the underlying 2:1 layer in one of six ways, but the high symmetry of the interlayer makes this positioning reduce to two, labelled as position *a* or position *b* (part ***b***). Note that in position *a*, interlayer cations project on to the underlying Si and the OH group in the centre of the silicate ring. Thus, a symbol, such as I*a* or II*b*, can be used to designate the octahedral set occupied and the positioning of the interlayer. A similar arrangement of stacking the next 2:1 layer above the interlayer occurs, but because of the lower symmetry of the 2:1 layer, these positions are given individual numbers (part ***c***), 1 to 6. These represent six positions of the OH group (silicate ring centre) of the bottom tetrahedral sheet in the upper 2:1 layer in projection onto the interlayer OH groups. Other forms of these symbols have been used, such as repeating the *a* or *b*, but to designate the positioning of the upper 2:1 layer, and/or using the 1 to 6 numbers to further designate monoclinic or triclinic symmetry (modified slightly from Bailey, 1988b; with permission of The Clay Minerals Society).

3.5. Vermiculite

Layer stacking in vermiculite has similar nomenclature to that of chlorite, but vermiculites have various stacking sequences that depend on relative humidity in addition to the position of the interlayer cations. Different degrees of stacking order/disorder of samples from different localities or even from samples within the same locality are common. Shirozu & Bailey (1966) used a sample from Llano, Texas (USA), that showed a two-layer structure with shifts within each layer of $-a_1$, an interlayer sheet orientation of I*a*, and the displacement of successive 2:1 layers that alternate by

$+b/3$ and $-b/3$. The interlayer Mg cations partially occupy each of the trioctahedral sites statistically by 0.41 Mg atoms and H_2O molecules statistically occupy 3.72 of the six possible coordinating sites at the studied relative humidity with H bonding to the basal oxygen plane. Interestingly, partial Al ordering occurs in the tetrahedral sites and the occupied Mg interlayer site preferentially associates with the adjacent Al-rich tetrahedral site. In chlorite, the I*a* structure is considered less stable than the more common II*b* and I*b* structures because of the repulsions between Si-rich tetrahedra and the interlayer cations, but these repulsions are minimized by the ordering in vermiculite and the partial interlayer occupancy. De la Calle *et al.* (1988), and later confirmed by Argüelles *et al.* (2010), showed that vermiculite from Santa Olalla, Spain, has a large number of stacking faults and that the regular alternation of $+b/3$ and $-b/3$ shifts is completely disrupted and approaches a stacking sequence where the $b/3$ shifts are random. This is referred to as semi-random stacking; see section 4.1.

3.6. Turbostratic (disorder) effects: smectite

Smectite minerals are generally recognized as having turbostratic disorder, where stacking of each layer parallels adjacent layers, but the layers are displaced and/or rotated in random amounts to adjacent layers. Turbostratic disorder does not occur where the layer-to-layer bonding is strong, such as in the micas where there is an interlayer cation in each ditrigonal ring and high layer charge or in chlorite where hydrogen bonding and electrostatic interactions occur between the 2:1 layer and the interlayer. In addition, in part also because of the low net charge on the layer, smectite stacking may involve different layer-to-layer distances unless care is taken to maintain a single type of exchangeable cation under controlled environmental conditions. However, particles having different layer-to-layer distances affect the diffraction pattern differently from turbostratic stacking (see section 4.2.1).

3.7. Talc and pyrophyllite

Talc/pyrophyllite are unlike the 1:1 (*e.g.* serpentines) or some of the 2:1 layer structures (*e.g.* chlorite), which require optimal hydrogen bonding between layers, or 2:1 layer silicates that have layers anchored by interlayer cations (*e.g.* micas). Instead, regular stacking in the talc/pyrophyllite group appears related to the minimization of Si^{4+} to Si^{4+} repulsive forces across the interlayer (Zvyagin *et al.*, 1969) and to van der Waals interactions to attract the layers (and a small electrostatic attraction as well; Giese, 1975). Single-crystal studies of talc-1*A* (Perdakatsis & Burzlaff, 1981) and pyrophyllite-1*A* (= 1*Tc*, Lee & Guggenheim, 1981) have confirmed that the 2:1 layers are offset so that Si atoms do not superpose across the interlayer over Si atoms of adjacent layers; the related shift is approximately $1/3a$ along one of the three pseudohexagonal axes. This shift also projects four of the six basal oxygen atoms within the hexagonal ring of one sheet onto the middle of the tetrahedral edges on the adjacent sheet (Bailey, 1984). Unlike the micas, there are no six- or twelve-fold interlayer sites in these minerals, and thus talc and pyrophyllite should not be considered as interlayer-cation-deficient analogues of the micas. A two-layer, monoclinic form of pyrophyllite occurs naturally, and stacking sequences based on TEM analysis indicate that other forms can occur; see Evans & Guggenheim (1988) for a summary.

3.8. Interstratifications, including intercalations

The terms 'interstratification', 'mixed layering', and 'interlayering' are used collectively to describe the stacking of two or more assemblages of layers (*e.g.* 1:1 layer of kaolinite composition and 2:1 layer of smectite composition; 2:1 layer of muscovite composition and 2:1 layer of montmorillonite composition). These terms involve how different types (1:1, 2:1) of layers or different chemical compositions of the same type of layer form a stacking sequence or assemblage. For cases where 'chlorite' is being considered as a component, both the 2:1 layer and the interlayer are considered as a single unit. Although the interlayer is never properly referred to as 'brucite' nor is the 2:1 layer referred to as 'talc' in chlorite (appropriate usage is 'brucite-like' or 'talc-like') it is permissible to describe the components of an interstratified stacking sequence by mineral names, as in 'illite-smectite' interstratifications. Many, if not all of these interstratification assemblages, probably originate *via* an Ostwald step rule as a (kinetic) response to the progress of a reaction (*e.g.* see Essene & Peacor, 1995) rather than a simple effect of temperature. Clearly, time, temperature and availability of chemical components determine transitional series. Interstratification assemblages involve random or semi-random patterns of interleaving, whereas an ordered, regular pattern of two layers is considered a mineral and a mineral name may be assigned (Table 2) following the criteria of Bailey (1981). A recent summary of the formation mechanisms of interstratified clay minerals, especially from TEM studies and descriptions of disorder, may be found in Meunier (2010) and other chapters in that volume and are not considered further here.

Table 2. Regularly interstratified phyllosilicates (from Guggenheim *et al.*, 2006).

Layer types	Group affiliation of layers	Octahedral character	Species
	Interstratifications of alternating layers in 50-50 proportions		
1:1	none		
2:1	pyrophyllite-smectite[1]	dioctahedral-dioctahedral	brinrobertsite
	talc-smectite	trioctahedral-trioctahedral	aliettite
	talc-chlorite	trioctahedral-trioctahedral	kulkeite
	mica-smectite	dioctahedral-dioctahedral	rectorite
	biotite-vermiculite	trioctahedral-trioctahedral	hydrobiotite
	chlorite-smectite	trioctahedral-trioctahedral	low-charge corrensite
	chlorite-vermiculite	trioctahedral-trioctahedral	high-charge corrensite
	chlorite-smectite	dioctahedral on average[2]	tosudite
1:1 and 2:1	serpentine-chlorite	trioctahedral-trioctahedral	dozyite

[1] group affiliations are given from the smallest d_{001} component in the natural state to the larger

[2] rigorously defined as having a total octahedral population between 6.0 and 7.0 on the basis of $O_{20}(OH)_{10}$. Thus, possible combinations are: di,dioctahedral chlorite/dioctahedral smectite, di,dioctahedral chlorite/trioctahedral smectite, di,trioctahedral chlorite/dioctahedral smectite, and tri,dioctahderal chlorite/dioctahedral smectite. There are no known occurrences of tri,dioctahedral chlorite.

3.8.1. Reichweite

The concept of Reichweite is based on both the probability of finding layer B after layer A for a two-component layer assemblage and the influence of the preceding layer, and this represents a short-range order parameter. However, a basic assumption for the statistics involved in determining Reichweite is that no influence from a previous layer type needs to be considered in this analysis, *i.e.* any influence, such as atom-to-atom interactions, may be ignored. Two parameters define the distribution of layers, the Reichweite, R, and a bulk composition term. The Reichweite concept can be useful in numerically identifying the ordering types of two-component assemblages and was popularized by Reynolds (1980). For example, the probability based on a resultant coin toss of either heads or tails is 0.5 because it is a two-sided coin and there is no influence of a preceding result. In this case, the Reichweite is given as 0.0 (random). In contrast, for a three-unit assemblage ABBABB . . ., the first layer has an effect on two additional layers and R is equal to 2 or $R2$; a four-unit assemblage ABAAABAA . . . has an R value of 3 (or $R3$). The ABAB . . . sequence is $R = 1$ or $R1$. Thus, the R indicates the most distant layer of an assemblage that affects the probability of finding a final layer which is influenced by the first (not necessarily identical to the first, however).

Moore & Reynolds (1997) noted that randomness can occur if the proportions of each layer differ from the average of the above ideal sequences, and this represents the compositional aspect. For example, if 60% of the layers are A layers and $R = 1$, some disorder occurs because there is an insufficient number of B layers to allow for perfect alternation. In this case, the B layers are separated by one or more A layers. Probabilities may still be determined. For a two-component system, fractions of layers are defined as:

$$P_A + P_B = 1 \qquad (1)$$

Now, junction probabilities can be defined (Reynolds, 1980) where there is a total of four possibilities: $P_{A.B}$, $P_{B.A}$, $P_{A.A}$ and $P_{B.B}$. For example, $P_{A.B}$ is read as the probability of layer B following layer A. To find an actual AB pair in the crystal, the term is designated P_{AB}, and this term is equal to $P_A P_{A.B}$. Reynolds (1980) gives the following junction probability equations:

$$P_{A.A} + P_{A.B} = 1 \qquad (2)$$

$$P_{B.B} + P_{B.A} = 1 \qquad (3)$$

$$P_{AB} = P_{BA} = P_A P_{A.B} = P_B P_{B.A}; \quad P_{A.B} = (P_B P_{B.A})/P_A \qquad (4)$$

Equation 2 states that an A or a B must follow an A layer, and equation 3 is analogous for a B layer. Thus, if a compositional parameter is deduced, *e.g.* from an energy dispersive system (EDS) analysis, such as $P_A = 0.3$, and a junction probability of $P_{B.B}$ of 0.8 is used, the probability of two A layers next to each other can be determined: for example, then $P_B = 0.7$ (eq. 1), $P_{B.A} = 1 - 0.8 = 0.2$ (equation 3), $P_{A.B} = [(0.7)\ (0.2)]/0.3 = 0.467$ (equation 4), and $P_{A.A} = 1 - 0.467 = 0.533$ (equation 2), which is the probability of layer A following layer A. Reynolds (1980) and Drits (1997) offer additional details and examples.

3.8.2. Exchange and solvation in swelling clays

In addition to interstratifications of layers and layer sequences, processes may also involve interlayer chemistry: 'intercalation' is the general term for the movement of an atom, ion, or molecule into a host phase, such that it leaves the host layers essentially unchanged and the stacking generally intact. For the purpose of this chapter, a swelling clay mineral may be considered the host, with the intercalation reaction involving the interlayer. Intercalation can be either an exchange reaction or a 'solvation' reaction (the union of a dissolved substance, a liquid such as ethylene glycol, and a solvent, the clay solid portion) without exchange. Because the position, intensity, and shape of diffraction peaks are dependent on the assemblage and periodicity of 'layers', whether those assemblages involve 1:1 or 2:1 layers or layers defined in part by the interlayer spacings, the applications discussed above that relate to interstratifications can be applied also to layer spacings defined in part by an inhomogeneous distribution of H_2O molecules in the interlayers of a clay mineral. Exchange and solvation reactions and their effects on the colloid properties of clay in solutions have been reviewed partially by MacEwan & Wilson (1980), and more completely in volumes by Güven & Pollastro (1992), Mermut (1994) and Sawhney (1996), and this topic is beyond the scope of the present chapter. Bergaya & Lagaly (2011) consider intercalations in swelling and non-swelling clay minerals, including pillared clay minerals and nanocomposite polymers, in this volume, and Mottana & Aldega (2011) discuss intercalation processes by using X-ray absorption spectroscopic (XAS) techniques in a chapter elsewhere in this volume; see also Putnis (1992) for a salient, but short description of XAS and Gates (2006) for details of the XAS technique and applications to clays and clay minerals.

4. Order and disorder effects on diffraction

4.1. Order/disorder in stacking

Stacking in most phyllosilicates (smectite is an important exception) is somewhat fixed by electrostatic interactions of the interlayer cation (*e.g.* mica), van der Waals interactions (*e.g.* talc), hydrogen bonding (*e.g.* 1:1 layer silicates, chlorite), or the combination of these. Therefore, the registry of adjacent layers parallel to the (001) plane is not truly random, and the term used to describe this is 'semi-random' stacking.

4.2. General diffraction characteristics

Classes of X-ray reflections for the hydrous phyllosilicates can be used to identify types of disorder in stacking (Fig. 9). The basic 2:1, trioctahedral, platy (ideal) phyllosilicate structure involves one or more octahedral sheets, and these sheets contain octahedral cations and anions that repeat at $b/3$ intervals (based on an orthohexagonal cell of

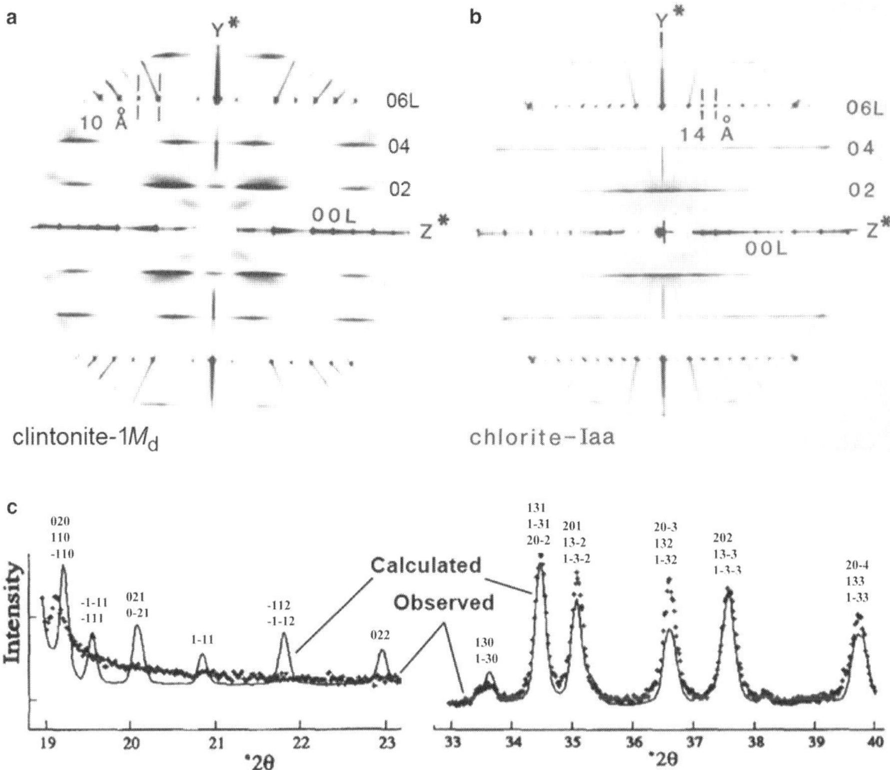

Fig. 9. XRD patterns illustrating stacking disorder effects in mica and chlorite: (***a***) the effect of stacking disorder in a precession photograph of clintonite (mica); and (***b***) semi-random stacking in chlorite (from Bailey, 1988b). Note that both overexposed photographs show sharp reflections of only the $k = 3n$ reflections. The streaked reflections will not appear in powder diffractometer tracings. Note the difference in the calculated and observed patterns in (***c***). Precession photographs use Mo radiation; powder diffraction tracings use Cu radiation. (Parts ***a*** and ***b*** from Bailey, 1988b; part **c** modified from Walker & Bish, 1992; with permission of The Clay Minerals Society.)

$a \approx 5$ Å, $b \approx 9.5$ Å). In micas, the anions of this sheet also either link to the tetrahedral cations (apical oxygen atoms) or reside in the centre of the silicate ring (OH groups), and thus they also repeat at $b/3$ intervals. In contrast, the basal oxygen atoms do not repeat at $b/3$ intervals. Although all these atoms contribute to the intensity of $k = 3n$ reflections, those atoms that repeat at $b/3$ intervals act in phase and dominate to produce relatively strong reflections (Bailey, 1988b). In addition to being relatively strong, these reflections are sharp owing to the strong $b/3$ spacing (periodicity) within the sheet components and because of the semi-random nature involving bonding connecting the layers. Workers involved in polytype derivation refer to these reflections as 'family' reflections, *e.g.* see Nespolo & Ďurovič (2002) and Ďurovič (1997). Basal oxygen atoms, which do not repeat at $b/3$ intervals, and tetrahedral cations contribute to

$k \neq 3n$ reflections ('non-family' reflections), whereas the octahedral cation sheet does not. Thus, the tetrahedral sheet dominates these reflections. The $k \neq 3n$ reflections provide important information about the details of the stacking and thus the polytype, whereas the $k = 3n$ reflections provide general symmetry information. Therefore, stacking disorder affects the $k \neq 3n$ reflections (see Fig. 9), at one extreme where perfect order occurs to produce sharp reflections, to the opposite extreme of complete disorder to produce continuous rods parallel to the [001]* direction. For the most part, the general arguments hold for 1:1 platy phyllosilicates, as well as for dioctahedral varieties. Thus, a qualitative interpretation of the diffraction pattern is possible for the easy separation of the effects from the octahedral sheet vs. the tetrahedral sheet.

The (00*l*) reflections remain sharp regardless of stacking errors because there is no lateral-displacement information about the layers where *h* and *k* are equal to zero. Thus, the 00*l* reflections define the basic stacking unit spacing, *e.g.* 10 Å for micas, 14 Å for chlorites. The $k \neq 3n$ reflections are useful to define the repeat distance of the polytype. Therefore, a powder diffractometer with an oriented aggregate sample mount shows the type of phyllosilicate present (*i.e.* 00*l* information only) but there is no polytype information available in the data; only a random powder sample mount with *h* and *k* information provides data on the polytype.

The space-group determination of all structures uses the systematic presence (or systematic absence) of reflections. The lack of information about essential structural features, such as lateral displacements of the layers, when considering (00*l*) reflections may produce a systematic absence that is not necessarily related to the space group symmetry and these absences are referred to as 'non-space group absences'. Although common in phyllosilicate structures, non-space group absences are not a common characteristic of non-layer structures.

4.2.1. *Turbostratic effects*

Details about the origin of two-dimensional (X-ray) diffraction bands are given in Brindley (1980) and reviewed by Reynolds (1989) and Moore & Reynolds (1997), and these details are beyond the scope of this overview. The bands occur from turbostratic disorder, and differ in appearance from effects of semi-random order. Diffraction bands affect those reflections originating from the layer, *i.e.* *h*00, 0*k*0 and *hk*0 type reflections. Like semi-random stacking effects, the reciprocal lattice reflection is represented by a rod along [001]* rather than a point, but different reflections are involved when compared to semi-random stacking effects. The two-dimensional diffraction band has a characteristic shape with a sharp low-angle intensity and a diminishing-intensity tail at higher angles (Fig. 10). The *hk* position is located approximately at the low-angle position of the maximum intensity within the band, and the diminishing-intensity tail approximately correlates to intensity of the rod along [001]*. Therefore, diffraction bands are designated by their (Miller index) *h* and *k* value which describes the approximate *hk* position, such as 02 or 13. For an orthohexagonal cell, *h*,*k* index pairs with the same *d* values follow the relationship: $3h^2 + k^2$ (this equation is derived from the general formula for the orthorhombic system of $1/d_{hkl2} = h^2/a^2 + k^2/b^2 + l^2/c^2$ and $b = a\sqrt{3}$ and $l = 0$). For example, where $h = 1$ and $k = 1$,

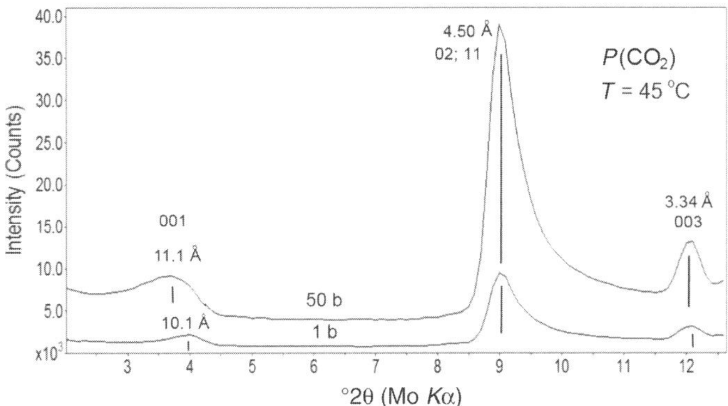

Fig. 10. Powder diffraction patterns show the same montmorillonite (Na-exchanged smectite, SWy-2) under two environmental conditions: one at $P(CO_2) = 50$ bars, $T = 45°C$, at near-dry conditions, and the other at ambient pressure with other conditions identical. Labels for each peak include index and d value. Note that although the interlayer spacing may change, the 02; 11 diffraction band remains fixed at $d = 4.50$ Å. The diffraction band originates from within the layer, and thus is not influenced by layer-spacing changes (unpublished data, Giesting, Guggenheim & Koster van Groos).

the diffraction band is designated as 02; 11; $1\bar{1}$ (or more often, using only the positive indices, 02; 11 or less commonly, $02l$; $11l$) as each designation has a solution of $3h^2 + k^2 = 4$. Note that variations in layer-to-layer distances do not affect the two-theta positions of these bands because diffraction bands originate from the layer only.

4.2.2. Diffraction effects of interstratifications: rational and irrational effects

A regular interstratification of layer A and layer B produces a superstructure with d_{00l} values, d_{AB}, of $d_A + d_B$ along the stacking direction. The resulting superstructure produces additional X-ray reflections in comparison to the simple mixture. If the spacing is sufficiently regular to produce a large number of relatively sharp (00l) reflections (>10) based on a coefficient of variability (see below), the phase can be given mineral status (Bailey, 1981). These reflections are essentially 'rational' (integral) and follow the Bragg equation.

Ideally, if layer A is physically mixed (*i.e.* with separate phase boundaries) with layer B, the resulting diffraction pattern will be approximately the sum of the diffraction patterns of each component (if absorption effects of B on A or A on B are ignored or are small). In contrast to physical mixtures, Mering (1949) qualitatively described the effect of stacking by considering the randomness of an interstratification on the diffraction pattern. For these materials, domains of layer A have parallel boundaries with B layers and the domains are well aligned along the [001] direction. Therefore, X-ray waves from the components interfere with each other and produce a pattern that differs from a simple physical mixture. Nonetheless, the resultant pattern can be derived as a function of the peak positions of the end members, layer A and layer B,

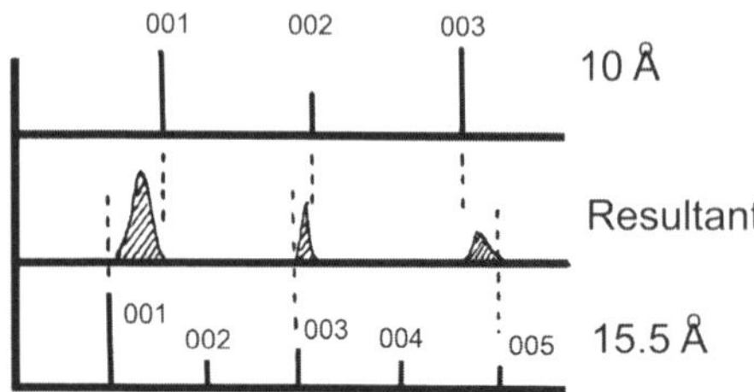

Fig. 11. Mering principle illustrated in a sketch of a powder pattern by Brindley (1981). The top shows an XRD 'pattern' of a pure 10-Å (mica) phase, and the bottom shows a pattern for a pure 15.5-Å (smectite) phase. Each vertical 'stick' represents a sharp reflection. The resultant pattern shows the case where these phases are randomly interstratified with equal numbers of each layer type. See text for explanation (modified from Brindley, 1981; with permission of the Mineralogical Association of Canada).

and their proportion in the mixture. For example, for a completely random mixture of an interstratification of A and B: $d_{ave} = x\,d_A + (1 - x)\,d_B$, where x is the proportion of A and $(1 - x)$ is the proportion of B layers, and in this case $x = 0.5$. The intensity distribution of the resultant peak involves the intensity of the interacting peaks from A and B and the volume of each component. Brindley (1981) illustrated the concept (Fig. 11) by considering two interstratified phases with d_{001} values of 10 Å (mica, top pattern) and 15.5 Å (smectite, bottom pattern). The resultant (middle pattern) shows 'compromise reflections' derived from the patterns of mica and smectite. When such resultant patterns are indexed, the peaks are labelled by the contribution of each component with the smaller spacing first (not shown in the figure). Thus, 002/003 implies the compromise between the 002 of mica and the 003 of smectite. Widely separated peaks between mica and smectite would not produce a discernible compromise peak (*e.g.* 002 of smectite and either 001 or 002 of mica) as the resultant would merge into background (*i.e.* very broad). Broad compromise peaks result where two peaks occur and they are moderately separated in 2θ. A resultant compromise peak (*e.g.* 002/003) between two closely spaced peaks may be sharp, and an isolated but very strong peak would produce a resultant peak that is not displaced from the original phase. Therefore, a pattern derived from a random interstratification may appear to have both 'rational' (integral) and 'irrational' (non-integral) reflections and variations in peak widths. The Méring approach is especially useful as a first step in interpreting a diffraction pattern with irrational reflections. More precise determinations are possible using computer programs such as *NEWMOD* (Reynolds, 1985), *NEWMOD+* (Yuan & Bish, 2010), and a program to assist in obtaining structure parameters by Plançon & Roux (2010).

Bailey (1981) defined a coefficient of variation (CV) for d_{00l} reflections which is defined as $CV = 100s/x_{ave}$, where the standard deviation, s, is equal to $\{\Sigma(x_i - x_{ave})^2/(n - 1)^2\}$, x_i is an individual value for the observed $l \times d_{001}$ reflection, x_{ave} is the mean of the x_i values, l is the order of the reflection, and n is the number of x_i values. The value of CV <0.75 defines the mineral as a new species because it has sufficient regularity of the alternation pattern of A and B layers based on peak position. However, the use of peak location and peak broadening can be used semi-quantitatively to detect small numbers of interstratified layers. Moore & Reynolds (1997) devised a broadening descriptor or a prediction for the peak width, Q, that is more sensitive to interstratifications (to ~<15% of one component) than simple peak location. If the interstratification is dominated by layer A, then Q is equal to the deviation of $l_A\{d_{001B}/d_{001A}\}$ from n, where l_A is the integer value of the order of the reflection as dominated by layer A, d_{001} is the first order spacing of either an ideal A or B, and n is an integer closest to the value of $l_A\{d_{001B}/d_{001A}\}$. Q values for $00l$ reflections can range from 0.0 (a reflection with no contribution from layer B and essentially B and A contributions superpose) to 0.5 (maximum effect of B layer contributions because moderately separated peaks produce a broad compromise reflection). Comparisons to a sample under study involving A and B layering can then be made based on expected peak broadening from Q. To avoid broadening effects from particle size (relating to particle thickness where a thin crystal shows peak broadening), which is related to θ, line breadths may be multiplied by cos θ. Moore & Reynolds (1997) discussed also how Q may relate to the quantitative analysis of small numbers of interstratified layers, which is beyond the scope of the present chapter.

Interstratifications have been discussed in terms of layer A and layer B, and the implication is that the layers refer to clay mineral spacings similar to those of Tables 1 and 2. However, the discussion applies equally to any two layers with different layer spacings, including swelling clays with spacings derived after sample preparation, such as after (ethylene glycol) solvation or heating. Lanson (2011, this volume) covers additional details of the effect of defects and layer stacking.

5. Modulated phyllosilicates, polysomatic structures, and cylindrical structures

5.1. Overview

A modulated phyllosilicate is a layer silicate structure where a periodic perturbation exists (Guggenheim & Eggleton, 1987, 1988; Martin *et al.*, 1991), and this perturbation usually involves the formation of a superstructure. Perturbations may involve discontinuities in the tetrahedral sheets, with tetrahedral inversions that connect sheets across the interlayer region, but with a continuous octahedral sheet to maintain the planar- or layer-like character of the phyllosilicate. Guggenheim & Eggleton (1987, 1988) subdivided this group with linkage configurations that are 'island-like' (*e.g.* for 2:1 layers, stilpnomelane; for 1:1 layers, greenalite) and 'strip-like' (*e.g.* for 2:1 layers, minnesotaite; for 1:1 layers, antigorite), although there are more complex examples (*e.g.* for 2:1 layers, bannisterite) that do not fall under either linkage configuration. Another type of modulation/perturbation involves vacant regions where the octahedral sheet is discontinuous and out-of-plane displacements occur. This type is more pyribole-like (*i.e.* palygorskite-sepiolite), and for these structures to maintain their layer-like characteristics, the structure requires a continuous plane of basal oxygen atoms. Table 3 (Guggenheim *et al.*, 2006) summarizes the structural varieties.

5.2. Modulated 2:1 phyllosilicates involving a continuous octahedral sheet

The origin of modulated 2:1 phyllosilicates, those with continuous octahedral sheets, involves a misfit between a tetrahedral sheet that is too small and a larger octahedral sheet, and the structural response is usually dramatic with tetrahedral inversions. Compositions that favour these modulated phyllosilicates tend to be ferrous iron and/or

Table 3. Classification of non-planar hydrous phyllosilicates (from Guggenheim *et al.*, 2006).

Layer type	**Modulated component**	**Linkage configuration**	**Unit layer, c sin β value**	**Traditional affiliation**	**Species**
			Modulated structures		
1:1 layer	Tet. Sheet	Strips	7 Å	Serpentine	antigorite, bemenitite
		Islands	7 Å	Serpentine	greenalite, caryopilite, pyrosmalite, manganpyrosmalite, ferropyrosmalite, friedelite, megillite, schallerite, nelenite
		Other		None	None
2:1 layer	Tet. Sheet	Strips	9.5 Å	Talc	minnesotaite
			12.5 Å	Mica	ganophyllite, eggletonite
		Islands	9.6–12.5 Å	Mica/ complex	zussmanite, parsettensite, stilpnomelane, ferrostilpnomelanene, ferristilpnomelane, lennilenapeite
		Other	12.3 Å	None	bannisterite
			14 Å	Chlorite	gonyerite
	Oct. Sheet	Strips	12.7–13.4 Å	Pyribole	sepiolite, loughlinite, falcondoite, palygorskite, yofortierite
		Rolled and spheroidal structures			
1:1 layer	None	Trioctahedral	–	Serpentine	chrysotile, pecoraite
		Dioctahedral	–	Kaolin	halloysite (non-planar)

manganese rich (large octahedra) and silicon rich (small tetrahedra). Thus, most of these structures occur in environments that are Al poor, because Al substitutions in either the octahedral sites or the tetrahedral sites will produce a congruency of sheets. Figure 12 summarizes schematically possible mechanisms available to a structure to reduce or minimize misfit. Only a short discussion regarding strip structures is given here.

Guggenheim & Eggleton (1987, 1988) have shown that modulation diversity can be directly related to misfit, as long as layer-to-layer interactions are minimal as is generally the case for 2:1 varieties and some 1:1 structures. For example, minnesotaite (Guggenheim & Eggleton, 1986), crystallizes in two forms relating to the size of the octahedral sheet (Fig. 12e): Iron- and Mg-rich specimens produce a primitive cell (*P*) with strip-widths of four tetrahedra along the [100]* direction; in contrast, a *C*-centred cell is produced where the octahedral sheet is composed mostly of Fe and strips containing three tetrahedra alternate with widths of four tetrahedra. Minnesotaite-*P* has potentially less misfit and thus ten tetrahedra $(4 + 1 + 4 + 1)$ spanning nine octahedra and a chemical formula of $(Fe,Mg)_{30}Si_{40}O_{96}(OH)_{28}$. Single tetrahedra are inverted. Minnesotaite-*C*, with potentially greater misfit, has a $(4 + 1 + 3 + 1)$ tetrahedral strip sequence with a resultant chemical formula of $(Fe,Mg)_{27}Si_{26}O_{86}(OH)_{26}$. Samples with greater amounts of Mg can probably produce greater spans, and Guggenheim & Eggleton noted that they did observe the occurrence of strips of five tetrahedra before an inversion. Tetrahedral inversions usually involve additional tetrahedra (when compared to ideal planar structures) that can span the octahedral sheet, and the configurational change affects the tetrahedral cation to octahedral cation ratios. Potential misfit between a large octahedral sheet and a small tetrahedral sheet has been used to successfully predict which layer silicate structures may be modulated; refer to Guggenheim & Eggleton (1987) for more information.

The modulation in strip structures (extra tetrahedra) occurs in only one direction, and thus extra tetrahedra relative to an ideal structure do not explain how misfit is relieved in directions perpendicular to the modulation direction (in the case of minnesotaite, along the [010]* direction). Clearly, more numerous strips per given set of octahedra suggest that octahedra are less constrained where the coordinating oxygen atoms are not shared with tetrahedra; tetrahedral inversions require that OH groups complete nearest-neighbour coordination of octahedra instead of apical oxygen atoms. Strain is also minimized by the offset of tetrahedral sheet sections (as shown in Fig. 12d). Other factors, such as cation order (*e.g.* in ganophyllite, Fig. 12f, where ordering of the larger Al cation occurs), allow tetrahedra to span the octahedra along the strip direction.

5.3. Modulated 2:1 phyllosilicates involving a discontinuous octahedral sheet

The palygorskite-sepiolite mineral group has an affinity to 2:1 layer silicates with the following essential structural features (Guggenheim & Krekeler, 2011): a continuous basal oxygen plane, an inverted tetrahedral arrangement that forms pyribole-like ribbons (laterally connected pyroxene-like chains), and a discontinuous octahedral sheet. The discontinuous octahedral sheet (Fig. 13), which is part of the ribbon along with the chains, is terminated by OH_2 anions to complete octahedral coordination around the *M* cations. Although the 2:1 modulated phyllosilicates appear to result from misfit between the tetrahedral sheet and the octahedral sheets, the same is not true for palygorskite and sepiolite. Guggenheim & Krekeler (2011) noted that the average octahedral and tetrahedral bond distances are equal to values found in the micas, which do not have modulations. They reviewed natural occurrences and synthesis conditions of the palygorskite-sepiolite group minerals and found that these minerals form in high-alkali environments. Natural environments vary from salt lakes (*i.e.* low-temperature, near-surface aqueous solutions) to agpaitic conditions, where $(Na + K)/Al > 1$ at hydrothermal temperatures of $<350°C$. The large number of OH and OH_2 anions in the structures of the palygorskite-sepiolite group minerals is consistent with an aqueous environment and high OH activity (a_{OH}). This observation suggests that parameters that affect a_{OH} also affect the continuity of the octahedral sheet at high a_{OH}. These parameters possibly include a_{OH}, a_{alkali}, a_{SiO_2} and/or a_{Mg}. Palygorskite and sepiolite deposits commonly occur with trioctahedral smectite (stevensite, saponite), which also requires high a_{OH}, although not necessarily identical values to palygorskite and sepiolite. More data are required to define the importance of a_{OH} in disrupting the continuity of the octahedral sheet in these minerals.

5.4. Modulated 1:1 phyllosilicates

Some modulated 1:1 phyllosilicates (*e.g.* greenalite) result from misfit between tetrahedral and octahedral sheets, similar to the modulated 2:1 silicates. However, Bates (1959; see discussion also in Guggenheim and Eggleton, 1987) recognized that misfit

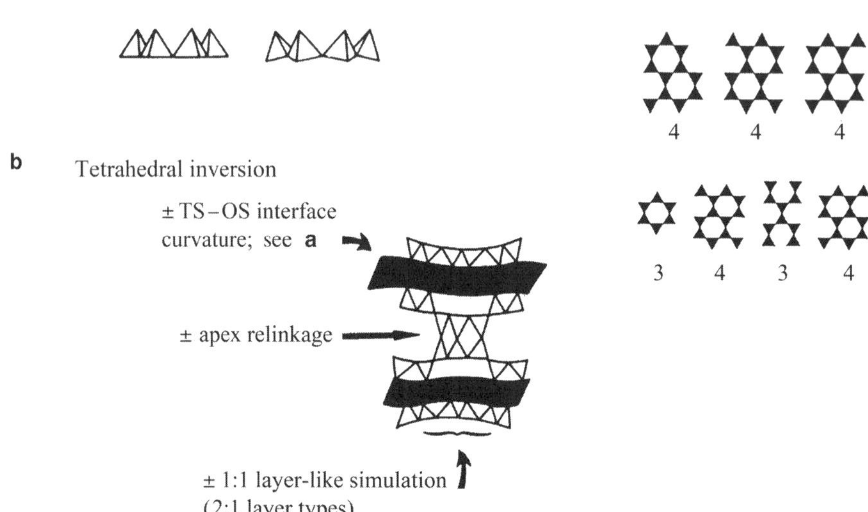

Fig. 12. Chemical and structural methods for tetrahedra and octahedra to obtain congruency for structures with continuous octahedral sheets. The top portion notes that Al substitutions in the octahedral sheet (OS) will generally reduce the octahedral sheet size. In contrast, Al substitutions for Si will increase the tetrahedral sheet (TS) size. Structural mechanisms to obtain congruency include (***a***) limited curving of the octahedral-tetrahedral interface by out-of-plane tetrahedral tilting (*e.g.* minnesotaite), (***b***) tetrahedral inversion, (***c***) tetrahedral ring configuration change from 6-fold to another ring type (*e.g.* zussmanite, with 6- and 3-fold rings, Lopes-Vieira & Zussman, 1969), (***d***) offsets in the tetrahedral and octahedra sheet linkages, (***e***) frequency of modulation (*e.g.* two types of minnesotaite showing a primitive cell of strips four tetrahedra wide and a C-centred cell with 4 + 3 alternating strip widths), (***f***) cation ordering in the tetrahedral sites with the larger Al sites bridging the larger octahedra in one direction and offsets in another (as found in ganophyllite) and (***g***) sheet-thickness adjustments (after Guggenheim & Eggleton, 1988; with permission of the Mineralogical Society of America).

is not a sufficient explanation for the occurrences of rolled, coiled, cylindrical or tubular serpentines, and Wicks & Whitaker (1975) showed compositional overlap between platy serpentine (lizardite) and rolled/coiled serpentine (chrysotile). Unlike the 2:1 modulated phyllosilicates, 1:1 layers are not semi-independent because layers are linked by H bonding. In general, because of cation substitutions, electrostatic interactions occur between the net negative charge of the tetrahedral sheet and the net positive charge of the octahedral sheet across the interlayer region (*e.g.* Gillery, 1959). Recent work

c TS configurational change
± apex relinkage

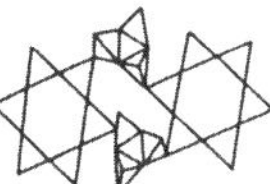

d Displacement of TS sections
across OS (2:1 layer types)

f Cation ordering

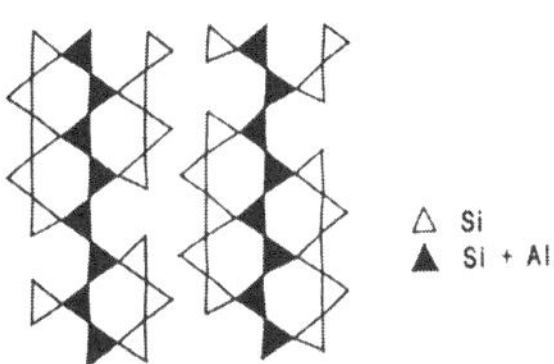

g Sheet thickness adjustment

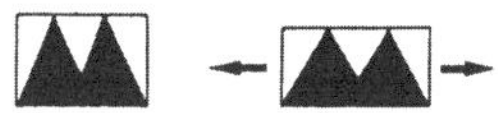

Fig. 12. Continued.

on antigorite (*e.g.* Capitani & Mellini, 2004, 2006; Capitani *et al.*, 2009), which has a wave-like 1:1 layer structure and the reversal of tetrahedral directions at the half-wave, does not show significant evidence of misfit, although different curvature radii

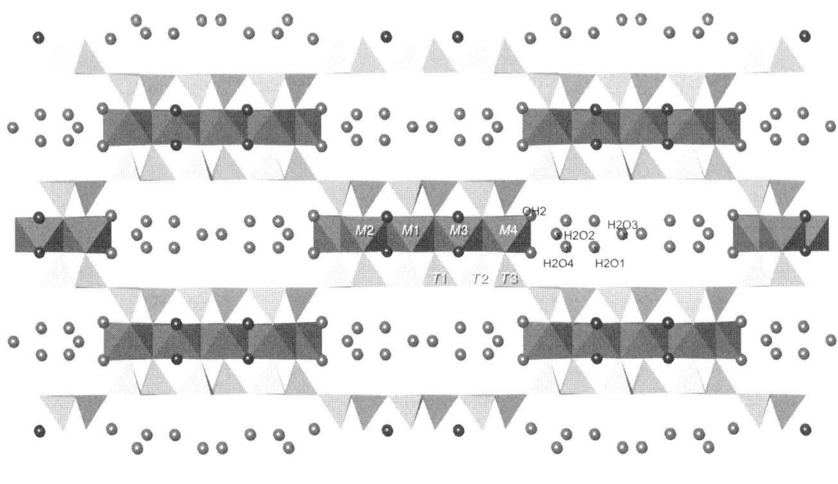

Fig. 13. Palygorskite-sepiolite minerals have a discontinuous octahedral sheet, as shown by sepiolite projected down [001]. Exchangeable zeolite-like H_2O molecules reside in the channels. The OH_2 are coordinated to *M*4 octahedra at the edges of the channels (after Guggenheim & Krekeler, 2011 based on atomic coordinates from Post *et al.*, 2007; with permission).

of the half waves suggest an absence of continuous hydrogen bonding across the interlayer. The crystal chemical causes of some of the morphologies of the kaolin/serpentine group are unknown, although the activity of OH probably plays an important role; see also section 5.5.

5.5. Rolled, tubular, cylindrical, curved 1:1 structures

Non-planar lattices of this type are generally not considered 'modulated' because of the lack of periodicity of any perturbation, and thus are treated here as a special case. Chrysotile occurs in many forms. Cylinders may be in the form of concentric 1:1 layers or hollow fibres generated as a spiral of 1:1 layers, much like a carpet being rolled in spiral form, or cone-in-cone structures (*e.g.* Yada, 1971), and the fibre axis is the *a* axis. The octahedral sheet is always located on the outside of the convex cylinder-like form, suggesting that a smaller tetrahedral sheet is being tilted to accommodate a larger octahedral sheet, although this explanation is probably too simplistic (see above). Based on misfit arguments (Wicks & Whittaker, 1975), the optimal diameter for chrysotile fibres is ~176 Å, and many chrysotile fibres have an outside dimension of ~250 Å (Yada, 1971), with a range from 80 to 280 Å. Brindley (1980) suggested that the range of radii of 40 to 140 Å (diameter = 80 – 280 Å) averages to 90 Å (diameter = 180 Å), and the average seems consistent with misfit arguments. The core of the tubes may be hollow or composed of amorphous material (Martinez & Comer, 1964). Other chrysotile fibres are known to be polygonal in cross section and much larger in diameter, commonly greater than 1000 Å (Cressey & Zussman, 1976). Polygonal chrysotile may have 15 or 30 sectors and 1:1 layers that are flat in each sector, but curved sector separators. Cressey *et al.* (2008, 2010) reported the occurrence of large lizardite-comprised spheres with possible curved layers, and modelled these forms by comparisons to face-centred buckeyballs. Cressey *et al.* (2010) suggested that a disc with five-fold symmetry is the fundamental seed structure for fibres, polygonal fibres, and spheres.

Halloysites are similar in composition to kaolinite but with H_2O between the 1:1 layers. The state of hydration is addressed by adding a suffix based on the d_{001} value, *e.g.* halloysite-7Å, which is dehydrated halloysite, halloysite-7.9Å, halloysite-10Å, *etc.* Morphologies of halloysites are planar, curved, cylindrical, and near spherical, with fibres commonly elongated along the *b* axis and less commonly along the *a* axis. For halloysite, the octahedral sheet is located on the inner side of the rolled layers (in contrast to chrysotile), with the larger tetrahedral sheet presumably attached to the smaller octahedral sheet. Evidence indicates that some fibres are either primary (*e.g.* Kohyama *et al.*, 1978) or secondary as a result of hydration (*e.g.* Singh & Mackinnon, 1996). The experimental study of Singh & Mackinnon, where platy kaolinite is transformed to rolled halloysite, suggests that H_2O disrupts the H bond network linking the 1:1 layers to allow rolling (see also Costanzo & Giese, 1985). Singh (1996) suggested that the rolled form is more stable than an adjustment by tetrahedral rotation to account for lateral misfit because Si–Si repulsions are minimized more effectively by rolling.

5.6. Polysomes, order/disorder and diffraction effects

Whereas polytypes involve building units of one type, polysomes involve units of two (or possibly more) building blocks or modules to form an atomic structure (*e.g.* Baronnet, 1997). Baronnet (1997) noted that compositions and structures of the modules may or may not be different (*e.g.* interstratifications of chlorite/lizardite of the same composition; interstratifications of muscovite/paragonite, both mica structures but with different compositions). Several modulated phyllosilicate structures, especially those belonging to strip structures, are composed of polysome arrangements. For example, minnesotaite (Guggenheim & Eggleton, 1986) forms in two modifications (see above, Fig. 12e), (a) with strips (*i.e.* modules) that are four tetrahedra wide between tetrahedral inversions or (b) strips that alternate in widths of three and four tetrahedra. Regularly ordered arrangements of strips consisting of widths of only four tetrahedra produce a unit cell that is primitive (no systematic absences), whereas alternating strips of widths of three and four tetrahedra produce a unit cell that is C centred (reflections occur in the $hk0$ plane of the $h+k=2n$ type). Transmission electron microscopy (TEM), selected area diffraction (SAD) patterns of the $hk0$ planes of the two types of structures are superficially similar (Fig. 14), but there is an apparent offset caused by the P *vs.* C centering when comparing patterns. Disorder of strip linkages produces streaking parallel to the [100]* direction (note slight streaking in Fig. 14b), but stacking disorder has a similar effect on the diffraction pattern of the $0kl$ net (Fig. 14c) with pronounced streaking along the [001]* direction (see general diffraction effects). Modulated phyllosilicates are readily identified by TEM by examining the superlattice reflections found in SAD patterns of grains lying on the (001) plane; see Guggenheim & Eggleton (1988). Because the superlattice reflections are generally weak in intensity, powder XRD traces do not produce patterns that are very informative about the superstructure,

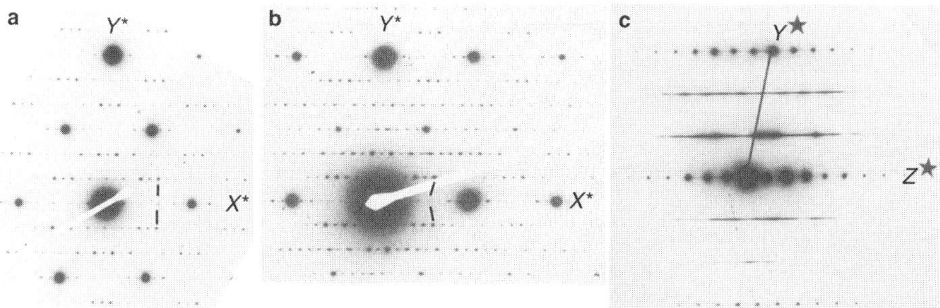

Fig. 14. Minnesotaite diffraction patterns of two polysomes are shown: a P cell is illustrated in (***a***) and a C-centred cell is shown in (***b***). Both are $hk0$ diffraction nets. Note the lines connecting points; vertical (no systematic absences) in (***a***) and offset ($h+k=2n$ reflection present) in (***b***). The primitive cell results from a structure that only has strips four tetrahedra wide along the a axis, whereas the C cell has strips four tetrahedra wide alternating with strips three tetrahedra wide along a. Any strip-width disorder is observed by streaking. Note the weak streaking in (***b***) along the [100]* direction. Part (***c***) shows stacking disorder, observed by streaking along [001]* for $k \neq 3n$ reflections in the $0kl$ net (from Guggenheim & Eggleton, 1986; with permission of the Mineralogical Association of Canada).

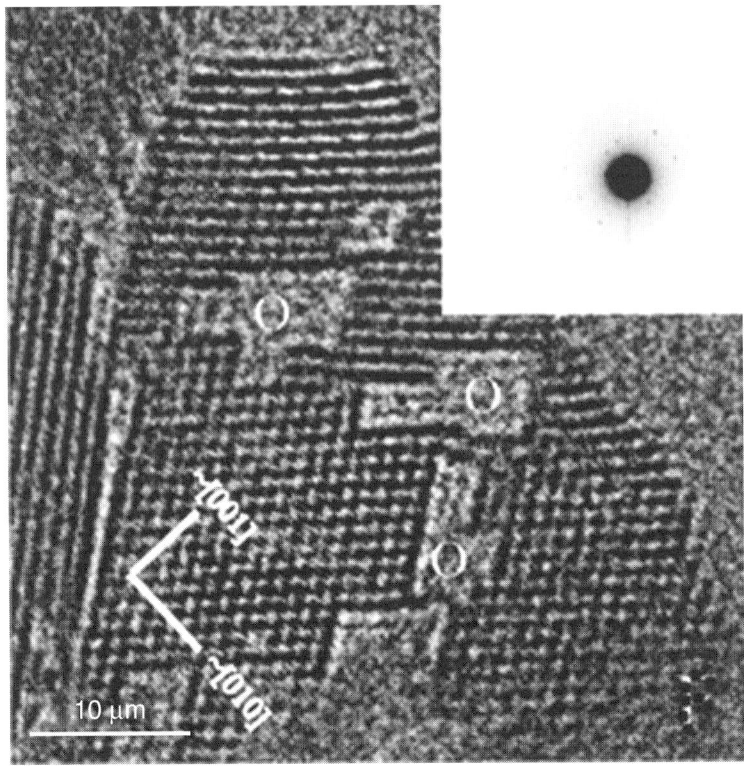

Fig. 15. TEM image approximately along the [100] direction of yofortierite (Mn-rich palygorskite) showing open channel defects, labelled 'O'. These types of defects may affect the adsorption properties of these minerals. The scale is shown in lower left corner of the figure (reprinted from Figure 2D of Krekeler & Guggenheim, 2008, *Applied Clay Science*; with permission from Elsevier).

but such patterns can be used to show the traditional, substructure affiliation (*e.g.* mica, serpentine).

The palygorskite and sepiolite mineral group shows variations in width of the polysomes, polysome omissions (open channel defects), and stacking errors, and these defects can be monitored by TEM. Krekeler & Guggenheim (2008) showed (Fig. 15) that open channel defects occur in sepiolite and yofortierite (= Mn-rich palygorskite) and that these defects are common and vary in size from 3.9 to 75 nm^2. Because of their large size, open channel defects may influence the adsorption of large organic molecules, which normally cannot be exchanged in the regions at the edges of the ribbons owing to steric constraints.

Acknowledgements

I thank P. Giesting for reviewing an early version of the manuscript.

References

Abad-Ortega, M. & Nieto, F. (1995) Extension and closure of the compositional gap between Mn- and Mg-rich chlorites toward Fe-rich compositions. *European Journal of Mineralogy*, **7**, 363–367.

Anderson, C.S. & Bailey, S.W. (1981) A new cation ordering pattern in amesite-$2H_2$. *American Mineralogist*, **66**, 185–195.

Antognozzi, M., Wotherspoon, A., Hayes, J.M., Miles, M.J., Szczelkun, M.D. & Valdré, G. (2006) A chlorite mineral surface actively drives the deposition of DNA molecules in stretched conformations. *Nanotechnology*, **17**, 3897–3902.

Argüelles, A., Leoni, M., Blanco, J.A. & Marcos, C. (2010) Semi-ordered crystalline structure of the Santa Olalla vermiculite inferred from X-ray powder diffraction. *American Mineralogist*, **95**, 126–134.

Bailey, S.W. (1963) Polytypism of the kaolin minerals. *American Mineralogist*, **48**, 1196–1209.

Bailey, S.W. (1969) Polytypism of trioctahedral 1:1 layer structures. *Clays and Clay Minerals*, **17**, 355–371.

Bailey, S.W. (1980) Structures of layer silicates. In: *Crystal Structures of Clay Minerals and their X-ray Identification* (G.W. Brindley & G. Brown, editors). Mineralogical Society, London, pp. 1–123.

Bailey, S.W. (1981) A system of nomenclature for regular interstratifications. *The Canadian Mineralogist*, **19**, 651–655.

Bailey, S.W. (1984) Classification and structures of the micas. In: *Micas* (S.W. Bailey, editor). Reviews in Mineralogy, **13**, Mineralogical Society of America, Washington, D.C., pp. 1–12.

Bailey, S.W. (1988a) Chlorites: Structures and crystal chemistry. In: *Hydrous Phyllosilicates (Exclusive of Micas)* (S.W. Bailey, editor). Reviews in Mineralogy, **18**, Mineralogical Society of America, Washington, D.C., pp. 347–403.

Bailey, S.W. (1988b) X-ray diffraction identification of the polytypes of mica, serpentine, and chlorite. *Clays and Clay Minerals*, **36**, 193–213.

Baronnet, A. (1997) Equilibrium and kintetic processes for polytype and polysome generation. In: *Modular Aspects of Minerals* (S. Merlino, editor). EMU Notes in Mineralogy **1**, European Mineralogical Union, Eötvös University Press, Budapest, pp. 119–152.

Bassett, W.A. (1960) Role of hydroxyl orientation in mica alteration. *Bulletin of the Geological Society of America*, **71**, 449–456.

Bates, T.F. (1959) Morphology and crystal chemistry of 1:1 layer lattice silicates. *American Mineralogist*, **44**, 78–114.

Bergaya, F. & Lagaly, G. (2011) Intercalation processes of layered minerals. In: *Layered Mineral Structures and their Application in Advanced Technologies* (M.F. Brigatti & A. Mottana, editors). EMU Notes in Mineralogy, **11**, European Mineralogical Union and the Mineralogical Society of Great Britain & Ireland, pp. 259–284.

Bickmore, B.R., Rosso, K.M., Nagy, K.L., Cygan, R.T. & Tadanier, C.J. (2003) *Ab initio* determination of edge surface structures for dioctahedral 2:1 phyllosilicates: implications for acid-base reactivity. *Clays and Clay Minerals*, **51**, 359–371.

Bleam, W.F. (1990) The nature of cation-substitution sites in phyllosilicates. *Clays and Clay Minerals*, **38**, 527–536.

Brigatti, M.F. & Guggenheim, S. (2002) Mica crystal chemistry and the influence of pressure, temperature, and solid solution on atomistic models. In: *Micas: Crystal Chemistry and Metamorphic Petrology* (A. Mottana, F.P. Sassi, J.B. Thompson, Jr. & S. Guggenheim, editors). Reviews in Mineralogy and Geochemistry, **46**, Mineralogical Society of America, Washington, D.C. and Accademia Nationale dei Lincei, Roma, Italy, pp. 1–98.

Brigatti, M.F., Galán, E. & Theng, B.K.G. (2006) Structures and mineralogy of clay minerals. In: *Handbook of Clay Science* (F. Bergaya, B.K.G. Theng & G. Lagaly, editors). Developments in Clay Science, **1**, Elsevier, Oxford, UK, pp. 19–86.

Brigatti, M.F., Malferrari, D., Laurora, A. & Elmi, C. (2011) Structure and mineralogy of layer silicates: recent perspectives and new trends. In: *Layered Mineral Structures and their Application in Advanced Technologies* (M.F. Brigatti & A. Mottana, editors), EMU Notes in Mineralogy, **11**, European Mineralogical Union and the Mineralogical Society of Great Britain & Ireland, pp. 1–71.

Brindley, G.W. (1980) Order-disorder in clay mineral structures. In: *Crystal Structures of Clay Minerals and their X-ray Identification* (G.W. Brindley & G. Brown, editors). Monograph **5**, Mineralogical Society, London, pp. 125–195.

Brindley, G.W. (1981) X-ray identification (with ancillary techniques) of clay minerals. In: *Clays and the Resource Geologist* (F.J. Longstaffe, editor). Mineralogical Association of Canada, Toronto, Canada, pp. 22–38.

Bujnowski, T.J., Guggenheim, S. & Brigatti, M.F. (2010) Al-saturated phlogopite: Charge considerations and crystal chemistry. *Clays and Clay Minerals*, **57**, 673–685.

Capitani, G. & Mellini, M. (2004) The modulated crystal structure of antigorite: The $m = 17$ polysome. *American Mineralogist*, **89**, 147–158.

Capitani, G. & Mellini, M. (2006) The crystal structure of a second antigorite polysome ($m = 16$), by single crystal synchrotron diffraction. *American Mineralogist*, **91**, 394–399.

Capitani, G., Stixrude, L. & Mellini, M. (2009) First-principles energetics and structural relaxation of antigorite. *American Mineralogist*, **94**, 1271–1278.

Cheng, L., Fenter, P., Nagy, K.L., Schlegel, M.L. & Sturchio, N.C. (2001) Molecular-scale density oscillations in water adjacent to a mica surface. *Physical Review Letters*, **87**, 156103(4).

Chernosky, J.V., Jr. (1975) Aggregate refractive indices and unit cell parameters of synthetic serpentine in the system MgO-Al_2O_3-SiO_2-H_2O. *American Mineralogist*, **60**, 200–208.

Chevychelov, V.Yu., Botcharikov, R.E. & Holtz, F. (2008) Experimental study of fluorine and chlorine contents in mica (biotite) and their partitioning between mica, phonolite melt, and fluid. *Geochemistry International*, **46**, 1081–1089 (translated from Russian).

Christidis, G.E. (2011) The concept of layer charge of smectites and its implications for important smectite-water properties. In: *Layered Mineral Structures and their Application in Advanced Technologies* (M.F. Brigatti & A. Mottana, editors), EMU Notes in Mineralogy, **11**, European Mineralogical Union and the Mineralogical Society of Great Britain & Ireland, pp. 237–258.

Christidis, G.E. & Eberl, D.D. (2003) Determination of layer-charge characteristics of smectites. *Clays and Clay Minerals*, **51**, 644–655.

Costanzo, P.M. & Giese, R.F., Jr. (1985) Dehydration of synthetic hydrated kaolinite: a model for dehydration of halloysite (10 Å). *Clays and Clay Minerals*, **33**, 415–423.

Cressey, B.A. & Zussman, J. (1976) Electron microscopic studies of serpentinites. *The Canadian Mineralogist*, **14**, 307–313.

Cressey, G., Cressey, B.A. & Wicks, F.J. (2008) Polyhedral serpentine: a spherical analogue of polygonal serpentine? *Mineralogical Magazine*, **72**, 1229–1242.

Cressey, G., Cressey, B.A., Wicks, F.J. & Yada, K. (2010) A disc with fivefold symmetry: the proposed fundamental seed structure for the formation of chrysotile asbestos fibres, polygonal serpentine fibres and polyhedral lizardite spheres. *Mineralogical Magazine*, **74**, 29–37.

de la Calle, C., Suquet, H. & Pons, C.H. (1988) Stacking order in a 14.30 Å Mg-vermiculite. *Clays and Clay Minerals*, **36**, 481–490.

Drits, V. A. (1997) Mixed-layer minerals. In: *Modular Aspects of Minerals* (S. Merlino, editor). EMU Notes in Mineralogy, **1**, European Mineralogical Union, Eötvös University Press, Budapest, pp. 153–192.

Ďurovič, S. (1994) Classification of phyllosilicates according to the symmetry of their octahedral sheets. *Ceramics-Silikàty*, **38**, 81–84.

Ďurovič, S. (1997) Fundamentals of the OD theory. *Modular Aspects of Minerals* (S. Merlino, editor). EMU Notes in Mineralogy, **1**, European Mineralogical Union, Eötvös University Press, Budapest, pp. 3–28.

Eggleton, R.A. & Ashley, P.M. (1989) Norrishite, a new manganese mica, $K(Mn^{3+}_2Li)Si_4O_{12}$, from the Hoskins mine, New South Wales, Australia. *American Mineralogist*, **74**, 1360–1367.

Essene, E.J. & Peacor, D.R. (1995) Clay mineral thermometry – A critical perspective. *Clays and Clay Minerals*, **43**, 540–553.

Evans, B.W. & Guggenheim, S. (1988) Talc, pyrophyllite and related minerals. In: *Hydrous Phyllosilicates (Exclusive of Micas)* (S.W. Bailey, editor). Reviews in Mineralogy, **18**, Mineralogical Society of America, Washington, D.C., pp. 225–294.

Farmer, V.C. (1978) Water on particle surfaces. In: *The Chemistry of Soil Constituents* (D.J. Greenland & M.H.B. Hays, editors). Wiley and Sons, New York, pp. 405–448.

Ferraris, C., Grobety, B. & Wessicken, R. (2001a) Phlogopite exsolutions within muscovite: a first evidence for a higher-temperature re-equilibrium, studied by HRTEM and AFM techniques. *European Journal of Mineralogy*, **13**, 15–26.

Ferraris, G., Gula, A., Ivaldi, G., Nespolo, M., Sokolova, E., Uvarova, Yu. & Khomyakov, A.P. (2001b) First structure determination of an MDO-2*O* mica polytype associated with a 1*M* polytype. *European Journal of Mineralogy*, **13**, 1013–1023.

Ferrow, E.A., London, D., Goodman, K.S. & Veblen, D.R. (1990) Sheet silicates of the Lawler Peak granite, Arizona: chemistry, structural variations, and exsolution. *Contributions to Mineralogy and Petrology*, **105**, 491–501.

Filut, M.A., Rule, A.C. & Bailey, S.W. (1985) Crystal structure refinement of anandite-2Or, a barium- and sulfur-bearing trioctahedral mica. *American Mineralogist*, **70**, 1298–1308.

Foster, M.D. (1962) Interpretation of the composition of and a classification of the chlorites. *United States Geological Survey Professional Paper*, **414-A**, 1–33.

Gates, W.P. (2006) X-ray absorption spectroscopy. In: *Handbook of Clay Science* (F. Bergaya, B.K.G. Theng & G. Lagaly, editors). Developments in Clay Science, **1**, Elsevier, Oxford, UK, pp. 789–864.

Giese, R.F. (1975) Interlayer bonding in talc and pyrophyllite. *Clays and Clay Minerals*, **23**, 165–166.

Giese, R.F. & van Oss, C.J. (1993) The surface thermodynamic properties of silicates and their interactions with biological materials. In: *Health Effects of Mineral Dusts* (G.D. Guthrie & B.T. Mossman, editors). Reviews in Mineralogy, **28**, Mineralogical Society of America, Washington, D.C., pp. 327–346.

Gillery, F.H. (1959) The X-ray study of synthetic Mg-Al serpentines and chlorites. *American Mineralogist*, **44**, 143–152.

Guggenheim, S. & Eggleton, R.A. (1986) Structural modulations in iron-rich and magnesium-rich minnesotaite. *The Canadian Mineralogist*, **24**, 479–497.

Guggenheim, S. & Eggleton, R.A. (1987) Modulated 2:1 Layer Silicates: Review, Systematics and Predictions. *American Mineralogist*, **72**, 726–740.

Guggenheim, S. & Eggleton, R.A. (1988) Crystal chemistry, classification, and identification of modulated layer silicates. In: *Hydrous Phyllosilicates (Exclusive of the Micas)* (S.W. Bailey, editor). Reviews in Mineralogy, **19**, Mineralogical Society of America, Washington, D.C., pp. 675–725.

Guggenheim, S. & Eggleton, R.A. (1998) Modulated crystal structures of greenalite and caryopilite: a system with long-range, in-plane structural disorder in the tetrahedra sheet. *The Canadian Mineralogist*, **36**, 163–179.

Guggenheim, S. & Krekeler, M.P.S. (2011) The structures and micro-structures of the palygorskite-sepiolite group minerals. In: *Advances in the Structure and Chemistry of Sepiolite and Palygorskite* (E. Galán, editor). Developments in Clay Science, Vol. **3**. Elsevier, Amsterdam (in press).

Guggenheim, S., Chang, Y.-H. & Koster van Groos, A.F. (1987) Muscovite dehydroxylation: High-temperature studies. *American Mineralogist*, **72**, 537–550.

Guggenheim, S., Bain, D.C., Bergaya, F., Brigatti, M.F., Drits, V., Eberl, D.D., Formoso, M., Galán, E., Merriman, R.J., Peacor, D.R., Stanjek, H. & Watanabe, T. (2002) Report of the Association Internationale pour L'etude des Argiles (AIPEA) Nomenclature Committee for 2001: Order, disorder, and crystallinity in phyllosilicates and the use of the "crystallinity" index. *Clays and Clay Minerals*, **50**, 406–409, and *Clay Minerals*, **37**, 389–393.

Guggenheim, S., Adams, J.M., Bain, D.C., Bergaya, F., Brigatti, M.F., Drits, V.A., Formoso, M.L.L., Galán, E., Kogure, T. & Stanjek, H. (2006) Summary of recommendations of the nomenclature committees relevant to clay mineralogy: Report of the Association Internationale pour l'Etude des Argiles (AIPEA) Nomenclature Committee for 2006. *Clay Minerals*, **41**, 863–877.

Guggenheim, S., Adams, J.M., Bain, D.C., Bergaya, F., Brigatti, M.F., Drits, V.A., Formoso, M.L.L., Galán, E., Kogure, T. & Stanjek, H. (2007) Summary of recommendations of the nomenclature committees relevant to clay mineralogy: Report of the Association Internationale pour l'Etude des Argiles (AIPEA) Nomenclature Committee for 2006: Corrigendum. *Clay Minerals*, **42**, 575–577.

Guinier, A., Bokij, G.B., Boll-Dornberger, K., Cowley, J.M., Jagodzinski, H., Krishna, P., Zvyagin, B.B., Cox, D. E., Goodman, P., Hahn, Th., Kuchitsu, K. & Abrahams, S.C. (1984) Nomenclature of polytype structures. Report of the International Union of Crystallography Ad-Hoc Committee on the Nomenclature of Disordered, Modulated, and Polytype Structures. *Acta Crystallographica*, **A40**, 399–404.

Güven, N. (1992) Molecular aspects of clay/water interactions. In: *Clay-Water Interface and its Rheological Implications* (N. Güven & R.M. Pollastro, editors). CMS Workshop Lectures, **4**, The Clay Minerals Society, Boulder, Colorado, USA, pp. 1–79.

Güven, N. & Pollastro, R.M., editors (1992) *Clay-Water Interface and its Rheological Implications*. CMS Workshop Lectures, **4**, The Clay Minerals Society, Boulder, Colorado, USA.

Hybler, J., Petricek, V., Fabry, J. & Ďurovič, S. (2002) Refinement of the crystal structure of cronstedtite-$2H_2$. *Clays and Clay Minerals*, **50**, 601–613.

Inoue, A., Meunier, A., Patrier-Mas, P., Rigault, C., Beaufort, D. & Viellard, P. (2009) Application of chemical geothermometry to low-temperature trioctahedral chlorites. *Clays and Clay Minerals*, **57**, 371–382.

Johnston, C. T. (1996) Sorption of organic compounds on clay minerals: A surface functional group approach. In: *Organic Pollutants in the Environment* (B. L. Sawhney, editor). CMS Workshop Lectures, **8**, The Clay Minerals Society, Boulder, Colorado, USA, pp. 1–44.

Johnston, C.T. (2010) Probing the nanoscale architecture of clay minerals. *Clay Minerals*, **45**, 245–280.

Kager, P.C.A. & Oen, I.S. (1983) Iron-rich talc-opal-minnesotaite spherulites and crystallochemical relations of talc and minnesotaite. *Mineralogical Magazine*, **47**, 229–231.

Kogure, T. & Inoue, A. (2005) Determination of defect structures in kaolin minerals by high-resolution transmission electron microscopy (HRTEM). *American Mineralogist*, **90**, 85–89.

Kogure, T., Miyawaki, R. & Banno, Y. (2005) The true structure of wonesite, and interlayer-deficient trioctahedral sodium mica. *American Mineralogist*, **90**, 725–731.

Kohyama, N., Fukushima, K. & Fukami, A. (1978) Observation of the hydrated form of tubular halloysite by an electron microscope equipped with an environmental cell. *Clays and Clay Minerals*, **26**, 25–40.

Krekeler, M.P.S. & Guggenheim, S. (2008) Defects in microstructure in palygorskite-sepiolite minerals: A transmission electron microscopy (TEM) study. *Applied Clay Science*, **39**, 98–105.

Lanson, B. (2011) Modelling of X-ray diffraction profiles: Investigation of defective lamellar structure crystal chamistry. In: *Layered Mineral Structures and their Application in Advanced Technologies* (M.F. Brigatti & A. Mottana, editors), EMU Notes in Mineralogy, **11**, European Mineralogical Union and the Mineralogical Society of Great Britain & Ireland, pp. 151–202.

Lee, J.H. & Guggenheim, S. (1981) Single crystal X-ray refinement of pyrophyllite-1*Tc*. *American Mineralogist*, **66**, 350–357.

Lee, S.S., Guggenheim, S., Dyar, M.D. & Guidotti, C.V. (2007) Chemical composition, statistical analysis of the unit cell, and electrostatic modeling of the structure of Al-saturated chlorite from metamorphosed rocks. *American Mineralogist*, **92**, 954–965.

Levinson, A.A. (1953) Studies in the mica group: Relationship between polymorphism and composition in the muscovite-lepidolite series. *American Mineralogist*, **38**, 88–107.

Levinson, A.A, (1955) Studies in the mica group: polymorphism among illites and hydrous micas. *American Mineralogist*, **40**, 41–49.

Lin, J.-C. & Guggenheim, S. (1983) The crystal structure of a Li,Be-rich brittle mica: A dioctahedral-trioctahedral intermediate. *American Mineralogist*, **68**, 130–142.

Lister, J. & Bailey, S.W. (1967) Chlorite polytypism: IV. Regular two-layer structures. *American Mineralogist*, **52**, 1614–1631.

Lopes-Vieira, A. & Zussman, J. (1969) Further detail on the crystal structure of zussmanite. *Mineralogical Magazine*, **37**, 49–60.

Lyklema, J. (1991) *Fundamentals of Interface and Colloid Science. Volume 1. Fundamentals.* Academic Press, London.

MacEwan, D.M.C. & Wilson, M.J. (1980) Interlayer and intercalation complexes of clay minerals. In: *Crystal Structures of Clay Minerals and their X-ray Identification* (G.W. Brindley & G. Brown, editors). Monograph **5**, Mineralogical Society, London, pp. 197–248.

Martin, R.T., Bailey, S.W., Eberl, D.D., Fanning, D.S., Guggenheim, S., Kodama, H., Pevear, D.R., Środoń, J. & Wicks, F.J. (1991) Report of the Clay Minerals Society Nomenclature Committee: Revised classification of clay materials. *Clays and Clay Minerals*, **39**, 333–335.

Martin, F., Micoud, P., Delmotte, L., Marichal, C., Le Dred, R., de Parseval, P., Mari, A., Fortuné, J-P., Salvi, S., Béziat, D., Grauby, O. & Ferret, J. (1999) The structural formula of talc from the Trimouns deposit, Pyrenées, France. *The Canadian Mineralogist*, **37**, 997–1006.

Martinez, E. & Comer, J.J. (1964) The concentration and study of the interstitial material in chrysotile asbestos. *American Mineralogist*, **49**, 153–157.

Massa, W. (2004) *Crystal Structure Determination*. Springer, Berlin, 210 pp.

Mellini, M. (1982) The crystal structure of lizardite 1*T*: Hydrogen bonds and polytypism. *American Mineralogist*, **67**, 587–598.

Mellini, M., Cressey, G., Wicks, F.J. & Cressey, B.A. (2010) The crystal structure of Mg end-member lizardite-1*T* forming polyhedral spheres from the Lizard, Cornwall. *Mineralogical Magazine*, **74**, 277–284.

Mering, J. (1949) L'interference des rayons X dans les systèmes à stratification désordonnée. *Acta Crystallographica*, **2**, 371–377.

Mermut, A.R., editor (1994) *Layer Charge Characteristics of 2:1 Silicate Clay Minerals*. CMS Workshop Lectures, **6**. The Clay Minerals Society, Boulder, Colorado, USA.

Meunier, A. (2010) Formation mechanisms of mixed-layer clay minerals. In: *Interstratified Clay Minerals: Origin, Characterization & Geochemical Significance* (S. Fiore, J. Cuadros & F.J. Huertas, editors). AIPEA Educational Series, **1**, Digilabs, Bari, Italy, pp. 53–71.

Moore, D.M. & Reynolds, R.C., Jr. (1997) *X-ray Diffraction and the Identification and Analysis of Clay Minerals*. Oxford University Press, New York, 378 pp.

Mottana, A. & Aldega, L. (2011) Advanced techniques to define intercalation processes. In: *Layered Mineral Structures and their Application in Advanced Technologies* (M.F. Brigatti & A. Mottana, editors), EMU Notes in Mineralogy, **11**, European Mineralogical Union and the Mineralogical Society of Great Britain & Ireland, pp. 285–312.

Nazzareni, S., Comodi, P., Bindi, L., Safonov, O.G., Litvin, Y.A. & Perchuk, L.L. (2008) Synthetic hypersilicic Cl bearing mica in the phlogopite celadonite join: A multimethodical characterization of the missing link between di- and tri-octahedral micas at high pressures. *American Mineralogist*, **93**, 1429–1436.

Nelson, D.O. & Guggenheim, S. (1993) Inferred limitations to the oxidation of Fe in chlorite: A high-temperature single-crystal X-ray study. *American Mineralogist*, **78**, 1197–1207.

Nespolo, M. & Ďurovič, S. (2002) Crystallographic basis of polytypism and twinning in micas. In: *Micas: Crystal Chemistry and Metamophic Petrology* (A. Mottana, F.P. Sassi, J.B. Thompson Jr. & S. Guggenheim, editors). Reviews in Mineralogy and Geochemistry, **46**, Mineralogical Society of America, Chantilly, Virginia, USA, pp. 155–272.

Nespolo, M., Takeda, H. & Ferraris, G. (1997) Crystallography of mica polytypes. In: *Modular Aspects of Minerals* (S. Merlino editor). EMU Notes in Mineralogy, **1**, Eötvös University Press, Budapest, pp. 81–118.

Newman, A.C.D. & Brown, G. (1987) The chemical constitution of clays. In: *Chemistry of Clays and Clay Minerals* (A.C.D. Newman, editor). Monograph **6**, Mineralogical Society, London, pp. 1–128.

Park, C., Fenter, P.A., Sturchio, N.C. & Nagy, K.L. (2008) Thermodynamics, interfacial structure, and pH hysteresis of Rb^+ and Sr^{2+} adsorption at the muscovite (001) – solution interface. *Langmuir*, **24**, 13993–14004.

Park, S.-H. & Sposito, G. (2002) Structure of water adsorbed on a mica surface. *Physical Review Letters*, **89**, 085501-1, 3.

Pekov, I.V., Chukanov, N.V., Ferraris, G., Ivaldi, G., Pushcharovsky, D. & Zadov, A.E. (2003) Shirokshinite, $K(NaMg_2)Si_4O_{10}F_2$, a new mica with octahedral Na from Khibiny massif, Kola Peninsula: descriptive data and structural disorder. *European Journal of Mineralogy*, **15**, 447–454.

Perdakatsis, B. & Burzlaff, H. (1981) Strukturverfeinerung am Talk $Mg_3[(OH)_2Si_4O_{10}]$. *Zietschrift für Kristallographie*, **156**, 177–186.

Petit, S., Martin, F., Wiewióra, A., de Parseval, P. & Decarreau, A. (2004) Crystal-chemistry of talc: A near infrared (NIR) spectroscopic study. *American Mineralogist*, **89**, 319–326.

Plançon, A. & Roux, J. (2010) Software for the assisted determination of the structural paramters of mixed-layer phyllosilicates. *European Journal of Mineralogy*, **22**, 733–740.

Post, J.E., Bish, D.L. & Heaney, P.J. (2007) Synchrotron powder X-ray diffraction study of the structure and dehydration behavior of sepiolite. *American Mineralogist*, **92**, 91–97.

Putnis, A. (1992) *Introduction to Mineral Sciences*, Cambridge University Press, Cambridge, England.

Ramsdell, L.S. (1947) Studies on silicon carbide. *American Mineralogist*, **32**, 64–82.

Reynolds, R.C. (1980) Interstratified clay minerals. *Crystal Structures of Clay Minerals and their X-ray Identification* (G.W. Brindley & G. Brown, editors). Mineralogical Society, London, pp. 249–304.

Reynolds, R.C., Jr. (1985) *NEWMOD, a computer program for the calculation of the basal diffraction intensities of mixed-layer clay minerals*. R. C. Reynolds, Hanover, New Hampshire, USA.

Reynolds, R.C., Jr (1989) Diffraction by small and disordered crystals. In: *Modern Powder Diffraction* (D.L. Bish & J.E. Post, editors). Reviews in Mineralogy, **20**, Mineralogical Society of America, Washington, D.C., pp. 145–182.

Rieder, M., Cavazzini, G., D'Yakonov, Yu. S., Frank-Kamenetskii, V.A., Gottardi, G., Guggenheim, S., Koval, P.V., Müller, G., Neiva, A.M.R., Radoslovich, E.W., Robert, J.-L., Sassi, F.P., Takeda, H., Weiss, Z. & Wones, D.R. (1998) International Mineralogical Association Report: Nomenclature of the micas. *Clays and Clay Minerals*, **46**, 586–595.

Robinson, G.W. & Chamberlain, S.C. (1984) The Sterling mine, Antwerp, New York, *Mineralogical Record*, **15**, 199–216.

Ruiz Cruz, M.D. & Sanz de Galdeano, C. (2009) Exsolution microstructures in NH_4-bearing muscovite and annite in gneisses from the Torrox area, Betic Cordillera, Spain. *The Canadian Mineralogist*, **47**, 107–128.

Sainz-Diaz, C.I. (2011) Applications of computational atomistic methods to phyllosilicates. In: *Layered Mineral Structures and their Application in Advanced Technologies* (M.F. Brigatti & A. Mottana, editors), EMU Notes in Mineralogy, **11**, European Mineralogical Union and the Mineralogical Society of Great Britain & Ireland, pp. 203–236.

Schlegel, M.L., Nagy, K.L., Fenter, P., Cheng, L., Sturchio, N.C. & Jacobsen, S.D. (2006) Cation sorption on the muscovite (001) surface in chloride solutions using high-resolution X-ray reflectivity. *Geochimica et Cosmochimica Acta*, **70**, 3549–3565.

Schoonheydt, R.A. & Johnston, C.T. (2011) The surface properties of clay minerals. In: *Layered Mineral Structures and their Application in Advanced Technologies* (M.F. Brigatti & A. Mottana, editors), EMU Notes in Mineralogy, **11**, European Mineralogical Union and the Mineralogical Society of Great Britain & Ireland, pp. 335–370.

Schreyer, W., Fransolet, A.-M. & Abraham, K. (1986) A miscibility gap in trioctahedral Mn-Mg-Fe chlorites: Evidence from the Lienne Valley manganese deposit, Ardennes, Belgium. *Contributions to Mineralogy and Petrology*, **94**, 333–342.

Sawhney, B.L., editor (1996) *Organic Pollutants in the Environment*. CMS Workshop Lectures, **8**, The Clay Minerals Society, Boulder, Colorado, USA.

Shirozu, H. & Bailey, S.W. (1966) Crystal structure of a two-layer Mg-vermiculite. *American Mineralogist*, **51**, 1124–1143.

Singh, B. (1996) Why does halloysite roll? – A new model. *Clays and Clay Minerals*, **44**, 191–196.

Singh, B. & Mackinnon, I.D.R. (1996) Experimental transformation of kaolinite to halloysite. *Clays and Clay Minerals*, **44**, 825–834.

Skipper, N.T., Soper, A.K. & McConnell, J.D.C. (1991) The structure of interlayer water in vermiculite. *Journal of Physical Chemistry*, **94**, 5751–5760.

Skipper, N.T., Soper, A.K. & Smalley, M.V. (1994) Neutron diffraction study of calcium vermiculite: Hydration of calcium ions in a confined environment. *Journal of Physical Chemistry*, **98**, 942–945.

Skipper, N.T., Smalley, M.V., Williams, G.D., Soper, A.K. & Thompson, C.H. (1995) Direct measurement of the electric double layer structure in hydrated lithium vermiculite clays by neutron diffraction. *Journal of Physical Chemistry*, **99**, 14201–14204.

Smith, J.V. & Yoder, H.S., Jr. (1956) Experimental and theoretical studies of the mica polymorphs. *Mineralogical Magazine*, **31**, 209–235.

Spear, F.S., Hazen, R.M. & Rumble, D., III (1981) Wonesite: a new rock-forming silicate from the Post Pond Volcanics, Vermont. *American Mineralogist*, **66**, 100–105.

Sposito, G. (1984) *The Surface Chemistry of Soils*. Oxford University Press, New York, 234 pp.

Sposito, G., Skipper, N.T., Sutton, R., Park, S.-H., Soper, A.K. & Greathouse, J.A. (1999) Surface geochemistry of the clay minerals. *Proceedings of the National Academy of Sciences, USA*, **96**, 3358–3364.

Sykes, M.L. & Moody, J.B. (1978) Pyrophyllite and metamorphism in the Carolina slate belt. *American Mineralogist*, **63**, 96–108.

Takeda, H. & Ross, M. (1975) Mica polytypism: Dissimilarities in the crystal structures of coexisting 1*M* and $2M_1$ biotite. *American Mineralogist*, **60**, 1030–1040.

Toraya, H., Iwai, S., Marumo, F., Daimon, M. & Kondo, R. (1976) The crystal structure of tetrasilicic potasium fluor mica, $KMg_{2.5}Si_4O_{10}F_2$. *Zeitschrift für Kristallographie*, **144**, 42–52.

Toraya, H., Iwai, S. & Marumo, F. (1978) The crystal structures of germanate micas, $KMg_{2.5}Ge_4O_{10}F_2$ and $KLiMg_2Ge_4O_{10}F_2$. *Zeitschrift für Kristallographie*, **148**, 65–81.

Tyrna, P.L. & Guggenheim, S. (1991) The crystal structure of norrishite, $KLiMn^{3+2}Si_4O_{12}$: an oxygen based mica. *American Mineralogist*, **76**, 266–271.

Valdré, G. (2007) Natural nanoscale surface potential of clinochlore and its ability to align nucleotides and drive DNA conformational change. *European Journal of Mineralogy*, **19**, 309–319.

Valdré, G., Antognozzi, M., Wotherspoon, A. & Miles, M.J. (2004) Influence of properties of layered silicate minerals on adsorbed DNA surface affinity, self-assembly and nanopatterning. *Philosophical Magazine Letters*, **84**, 539–545.

Valdré, G., Moro, D. & Ulian, G. (2011) Interaction of organic molecules with layer silicates, oxides and hydroxides and related surface-nano-characterization techniques. In: *Layered Mineral Structures and their Application in Advanced Technologies* (M.F. Brigatti & A. Mottana, editors). EMU Notes in Mineralogy, **11**, European Mineralogical Union and the Mineralogical Society of Great Britain & Ireland, pp. 313–334.

Veblen, D.R. (1983) Exsolution and crystal chemistry of the sodium mica wonesite. *American Mineralogist*, **68**, 554–565.

Wahle, M.W., Bujnowski, T.J., Guggenheim, S. & Kogure, T. (2010) Guidottiite, the Mn-analogue of cronstedtite: a new serpentine-group mineral from South Africa. *Clays and Clay Minerals*, **58**, 364–376.

Walker, J.R. & Bish, D.L. (1992) Application of Rietveld refinement techniques to a disordered II*b* Mg-chlorite. *Clays and Clay Minerals*, **40**, 319–322.

Wicks, F.J. & Whittaker, E.J.W. (1975) A reappraisal of the structures of the serpentine minerals. *The Canadian Mineralogist*, **13**, 227–243.

Yada, K. (1971) Study of microstructure of chrysotile asbestos by high resolution electron microscopy. *Acta Crystallographica*, **A27**, 659–664.

Yariv, S. (1992) The effect of tetrahedral substitution of Si by Al on the surface acidity of the oxygen plane of clay minerals. *International Reviews in Physical Chemistry*, **11**, 345–375.

Yuan, H. & Bish, D.L. (2010) NEWMOD+, a new version of the NEWMOD program for interpreting X-ray powder diffraction patterns from interstratified clay minerals. *Clays and Clay Minerals*, **58**, 318–326.

Zvyagin, B.B., Mishchenko, K.S. & Soboleva, S.V. (1969) Structure of pyrophyllite and talc in relation to the polytypes of mica-type minerals. *Soviet Physics Crystallography*, **13**, 511–515.

EMU Notes in Mineralogy, Vol. 11 (2011), Chapter 3, 123–149

Layered titanosilicates

ZHI LIN, FILIPE A. ALMEIDA PAZ and JOÃO ROCHA

Department of Chemistry, CICECO, University of Aveiro,
3810-193 Aveiro, Portugal
e-mail: rocha@ua.pt

Layered materials have much potential for applications as ion exchangers, gas sorbents and catalysts. Layered titanosilicates are of particular interest because on the one hand they possess large amounts of titanium and, on the other, they exhibit a wide variety of crystal structures. This chapter summarizes the main structural features in the family of heteropolyhedral layered titanosilicates and presents briefly the synthesis and application studies on some of these materials. The most representative members have been described in terms of the construction of each individual sheet from the individual building blocks. In general, the structures of these materials encompass single layers, mica-type layers (heterophyllosilicates) and perforated 'microporous' layers. The most commonly used synthesis process and potential applications of some layered titanosilicates are reviewed briefly. The materials with perforated 'microporous' sheets are particularly interesting because they can be considered as microporous materials with one of the dimensions in the nano size regime, meaning the materials have fast transport properties. Most of the structures reviewed are known only as minerals. Both the synthesis and potential applications of layered titanosilicates have been significantly underexplored and still present a great number of challenges and opportunities for future research.

1. Introduction

The most commonly employed inorganic functional materials are undoubtedly zeolites, which are, in general, defined as anionic three-dimensional, microporous, networks assembled from smaller $(SiO_4)^{4-}$ and $(AlO_4)^{5-}$ building units. However, layered (*i.e.* two-dimensional) inorganic materials have also found many uses as ion-exchangers, in adsorption processes and as catalysts (Centi & Perathoner, 2008; Corma *et al.*, 1998, 2000).

The infinite two-dimensional nature of the latter materials comes with a number of advantages, in particular those associated with the ability to intercalate in the interlamellar space a great number and variety of chemical species. For example, a recent use of inorganic layered structures includes the design and construction of organic-inorganic nanocomposites *via* the intercalation of functional organic species (Carrado, 2000; Choi *et al.*, 2008; Kwon & Shin, 2000; Ogawa *et al.*, 1998; Shimojima *et al.*, 2001). Layered silicates may also be transformed into mesoporous materials by pillaring processes. These encompass the intercalation of surfactant molecules into the interlayer space, followed by a pH adjustment, which induces the connection of the silicate sheets around the surfactant phase and, ultimately, the formation of a mesoporous

DOI: 10.1180/EMU-notes.11.3

structure (Kan *et al.*, 2000). Note that if the layers are perforated (*i.e.* microporous) *a priori* the resulting materials will exhibit dual-level porosity (microporous and mesoporous porosity).

Layered materials may allow the fabrication of permselective organic-inorganic composite membranes with a desirable combination of mechanical, thermal and barrier properties (Yano *et al.*, 1993; Krishnamoorti *et al.*, 1996; Wang & Pinnavaia, 1998; Umemura *et al.*, 2001; Choi *et al.*, 2008). The chemical moieties lining the interlayer surfaces can act as specific anchoring sites to interact with the active groups of the polymers (Weimer *et al.*, 1999), or promote the storage of drugs and enzymes. Perforated (*i.e.* 'microporous') sheets may further improve the properties of the polymer membranes.

An interesting recent development in zeolite synthesis is the preparation of, *e.g.* ZSM-5 crystals with one crystal dimension confined to the nanometer scale ($\sim$2 nm), meaning that only a small number of unit cells is aligned to form that specific crystal face (Choi *et al.*, 2009). This nanosheet catalyst overcomes diffusion limitations, thus providing high catalytic activity and a long lifetime of the catalyst. In a similar fashion, the delamination of the layered zeolite MCM-22 affords a material with large surface area, and composed of thin and ordered silicate sheets which allow improved access to the catalytic sites (Corma *et al.*, 1998). Individual layers exfoliated from layered-type materials with perforated sheets may also have properties very similar to zeolite nanosheets.

This chapter summarizes the main structural features of a specific family of layered inorganic materials: heteropolyhedral layered titanosilicates. We shall describe the most representative members, focusing on the step-wise construction of each individual sheet starting from the individual building blocks. The most commonly employed synthesis methods and potential applications of some layered titanosilicates are also reviewed briefly.

2. Construction of layered titanosilicates

2.1. Building blocks

The construction of heteropolyhedral layered titanosilicates can be envisaged by the use of rigid building blocks which interconnect in different ways, thus resulting in different crystal structures (corresponding to minerals or mineral analogues).

The most important building block is the SiO_4^{4-} tetrahedron, which can appear in different forms in minerals as depicted in Figure 1. For example, the SiO_4^{4-} unit can appear isolated as observed in the mineral natisite and in nesosilicates. However, it also has the ability to condense with itself, leading to the formation of secondary units such as $Si_2O_7^{6-}$, which is observed in the minerals composing the bafertisite group and the sorosilicates. Condensation can be more extensive leading to the formation of chains (as in the lintisite-type structures and in inosilicates), dendritic structures (as in the astrophyllite-group minerals), $Si_6O_{19}^{14-}$ belts (as in the nafertisite group), layers or double layers (as in AM-1 and the mineral jonesite, respectively).

In titanosilicate layers, the titanium polyhedra: (1) are typically isolated from each other (by silicate units), with this being the most common occurrence; or (2) form

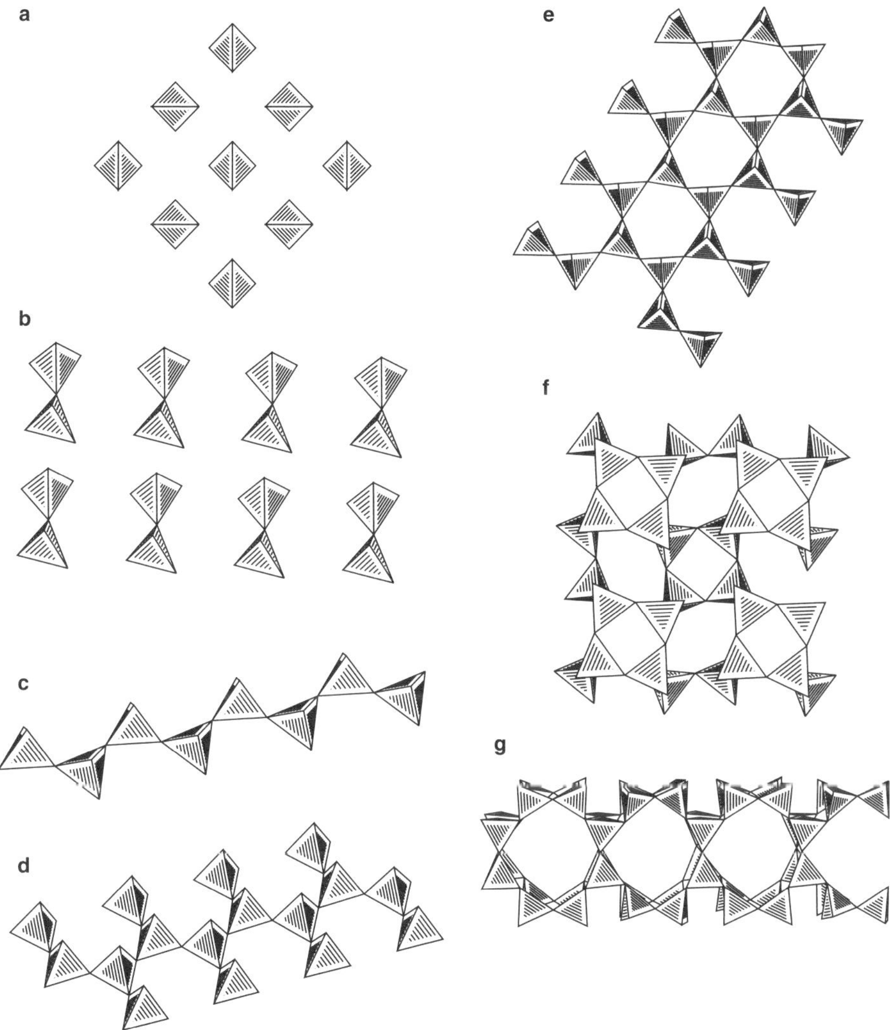

Fig. 1. The connections between SiO_4^{4-} tetrahedra in layered titanosilicates: (*a*) isolated SiO_4^{6-}; (*b*) $Si_2O_7^{6-}$ dimers; (*c*) SiO_2 single chain; (*d*) dentritic $Si_4O_{13}^{10-}$ chain; (*e*) $Si_6O_{19}^{4-}$ belt; (*f*) silicate single layer; and (*g*) silicate double layer.

brookite-type $[Ti_2O_8]^{8-}$ zigzag chains *via* edge-sharing as found in the mineral lintisite (Fig. 2).

2.2. Layer construction

The combination of the two types of building blocks highlighted in the previous subsection leads to the formation of extended structures. In this chapter we shall deal only with the layered structures which usually result from this self-assembly.

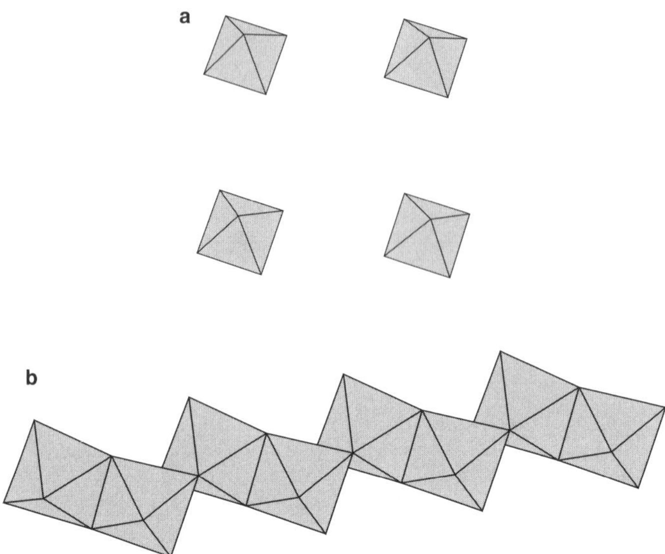

Fig. 2. Typical appearances of TiO_5 or TiO_6 polyhedra in layered titanosilicates: (***a***) isolated polyhedra; and (***b***) brookite-type $[Ti_2O_8]^{8-}$ zigzag chain.

In general, layered titanosilicates may be composed of single layers, mica-type layers or perforated 'microporous' layers. Materials having mica-type layers form a large family of compounds usually referred to as heterophyllosilicates. These structures have been reviewed by Ferraris & Gula (2005), Ferraris (2008) and Ferraris *et al.* (2008). For additional structural details we refer to the work by Ferraris and collaborators, and Sokolova (2006). In these structures, the SiO_4 connectivities depicted in Figures 1b, d, e are those more commonly observed.

3. Representative structures

3.1. AM-1: $Na_4[Ti_2Si_8O_{22}]·4H_2O$

Roberts *et al.* (1996) solved *ab initio* the structure of AM-1 from powder X-ray diffraction (XRD) data in combination with X-ray absorption spectroscopy (XAS) in the tetragonal $P42_12$ space group. In a recent report, Ferdov *et al.* (2007) confirmed the structural features using single-crystal XRD data. The latter study permitted the unequivocal location of the hydrogen atoms associated with the intercalated water molecules, ultimately describing in detail the hydrogen-bonding sub-network present in the crystal structure.

One of the most striking structural features of AM-1 (Fig. 3a) concerns the fact that four primary SiO_4^{4-} units (related by a fourfold rotation axis) are joined together *via* corner sharing to construct tetrameric $(Si_4O_{12})^{8-}$ units as depicted in Figure 3

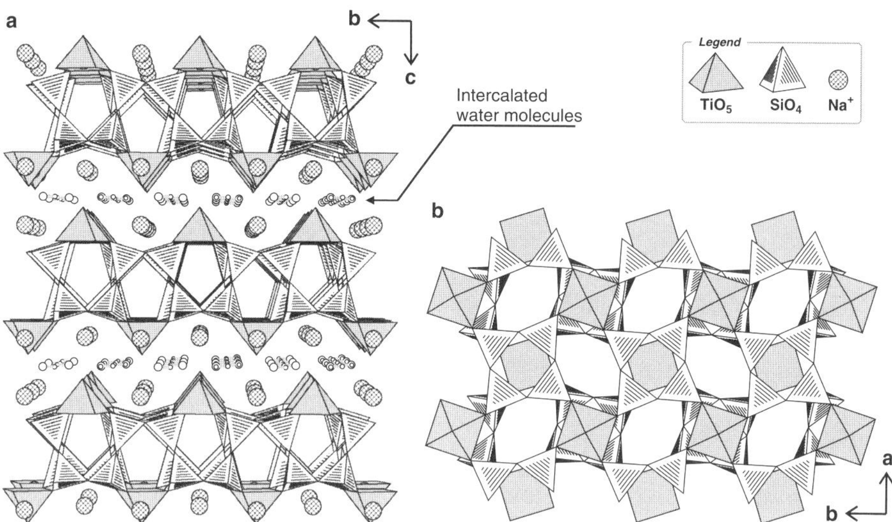

Fig. 3. (***a***) Crystal packing of the AM-1 structure viewed in perspective along the [100] direction of the unit cell. (***b***) Top view of the anionic titanosilicate layer.

(inter Si···Si within the tetramer of ~3.13 Å). These units are interconnected (also *via* corner sharing), in a zigzag fashion, along the [001] direction of the unit cell, leading to an infinite two-dimensional siliceous layer (inter Si···Si along the *c* axis of ~3.19 Å). Connections among adjacent tetrameric $(Si_4O_{12})^{8-}$ moieties in the *ab* plane of the unit cell are ensured by $(TiO{\cdot}O_4)^{4-}$ units (with coordination geometry resembling a distorted square pyramid), having the terminal Ti═O oxido group (occupying the apex of the pyramid) pointing to the interlayer space. The shortest Ti···Si distance (*via* an oxo bridge) is ~3.36 Å. The combination of the aforementioned connections leads to the formation of anionic $[Ti_2Si_8O_{22}]^{4-}$ three-tier layers which have five-membered ring apertures along the [100] and [010] direction of the unit cell, and distorted six-membered rings crossing the layers along the [001] direction (Fig. 3b). The latter apertures are comparable to those found in zeo-type materials and only large enough to allow the passage of small molecules such as H_2. The anionic $[Ti_2Si_8O_{22}]^{4-}$ layers can also be envisaged as constructed from two individual $TiSi_4O_{11}$ layers, related by 2_1 screw axes, mutually shifted along the [110] direction by $(\sqrt{2}/2)a$. This type of construction permits the visualization of small cages composed of eight SiO_4 units and one TiO_5.

The structure of AM-1 is clearly distinct from that of the mineral narsarsukite ($Na_2[TiSi_4O_{11}]$ – $I4/m$ space group, Peacor & Buerger, 1962). In the latter structure titanium appears with octahedral coordination environment, with the TiO_6 units being corner-shared forming a titanium oxide chain along the [001] direction of the unit cell. The connections between adjacent tetrameric $(Si_4O_{12})^{8-}$ units are completely distinct, forming a one-dimensional arrangement instead of the layered structure observed in AM-1. These considerable structural differences result in a framework-type structure for narsarsukite with the charge-balancing Na^+ being located in the cavities inside the structure.

The anionic charge of the $[Ti_2Si_8O_{22}]^{4-}$ titanosilicate layers is balanced by the interlamellar Na^+ cations (Fig. 3a), structurally located over the six-membered rings (shortest Na···Ti and Na···Si distances of ~3.73 and 3.08 Å), which effectively blocks the entrance to the intralamellar space. These cations can be exchanged by, *e.g.* protonated alkylamines. Farther from the titanosilicate layers there is a layer of water molecules, sandwiched between the Na^+ cations close to each $[Ti_2Si_8O_{22}]^{4-}$ layer. The crystallographically independent water molecule is engaged in relatively strong O—H···O hydrogen bonding interactions with the terminal Ti=O oxido groups from two adjacent anionic layers: while the $D \cdots A$ distance (D = donor and A = acceptor) is ~2.90 Å, the interactions angle [$<(DHA)$] is ~145°. Besides the electrostatic contacts, these supramolecular interactions are the only ones establishing direct links between neighbouring anionic layers.

Even though only one type of chemical environment was found for the water molecules of crystallization (*i.e.* a single crystallographic independent moiety), dehydration does not occur in a simple fashion. Indeed, AM-1 loses water in two consecutive steps: the first between 100 and 180°C, and the second in the 180–400°C range (Du *et al.*, 1996; Lin *et al.*, 1997; Ferdov *et al.*, 2007; Veltri *et al.*, 2006). This dehydration process can also be accompanied by differential scanning calorimetry (DSC)/differential thermal analysis (DTA) which clearly shows the presence of two peaks at ~150 and 245–270°C. This complex process cannot be explained by a simple structural disorder because the most recent crystal-structure determination clearly showed a small isotropic displacement parameter for the oxygen atom of the water molecule (Ferdov *et al.*, 2007). The structural reasons for the origin of such differences remain unclear.

3.2. Lintisite-type structures

Lintisite was discovered in Mount Alluaiv, Lovozero massif, Kola peninsula, Russia (Khomyakov *et al.*, 1990), and in Mont Saint-Hilaire, Canada (Chao *et al.*, 1990). Its crystal structure was determined in the monoclinic $C2/c$ space group by Merlino *et al.* (1990) from single-crystal XRD data and formulated as $Na_3Li[Ti_2Si_4O_{14}] \cdot 2H_2O$ (Fig. 4a).

The anionic $[Ti_2Si_4O_{14}]^{4-}$ layer of lintisite is formed by a single crystallographically independent Ti^{4+} metallic centre and two Si tetrahedra. The coordination geometry of Ti^{4+} can be envisaged as a highly distorted octahedron because the metal is displaced from the geometrical centre of the octahedron due to repulsion (thus resulting in a wide range of Ti–O bond lengths). Adjacent TiO_6 units are interconnected along the [001] direction of the unit cell (*via* edge sharing) to form brookite-type $[Ti_2O_8]^{8-}$ zigzag single chains (Fig. 4). Conversely, SiO_4^{4-} tetrahedra are, instead, interconnected *via* corner sharing to also form zigzag pyroxene-type chains running parallel to the same direction. The closest Ti···Ti and Si···Si internuclear distances along the chains are ~3.08 and 3.06 Å, respectively. Each crystallographically independent Si forms a distinct pyroxene-type chain. In this context, the anionic $[Ti_2Si_4O_{14}]^{4-}$ layer placed

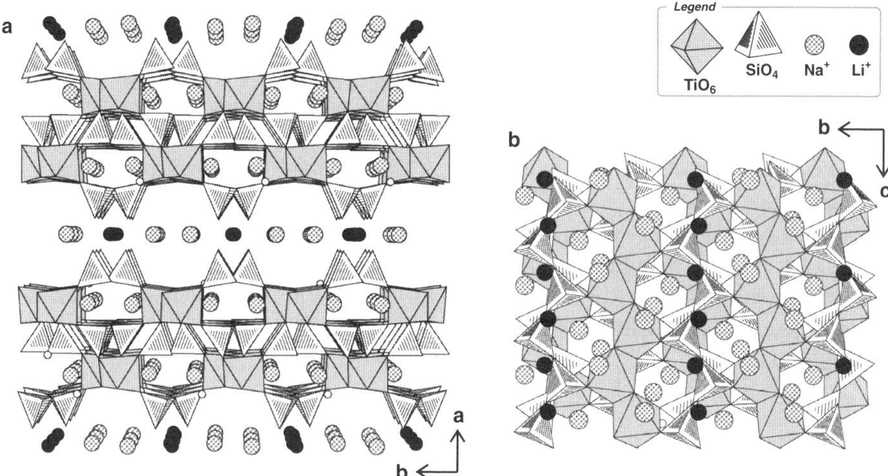

Fig. 4. (*a*) Crystal packing of the structure of lintisite viewed in perspective along the [001] direction of the unit cell. (*b*) Top view of the anionic titanosilicate layer emphasizing the location of the Li^+ cations over the pyroxene-type chains.

in the *bc* plane of the unit cell can also be envisaged as a five-tier sheet composed of chains of SiO_4:TiO_6:SiO_4:TiO_6:SiO_4 (Fig. 4a). The connectivity between the SiO_4 tetrahedral and TiO_6 octahedral chains is similar to that of the aegirine–augite series (Ghose, 1986).

The anionic charge of the layers is neutralized by the inclusion of Na^+ and Li^+ cations in the structure. While the former cations (along with the water molecules of crystallization) are distributed regularly throughout the entire crystal structure, being located both within the cavities of the anionic layer and also in the interlamellar space, the Li^+ cations are specifically located over the external pyroxene-type chains of $[Ti_2Si_4O_{14}]^{4-}$ (Fig. 4). The interactions of Li^+ with the apexes of the SiO_4 units lead to the formation of a chain of distorted LiO_4 tetrahedra (Li—O and O—Li—O length and angles of ~2.01 Å and ~114.8–121.2°, respectively), with the closest Li···Li distance within the chain being ~2.62 Å.

The crystal structure of lintisite is reminiscent of that of the mineral vinogradovite (Rastsvetaeva and Andrianov, 1984). Similar titanate and silicate chains (along with the octahedral distortions) are also observed for the latter mineral. If neighbouring layers of lintisite shift fractionally by 0.4613 or 0.5387 along the *c* axis, the SiO_4 tetrahedra could thus interconnect, ultimately generating the crystal structure of vinogradovite. This modular relation was also discussed by Merlino *et al.* (1990) and Ferraris *et al.* (2004). The hydrothermal formation of vinogradovite at 200°C has been reported by Chapman & Roe (1990).

The crystal structures of AM-4, ($Na_3(Na,H)[Ti_2Si_4O_{14}]\cdot 2H_2O$; *ab initio* from powder XRD data; Dadachov *et al.*, 1997), kukisvumite ($Na_6ZnTi_4[Si_2O_6]_4O_4\cdot 4H_2O$; mineral found in the Khibina massif, Kola peninsula, Russia, in 1991; Yakovenchuk *et al.*,

1991; crystal structure determined and discussed in detail by Merlino *et al.*, 2000) and manganokukisvumite ($Na_6Mn[Ti_4Si_8O_{28}]\cdot 4H_2O$; mineral discovered in Mont Saint-Hilaire, Canada, in 1987; Gault *et al.*, 2004) share striking similarities with that of lintisite. In short, all titanosilicate layers are identical except for the replacement of Li^+ by other elements. For example, in AM-4, Na^+ or H^+ ions occupy the Li^+ sites (Dadachov *et al.*, 1997), while in kukisvumite and manganokukisvumite, the same sites are occupied by zinc and manganese cations (sites partially occupied) instead. Cation replacement is not always straightforward and can lead to more complex structural arrangements, *e.g.* the ^{23}Na MAS NMR spectrum of AM-4 displays at least eight resonances (Lin *et al.*, 1997) while the structure model of lintisite only requires three independent sites (Dadachov *et al.*, 1997). This discrepancy remains unexplained.

3.3. Natisite: $Na_2[TiSiO_5]$

The layered mineral natisite (Fig. 5) crystallizes in the tetragonal $P4/nmm$ space group (Nyman *et al.*, 1978). The crystal structure comprises a single Ti^{4+} centre (located at a 4mm special position within the unit cell) bound to four symmetry-related isolated SiO_4 tetrahedra (*via* corner sharing) and a terminal Ti=O oxido group. This mineral is, therefore, a nesosilicate. The TiO_5 coordination environment resembles a square pyramid, with the basal Ti—O bond lengths of ~1.99 Å while the apical Ti=O distance is ~1.70 Å. The metallic centre is raised slightly from the basal plane of the pyramid (by ~0.60 Å). Because of this simple corner-sharing connectivity, both SiO_4 and TiO_5 units appear isolated in the crystal structure with the minimum Ti···Si distance within the anionic $[TiSiO_5]^{2-}$ layer always being ~3.26 Å. As depicted in Figure 5b, within the anionic layer mutually close SiO_4 and TiO_5 units are related by inversion which leads to the presence of Ti=O oxido groups on both sides of the titanosilicate layer.

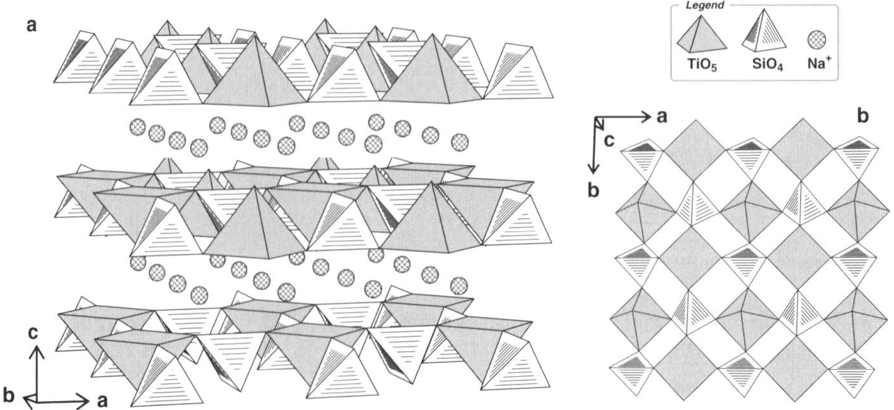

Fig. 5. (***a***) Crystal packing of the mineral natisite. (***b***) Top view of the anionic titanosilicate layer.

The crystal charge is balanced by a crystallographically independent Na^+ cation, structurally located between the anionic $[TiSiO_5]^{2-}$ layers (Fig. 5a). These cations have the additional structural function of establishing physical connections between adjacent anionic $[TiSiO_5]^{2-}$ layers, interacting with six oxygen atoms in a slightly distorted octahedral coordination fashion. While the equatorial plane is composed of oxygen atoms from SiO_4 units (Na—O distance ≈ 2.31 Å), the apical positions are occupied by interactions with the terminal oxido groups (Na—O distance ~2.58 Å) instead. The closest Na···Na distance within the interlamellar space is ~3.24 Å.

3.4. Jonesite: $Ba_2(K,Na)[Ti_2(Si_5Al)O_{18}(H_2O)]{\cdot}nH_2O$

The mineral jonesite, $Ba_2(K,Na)[Ti_2(Si_5Al)O_{18}(H_2O)]{\cdot}nH_2O$ (monoclinic $P2_1/n$), was first discovered in the Benitoite Gem Mine, San Benito County, California (Wise *et al.*, 1977). Its crystal structure was, however, fully elucidated only recently by Krivovichev & Armbruster (2004) (Fig. 6).

The structure consists of anionic double layers, $[Ti_2(Si_5Al)O_{18}(H_2O)]^{5-}$, having eight-membered apertures running parallel to the [100] direction of the unit cell (Fig. 6). The heterosilicate layers in contact with the interlamellar space are of bafertisite-type (see also Fig. 9), formed of TiO_6 octahedra (two crystallographically distinct

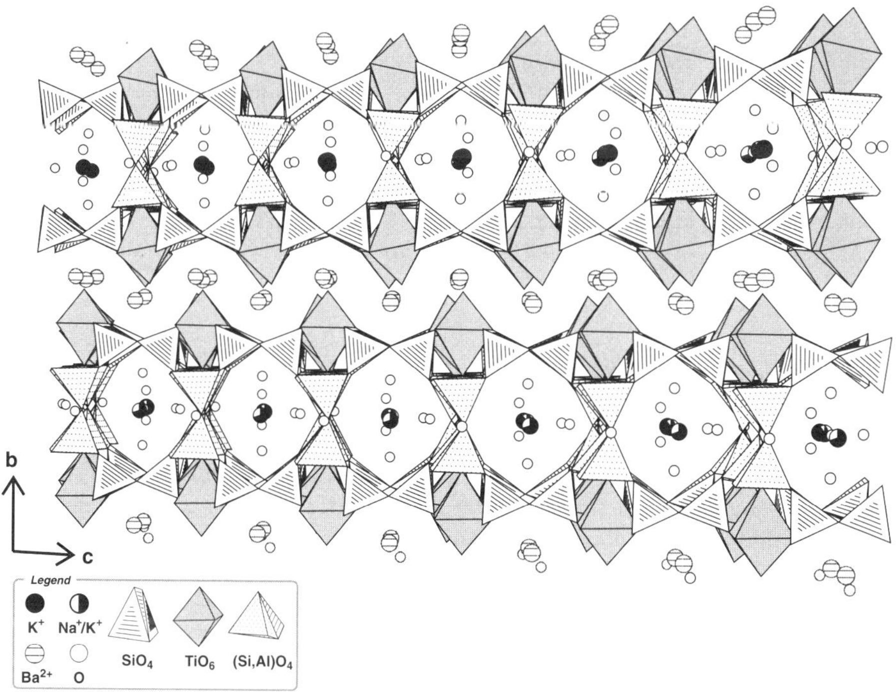

Fig. 6. Crystal structure of jonesite viewed in perspective along the [100] direction of the unit cell.

types) joined (*via* corner sharing) by individual Si_2O_7 units. The shortest Ti···Ti distance within this layer occurs along the *a* axis of the unit cell, being ~5.36 Å. A feature worth emphasizing concerns the fact that in jonesite the molecular geometry of the $Si_2O_7^{6-}$ units is significantly distorted from that observed in other related minerals described in this chapter, with the Si—O—Si angle being nearly 150°. This feature seems to result from the type of interlayer connectivity described in detail below. The two distinct TiO_6 polyhedra exhibit considerable distorted octahedral coordination environments, having very short Ti=O bonds (~1.71 Å), pointing towards the interlamellar space, *trans*-coordinated to a long Ti—O bond (of ~2.40 Å). Remarkably, the metallic centres are not displaced significantly from the average equatorial plane, thus suggesting that this long bond is instead the result of a weaker interaction with the neighbouring $(Si,Al)O_4$ units (hereafter denominated by TO_4 with Si and Al occurring in the same proportion in the crystal structure).

Bafertisite-type sheets are pillared along the [010] direction by T_4O_{12} units which create four-membered apertures to the inner portion of the anionic layer. These units are disordered over two distinct crystallographic positions, mutually related by $\pm a/2$, having rates of occupancy of 79 and 21%. The disorder was ascribed to stacking faults while connecting the two titanosilicate sheets which can shift along the *a* axis by $\pm a/2$. Thus, in the absence of these units, the TiO_6 unit from the bafertisite-type sheets has coordinated water molecules in the apical positions. Remarkably, jonesite has a very similar structure to the mineral bafertisite but, instead of T_4O_{12} groups, the latter contains edge-sharing FeO_6 octahedra instead joining together the two sheets, ultimately leading to a dense, non-porous compound.

The pillaring process leads to the formation of the aforementioned eight-membered apertures, having cross-section diameters of ~3.3–3.4 Å. These inner cavities of the anionic layer house the charge-balancing Na^+ and K^+ cations, and also a large number of water molecules of crystallization. Ba^{2+} cations are only located in the interlamellar space (see Fig. 6) with the shortest Ba···Ba distance being 4.37 Å. The silicate double layers with eight-membered channels are also observed in the rhodesite-type structure (Ferraris & Gula, 2005).

3.5. Lamprophyllite: $(Sr,Ba)_2[Na_3Ti_3O_2(Si_2O_7)_2(OH)_2]$

Lamprophyllite, $(Sr,Ba)_2[Na_3Ti_3O_2(Si_2O_7)_2(OH)_2]$, belongs to the bafertisite group which includes at least 25 members (Ferraris & Gula, 2005; Ferraris *et al.*, 2008). This mineral was first discovered 100 years ago in Kola peninsula, Russia, and later found in other alkaline massifs around the world (Krivovichev *et al.*, 2003 and references therein). The monoclinic (space group $C2/m$) and orthorhombic (space group *Pnmn*) forms of lamprophyllite are commonly known as polytypes 2*M* and 2*O* (Fig. 7).

The two polymorphic forms of lamprophyllite are formed by a three-tier titanosilicate layer in the *bc* plane of the unit cells (Fig. 7a, b and d, e), having two markedly distinct coordination environments for Ti^{4+}. The surfaces of these layers consist of square pyramidal TiO_5 units corner-shared to $Si_2O_7^{6-}$ groups, leading to the formation of a bafertisite-type $TiOSi_2O_7$ sheet. Unlike that described for AM-1 and jonesite, the terminal

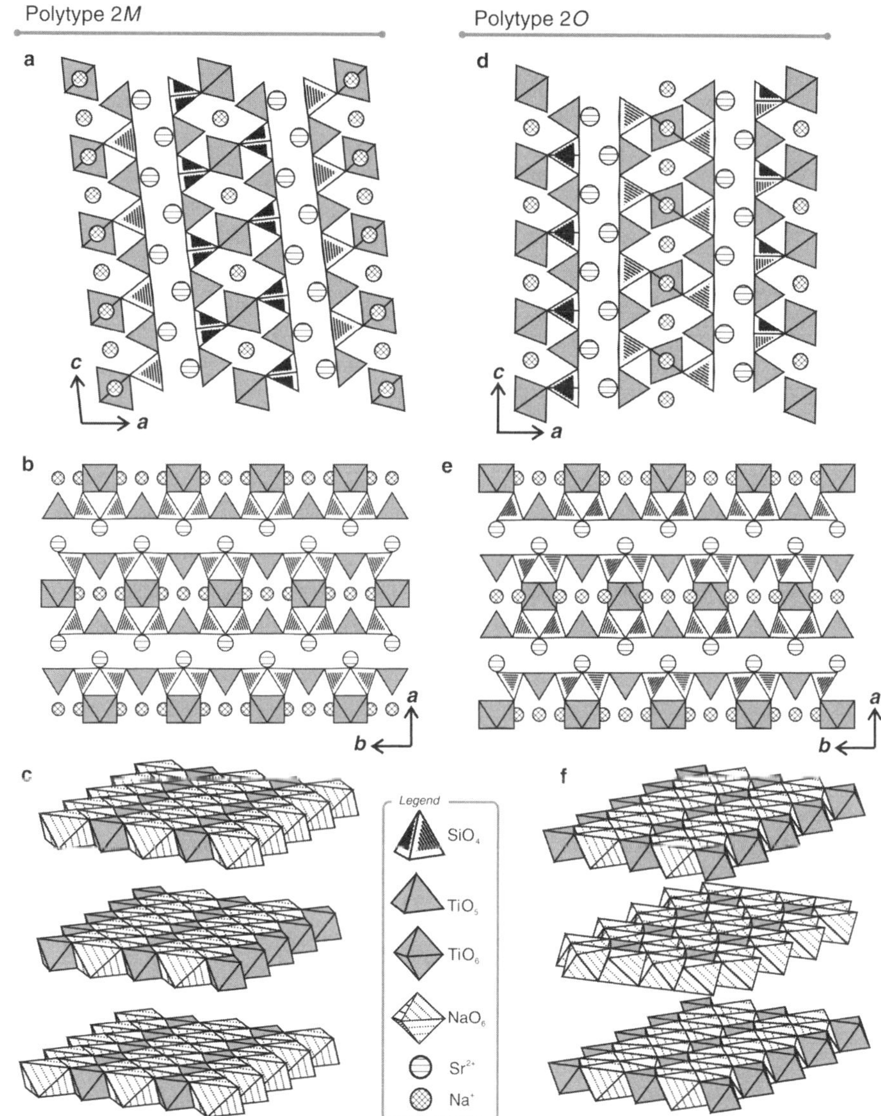

Fig. 7. Polytypes 2*M* and 2*O* of lamprophyllite.

Ti=O oxido group (distances of ~1.68 and 1.69 Å for polytypes 2*M* and 2*O*, respectively) in the two polymorphic forms does not point to the interlamellar space. As expected, the Ti^{4+} centres are raised slightly (by ~0.47 Å) from the basal plane which is composed of four SiO_4 units (Ti—O distances of 1.96 and 1.95 Å for 2*M* and 2*O*, respectively). Within this sheet, the shortest $Ti_{pyramidal}\cdots Ti_{pyramidal}$ distance is 5.37 and 5.38 Å for polytypes 2*M* and 2*O*, respectively.

The aforementioned bafertisite-type sheet is identical to the two polytypes, with the difference between the two structures residing in the orientation of the intermediate Na,Ti octahedral sheet. In the polytypte 2*M* structure, the NaO_6 and TiO_6 octahedra between adjacent layers are aligned in the same direction (because of their relation *via* a twofold screw axis – see Fig. 7c); in the polytype 2*O* structure, adjacent layers are mutually rotated instead (related by intersecting 2_1 screw rotation axes parallel and perpendicular to the layers) as depicted in Figure 7f. This second crystallographic type of Ti^{4+} appears with a typical octahedral coordination environment, with the equatorial plane being composed of four SiO_4 tetrahedra (two from each bafertisite-type sheet) while the apical positions are occupied by hydroxyl groups. As a consequence, the Ti—O distances are slightly shorter for the latter (1.95 and 1.93 Å for 2*M* and 2*O*, respectively); for the two polytypes, the Ti—O distances to SiO_4 are ~2.00 Å. Within this octahedral sheet, the shortest $Ti_{octahedral}\cdots Ti_{octahedral}$, $Ti_{octahedral}\cdots Na$ and Na···Na distances are 5.37/5.38 Å, 3.18/3.19 Å and 3.25/3.26 Å for polytypes 2*M*/2*O*, respectively.

The charge of these three-tier anionic $[Na_3Ti_3O_2(Si_2O_7)_2(OH)_2]^{4-}$ layers is balanced by the presence of strontium cations (partially replaced by sodium) residing in the interlayer space (see (Fig. 7a, b and 7d, e). Strontium-free lamprophyllites have also been reported, *e.g.* barytolamprophyllite, $(Ba,K)_2[Na_3Ti_3(Si_2O_7)_2O_2(OH)_2]$, (Sokolova & Cámara, 2008a) and nabalamprophyllite, Ba(Na,Ba) $[Na_3TiTi_2O_2Si_4O_{14}(OH,F)_2]$, (Chukanov *et al.*, 2004).

3.6. Delindeite: $Ba_2(Na,K,\square)_3(Ti,Fe)[Ti_2(O,OH)_4Si_4O_{14}](H_2O,OH)_2$

The mineral delindeite (named after Henry S. deLinde), $Ba_2(Na,K,\square)_3(Ti,Fe)$ $[Ti_2(O,OH)_4Si_4O_{14}](H_2O,OH)_2$ (monoclinic $A2/m$), belongs to the general class of sorosilicates and was discovered in 1987 (Appeman *et al.*, 1987). Its crystal structure was unveiled by Ferraris *et al.* (2001a), showing that it belongs to the general bafertisite group (Fig. 8).

The basic building blocks of delindeite are $Si_2O_7^{6-}$ dimeric units and TiO_6 octahedra which interconnect in the *ab* plane of the unit cell, *via* corner-sharing, to form a bafertisite-type sheet (as depicted in Fig. 9). For this type of layer, the shortest Ti···Ti distance occurs across the layer (in this case parallel to the [100] direction) and is ~5.33 Å. Two bafertisite-type sheets are joined together by a mixed octahedral $(Ti,Fe)O_6$ site with rates of occupancy of 2/3:1/3 for Ti:Fe, respectively. This octahedron consists of four $Si_2O_7^{6-}$ dimeric units making up the equatorial plane (contrasting with the only two observed for lamprophyllite), while the apical positions (bound to the bafertisite-type sheets) are occupied by oxido or hydroxyl groups instead. The shortest (Ti,Fe)···Ti distance to a metal from the bafertisite-type sheets is ~3.72 Å.

The initial description of the structure of delindeite also contemplated disorder located in its building blocks (as shown in Fig. 8). The interlamellar Ba^{2+} cations were modelled as disordered over two distinct crystallographic positions (rates of occupancy of 96% and two of 2% each) which are spatially close: the greater occupancy site (96%) is separated by only ~1.04 and 1.08 Å from the two remaining sites (2%

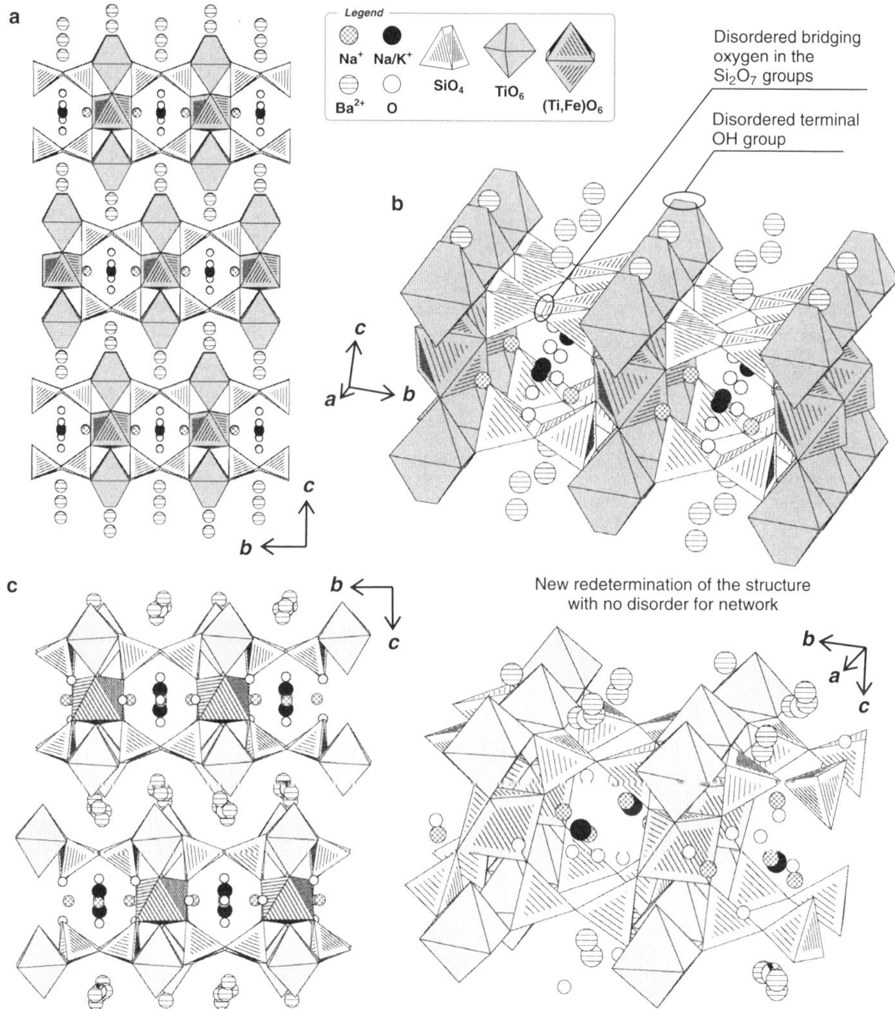

Fig. 8. Crystal packing and top view of the anionic titanosilicate layer of the mineral delindeite.

occupied). As depicted in Figure 8b, the apex of the TiO_6 polyhedra making up the bafertisite-type sheets appears disordered over two distinct crystallographic locations (50% rate of occupancy for each), with half of them being hydroxyl groups (based on charge-balancing reasons). These two sites are ~0.81 Å apart. In addition, the central bridging oxygen atom of the $Si_2O_7^{6-}$ units appears with two possible locations (rates of occupancy of 54 and 46%), separated from each other by ~1.25 Å. Within the octahedral layers (sandwiched between the bafertisite-type sheets), Na^+ has two distinct crystallographic sites. One site shares its occupancy with K^+ and is split by symmetry (the two are ~0.61 Å apart). The second site is coordinated in a typical octahedral

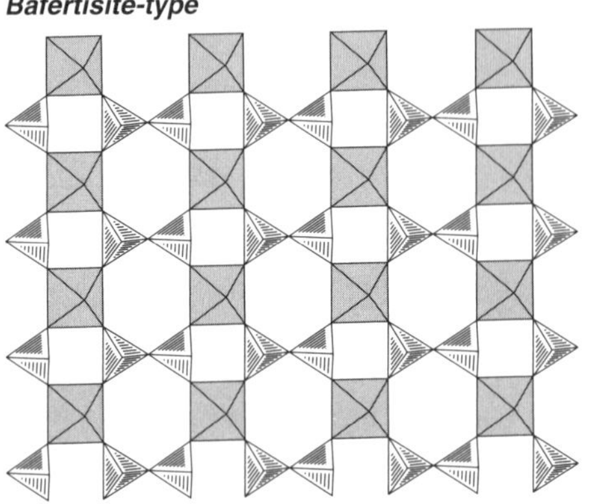

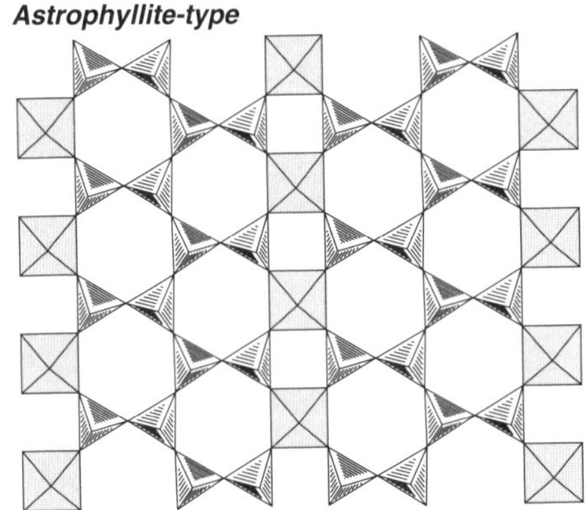

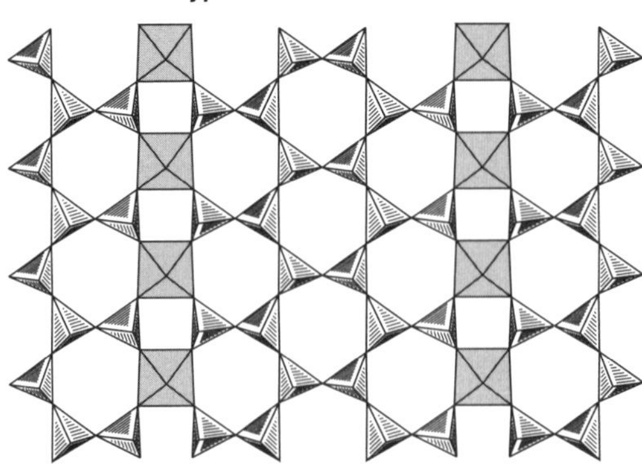

Fig. 9. Different types of heterosilicate layers.

fashion with water molecules dispersed over two positions (separated by ~0.81 Å) with rates of occupancy of 38 and 58%. This extensive disorder and the presence of diffraction streaks in the patterns motivated Sokolova & Cámara (2007) to reinvestigate the structure of delindeite. They found that by doubling the *a* and *b* axes of the unit cell (concomitantly the unit-cell volume increases by four times) the structure could be solved in the monoclinic $C2/c$ space group. In this new model, the bafertisite-type sheet, formed by the TiO_6 octahedra and $Si_2O_7^{6-}$ groups, now appears completely ordered (Fig. 8c).

3.7. Heterophyllosilicates

Heterophyllosilicates are layered titanosilicates in which the SiO_4 layer is interrupted by titanium polyhedra. This substitution is usually periodic leading to the formation of bafertisite-, astrophyllite- or nafertisite-type layers (Fig. 9). This subject was reviewed extensively by Ferraris and colleagues (Ferraris & Gula, 2005; Ferraris, 2008; Ferraris *et al.*, 2008). Here we shall provide some examples of minerals belonging to each of the aforementioned classes.

3.7.1. Bafertisite-type

Bafertisite-type layers are very common and are present in, at least, 25 minerals (Ferraris & Gula, 2005; Sokolova, 2006) including lamprophyllite and delindeite (see previous subsections for structural details). Titanium may be replaced by other elements. For example, in surkhobite (monoclinic *C*2), $(Ba,K)_2\,CaNa(Mn,Fe^{2+},Fe^{3+})_8$-$Ti_4(Si_2O_7)_4\,O_4(F,OH,O)_6$ (Rastsvetaeva *et al.*, 2008), Ti is partially replaced by Nb and Zr. Vuonnemite (triclinic $P\bar{1}$), $Na_{11}[TiNb_2(Si_2O_7)_2(PO_4)_2O_3(F,OH)]$ (Ercit *et al.*, 1998), is a sorosilicate consisting of a five-tier anionic layer: $[TiNb_2(Si_2O_7)_2(PO_4)_2O_3(F,OH)]^{11-}$. The heterosilicate sheet consists solely of isolated $Si_2O_7^{6-}$ dimeric units and NbO_6 octahedra (Fig. 10), with the shortest Nb···Nb distance being ~5.50 Å (along the *a* axis of the unit cell). Only the O atoms in the equatorial plane of the NbO_6 octahedra are shared with $Si_2O_7^{6-}$ units. The apical positions are occupied by F^-/OH^- groups (pointing to the inner section of the anionic layer) and phosphate moieties (making up the outer tier of the anionic layer). Heterosilicate sheets are joined together along the *c* axis *via* intermediate octahedral TiO_6 units, with the shortest Nb···Ti and Ti···Ti distances being ~4.95 and 5.50 Å, respectively. The crystal charge is balanced by the presence of Na^+ cations, both between and inside the anionic layers (Fig. 10). The shortest Na···Na distance occurs, however, in the former location, ~3.16 Å.

Variation in the chemical composition of the octahedral layer joining together the heterosilicate sheets leads to distinct mineral phases: instead of Ti, these polyhedral layers may contain Fe or Mn, forming layers with or without Na^+. A striking example concerns bafertisite, $BaFe_2TiO[Si_2O_7](OH)_2$ (monoclinic *Cm* – see Fig. 11). In this mineral, the heterosilicate layers (formed by isolated $Si_2O_7^{6-}$ units interconnecting the TiO_6 octahedra) are joined together by a compact octahedral layer consisting solely of

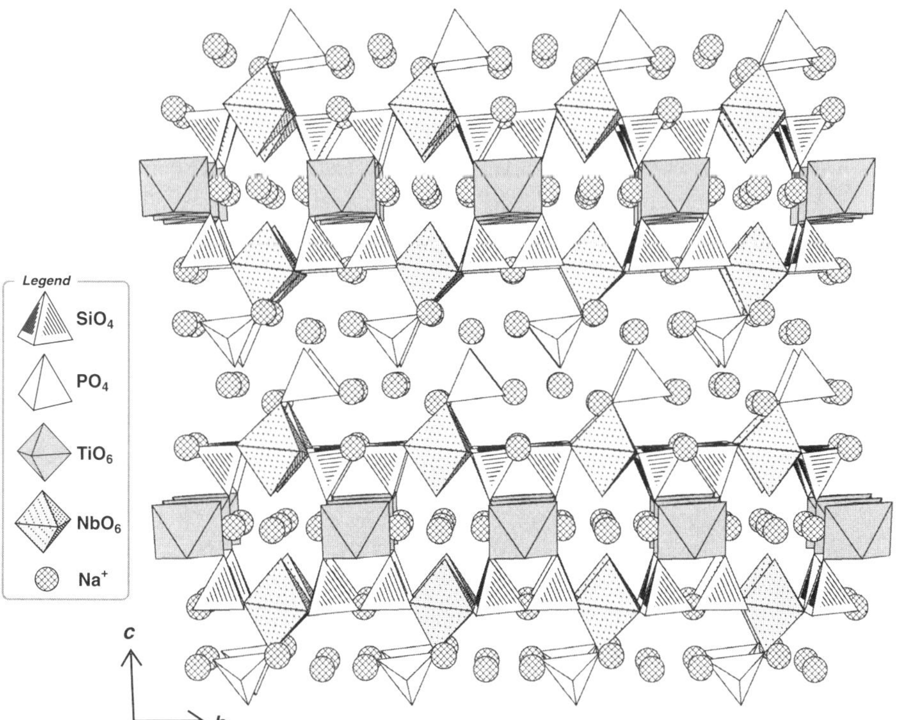

Fig. 10. Crystal packing of vuonnemite viewed in perspective along the [100] direction of the unit cell.

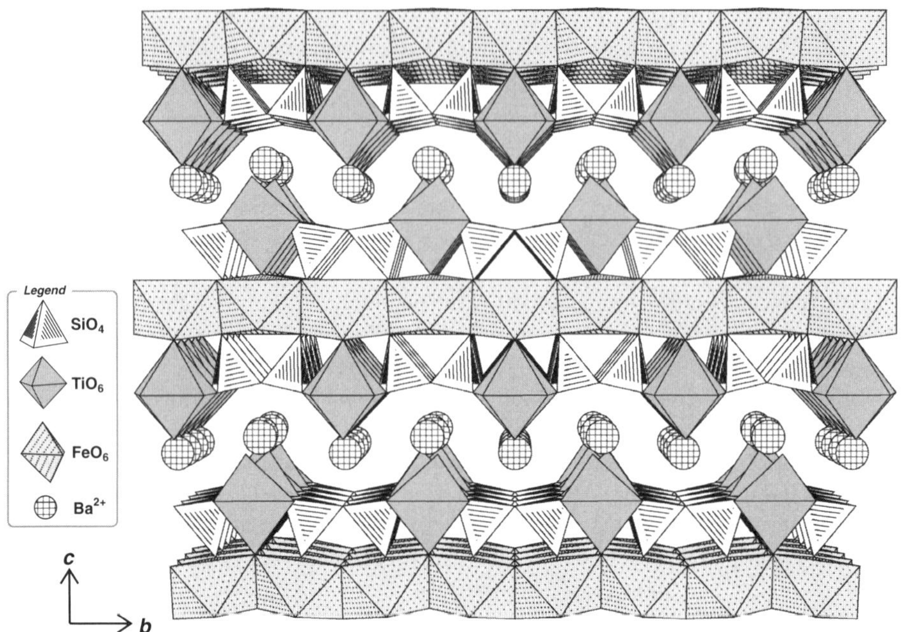

Fig. 11. Crystal packing of bafertisite viewed in perspective along the [100] direction of the unit cell.

edge-shared FeO_6 units (Fig. 11), with the shortest Fe···Fe distance of ~3.16 Å. Charge-balancing Ba^{2+} cations occupy the interlamellar space (minimum inter-cation distance of ~4.43 Å), being interrupted periodically by the terminal apical hydroxyl groups of the TiO_6 polyhedra.

In murmanite, $Na_2MnTi_3(Si_2O_7)_2(OH)_4{\cdot}4H_2O$ (Khalilov *et al.*, 1965) (triclinic *P*1 – see Figure 12), the intermediate octahedral layer is composed of Ti, Mn and Na instead, which is formed by edge-sharing of $(Ti,Mn)O_6$ and NaO_6 octahedral chains running parallel to the *a* axis of the unit cell. The —OH content is dependent on the amount of Mn present in the structure with Ti and Mn octahedra being distributed in an alternating fashion along the chain. The intra-chain (Ti,Mn)···(Ti,Mn) and Na···Na intermetallic distances were found to be in the 3.05–3.50 and 3.28–3.36 Å ranges, respectively. The layers are neutral with the interlamellar space being occupied solely by water molecules of crystallization engaged in O—H···O hydrogen bonding interactions (O···O distances in the ~3.01–3.28 Å range). Murmanite is amenable to isomorphic substitution: the TiO_6 octahedra making up the heterosilicate sheets may be partially replaced by Nb, while within the $(Ti,Mn)O_6$ octahedral chains, Mn can be replaced by Fe to form $(Ti,Fe)O_6$ octahedral chains instead (Rastsvetaeva & Andrianov, 1986). The apexes of the TiO_6 octahedra making up the heterosilicate sheets are typically occupied by water molecules. If the apical positions of TiO_6 are occupied by phosphate moieties instead, there is an expansion of the crystal structure through the *c* axis with the concomitant formation of a crystal structure similar to that

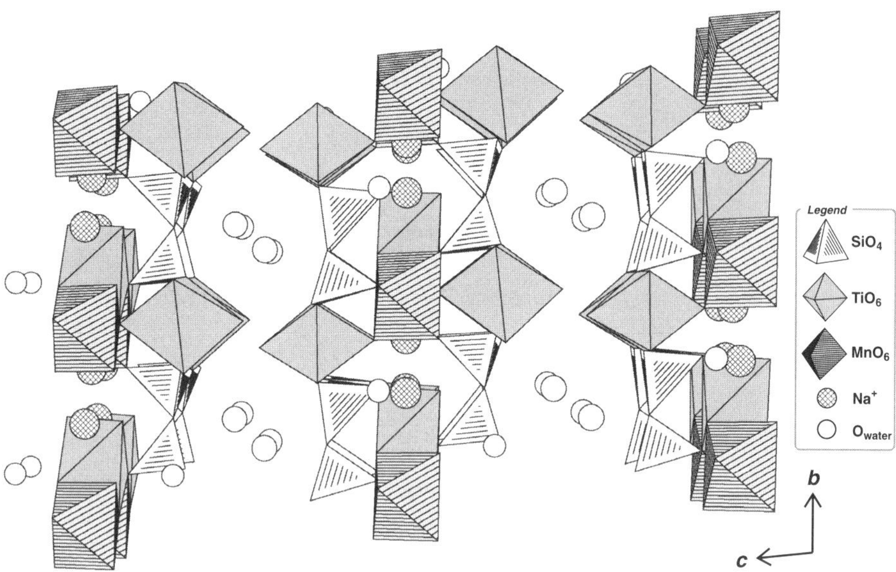

Fig. 12. Crystal packing of murmanite viewed in perspective along the [100] direction of the unit cell.

of the mineral lomonosovite [ideally $Na_{10}Ti_4O_4(Si_2O_7)_2(PO_4)_2$, Belov *et al.*, 1977; Cámara *et al.*, 2008].

Layers may also be more complex, *e.g.* those encountered in the mineral bornemanite ($BaNa_3\{(Na,Ti)_4[(Ti,Nb)_2O_2Si_4O_{14}](F,OH)_2\}PO_4$, monoclinic *Ib*, Ferraris *et al.*, 2001b, ideally $Na_6\square BaTi_2Nb(Si_2O_7)_2(PO_4)O_2(OH)F$, triclinic $P\bar{1}$, Cámara & Sokolova, 2007) and nechelyustovite ($Na_4Ba_2Mn_{1.5}\square_{2.5}Ti_5Nb(Si_2O_7)_4O_4$ $(OH)_3F(H_2O)_6$, Cámara & Sokolova, 2009, Nèmeth *et al.*, 2009).

3.7.2. Astrophyllite- and Nafertisite-type

At least nine minerals exhibit astrophyllite-type crystal structures (Piilonen *et al.*, 2003, 2004; Ferraris & Gula, 2005): astrophyllite [$(K,Na)_3(Fe,Mn)_7Ti_2Si_8O_{24}(O,OH)_7$]; magnesioastrophyllite [$(Na,K)_4Mg_2(Fe,Fe,Mn)_5Ti_2Si_8O_{24}(O,OH,F)_7$]; hydroastrophyllite [$(H_3O,K,Ca)_3(Fe,Mn)_{5\text{-}6}Ti_2Si_8(O,OH)_{31}$]; kupletskite [$(K,Na)_3(Mn,Fe)_7Ti_2Si_8O_{26}(OH)_4F$]; kupletskite-(Cs) [$(Cs,K,Na)_3(Mn,Fe^{2+})_7(Ti,Nb)_2Si_8O_{24}(O,OH,F)_7$]; niobokupletskite [$K_2Na(Mn,Zn,Fe)_7(Nb,Zr,Ti)_2Si_8O_{26}(OH)_4(O,F)$]; nalivkinite [$Li_2Na(Fe^{2+},Mn^{2+})_7Ti_2Si_8O_{26}(OH)_4F$]; niobophyllite [$(K,Na)_3(Fe,Mn)_6(Nb,Ti)_2Si_8(O,OH,F)_{31}$]; and zircophyllite [$(K,Na)_3(Mn,Fe^{2+})_7(Zr,Ti)_2Si_8O_{24}(O,OH,F)_7$].

The heterosilicate sheets of nafertisite and astrophyllite-type compounds (in contact with the interlayer space – see Fig. 13) consist of a one-dimensional zigzag chain of corner-shared $Si_2O_7^{6-}$ units (inosilicate) (see Fig. 9) which establish physical links between pyramidal Ti (or Nb in niobophyllite). The shortest intermetallic distance within this layer occurs along the [100] direction of the unit cell and is 5.33 Å. These heterosilicate layers are joined together by an octahedral sheet which, in astrophyllite

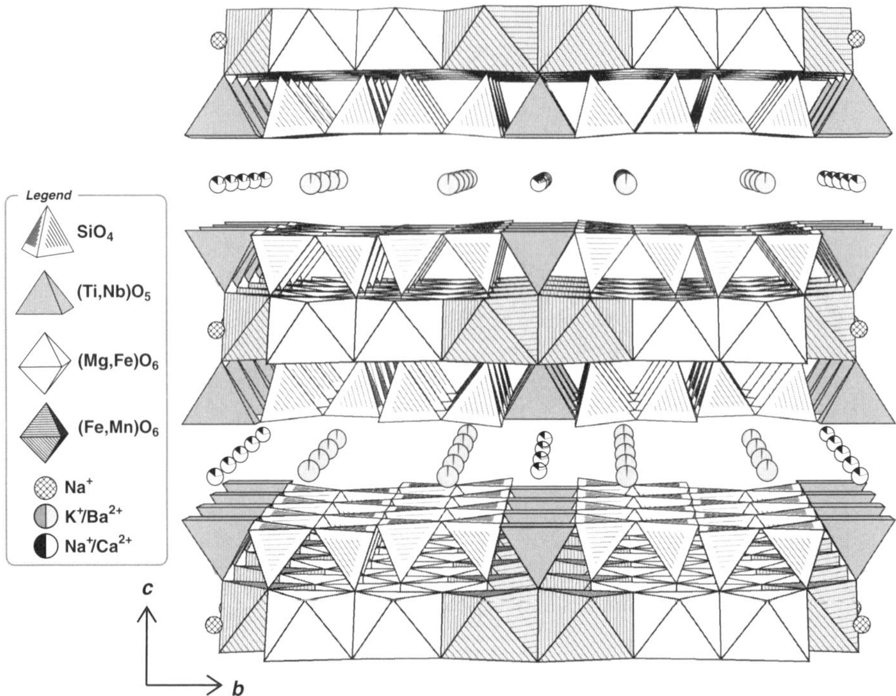

Fig. 13. Crystal structure of magnesioastrophyllite viewed in perspective along the [100] direction of the unit cell.

and magnesioastrophyllite (Sokolova & Cámara 2008b), may contain Fe^{2+}, Mn^{2+} and Mg^{2+}. The shortest intermetallic distance within this octahedral layer is ~3.11 Å. In other members of the series (see above), this octahedral layer may be partially replaced by other cations such as Zr (in zircophyllite, Kapustin, 1972), Zn or Nb (in niobokupletskite, Piilonen *et al.*, 2000). The three-tier anionic layers are separated by a layer of charge-balancing cations, typically highly disordered and belonging to the first and second groups of the periodic table.

The phyllosilicate nafertisite, $Na_3Fe_8(Ti_2Si_{12}O_{37})(OH)_6$, (Ferraris *et al.*, 1996) only contains Fe^{2+} and Fe^{3+} making up the octahedral layers sandwiched between the heterosilicate sheets as depicted in Figure 14. The shortest Fe···Fe distance within this layer is very comparable to that of astrophyllite-type structures being calculated at ~3.13 Å. A striking distinction between nafertisite and astrophyllite is the coordination geometry of the heteropolyhedra making up the heterosilicate sheets: while in the latter family, and as mentioned above, Ti (or Nb) appears with a typical distorted square pyramidal coordination environment (probably due to the presence in the apical position of a charge-balancing cation), in nafertisite Ti is coordinated to a hydroxyl group, thus leading to an octahedral environment. As in astrophyllite type compounds, the shortest Ti···Ti distance within the heterosilicate sheets also occurs along the *a* axis of the unit

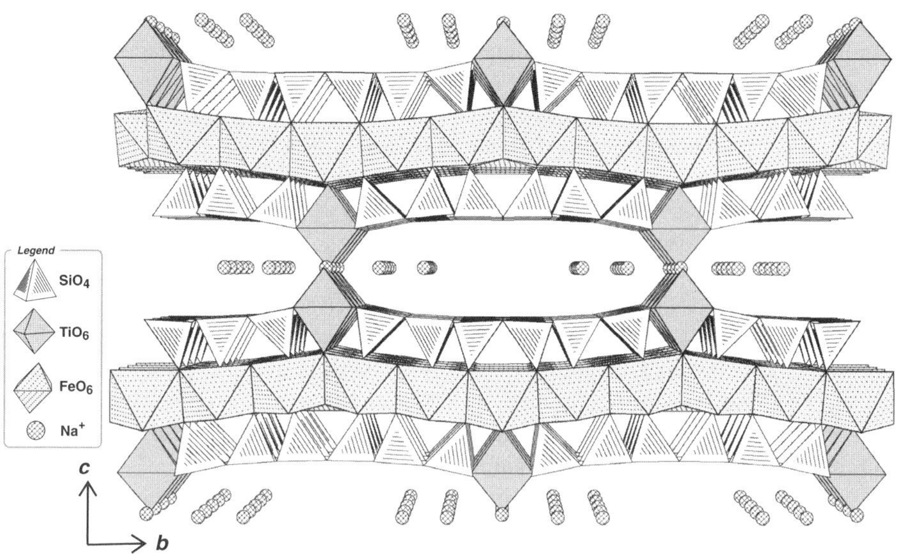

Fig. 14. Crystal packing of nafertisite viewed in perspective along the [100] direction of the unit cell.

cell being ~5.36 Å. A mineral structurally related to nafertisite is the sorosilicate caryochroite: $(Na,Sr)_3(Fe^{3+},Mg)_{10}[Ti_2Si_{12}O_{37}](H_2O,O,OH)_{17}$ (Kartashov *et al.*, 2006).

4. Prcparation mcthods

Comprehensive investigations throughout the years, focused on the preparation of tita nosilicates, have led to the isolation of the representative layered materials highlighted in the previous sections. The following paragraphs summarize the most significant advances concerning the synthesis of AM-1, AM-4 and natisite.

Anderson *et al.* (1995) first reported AM-1 as a new phase isolated during their study of the synthesis of ETS-10. Du *et al.* (1996) and Roberts *et al.* (1996) reported the synthesis and structural determination of the layered titanosilicate JDF-L1. Later, it was found that AM-1 and JDF-L1 possess the same structure (Lin *et al.*, 1997). Veltri *et al.* (2006) also investigated this phase and named their material NTS. They mainly studied the formation of this compound in pellet form and described its crystallization field and kinetic curves. Ferdov *et al.* (2002a) reported a rapid synthesis, using $TiCl_4$ as the titanium source, of a pure AM-1 phase in <16 hours. Rubio *et al.* (2009) studied the seeding effect on the formation of AM-1 and concluded that seeding decreases significantly the crystallization time to just 6 hours and also the crystal size (~100 nm thick). AM-1 was synthesized at 180–230°C using $TiCl_3$, $TiCl_4$ or tetrabutyl orthotitanate as titanium sources, and fumed silica and sodium silicate solution as Si sources (Du *et al.*, 1996; Lin *et al.*, 1997; Ferdov *et al.*, 2002a). Although the formation of AM-1 does require organic additives, crystal habit does depend on their presence or absence. The addition of triethylamine results in samples with intergrown or twinned aggregates, whereas tetrapropylammonium bromide induces square plate-like crystals (Du *et al.*, 1996).

The synthesis and characterization of the novel layered titanosilicate AM-4 under mild hydrothermal conditions were reported by Lin *et al.* (1997). This phase was synthesized at 230°C (over 4 days) using $TiCl_3$ as the titanium source and sodium silicate solution as the silicon source. The crystal morphology of AM-4 was found to be similar to that of manganokukisvumite.

The reactive Na_2O-ZnO-SiO_2-H_2O system was studied at 550°C while sealed in a titanium tube. The large NaOH content (25–30%) dissolved the titanium tube forming a titanosilicate structurally similar to the mineral natisite without zinc (Nikitin *et al.*, 1964). Later, Nyman *et al.* (1978) prepared natisite by sealing a water solution containing NaOH, TiO_2 and SiO_2 in a gold tube, at 350°C and 2 kbar. This phase was prepared more recently at 200°C (Kostov-Kytin *et al.*, 2002, 2007; Ferdov *et al.*, 2002b) using SiO_2 and $TiCl_4$ as the silicon and titanium sources, respectively.

AM-4 and natisite may also be obtained from the transformation of other microporous titanosilicates by increasing the reaction time. For example, the former was prepared from GTS-1 or ETS-4, while the latter was prepared from sitinakite and paranatisite (Kostov-Kytin *et al.*, 2007).

The synthetic analogues of jonesite, delindeite and lamprophyllite have not yet been prepared. Ferraris *et al.* (2008) reported some pioneering work on the synthesis of bafertisite at 250–450°C for 160–340 h. The best material was obtained at 380°C, 1 kbar, pH 8.5 for 160 hours and contained ~20% bafertisite.

5. Applications

Some of the compounds highlighted in the previous subsections can be employed as functional materials, either used in their as-prepared forms or by modifying the crystal structures post-synthesis. For example, AM-1 has a number of structural features which may direct interesting functional properties: (1) the TiO_5 units in the interlayer surface of AM-1 may act as the active sites for oxidation catalysis; (2) the hydrogen-bonded water molecules and Na^+ cations occupying the interlayer space (Ferdov *et al.*, 2007) may be replaced by large organic or inorganic molecules; (3) five-membered ring apertures along the [100] direction of the unit cell suggest some level of porosity. These features have already been explored and a handful of remarkable studies are already available in the literature. For example, as-prepared AM-1 is inactive in the epoxidation of cyclohexene. However, after treatment with mixed hydrogen peroxide and diluted HCl, it selectively oxidizes phenol to quinine (Roberts *et al.*, 1996). In similar fashion to the AMH-3 material (Choi *et al.*, 2008), the porous layers of AM-1 may be particularly interesting for the preparation of permselective membranes (for hydrogen separation) using a polymer-inorganic layer precursor. Rubio *et al.* (2010) studied the delamination of AM-1 and its use as additives to polymer membranes. AM-1 was first proton-exchanged using the amino acid, histidine, intercalated with

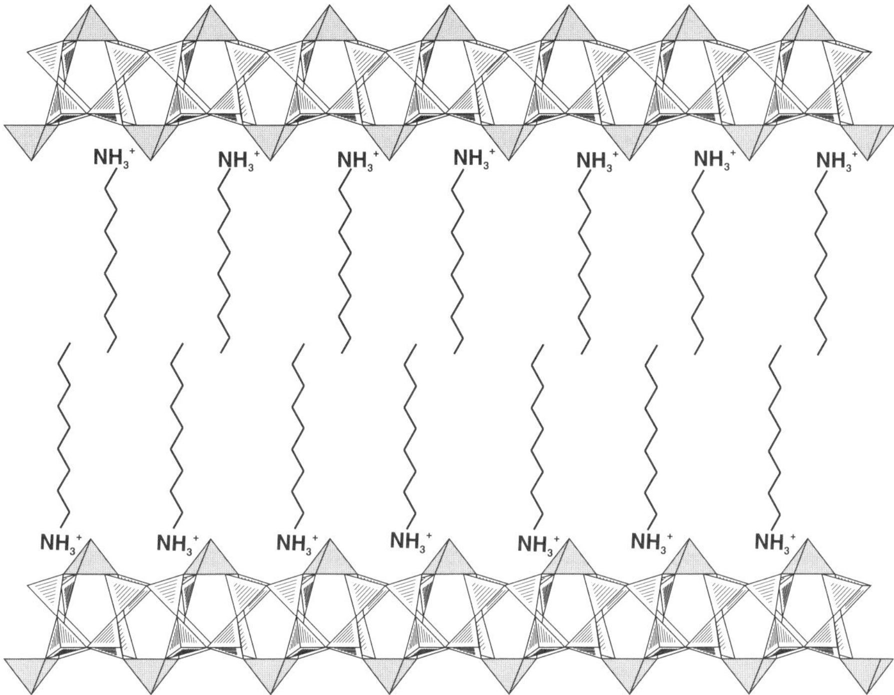

Fig. 15. Schematic representation of the intercalation of nonylammonium cations between the anionic $[Ti_2Si_8O_{22}]^{4-}$ layers of AM-1.

nonylamine and finally extracted in a mixture of HCl/water/ethanol, forming an exfoliated material composed of sheets with nanosized thickness. The addition of the delaminated AM-1 to a polysulfone membrane improved the hydrogen selectivity.

AM-1 may also be intercalated using an aqueous solution of nonylammonium salt. The nonylammonium cations form double layers with the protonated $-NH_3^+$ heads pointing towards the titanosilicate layers (Du *et al.*, 1996) (Fig. 15). The physical and chemical properties of active AM-1 materials may be modified by controlling the morphology of crystals. It is known, for example, that proton exchange of AM-1 is significantly increased as the crystals become thinner (Rubio *et al.*, 2009).

The structure of AM-1 collapses after heating the material at 580°C for just 1 hour. Nevertheless, controlled thermal treatment of AM-1 can promote a phase transition, *via* a mixture of amorphous and microcrystalline silica, into narsarsukite. This transformation occurs at 700°C (Kostov-Kytin *et al.*, 2004; Veltri *et al.*, 2006). ETS-10 and AM-3 can also be transformed into narsarsukite at ~800°C (Rocha *et al.*, 2000; Lin *et al.*, 2006).

Park *et al.* (2009) pillared AM-1 in H-AM-1/dodecylamine/TEOS(tetraethylorthosilicate) and obtained mesoporous titanosilicate materials after calcination (in air) at 500°C. The pillared material was stable at 700°C for 5 hours. Park (2010) also reported that the simultaneous intercalation of octyltriethoxysilane and dodecylamine into the titanosilicate layers of AM-1 without a pre-swelling treatment may be achieved in ethanol, providing a way to prepare the material under atmospheric conditions at low cost and with little waste.

Decaillon *et al.* (2002) studied the sorption behaviour of AM-4. Those authors noted that this material has non-bridging oxygen atoms (O_{nbo}) of the Si-O_{nbo} type, which have a high anionic charge thus favouring strontium uptake. It was found that for AM-4, among Na^+, K^+, Ca^{2+} and Mg^{2+}, Ca^{2+} is clearly the main ion competing for strontium. Thus, the material has a greater affinity for ions with high charge, with the uptake depending on the pH: in acidic and neutral conditions the uptake of strontium is exceedingly low; between pH 8 and 10, the uptake increases rapidly to 99.6%. The strontium exchange sites seem to be weakly acidic. AM-4 was also tested in natural waters showing some efficiency in strontium removal.

AM-4 has also been tested for the sorption of radionuclides (^{60}Co, ^{110m}Ag, ^{125}Sb, ^{134}Cs, ^{137}Cs, ^{235}U, ^{236}P and ^{241}Am) (Al-Attar & Dyer, 2001, Al-Attar *et al.*, 2003a, 2003b). The compound shows little affinity for Cs, Co, Sb or Ag (Al-Attar *et al.*, 2003b). Amin *et al.* (2000) reported, however, that AM-4 is a promising material for the removal of Sr and Cs, and is also suitable for actinide removal from sodium salt solutions (Al-Attar & Dyer, 2001; Al-Attar *et al.*, 2003a).

6. Conclusions and outlook

The host-guest chemistry of layered materials affords them much potential for applications in fields such as ion exchange, gas sorption and separation, and catalysis, among others. Layered titanosilicates are of particular interest because they contain large amounts of a transition metal and often exhibit a large variety of complex crystal structures. Moreover, much of the research carried out by the materials-science community, working on layered titanosilicates, is prompted by the amazing examples provided by the considerable diversity of fascinating minerals found in nature.

Here, we have reviewed the archetypal members of the family of heteropolyhedral layered titanosilicates, focusing on the step-wise construction of discrete sheets from the individual building blocks. In general, the structures of these materials encompass single layers, mica-type layers (heterophyllosilicates) and perforated 'microporous' layers.

The commonly employed synthesis methods and potential applications of layered titanosilicates are also highlighted briefly. Neither area has been explored extensively and both present many challenges and opportunities for future research. In particular, the layers of exfoliated materials may be used to build up new materials with useful applications. In this context, perforated 'microporous' sheets are particularly interesting because they are reminiscent of microporous materials with one of the dimensions at the nano scale. Thus, they may be used in catalysis, adsorption and ion exchange, displaying fast transport properties, while preserving important features, such as well defined pore sizes, catalytic sites and ion-exchange capability (although having lower structural rigidity). Clearly, if the layers possess pores parallel to the sheet, the selectivity may be improved. For example, the structure of jonesite, having an [8]-membered ring along its sheets, fulfils this requirement.

Acknowledgements

The authors are grateful to Fundação para a Ciência e a Tecnologia (FCT, Portugal) for funding.

References

Al-Attar, L. & Dyer, A. (2001) Sorption of uranium onto titanosilicate materials. *Journal of Radioanalytical and Nuclear Chemistry*, **247**, 121–128.

Al-Attar, L., Dyer, A. & Harjula, R. (2003a) Uptake of radionuclides on microporous and layered ion exchange materials. *Journal of Materials Chemistry*, **13**, 2963–2968.

Al-Attar, L., Dyer, A., Paajanen, A. & Harjula, R. (2003b) Purification of nuclear wastes by novel inorganic ion exchanges. *Journal of Materials Chemistry*, **13**, 2969–2974.

Amin, S., Harjula, R., Möller, T., Dyer, A., Pillinger, M., Newton, J., Webb, M., Araya, A. & Tusa, E. (2000) Waste volume minimisation using highly selective inorganic ion exchange materials. Final Report, Contract No. F14W-CT95-0016, EUC

Anderson, M.W., Terasaki, O., Oshuna, T., O'Malley, P.J., Philippou, A., Mackay, S.P., Ferreira, A., Rocha, J. & Lidin, S. (1995) Microporous titanosilicate ETS-10: a structural survey. *Philosophical Magazine B*, **71**, 813–841.

Appeman, D.E., Evans, H.T., Nord, G.L., Dwornik, E.J. & Milton, C. (1987) Delindeite and lourenswalsite, two new titanosilicates from Magnet Cove region Arkansas. *Mineralogical Magazine*, **51**, 417–425.

Belov, N.V., Gavrilova, G.S., Soloveeva, L.P. & Khalilov, A.D. (1977) The refined structure of lomonosovite. *Soviet Physics Doklady*, **22**, 422–424.

Cámara, F. & Sokolova, E (2007) From structure topology to chemical composition. VI. Titanium silicates: the crystal structure and crystal chemistry of bornemanite, a group III Ti-disilicate mineral. *Mineralogical Magazine*, **71**, 593–610.

Cámara, F. & Sokolova, E. (2009) From structure topology to chemical composition. X. Titanium silicates: the crystal structure and crystal chemistry of nechelyustovite, a group III Ti-disilicate mineral. *Mineralogical Magazine*, **73**, 753–775.

Cámara, F., Sokolova, E., Hawthorne, F.C. & Abdu, Y. (2008) From structure topology to chemical composition. IX. Titanium silicates: revision of the crystal chemistry of lomonosovite and murmanite, Group-IV minerals. *Mineralogical Magazine*, **72**, 1207–1228.

Carrado, K.A. (2000) Synthetic organo- and polymer-clays: preparation, characterization, and materials applications. *Applied Clay Science*, **17**, 1–23.

Centi, G. & Perathoner, S. (2008) Catalysis by layered materials: A review. *Microporous and Mesoporous Materials*, **107**, 3–15.

Chao, G.Y., Conlon, R.P. & Van Velthuizen, J. (1990) Mont Saint-Hilaire unknowns. *The Mineralogical Record*, **21**, 363–368.

Chapman, D.M. & Roe, A.L. (1990) Synthesis, characterization and crystal-chemistry of microporous titanium silicate materials. *Zeolites*, **10**, 730–737.

Choi, S., Coronas, J., Jordan, E., Oh, W., Nair, S., Onorato, F., Shantz, D.F. & Tsapatsis, M. (2008) Layered silicates by swelling of AMH-3 and nanocomposite membranes. *Angewandte Chemie International Edition*, **47**, 552–555.

Choi, M., Na, K., Kim, J., Sakamoto, Y., Terasaki, O. & Ryoo, R. (2009) Stable single-unit-cell nanosheets of zeolite MFI as active and long-lived catalysts. *Nature*, **461**, 246–250.

Chukanov, N.V., Moiseev, M.M., Pekov, I.V., Lazebnik, K.A., Rastsvetaeva, R.K., Zayakina, N.V., Ferraris, G. & Ivaldi, G. (2004) Nabalamprophyllite $Ba(Na,Ba)\{Na_3Ti[Ti_2O_2Si_4O_{14}](OH,F)_2\}$, a new layer titanosilicate of the lamprophyllite group from the Inagli and Kovdor alkaline-ultrabasic massifs, Russia. [In Russian, English abstract]. *Zapiski Vserossiyskogo Mineralogicheskogo Obshchestva*, **133**, 59–72.

Corma, A., Fornes, V., Pergher, S.B., Maesen, T.L.M. & Buglass, J.G. (1998) Delaminated zeolite precursors as selective acidic catalysts. *Nature*, **396**, 353–356.

Corma, A., Diaz, U., Domine, M.E. & Fornes, V. (2000) New aluminosilicate and titanosilicate delaminated materials active for acid catalysis, and oxidation reactions using H_2O_2. *Journal of the American Chemical Society*, **122**, 2804–2809.

Dadachov, M.S., Rocha, J., Ferreira, A., Lin, Z. & Anderson, M.W. (1997) *Ab initio* structure determination of layered sodium titanium silicate containing edge-sharing titanate chains (AM-4) $Na_3(Na,H)Ti_2O_2[Si_2O_6]_2{\cdot}2H_2O$. *Chemical Communications*, 2371–2372.

Decaillon, J.G., Andrès, Y., Mokili, B.M., Abbé, J.C., Tournoux, M. & Patarin, J. (2002) Study of the ion exchange selectivity of layered titanium silicate $Na_3(Na,H)Ti_2O_2[Si_2O_6]_2{\cdot}2H_2O$, AM-4, for strontium. *Solvent Extraction and Ion Exchange*, **20**, 273–291.

Du, H., Chen, J. & Pang, W. (1996) Synthesis and characterization of a novel layered titanium silicate JDF-L1. *Journal of Materials Chemistry*, **6**, 1827–1830.

Ercit, T.S., Cooper, M.A. & Hawthorne, F.C. (1998) The crystal structure of vuonnemite, $Na_{11}TiNb_2(Si_2O_7)_2(PO_4)_2O_3(F,OH)$, a phosphate-bearing sorosilicate of the lomonosovite group locality: Ilimaussaq, Greenland. *The Canadian Mineralogist*, **36**, 1311–1320.

Ferdov, S., Kostov-Kytin, V. & Petrov, O. (2002a) A rapid method of synthesizing the layered titanosilicate JDF-L1. *Chemical Communications*, 1786–1787.

Ferdov, S., Kostov-Kytin, V. & Petrov, O. (2002b) Improved powder diffraction patterns for synthetic paranatisite and natisite. *Powder Diffraction*, **17**, 234–237.

Ferdov, S., Kolitsch, U., Lengauer, C., Tillmanns, E., Lin, L. & Sá Ferreira, R.A. (2007) Refinement of the layered titanosilicate AM-1 from single-crystal X-ray diffraction data. *Acta Crystallographica*, **E63**, i186.

Ferraris, G. (2008) Modular structures – the paradigmatic case of the heterophyllosilicates. *Zeitschrift für Kristallographie*, **223**, 76–84.

Ferraris, G. & Gula, A. (2005) Polysomatic aspects of microporous minerals – heterophyllosilicates, palysepioles and rhodesite – related structure. In: *Micro- and Mesoporous Materials* (G. Ferraris and S. Merlino, editors). Reviews in Mineralogy and Geochemistry, **57**, Mineralogical Society of America, Chantilly, Virginia, USA, pp. 69–104.

Ferraris, G., Ivaldi, G., Khomyakov, A.P., Soboleva, S.V., Belluso, E. & Pavese, A. (1996) Nafertisite, a layer titanosilicate member of a polysomatic series including mica. *European Journal of Mineralogy*, **8**, 241–249.

Ferraris, G., Ivaldi, G., Pushcharovsky, D.Y., Zubkova, N.V. & Perkov, I.V. (2001a) The crystal structure of delindeite, $Ba_2\{(Na,K,\square)_3(Ti,Fe)[Ti_2(O,OH)_4Si_4O_{14}](H_2O,OH)_2$, a member of the mero-plesiotype bafertisite series. *The Canadian Mineralogist*, **39**, 1307–1316.

Ferraris, G., Belluso, E., Gula, A., Soboleva, S.V., Ageeva, O.A. & Borutskii, B.E. (2001b) A structural model of the layer titanosilicate bornemanite based on seidozewrite and lomonosovite modules. *The Canadian Mineralogist*, **39**, 1665–1673.

Ferraris, G., Makovicky, E. & Merlino, S. (2004) *Crystallography of Modular Materials*. Oxford University Press, Oxford, UK, 266 pp.

Ferraris, G., Bloise, A. & Cadoni, M. (2008) Layered titanosilicates – A review and some results on the hydrothermal synthesis of bafertisite. *Microporous and Mesoporous Materials*, **107**, 108–112.

Gault, R.A., Ercit, T.S., Grice, J.D. & Van Velthuizen, J. (2004) Manganokukisvumite, a new mineral species from Mont Saint-Hilaire, Quebec. *The Canadian Mineralogist*, **42**, 781–785.

Ghose, S., Kersten, M., Langer, K., Rossi, G. & Ungaretti, L. (1986) Crystal field spectra and Jahn Teller effect of Mn^{3+} in clinopyroxene and clinoamphiboles from India. *Physics and Chemistry of Minerals*, **13**, 291–305.

Kan, Q.B., Fornes, V., Rey, F. & Corma, A. (2000) Transformation of layered aluminosilicates and gallosilicates with kanemite structure into mesoporous materials. *Journal of Materials Chemistry*, **10**, 993–1000.

Kartashov, P.M., Ferraris, G., Soboleva, S.V. & Chukanov, N.V. (2006) Caryochroite, a new heterophyllosilicate mineral species related to nafertisite, from the Lovozero massif, Kola Peninsula, Russia. *The Canadian Mineralogist*, **44**, 1331–1339.

Khalilov, A.D., Mamedov, K.S., Makarov, E.S. & Pyanzina, L.A. (1965) Crystal structure of murmanite. *Doklady Akademii Nauk SSSR*, **161**, 1409–1411.

Khomyakov, A.P., Polezhaeva, L.I., Merlino, S. & Pasero, M. (1990) Lintisite, $Na_3LiTi_2[Si_2O_6]_2O_2{\cdot}2H_2O$, a new mineral. [In Russian]. *Zapiski Vsesoyuznogo Mineralogicheskogo Obshchestvo*, **119**, 76–80.

Kapustin, Y.L. (1972) Zircophyllite, the zirconium analogue of astrophyllite. [In Russian]. *Zapiski Vsesoyuznogo Mineralogicheskogo Obshchestvo*, **101**, 459–463.

Kostov-Kytin, V., Ferdov, S. & Petrov, O. (2002) Hydrothermal synthesis and successive transformation of paranatisite into natisite. *Comptes-rendus de l'Académie Bulgare des Sciences*, **55**, 61–64.

Kostov-Kytin, V., Mihailova, B., Ferdov, S. & Petrov, O. (2004) Temperature-induced phase transformations of layered titanosilicate JDF-L1. *Solid State Sciences*, **6**, 967–972.

Kostov-Kytin, V., Ferdov, S., Kalvachev, Yu., Mihailova, B. & Petrov, O. (2007) Hydrothermal synthesis of microporous titanosilicates. *Microporous and Mesoporous Materials*, **105**, 232–238.

Krishnamoorti, R., Vaia, R.A. & Giannelis, E.P. (1996) Structure and dynamics of polymer-layered silicate nanocomposites. *Chemistry of Materials*, **8**, 1728–1734.

Krivovichev, S.V. & Armbruster, T. (2004) The crystal structure of jonesite, $Ba_2(K,Na)[Ti_2(Si_5Al)O_{18}(H_2O)](H_2O)_n$, a first example of titanosilicate with porous double layers. *American Mineralogist*, **89**, 314–318.

Krivovichev, S.V., Armbruster, T., Yakovenchuk, V.N., Pakhomovsky, Y.A. & Men'shiko, Y.P. (2003) Crystal structures of lamprophyllite-2*M* and lamprophyllite-2*O* from the Lovozero alkaline massif, Kola peninsula, Russia. *European Journal of Mineralogy*, **15**, 711–718.

Kwon, O.Y. & Shin, H.S. (2000) Preparation of porous silica-pillared layered phase: simultaneous intercalation of amine-tetraethylorthosilicate into the H^+-magadiite and intragallery amine-catalyzed hydrolysis of tetraethylorthosilicate. *Chemistry of Materials*, **12**, 1273–1278.

Lin, Z., Rainho, J.P., Rocha, J. & Carlos, L.D. (2006) Preparation of photoluminescent materials from a lanthanide-doped microporous titanosilicate precursor. *Materials Science Forum*, **514–516**, 123–127.

Lin, Z., Rocha, J., Brandão, P., Ferreira, A., Esculcas, A.P., Pedrosa de Jesus, J.D., Philippou, A. & Anderson, M.W. (1997) Synthesis and structural characterization of microporous umbite, penkvilksite and other titanosilicates. *The Journal of Physical Chemistry B*, **101**, 7114–7120.

Merlino, S., Pasero, M. & Khomyakov, A.P. (1990) The crystal structure of lintisite, $Na_3LiTi_2[Si_2O_6]_2O_2{\cdot}2H_2O$, a new titanosilicate from Lovozero (USSR). *Zeitschrift für Kristallographie*, **193**, 137–148.

Merlino, S., Pasero, M. & Ferro, O. (2000) The crystal structure of kukisvumite, $Na_6ZnTi_4(Si_2O_6)_4O_4{\cdot}4H_2O$. *Zeitschrift für Kristallographie*, **215**, 352–356.

Nèmeth, N., Khomyakov, A.P., Ferraris, G. & Menshikov, Yu. P. (2009) Nechelyustovite, a new heterophyllosilicate mineral, and new data on bykovaite: A comparative TEM study. *European Journal of Mineralogy*, **21**, 251–260.

Nikitin, A.V., Ilyuhin, V.V., Litvin, B.N., Mel'nikov, O.K. & Belov, N.V. (1964) Crystalline structure of synthetic titanium sodium silicate, $Na_2(TiO)[SiO_4]$. *Doklady Akademii Nauk SSSR*, **157(6)**, 1355–1357.

Nyman, H., O'Keeffe, M. & Bovin, J.O. (1978) Sodium titanium silicate, Na_2TiSiO_5. *Acta Crystallographica*, **B34**, 905–906.

Ogawa, M., Okutomo, S. & Kuroda, K. (1998) Control of interlayer microstructures of a layered silicate by surface modification with organochlorosilanes. *Journal of the American Chemical Society*, **120**, 7361–7362.

Park, K.W. (2010) Spectroscopy characteristics for interlamellar silylation of H^+-titanosilicate using dodecylamine and octyltriethoxysilane. *Microporous and Mesoporous Materials*, **127**, 142–146.

Park, K.W., Jung, J.H., Kim, J.D., Kim, S.K. & Kwon, O.Y. (2009) Preparation of mesoporous silica-pillared H^+-titanosilicates. *Microporous and Mesoporous Materials*, **118**, 100–105.

Peacor, D.R. & Buerger, M.J. (1962) The determination and refinement of the structure of narsarsukite, $Na_2TiOSi_4O_{10}$. *American Mineralogist*, **47**, 539–556.

Piilonen, P.C., Lalonde, A.E., McDonald, A.M. & Gault, R.A. (2000) Niobokupletskite, a new astrophyllite group mineral from Mont Saint-Hilaire, Québec, Canada: Description and crystal structure. *The Canadian Mineralogist*, **38**, 627–639.

Piilonen, P.C., McDonald, A.M. & Lalonde, A.E. (2003) Insights into astrophyllite-group minerals. II. Crystal chemistry. *The Canadian Mineralogist*, **41**, 27–54.

Piilonen, P.C., Rancourt, D.G., Evans, R.J., Lalonde, A.E., McDonald, A.M. & Shabani, A.A.T. (2004) The relationships between crystal-chemical and hyperfine parameters in astrophyllite-group minerals: A combined ^{57}Fe Mössbauer spectroscopy and single-crystal X-ray diffraction study. *European Journal of Mineralogy*, **16**, 989–1002.

Rastsvetaeva, R.K. & Andrianov, V.I. (1984) Refined crystal-structure of vinogradovite. *Soviet Physics – Crystallography*, **29**, 403–406.

Rastsvetaeva, R.K. & Andrianov, V.I. (1986) New data on the crystal structure of murmanite. *Soviet Physics – Crystallography*, **31**, 44–48.

Rastsvetaeva, R.K., Eskova, E.M., Dusmatov, V.D., Chukanov, N.V. & Schneider, F. (2008) Surkhobite: revalidation and redefinition with the new formula, $(Ba,K)_2CaNa(Mn,Fe^{2+},Fe^{3+})_8Ti_4(Si_2O_7)_4O_4(F,OH,O)_6$. *European Journal of Mineralogy*, **20**, 289–295.

Roberts, M.A., Sankar, G., Thomas, J.M., Jones, R.H., Du, H., Fang, M., Chen, J., Pang, W. & Xu, R. (1996) Synthesis and structure of a layered titanosilicate catalyst with five-coordinate titanium. *Nature*, **381**, 401–404.

Rocha, J., Carlos, L.D., Rainho, J.P., Lin, Z., Ferreira, P. & Almeida, R.M. (2000) Photoluminescence of new Er^{3+}-doped titanosilicate materials. *Journal of Materials Chemistry*, **10**, 1371–1375.

Rubio, C., Casado, C., Uriel, S., Tellez, C. & Coronas, J. (2009) Seeded synthesis of layered titanosilicate JDF-L1. *Materials Letters*, **63**, 113–115.

Rubio, C., Casado, C., Gorgojo, P., Etayo, F., Uriel, S., Téllez, C. & Coronas, J. (2010) Exfoliated titanosilicate material UZAR-S1 obtained from JDF-L1. *European Journal of Inorganic Chemistry*, 159–163.

Shimojima, A., Mochizuki, D. & Kuroda, K. (2001) Synthesis of silylated derivatives of a layered polysilicate kanemite with mono-, di-, and trichloro(alkyl)silanes. *Chemistry of Materials*, **13**, 3603–3609.

Sokolova, E. (2006) From structure topology to chemical composition. I. Structural hierarchy and stereochemistry in titanium disilicate minerals. *The Canadian Mineralogist*, **44**, 1273–1330.

Sokolova, E. & Cámara, F. (2007) From structure topology to chemical composition. II. Titanium silicates: revision of the crystal structural and chemical formula of deindeite. *The Canadian Mineralogist*, **45**, 1247–1261.

Sokolova, E. & Cámara, F. (2008a) From structural topology to chemical composition. III. Titanoslicates: the crystal chemistry of barytolamprophyllite. *The Canadian Mineralogist*, **46**, 403–412.

Sokolova, E. & Cámara, F. (2008b) Re-investigation of the crystal structure of magnesium astrophyllite. *European Journal of Mineralogy*, **20**, 253–260.

Umemura, Y., Yamagishi, A., Schoonheydt, R., Persoons, A. & De Schryver, F. (2001) Fabrication of hybrid films of alkylammonium cations ($C_nH_{2n+1}NH_3^+$; n = 4–18) and a smectite clay by the Langmuir-Blodgett method. *Langmuir*, **17**, 449–455.

Veltri, M., Vuono, D., De Luca, P., Nagy, J.B. & Nastro, A. (2006) Typical data of a new microporous material obtained from gels with titanium and silicon. *Journal of Thermal Analysis and Calorimetry*, **84**, 247–252.

Wang, Z. & Pinnavaia, T.J. (1998) Hybrid organic-inorganic nanocomposites: Exfoliation of magadiite nanolayers in an elastomeric epoxy polymer. *Chemistry of Materials*, **10**, 1820–1826.

Weimer, M.W., Chen, H., Giannelis, E.P. & Sogah, D.Y. (1999) Direct synthesis of dispersed nanocomposites by in situ living free radical polymerization using a silicate-anchored initiator. *Journal of the American Chemical Society*, **121**, 1615–1616.

Wise, W.S., Pabst, A. & Hinthorne, J.R. (1977) Jonesite, a new mineral from the Benitoite Gem Mine, San Benito County, California. *The Mineralogical Record*, **8**, 453–456.

Yakovenchuk, V.N., Pakhomovskii, Ya.A. & Bogdanova, A.N. (1991) Kukisvumite – a new mineral from the alkaline pegmatites of the Khibina massif (Kola Peninsula). [In Russian]. *Mineralogicheskiy Zhurnal*, **13**, 63–67.

Yano, K., Usuki, A., Okada, A., Kurauchi, T. & Kamigaito, O. (1993) Synthesis and properties of polyimide clay hybrid. *Journal of Polymer Science Part A: Polymer Chemistry*, **31**, 2493–2498.

EMU Notes in Mineralogy, Vol. 11 (2011), Chapter 4, 151–202

Modelling of X-ray diffraction profiles: Investigation of defective lamellar structure crystal chemistry

Bruno LANSON

Mineralogy & Environments, Institut des Sciences de la Terre, Université Joseph Fourier – CNRS, 38041 Grenoble cedex 9, France
e-mail: bruno.lanson@obs.ujf-grenoble.fr

Layered minerals and materials are ubiquitous and characterized by the frequent occurrence of stacking defects. In particular, interstratification (or mixed layering), which corresponds to the intimate intergrowth of layers differing in terms of their layer thickness and/or internal structure, and stacking faults, both random and well defined, are especially common. These defects impact heavily on the reactivity of the lamellar structures. In addition, they may record the conditions of mineral (trans)formation. Determining their nature, abundance and possibly their distribution is thus an essential step in their structural characterization leading to an understanding of their reactivity. Over recent decades, modelling of X-ray diffraction profiles has proved to be an important tool which allows detailed structural identification of defective lamellar structures. The present chapter will review the basic concepts of such identification and review the literature to outline how our understanding of defective structures and mixed layers has improved over the last decade or so and to describe some of the new perspectives opened by this improvement.

1. Introduction

Layered compounds are ubiquitous on Earth as minerals (*e.g.* clay minerals and layered oxides) but also as geogenic or synthetic materials (*e.g.* carbons, layered double hydroxides, layered dichalcogenides, high-temperature superconductors). In contrast to pseudo-layered compounds, they are characterized by interatomic bonds much stronger within two-dimensional fragments/layers than between these fragments. The sustained interest in these structures arises from their reactivity which is reinforced by

This chapter is dedicated to some of those who have initiated, guided, encouraged or helped the development of my scientific activity through the years. The long-term interaction with Victor A. Drits has been an essential and inexhaustible source of inspiration, knowledge and motivation. Curiosity and stimulation also resulted from reading Robert C. Reynolds' writings and listening to his talks. Alain Meunier and Bruce Velde guided my first steps into the magical world of clay science. Alain Manceau made possible to develop my academic activity and brought both scientific rigour and opportunities to it. Gérard Besson, Alain Plançon and Boris A. Sakharov shared their experience and knowledge of diffraction effects arising from defective structures. Finally, PhD students, especially Anne-Claire Gaillot, Francis Claret, Eric Ferrage and Sylvain Grangeon, carried out a significant part of the daily simulation work and brought innovative ideas to it. To all, thank you very much both for your scientific input to this work but also for the associated human dimension. To Martine, Bertrand, Meije and Robin, thank you for your love, patience and understanding.

DOI: 10.1180/EMU-notes.11.4

their anisotropic character. In particular, the intercalation properties of lamellar compounds have drawn the attention of the materials chemistry community for several decades in an effort to prepare polyfunctional materials. Similarly, the reactivity of natural layered silicates in response to their environmental conditions has been investigated thoroughly both for its potential ability to record temperature and/or pressure palaeo-conditions, and to assess the possible impact of these minerals on their environment, in particular in the context of waste storage.

Layered materials and minerals commonly exhibit a high density of structure defects. These defects range from local defects such as isomorphous substitutions, layer vacancies, or atomic displacements, to random or well defined stacking faults induced by non-periodic layer rotations, translations or twinning, and to interstratification (or mixed layering, both terms being used hereafter as synonymous) resulting from the coexistence within a given crystal of layers having different structure, thickness, or interlayer displacement. The occurrence of stacking faults in lamellar structures is favoured by the energetic similarity of the different stacking modes, owing to the weak interactions between adjacent layers. In addition, layered materials and minerals often exhibit minute crystal sizes that may be considered as an additional type of defect because of the induced disruption of the three-dimensional (3D) crystal periodicity.

The reactivity of lamellar structures actually depends, to a large extent, on these structure defects, on their nature, abundance and distribution within crystals. This influence is exemplified by the direct relationship between surface area and reactivity, or by the specific amphiphilic properties of layered silicates when intercalated with hydrophilic and hydrophobic species in successive interlayers (*e.g.* Ijdo & Pinnavaia, 1998a, 1998b). These structure defects also exert an essential influence on the actual physicochemical properties of these layered structures. Similarly, structure defects are often considered as indicators of the mineral conditions of formation. It is therefore essential to determine these parameters to gain additional insight into these physicochemical conditions in an effort to reconstruct reaction pathways, and/or to predict the reactivity of a given material/mineral. To establish these interpretations on scientifically sound bases, a detailed structural characterization of defective lamellar structures is thus essential, and X-ray diffraction (XRD) has been the preferred method of investigation.

However, as a result of their non-periodicity or their reduced periodicity, the diffraction maxima recorded from these compounds do not strictly obey Bragg's law, thus hampering the use of conventional diffraction approaches, such as the Rietveld method. Profiles and intensities of diffraction maxima are affected by the nature, abundance, and distribution of structure defects, however, thus allowing the determination of these parameters with diffraction techniques. Drits (1997, 2003) stressed the remarkable ability of diffraction techniques to extract average structural information from crystals deprived of 3D periodicity. This author also pointed out that the reliable determination of structural and chemical heterogeneity of layered structures depends essentially on the reliable interpretation of diffraction data.

In this chapter some of the literature from the last two decades is reviewed to: (1) provide a description of the different types of structure defects that may be encountered in layered structures and of their influence on XRD patterns; (2) provide insights into the

XRD data processing specific to these defective structures and into the structural information that can be retrieved; and (3) provide a review of the recent literature describing the nature, abundance and distribution of structure defects in layered silicates, layered oxides and layered double hydroxides.

2. General background on structure defects and induced diffraction effects

2.1. Different structure defects

2.1.1. Local defects

Layered structures commonly possess complex chemical compositions together with a wide spectrum of isomorphic substitutions of cations or anions, including the presence of vacant sites. It should be noted that local defects essentially affect the structure factor by modifying the average unit-cell content. As a consequence, this type of defect can be described by the usual structural methods allowing the calculation of diffracted intensity by a given structure, such as the Rietveld refinement, and will not be detailed here. An important side effect of the high density of local defects is the minute crystal size of these lamellar minerals and materials (Meunier, 2006). This finely divided character has essential implications both for the reactivity of these structures and for their XRD signature, inducing in particular a significant peak broadening that increases as a function of $1/\cos\theta$ (Scherrer, 1918; Guinier, 1964).

2.1.2. Stacking defects

In layered structures, the weak interactions between adjacent layers result in the energetic similarity of different stacking modes, and thus to the frequent occurrence of stacking faults. In particular, stacking faults may be due to random rotation and/or translation between consecutive layers. When the extent of the random translation is limited and does not totally disrupt the scattering coherence of adjacent layers, faults may be described as fluctuations of the translation vectors about a mean value. Depending on the nature of the interactions between the layers and on the physical reasons for the translations' variations, such fluctuations were shown to be of two types (Guinier, 1964; Drits & Tchoubar, 1990). In the first type (Fig. 1), fluctuations follow a unique law for any layer pair, and this law thus describes both short- and long-range order in a layer stack. As a result, the total translation between two *n*th-nearest neighbour layers is equal to *n* times the average translation, although the translation may vary from one layer pair to the other. The effect of this type of defect is similar to the thermal motion of atoms and is characterized by a loss of intensity when increasing the diffusion vector (**s**), which is directly related to the diffraction angle ($s = (2 \times \sin\theta)/\lambda$). In the second type of fluctuations, there is no long-range correlation between the defects and the total translation between two *n*th-nearest neighbour layers is then no longer equal to *n* times the average translation. The influence of this type of defect on the XRD patterns is stronger than that of the first type, leading both to a significant intensity decrease and to peak broadening when increasing the diffusion vector.

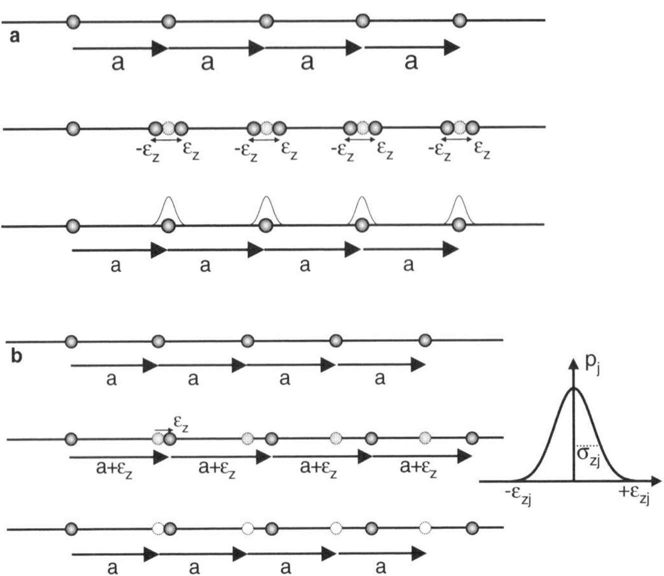

Fig. 1. Schematic description of translation vector fluctuations. (***a***) Fluctuations of the first type. (***b***) Fluctuations of the second type (see text for details – adapted from Ferrage, 2004). **a** represents the translation vector and ε_z the additional displacement that occurs with a p_j probability.

In contrast to fluctuations of the translation vectors, random stacking faults are characterized by a translation between adjacent layers the direction and amplitude of which are random, thus leading to a strikingly different distribution of deviation from the average translation (Drits & Tchoubar, 1990). When they occur, such random translations statistically disrupt the interferential coherence of the substacks between which they occur. As a result, these substacks behave independently from each other from a diffraction point of view, and the resolution of *hkl* reflections is progressively lost when increasing the proportion of such random stacking faults (Fig. 2). Random rotations induce the same effect. When random stacking faults occur systematically between successive layers (turbostratic stacking), unresolved *hkl* reflections lead to an asymmetric profile called *hk* diffraction band, characteristic of two-dimensional crystals, similar to that obtained from a single layer. These diffractions bands display a rapid increase of the diffracted intensity near the value $s_0 = |h\vec{a}^* + k\vec{b}^*|$ followed by a slow decrease.

Finally, well defined stacking faults may occur between successive layers with a given probability. In this case, it is possible to consider that different layers, characterized by their own unit-cell parameters and structure factors, coexist within the same crystals, and the diffraction effects are similar to those arising from mixed layers and described in the next section.

2.1.3. Interstratification

Mixed layers are remarkable examples of order-disorder observed in natural and synthetic lamellar crystals. They consist of the alternation either of layers exhibiting contrasting structures, compositions and basal distances or of layers having similar basal distances but differing by their internal structures or stacking mode, that is exhibiting different layer displacement or rotation between consecutive layers. The different layer types can coexist in variable proportions within the crystal and define a variety of layer stacking sequences. Mixed layering (or interstratification) has been widely recognized in natural and synthetic species such as layered silicates, layered manganates, hydrotalcites and synthetic layered double hydroxides, layered oxides in general, sulphides and intercalated graphites.

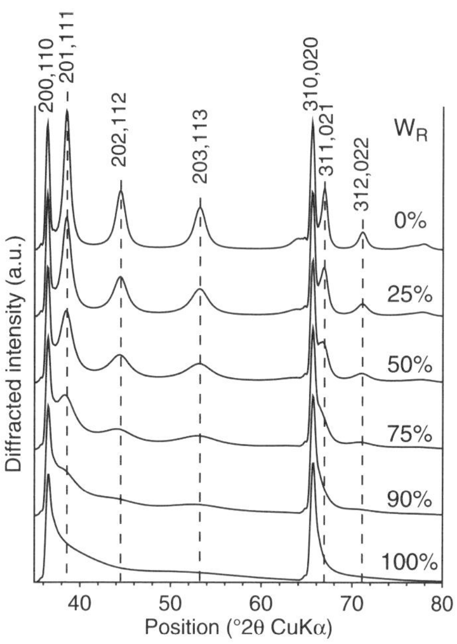

Fig. 2. Influence of random stacking faults on XRD patterns calculated for a layered manganese oxide (1*H* birnessite – adapted from Grangeon, 2008). W_R represents the occurrence probability of random stacking faults.

Two main categories of mixed layers can be established, depending on the actual distribution of interstratified layer types. The first type of mixed layers corresponds to regular structures in which different layer types alternate periodically, usually along the axis perpendicular to the layer plane (c^* axis). These mixed layers have, in many cases, been given mineral names, as they have strictly periodic structures and are often considered as distinct phases. Chlorites and corrensite are two examples of such structures that can be described as regular talc-brucite and chlorite-smectite, respectively. Within the materials-chemistry community these regular alternations of different layer types are known as the 'staging' phenomenon, the *n*th staging corresponding to the systematic occurrence of a given layer type at every *n*th layer (Fogg *et al.*, 1998a; Ijdo & Pinnavaia, 1998b). When the different layer types have the same structure but differ from their layer displacement, their regular alternation defines polytypic variants of a given mineral/species.

In the second type of mixed layers, the different layer types either alternate at random or tend to some sort of ordering (avoiding for example the existence of pairs of the minor layer type) or segregation (clustering layers of a given type). In this case and if interstratified layers have significantly different thicknesses and structures (Fig. 3), resulting peak positions do not obey Bragg's law but form a non-rational series ($d_{001} \neq l \times d_{00l}$) leading to the apparent lack of physical meaning for the observed peak positions. For a given composition, *i.e.* for given relative proportions of the different layer types, the actual position of the diffraction maxima corresponding to such a mixed layer depend on the actual layer sequences and on their relative abundances (see section 2.2.2.).

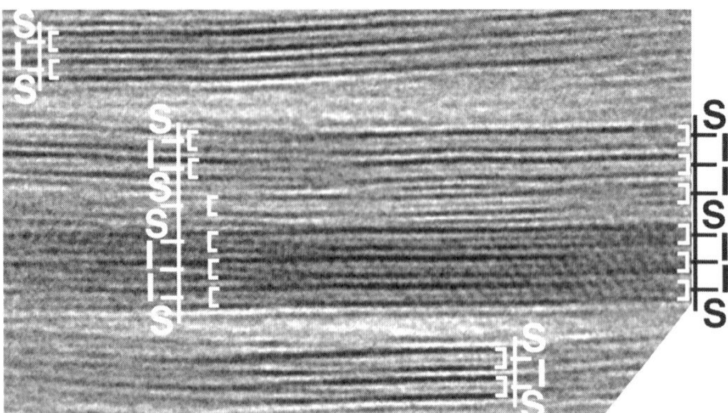

Fig. 3. Transmission election microscope (TEM) image of an illite-smectite sample (adapted from Murakami *et al.*, 2005). Brackets outline the T-O-T silicate layers; I and S accompanied by a pair of bars indicate illitic and smectitic layers, respectively. Bars near I or S are located at the centre of an octahedral cation plane. The distance between a pair of bars with I is 1.0 nm.

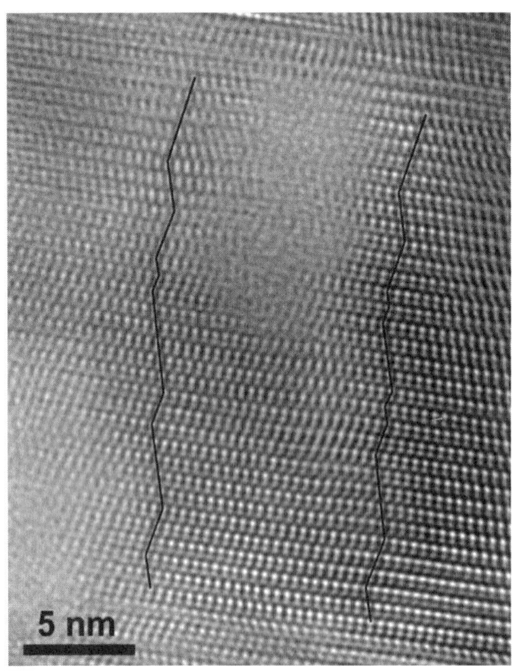

Fig. 4. Filtered high-resolution TEM image of a kaolinite crystal along one of the x_i directions (adapted from Kogure *et al.*, 2010). The solid lines connect equivalent points in successive layers and the direction changes correspond to the occurrence of well defined stacking faults.

The second type of mixed layers also includes structures in which the respective thicknesses of interstratified layers are multiples of each other (*e.g.* chlorite-serpentine). In this case, the positions of reflections corresponding to the mixed layer form a rational series, and the determination of the interstratified character of the structure then requires a more detailed analysis of peak position, profiles (especially width) and relative intensities for different reflections (Moore & Reynolds, 1997; Drits, 2003).

Finally, the second type of mixed layers includes structures in which interstratified layers have about the same thickness but distinct structures or layer displacement (Fig. 4). In this case, only the positions of non-basal reflections are affected. These reflections form non-rational series as basal reflections for mixed layers in which interstratified layers have significantly different thicknesses and structures. Within this last type of mixed layers, additional variety can arise from the possible incommensurability of the interstratified layers within the *ab* plane.

Although occurrences of interstratification reported in the literature correspond overwhelmingly to the coexistence of two layer types, both the occurrence and the description of mixed layers containing three, or more, layer types are possible. Over the last decade or so, the quantitative assessment of diffraction data has proven that such complex mixed layers may actually be extremely common; the main reasons for their rare description in the literature being: (1) the absence of a direct quantitative comparison between experimental and calculated patterns; and (2) the non-availability of adapted calculation routines.

2.2. Interpretation of diffraction data from disordered solids

Drits (2003) stressed the possibility of extracting average structural information from the diffraction patterns obtained on crystals lacking 3D periodicity. However, interpretation of XRD effects from defective compounds cannot be achieved satisfactorily with conventional XRD methods such as single-crystal diffraction and/or Rietveld structure refinement using powder diffraction data because of this reduced periodicity. This has led to the development of specific algorithms for the calculation of diffraction effects arising from defective structures.

2.2.1. Crystals exhibiting local defects and/or random stacking faults

The common occurrence of isomorphic cation or anion substitutions, including the presence of vacant sites, in layered structures essentially modifies the calculation of the structure factor. It is thus not specific to defective compounds and is accounted for by usual structural methods allowing the calculation of the intensity diffracted by a given structure. Although not described specifically in the present chapter, local defects should, however, be included in a comprehensive structural characterization of defective compounds.

Warren (1941) was among the first to propose a formalism that describes the diffraction effects arising from the presence of random stacking faults in lamellar crystals (Drits & Tchoubar, 1990; Leoni, 2008). This formalism allows us to extract structural information from the profile of *hk* bands, despite the absence of resolved *hkl* lines (Biscoe & Warren, 1942; Manceau *et al.*, 2000a; Villalobos *et al.*, 2006).

Recently, alternative approaches have been proposed to retrieve structural information from XRD patterns of such two-dimensional (2D) compounds. The first is based on the recursive description of the stacking of layers (Treacy *et al.*, 1991; Leoni *et al.*, 2004; Tsybulya *et al.*, 2004; Leoni, 2008); the second is based on the Rietveld algorithm (Rietveld, 1967; Ufer *et al.*, 2004). Both approaches have in common the ability to minimize 'automatically' the difference between experimental and calculated patterns. However, both rely on 'tricks' to mimic the diffraction behaviour of 2D periodic compounds. In the first case, as the stacking vector between successive layers must be defined, turbostratism is described as the random succession of different well defined

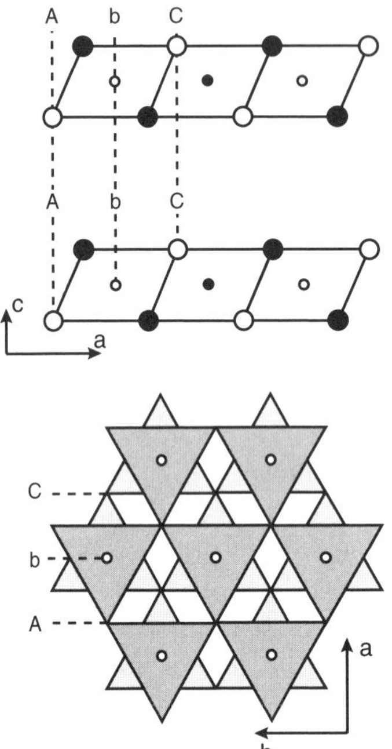

Fig. 5. Idealized structure model of layer stacking in a $1H$ polytype, or *hcp* (adapted from Drits *et al.*, 2007a). Top: projection along the **b** axis. Open and solid symbols indicate atoms at $y = 0$ and $y = \pm\frac{1}{2}$, respectively. Large circles represent O_{layer} atoms and small circles represent Mn_{layer} atoms in the phyllomanganate birnessite. Dashed lines outline specific positions of the close-packing formalism. Bottom: projection on the *ab* plane. The upper surface of the lower layer is shown as light-shaded triangles whereas the lower surface of the upper layer is shown as dark-shaded triangles. Mn_{layer} of the upper layer are shown as small open circles.

vectors (Leoni *et al.*, 2004). In the second case, the calculation of *hk* bands is performed for a large super-cell along the axis perpendicular to the layer plane, this super-cell containing a single layer. The resulting calculation is about equivalent to that performed for an equivalent super-cell containing several layers randomly shifted (Ufer *et al.*, 2004). However, in the latter case, several calculations need to be performed and averaged to smooth out the *hkl* oscillations. The two types of calculations lead to XRD patterns similar to those obtained assuming random stacking faults (Ufer *et al.*, 2004). Recently, recursive description of layer stacking has also been implemented in the *BGMN* code which aims to improve the description of turbostratic structures (Ufer *et al.*, 2008).

2.2.2. Well defined stacking defects/ interstratification

In addition to the above-described random stacking faults, lamellar compounds are characterized by the frequent occurrence of stacking faults corresponding to a well defined translation and/or rotation between adjacent layers. For example, using the close-packing notation, $1H$ layer stacking of an octahedral sheet may be described as AbC AbC AbC ... (Fig. 5). The presence of well defined faults can alter the periodicity of such stacking leading, for example, to an AbC AbC CbA CbA ... sequence (twinning – Fig. 6), or to an AbC AbC BcA BcA ... one ($+a/3$ layer displacement). Similarly $3R^-$ polytypes can be described as AbC CaB BcA AbC ... using the same close-packing notation. In this case, the occurrence of a '1H' fault corresponds to a null, rather than $-a/3$, layer displacement between two consecutive layers and results in an AbC AbC CaB BcA AbC ... sequence (Fig. 7). Such stacking faults can be described as the interstratification of different polytypic fragments.

From the diffraction point of view, the description of non-periodic crystals resulting from the presence of well defined stacking faults requires the definition of the fault nature, *i.e.* of the corresponding translation vector, of its abundance and of the probability law for its occurrence. The exact same set of parameters is required to describe the interstratification of layers having similar unit-cell parameters within the layer plane but different layer-to-layer distance. The intrinsic non-periodicity of crystals exhibiting such stacking faults, which include interstratification, hampers their structural characterization with conventional XRD methods such as single-crystal diffraction and/or Rietveld structure refinement using powder diffraction data. This has led to the development of specific algorithms for the calculation of diffraction effects arising from such defective structures consisting of layers having identical or different thicknesses.

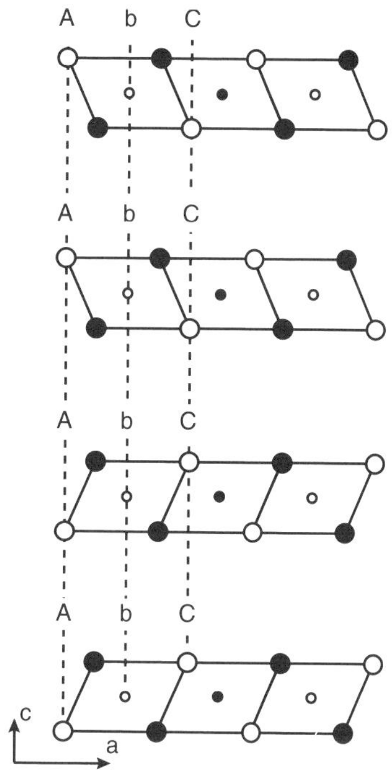

MacEwan (1956) and D'yakonov (1961, 1962) proposed direct methods for the structural analysis of mixed layers. These methods, especially that implemented by D'yakonov (1961, 1962), have been used, especially in the former USSR, to determine the thickness and relative proportions of the different layer types coexisting in mixed layers (Drits & Tchoubar, 1990 and references therein). They present several intrinsic limitations that prevent general use, however. In particular, Drits & Tchoubar (1990) outlined the inadequacy of this method to deal with mixed layers whose stacking sequences depart significantly from the maximum possible degree of ordering (MPDO) case (see section 2.3.1.). In addition and more importantly, these methods are inefficient when different phases coexist in a sample leading to overlapping reflections.

Fig. 6. Idealized structure model describing the occurrence of a twinning fault in a $1H$ polytype. The resulting sequence is described as AbC AbC CbA CbA ... using the close-packing formalism. Symbols as in Figure 5.

Alternatively, and at about the same time, indirect methods have been proposed that allow the calculation of diffraction effects arising from non-periodic interstratified structures. Hendricks & Teller (1942) were the first to propose an elegant matrix formalism allowing the calculation of diffraction effects arising from an infinite stack of interstratified layers. This formalism was further improved over the years, in particular to account for the limited extension of crystals and for a wider variety of stacking sequences (Méring, 1949; Kakinoki & Komura, 1952, 1954a, 1954b, 1965; Allegra, 1961, 1964; Cesari *et al.*, 1965; Cesari & Allegra, 1967). This formalism is still widely used, in particular among the clay community, to describe the intensities of basal ($00l$) and *hkl* reflections diffracted by a set of crystals containing different layer types (Drits & Sakharov, 1976; Plançon, 1981, 2002; Sakharov *et al.*, 1982a, 1982b; Watanabe, 1988; Drits & Tchoubar, 1990).

Another approach, which is based on the direct summation of the contributions to diffracted intensity coming from waves scattered by all possible layer subsequences in the interstratified crystals, was also developed for the calculation of XRD patterns from mixed layers (Reynolds, 1967, 1980). Finally, Treacy *et al.* (1991) proposed a recursive

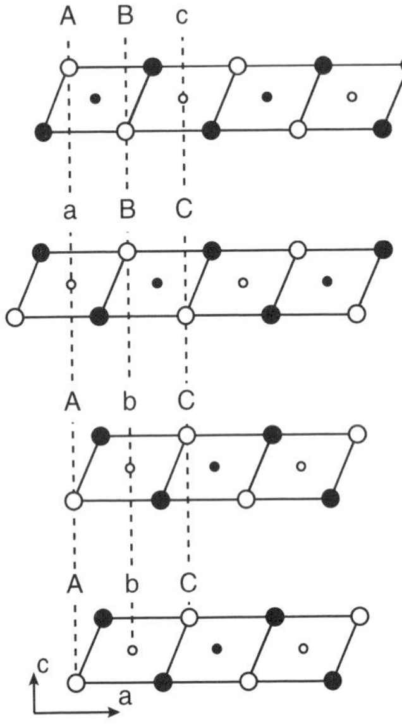

Fig. 7. Idealized structure model describing the occurrence of a $1H$ fault in a $3R^{-}$ polytype. This fault corresponds to a null, rather than $-a/3$, layer displacement between two consecutive layers and results in an AbC AbC AbC CaB BcA AbC ... sequence. Symbols as in Figure 5.

method for describing layer stacking in lamellar compounds exhibiting well defined stacking faults and/or interstratification and for calculating diffracted intensities from such compounds. This program was implemented in the *DIFFaX* code (Treacy *et al.*, 1991; Leoni *et al.*, 2004), which is used widely by the 'materials community'.

The determination of a structure model using one of these approaches usually relies on the comparison of XRD profiles calculated for a structure model with the data. The optimum fit to the data, and thus the selection of a structure model, is commonly obtained by a trial-and-error procedure. Attempts have been made to couple the above-described approaches with minimization routines in an effort to reduce the time needed to reach an optimum solution. Benefiting from the sustained interest of the oil industry for the potential of mixed layer as indicators of temperature palaeo-conditions, genetic algorithms were, for example, coupled to the Reynolds method (Pevear & Schuette, 1993) or more recently to the matrix formalism (Aplin *et al.*, 2006). Similarly, non-linear least-squares minimization routines have been coupled both to the *DIFFaX* code (Leoni *et al.*, 2004) and to the matrix approach (Plançon & Roux, 2010). The recursive approach has also been included in the general-purpose Rietveld code *BGMN* (Bergmann & Kleeberg, 1998; Ufer *et al.*, 2004, 2010). These efforts towards the coupling of routines allowing the calculation of diffraction effects from defective structures with minimization routines are beneficial, and probably necessary, for the development of these methods and of a more systematic quantitative assessment of the structure models proposed for such defective structures. Care must be taken, however, to avoid reducing the flexibility of the calculation routines while implementing the coupling.

2.3. XRD identification of defective structures

2.3.1. Calculation of XRD intensity from mixed layers

The calculation of diffraction effects arising from mixed layers would need a separate chapter of its own to enable a thorough description. It is beyond the scope of the present chapter. The interested reader will find relevant information on the direct summation method, on the recursive approach, and on the matrix formalism in Reynolds (1980, 1989) and Moore & Reynolds (1997), in Treacy *et al.* (1991) and Leoni *et al.* (2004), and in Drits & Tchoubar (1990), respectively.

Using the simple and elegant matrix formalism, the expression of the intensity diffracted by mixed layers has remained unchanged since the early works of Kakinoki & Komura (1952, 1954a, 1954b) and may be expressed as:

$$I = NSpur[V][W] + 2\mathrm{Re}\sum_{n=1}^{N-1}(N-n)Spur[V][W][Q]^n \quad (1)$$

where $[V]$ is the square matrix containing the structure factors, $[W]$ the diagonal matrix containing the occurrence probability of the different layers, layer pairs, layer triplets, . . . depending on the extent of the ordering parameter (S or Reichweite – Jagodzinski, 1949), and Q a square matrix of the product of junction probabilities describing layer stacking sequences by the corresponding phase term. S*pur* is the trace of a matrix, that is the sum of its diagonal elements, and N is the number of layers in a given crystal (Drits Tchoubar, 1990). The contents of the matrices $[V]$, $[W]$ and $[Q]$ depend on the structural model characterized by a given number of layer types and a short-range order factor, S (Reichweite, often reported in the literature as the R parameter).

To calculate diffraction effects arising from a mixed layer, the first requirement is thus to determine the number of the different layer types and their respective structures. Unit-cell parameters, together with atomic coordinates and occupancies of the different layer sites, are indeed necessary to define the structure factors and phase terms. For lamellar compounds exhibiting well defined stacking faults, the requirements are essentially the same, except that the different layer types are distinguished from their different stacking vectors within the *ab* plane. The next requirement is to describe the stacking sequences of the different layer types. For this, different sets of probability parameters are required. At first, it is necessary to determine the relative proportions (W_i) of the different layer types. Next, an essential parameter to be defined is the short-range order factor, S (Reichweite). This Reichweite parameter was originally proposed by Jagodzinski (1949) to describe how many of its neighbours will influence the presence of a given layer. Following this definition, the Reichweite parameter, S, can only have integer values. On the other hand, if this parameter defines the extent of the influence of a given layer, it provides no indication as to the nature of this influence, which is characterized by junction probability parameters. These probability parameters define the probability of finding a layer type (j) after a given layer type (i) and are usually denoted P_{ij}.

Although different models may be used to define these parameters (Plançon *et al.*, 1983, 1984a, 1984b), the quasi-homogeneous model that assumes Markovian statistics is overwhelmingly used in the literature (Kakinoki & Komura, 1952, 1954a, 1954b, 1965; Drits & Sakharov, 1976; Reynolds, 1980; Drits & Tchoubar, 1990). For illustration purposes, one may consider the case of mixed layers containing two layer types (A and B) and different sets of junction probability parameters. If one first assumes that S, the Reichweite parameter, is equal to 1 only four junction probability parameters (P_{AA}, P_{AB}, P_{BA} and P_{BB}) are required, together with the relative layer abundances (W_A and W_B) to thoroughly describe layer-stacking sequences.

These six parameters are linked by the following four relations:

$$W_A + W_B = 1 \quad (2)$$

$$P_{AA} + P_{AB} = 1, \text{ and } \quad P_{BA} + P_{BB} = 1 \quad (3)$$

$$W_A \times P_{AB} = W_B \times P_{BA} \quad (4)$$

The rationale for equation 2 is obvious. The two parts of equation 3 state that a given layer (A or B) can be followed only by A or B layers. From the first part of equation 3, it is clear that $W_A = W_A \times P_{AA} + W_A \times P_{AB}$. Similarly, the occurrence probability of A layers can be written as $W_A = W_A \times P_{AA} + W_B \times P_{BA}$, and equation 4 can easily be derived from the combination of the last two equations. As a result, there are only two independent parameters among the six probability parameters needed to describe stacking sequences in a two-component mixed layer with $S = 1$. One of these parameters should be W_A, or W_B. If $W_A > W_B$, it is convenient to choose P_{BA} or P_{BB} as the second independent parameter as their values may vary from 0 to 1. If, for example, P_{BB} is the independent parameter, then other probabilities can be defined as:

$$W_B = 1 - W_A \quad (5)$$

$$P_{BA} = 1 - P_{BB} \quad (6)$$

$$P_{AB} = (1 - W_A)/W_A \times (1 - P_{BB}) \quad (7)$$

$$P_{AA} = (2 \times W_A - 1)/W_A \times (1 - P_{BB}) \quad (8)$$

Plotting P_{AA} as a function of W_A, as proposed by Bethke *et al.* (1986), allows us to describe the whole range of possible stacking sequences for a two-component mixed layer with $S = 1$ (Fig. 8). When $P_{AA} = 1$ whatever the relative proportion of A layers, A and B layers are present in different crystals and this $P_{AA} = 1$ line corresponds to the physical mixture of periodic A and B phases. The opposite end of the diagram is defined by $P_{BB} = 0$ when $W_A > W_B$, meaning that it is impossible to find pairs of the minor layer. This corresponds to the maximum possible degree of ordering (MPDO) for $S = 1$. If $P_{BB} = 0$, then $P_{AA} = (2 \times W_A - 1)/W_A$ according to equation 8. The specific point of this line with $W_A = W_B = 0.5$ corresponds to the regular interstratification, or second staging, of A and B layers for which only ABAB … sequences are allowed. Minerals corresponding to such regular interstratification exhibit a rational series of reflections ($d_{001} = l \times d_{00l}$) corresponding to the AB sequence (super-periodicity) and can be given names if the periodiocity is sufficient (Bailey *et al.*, 1982). The domain shown to the right of this curve on Figure 3 is prohibited as it would require negative values of the P_{BB} parameter. In Figure 8 it is also possible to recognize the $P_{AA} = W_A$ line, along which the probability of finding an A layer following another A layer (P_{AA}) is equal to the relative abundance of A layers in the mixed layer (W_A). In this specific case of $S = 1$, there is no influence of the nature of a given layer on the occurrence probability of the next one and the interstratification is thus random,

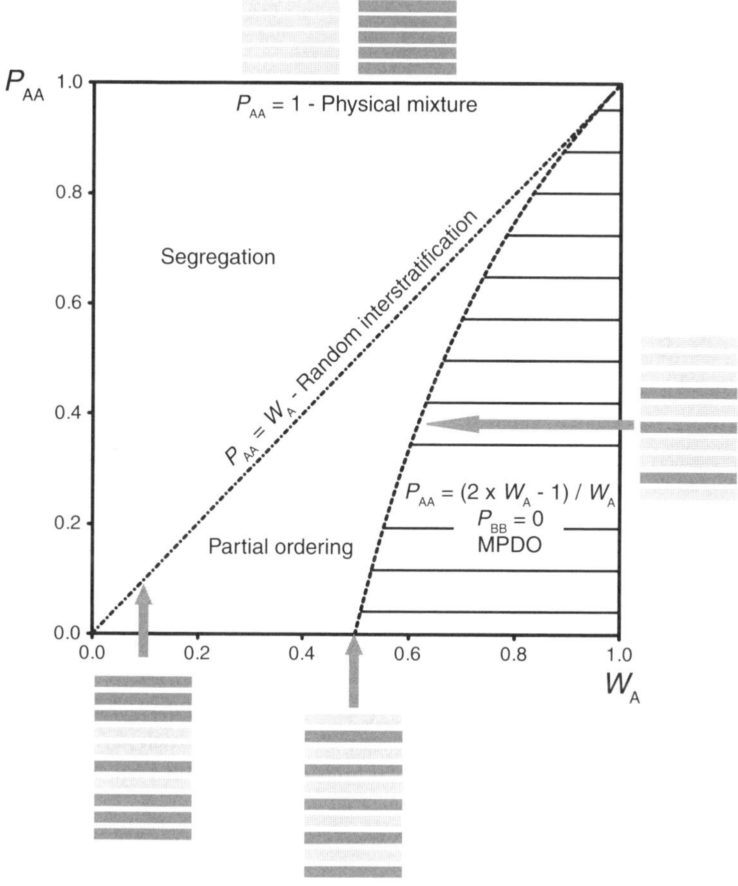

Fig. 8. Junction probability diagram for $S = 1$ two-component mixed layers (adapted from Bethke *et al.*, 1986). W_A and P_{AA} represent the relative proportion of A layers and the probability of finding an A layer after an A layer, respectively. Specific junction probabilities corresponding to random interstratification ($S = 0$) and maximum possible degree of ordering (MPDO) are shown as dotted-dashed and dashed lines, respectively. Domains corresponding to physical mixture, segregation and partial ordering are labelled. Possible stacking sequences corresponding to the different cases are schematized.

which is commonly described as the $S = 0$ case. The domains lying between physical mixture and random interstratification, on the one hand, and between random interstratification and maximum possible degree of ordering, on the other hand, are the domains of segregation and partial ordering, respectively. In the first domain, the presence of pairs of the minor layer is favoured ($1 > P_{AA} > W_A$ when $W_A < 0.5$), whereas this presence is hindered in the partial ordering domain ($0 < P_{AA} < W_A$ when $W_A < 0.5$), and prohibited ($P_{AA} = 0$) along the MPDO line. The relative proportions of the different stacking sequences, which are described by junction probability parameters, have an essential influence on diffraction patterns calculated for a given

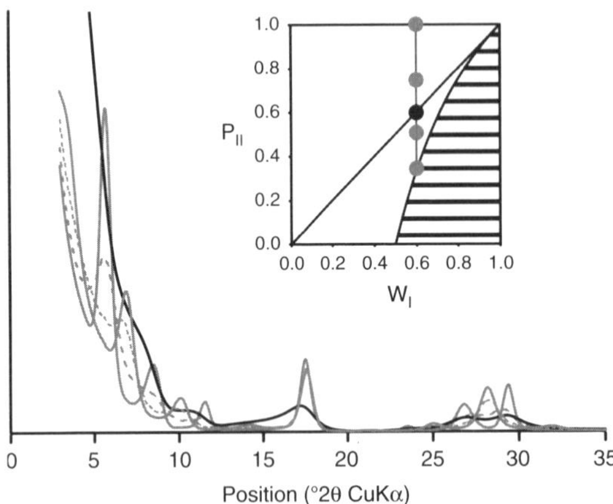

Fig. 9. XRD patterns calculated for the interstratification of illite and smectite layers as a function of junction probability parameters. Mixed-layer composition is constant ($W_I = 0.60$). Physical mixture ($P_{II} = 1.00$), segregation ($P_{II} = 0.733$), random interstratification ($P_{II} = 0.60$), partial ordering ($P_{II} = 0.50$), and MPDO ($P_{SS} = 0.00$, $P_{II} = 0.333$) cases are shown respectively as solid orange, dashed orange, solid black, dashed blue, and solid blue lines. The inset shows, in a diagram similar to Figure 8, the composition of the calculated mixed layer as a solid red line. Dots represent specific values of junction probability parameter P_{II} for which calculations were performed. Readers of the paper version of this chapter may wish to download a colour version of this figure from www.minersoc.org/emu-notes/emu-11/11-4-colour.pdf.

composition (relative proportions of the different layer types) and Reichweite parameter (Fig. 9). Occurrence and junction probability parameters are thus essential components of the structure model and should be reported as such (Bailey *et al.*, 1982; Guggenheim *et al.*, 2006).

This description provides a brief overview of the essential concepts and terms required for the description of stacking sequences in mixed layers and in lamellar compounds with well defined stacking faults. It should be noted however that the simplified view of stacking sequences offered by Figure 8 is hardly transposable for longer-range ($S > 1$) interactions that require first the definition of junction probability parameters describing the short-range ($S = 1$) order, which does not necessarily correspond to the MPDO case. Similarly, the description of multicomponent mixed layers requires the definition of additional probability parameters. A more complete, although not comprehensive, description of occurrence and junction probability parameters can be found in Reynolds (1980) and Drits & Tchoubar (1990).

2.3.2. *Usual identification methods*

In the middle of the 20th century, Jacques Méring proposed theoretical developments allowing the prediction of diffraction effects arising from the random interstratification of layers having different basal spacings (Méring, 1949). According to that author, basal reflections corresponding to a randomly interstratified mixed layer (two-component) are located between neighbouring 00*l* reflections corresponding to periodic structures whose layers are interstratified. The actual position of these reflections depends on the relative proportions of the interstratified layer types, and on the respective intensity of the neighbouring 00*l* reflections. In addition, the breadth of the mixed-layer reflections increases with the 'distance' between the 00*l* reflections corresponding to periodic structures whose layers are interstratified ('Q-rule' of Moore & Reynolds, 1997). More generally, positions of basal reflections corresponding to an ordered A-rich A-B with

$S = 1$ or $S = 2$ are located between neighbouring 00*l* reflections corresponding to periodic stacking sequences consisting of A and AB ($S = 1$), or AAB ($S = 2$), fragments (Drits *et al.*, 1994). More recently, Drits & McCarty (1996) showed that the so-called 'Méring's rules' can be extended to predict the positions of non-basal reflections arising from randomly interstratified two-component systems consisting of layers having identical thicknesses. They showed, in particular, that non-basal reflections from such mixed layers are located between the nearest *hkl* reflections of the periodic phases whose fragments are interstratified, the actual position of these non-basal reflections depending on the relative proportions of interstratified fragments.

The predicted shift of peak position as a function of the actual composition of the mixed layer has allowed the development of simplified identification for commonly reported mixed layers. In natural environments, interstratification is especially widespread among clay minerals (phyllosilicates) which differ in the type of interstratified layers and in their stacking sequences. Because of the reactivity of the frequently interstratified expandable layers and of their resulting ability to evolve as a function of physicochemical conditions, these mixed layers have drawn special attention for decades in an effort to use them as indicators of palaeo-temperature conditions and/or of reaction progress (Hower & Mowatt, 1966; Perry & Hower, 1970; Lanson *et al.*, 2009). Several simplified methods have thus been proposed over the years for the identification of illite-smectite (Środoń, 1980, 1984; Watanabe, 1981, 1988; Velde *et al.*, 1986; Inoue *et al.*, 1989; Drits & Plançon, 1994; Drits *et al.*, 1994; Plançon & Drits, 1994, 2000). These methods rely essentially on peak-migration curves linking the position of a given reflection (or of a given set of reflections) to the composition (relative proportion of the different layer types) of the mixed layers. The curves were derived essentially from the calculation of XRD patterns using either *Newmod*, developed by Reynolds, or the program based on the matrix formalism developed by Watanabe. The intensity ratio between some of these reflections, or between reflections and 'background', is used occasionally as an additional criterion to estimate the relative contents of the different layer types in the mixed layers.

Despite their widespread use, these simplified identification methods have major drawbacks. The first is the lack of direct comparison between experimental and calculated patterns, which is intrinsic to the approach. As a result, there is no way to assess the validity of the identification by using a parameter measuring the 'goodness of fit' as is usual in structural studies. Direct comparison of experimental XRD patterns with those calculated on the basis of the identification performed allows us to refute peak position as a valid criterion for mixed-layers structure characterization (Claret *et al.*, 2004; McCarty *et al.*, 2008), and thus these simplified identification methods. Another drawback of these methods comes from the use of a unique XRD pattern for identification purposes, which does not allow the validation of the proposed identification by independent XRD measurements on the same sample submitted to different treatments (see section 2.3.3 on the description of the multi-specimen method).

In addition, the profiles of the diffraction lines, which are strongly affected by interstratification effects, are not taken into account by these peak-position methods. Other drawbacks are the limitations of the programs used to calculate diffraction effects arising from mixed layers, and the limited range used for variable parameters in order to (over)simplify the identification process. Intrinsic limitations of the programs include, for example, their inability to calculate diffraction effects from multi-component mixed layers. Essential adjustable parameters which are insufficiently varied include the size and distribution of the coherent scattering domains, as the calculations are most often restricted to a single mean value, and junction probabilities. Most methods include calculations only for randomly interstratified mixed layers ($S = 0$, $P_{ii} = W_i$) and for the sole MPDO case for higher values of the Reichweite parameter, S, despite the key role of junction probabilities in the profiles of XRD patterns calculated for mixed layers (Fig. 9).

2.3.3. Multi-specimen method for XRD identification of mixed layers

Compared to usual structure determination and refinement methods, occurrence and junction probability parameters represent additional variable parameters required to describe the non-periodicity of stacking sequences. The quantitative comparison between experimental diffraction patterns and those calculated for the proposed structure model should thus be considered as a minimum requirement for the validation of this model in the case of mixed layers or of lamellar compounds exhibiting stacking faults. As the number of adjustable parameters is actually increased compared to periodic structures, it is valuable to consider additional constraints to the structure model.

These constraints are especially useful because of the low sensitivity of diffraction to the actual nature of structural disorder as illustrated by the common possibility of fitting the data equally well with different structure models (Drits, 1985). Experimental XRD patterns reported in Figure 10 are commonly attributed to randomly interstratified illite-smectite ($S = 0$, $W_S > W_I$) with a set of strong reflections whose positions are consistent with those of smectites, both in air-dried state (AD) and following ethylene-glycol solvation (EG), and a high low-angle shoulder for the 001 smectite peak (Inoue *et al.*, 1989). McCarty *et al.* (2009) and Lanson *et al.* (2009) reported that such XRD patterns could be described equally well assuming two distinct structure models (Fig. 10). The first assumes an illite-smectite in which smectite layers, the hydration and expansion behaviour of which is systematically heterogeneous, prevail. Junction probability parameters of this short-range ($S = 1$) mixed layer indicate the segregation of expandable layers ($P_{Exp\text{-}Exp} > W_{Exp}$). According to the alternative model, XRD patterns instead correspond to the physical mixture of discrete smectite, exhibiting again an heterogeneous hydration and expansion behaviour, and of a randomly interstratified illite-smectite in which illite layers prevail ($W_I > W_{Exp}$). Although the fit quality is improved slightly when using the mixture model (Lanson *et al.*, 2009), the two models provide satisfactory fits to the data, both in AD- and EG-solvated states (Fig. 10).

This example illustrates the need for (1) a quantitative comparison of XRD data with calculated patterns, and (2) additional constraints to assess the validity of structure models. Within this scope, and in contrast to most usual mixed-layer identification methods, the multi-specimen method requires the recording of XRD patterns for each sample after different treatments (*e.g.* Ca-saturated in AD- and/or EG-solvated states, and/or Na-saturated in AD-and/or EG-solvated states). For a given sample, XRD patterns

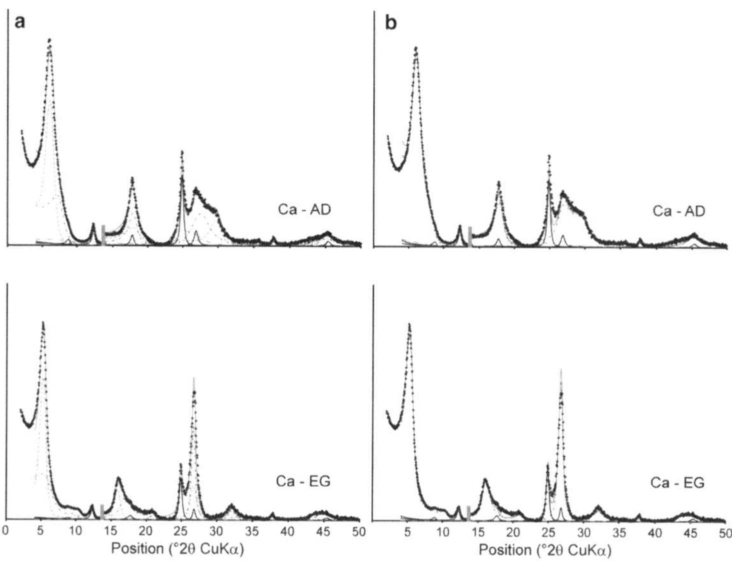

Fig. 10. Comparison between experimental and calculated XRD patterns for a diagenetically altered clayey sample from the Gulf Coast (adapted from Lanson *et al.*, 2009). XRD patterns were collected on Ca-saturated samples either air-dried and equilibrated at 40% relative humidity (Ca-AD – top) or solvated with ethylene glycol (Ca-EG – bottom). The grey bar indicates a modified scale factor for the high-angle region. (***a***) The structure model includes a randomly interstratified illite-smectite (57:43 ratio) and discrete smectite (dotted-dashed blue line and dashed red line, respectively) in addition to a randomly interstratified kaolinite-smectite (purple line), discrete kaolinite and discrete illite. (***b***) The structure model includes illite-smectite (45:55 ratio) with a tendency to segregation (dotted-dashed blue line) in addition to a randomly interstratified kaolinite-smectite (purple line), discrete kaolinite and discrete illite. Readers of the paper version of this chapter may wish to download a colour version of this figure from www.minersoc.org/emu-notes/emu-11/11-4-colour.pdf.

usually differ significantly after these treatments owing to the contrasting hydration/expansion properties of expandable layers. It is then possible to extract additional constraints on the actual structure of the different minerals present.

The method itself was proposed initially by Drits *et al.* (1997) and Sakharov *et al.* (1999a, 1999b), and consists of direct comparison of experimental XRD profiles with those calculated for a structure model. The optimum agreement between experimental and calculated XRD patterns is obtained by a trial-and-error procedure. Structure models include for each mixed layer, the number (not limited to 2), the nature and proportions of the different layer types and a statistical description of their stacking sequences (Reichweite parameter and junction probabilities). The different treatments may change the thickness and the scattering power (nature, amount and position of interlayer species) of the expandable interlayers but not the distribution of the different layer types. A consistent structure model is thus obtained for one sample when the stacking sequences of the different layer types obtained from all experimental XRD profiles of the same sample are nearly identical. In addition to these structural parameters, relative contributions of the various phases (including mixed layers) to the different XRD patterns recorded for the same sample must be similar for polyphasic clay parageneses (Sakharov *et al.*, 1999a; Claret *et al.*, 2004).

By constraining the structural characterization of a given sample from different experimental XRD patterns, the multi-specimen method allows us to overcome one major intrinsic limitation of the XRD identification of clay minerals. This limitation arises from the strong tendency of XRD to average structural parameters describing the periodicity of crystals. The resulting low sensitivity of XRD to variation in local disorder can allow for the possibility of several structure models giving rise to similar diffraction effects for a given set of experimental conditions. To determine the actual structure model, additional constraints obtained from the analysis of different XRD patterns obtained from the same sample after different treatments are thus essential (Sakharov *et al.*, 1999a; McCarty *et al.*, 2004). In addition, this method allows for a semi-quantitative phase analysis of the clay fraction including both discrete and mixed-layer clay phases in addition to the detailed structural characterization of these different components.

3. Structural characterization of defective layered structures

In the following sections, a variety of examples from the recent literature, in which defective lamellar compounds have been reported and/or the structure of which has been determined, is reviewed. Studies where the structure model was checked against the data have been favoured in the selection process required by the large number of examples available. Emphasis was placed on the variety of structural parameters required for a thorough and realistic description of their actual structure. The first objective of this review is to illustrate the potential of the different approaches for the structural characterization of non-periodic compounds. Limitations of the different approaches will also be assessed.

3.1. Layered silicates: structural characterization of mixed layers the elementary layers of which differ from their basal spacings

3.1.1. Layered silicates: hydration and expansion heterogeneity, intercalation

Because of their ability to incorporate and to exchange a variety of cations and molecules in their interlayer spaces, layers having different basal distances often coexist in expandable layered silicates (smectites), despite the overall homogeneity of layer charge. For example, Méring & Glaeser (1954) reported the interstratification of smectite layers with contrasting interlayer cation contents resulting from a partial exchange of Na by Ca. The random interstratification of Ca- and Na-saturated smectites was particularly visible under low relative humidity (RH) conditions from peak positions intermediate between those characteristic of the two end-members. At 5% RH, Na-saturated smectite was indeed dehydrated (0W smectite – $d_{001} \approx 9.6$ Å) whereas Ca-saturated interlayers incorporate a unique plane of H_2O molecules (monohydrated, or 1W, smectites – $d_{001} \approx 12.4$ Å).

The contrasting hydration energy of alkali and alkaline earth cations result, for a given RH value, in different hydration states for smectite layers depending on the interlayer cation. This specific behaviour allows us, for example, to prove the coexistence of different cations in smectite interlayers if the hydration contrast and/or the interlayer electronic density contrast is sufficient (Iwasaki & Watanabe, 1988; Ferrage *et al.*, 2005d). These authors used such a model to demonstrate that within a given interlayer, all cations are exchanged at the same time, leading to the segregation of the different cations in different interlayers. As a result, diffraction patterns indicate the interstratification of layer types whose contrasting hydration, and thus basal spacing, is characteristic of their cationic content. Oueslati *et al.* (2009) used a similar approach to assess the selectivity of smectite for different cations as a function of the ionic strength. In smectite samples saturated with bi-cationic solutions, they determined the relative proportion of mono- and divalent cations from the modelling of XRD patterns, each cation being characterized by its specific hydration behaviour. They deduced that at low ionic strength, smectites favour the presence of cations with low hydration over more hydrated cations.

Interstratification of smectite layers exhibiting different hydration states was reported also in homoionic smectite (Méring & Glaeser, 1954; Walker, 1956; Glaeser *et al.*, 1967; Sato *et al.*, 1992, 1996). More recently, the potential of smectite interlayer water as a vector for the migration of contaminants (*e.g.* radionuclides) or as a water 'reservoir' in deep geological formations has renewed the interest in its study. In particular, smectite hydration has been thoroughly described as a function of (1) the relative humidity under ambient pressure and temperature conditions, (2) the amount and location of layer charge, (3) the interlayer cation (Bérend *et al.*, 1995; Cases *et al.*, 1997; Calarge *et al.*, 2003; Meunier *et al.*, 2004). Ferrage *et al.* (2005a, 2005b, 2007c, 2010) and Karmous *et al.* (2009) showed that heterogeneity, rather than homogeneity, is the rule for smectite hydration even for homoionic specimens under controlled RH conditions. For such samples, these authors showed the systematic interstratification of layer types exhibiting contrasting hydration states. Moreover, the contributions of various randomly interstratified mixed layers are commonly required to account for the sample heterogeneity (Fig. 11). Although these contributions do not imply the actual presence of distinct (sub-)populations of particles in the sample, they indicate the partial segregation of the different hydration states within the sample.

A similar modelling approach could be implemented for a variety of samples in order to understand, for example, the link between smectite hydration and the amount and location of charge deficit. Ferrage *et al.* (2005b, 2007c) and Karmous *et al.* (2009) performed such a study on di- and trioctahedral smectites, respectively, suggesting in both cases the increased hydration heterogeneity in tetrahedrally substituted smectites. Ferrage *et al.* (2005b, 2007c) also proposed a conceptual model to reconcile the greater hydration of smectite, as estimated from the number of interlayer H_2O molecules, when increasing the layer charge deficit and the associated lower value of the basal distance. In a remarkable effort to overcome the intrinsic time-consuming character of the trial-and-error approach, Ferrage *et al.* (2007a) modelled hundreds of XRD traces obtained *in situ* while dehydrating smectite at different temperatures. The relative

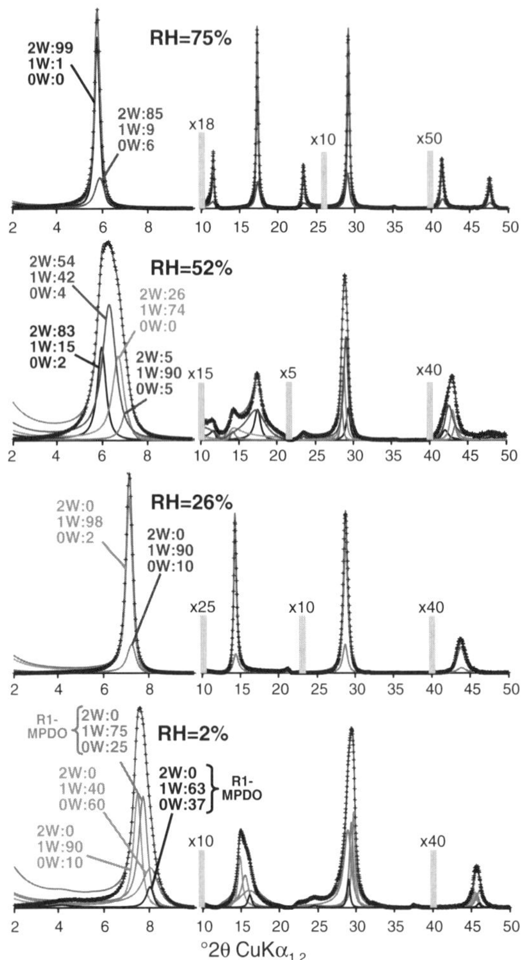

Fig. 11. Comparison between experimental and calculated XRD patterns as a function of RH for a low-charge saponite (trioctahedral smectite – adapted from Ferrage *et al.*, 2010). Respective contributions of the various mixed layers to the calculated profiles are shown in different colours. Experimental and calculated XRD patterns are shown as crosses and solid red lines, respectively. Other symbols as in Figure 10. Readers of the paper version of this chapter may wish to download a colour version of this figure from www.minersoc.org/emu-notes/emu-11/11-4-colour.pdf.

proportions of the different hydration states were thus determined as a function of temperature and dehydration time and served as the basis for the derivation of smectite dehydration kinetic parameters (Ferrage *et al.*, 2007b).

In most of the studies described above, interstratification of smectite layers with different cation content or hydration is essentially random, thus supporting the assertion that 'spontaneous segregation of diverse guest species into separate but stacked interlayers by a process comparable to staged heterostructures in graphite is rare' (Ijdo *et al.*, 1996). Walker (1956) reported, however, the ordered interstratification of 0W and 1W smectites layers from the presence of a 20.6 Å (9.0 + 11.6 Å) superperiodicity exhibiting >20 reflections as evidence of the high degree of regularity of the mixed layer. Additional reports of such ordered interstratification of smectite layers with different hydration states commonly describe this phenomenon as affecting only a minor part of the sample with much reduced regularity, at least for natural samples (Moore & Hower, 1986; Chipera & Bish, 2001; Ferrage *et al.*, 2007c).

Regular alternation, or interstratification, of organic and inorganic interlayer cations was, however, achieved successfully in an effort to prepare materials with amphiphilic properties from smectites (Ijdo & Pinnavaia, 1998a, 1998b). Second staging was identified from the presence of a reflection corresponding to the 1:1 regular interstratification ($S = 1$, MPDO) of layers with sodium and alkylammonium interlayer cations as a result of the partial exchange of sodium. Such regular interstratification was also achieved with a synthetic smectite for different inorganic cations having contrasting hydration energies (Möller *et al.*, 2010a, 2010b). In both cases, the regular interstratification probably originates from the increased charge homogeneity of the synthetic smectites used compared to natural varieties. Charge homogeneity favours the alternation of cations with different hydration properties in successive interlayers to optimize local charge compensation. Möller *et al.* (2010b) actually hypothesized, from ^{23}Na magic angle spinning nuclear magnetic resonance (MAS NMR) spectroscopy data, different contents of interlayer cations in successive interlayers despite the remarkably homogeneous layer-charge distribution (Breu *et al.*, 2001).

3.1.2. Natural occurrence of multi-component mixed layers

Even though mixed layers containing more than two layer types have been described only recently in the wealth of literature devoted to the structural characterization of these minerals, such multi-component (3- but also 4-component) mixed layers are most likely overwhelmingly present, and more especially so in natural samples. Heterogeneity rather than homogeneity is the rule for smectite hydration and expandability, even when working with homoionic samples under controlled RH conditions. Although "it is not certain that single layers of smectite or of vermiculite sandwiched between layers of a different type in an interstratification react in the same way as the non-interstratified minerals would react to solvation, hydration, and dehydration tests" (Bailey *et al.*, 1982), heterogeneity is probably common in mixed layers implying expandable layers. Consistently, all recent studies performed with calculation algorithms allowing the calculation of their XRD patterns have led to the identification of mixed layers that include more than two components as the result of the systematic heterogeneity of the swelling/hydration behaviour of expandable layers (Drits *et al.*, 1997, 2002a, 2002b, 2004, 2007b; Sakharov *et al.*, 1999a, 1999b, 2004a; Lindgreen *et al.*, 2000, 2002, 2008; Claret *et al.*, 2004; McCarty *et al.*, 2004, 2008; Inoue *et al.*, 2005; Aplin *et al.*, 2006; Hubert *et al.*, 2009; Lanson *et al.*, 2009). It should be noted that heterogeneity is systematically present whatever the chemistry (both Fe- and Al-rich) of expandable layers (McCarty *et al.*, 2004). In addition, hydration behaviour exhibited by expandable layers in smectite is not restricted to the usual 0W (d_{001} = 9.6–10.0 Å), 1W (d_{001} = 12.0–12.8 Å), and 2W (d_{001} = 14.8–15.6 Å) states and 'unusual' basal spacings have been reported for hydrated expandable layers (*e.g.* Lindgreen *et al.*, 2002; Lanson *et al.*, 2009; Fig. 12) as suggested by Bailey *et al.* (1982). This heterogeneity evidently proscribes the use of simplified peak-migration methods for the identification of mixed layers. Heterogeneity is actually not restricted to expandable layers, and the coexistence of different mica-like layers in a given mixed layer has also been reported, thus reinforcing the inability of these methods to provide a satisfactory identification of mixed layers. In particular, the coexistence of K^+ and NH_4^+ has been demonstrated repeatedly in the context of burial diagenesis in the vicinity of source rocks (Sakharov *et al.*, 1999a; Lindgreen *et al.*, 2000; Drits *et al.*, 2001, 2002a, 2005, 2007b).

In addition to the mixed layers whose multi-component character originates from the heterogeneity of the expandable or micaceous layers, mixed layers containing three really distinct layer types have been reported. In particular, complex mixed layers

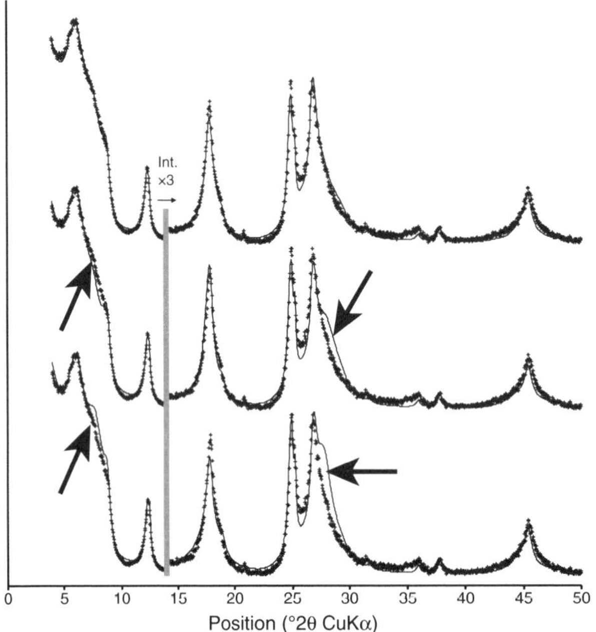

Fig. 12. Comparison between experimental and calculated XRD patterns for a diagenetically altered clayey sample from the Gulf Coast (adapted from Lanson *et al.*, 2009). XRD patterns were collected on Ca-saturated samples air-dried and equilibrated at 40% relative humidity. In addition to discrete illite, kaolinite, smectite, and chlorite, the optimum structure model includes a randomly interstratified kaolinite-smectite and a randomly interstratified illite-smectite (62:38 ratio). From top to bottom, the relative proportion of 15.00, 14.00, and 12.50 Å expandable layers in the latter mixed layer are 15:8:15 (optimum values), 23:0:15 and 15:0:23. Symbols as in Figure 10.

involving the coexistence of illite and expandable layers together with di-trioctahedral chlorite (sudoite) layers, seldom described as a discrete phase, have been described repeatedly especially in the context of burial diagenesis of clayey sediments (Lindgreen *et al.*, 2002, 2008; Lanson *et al.*, 2009) but also in soils (Hubert *et al.*, 2011). The interstratification of illite and expandable layers with kaolinite, rather than chlorite, layers has also been reported in a burial-diagenesis context (Drits *et al.*, 1997; Sakharov *et al.*, 1999b). More seldom-described complex mixed layers include berthierine-chlorite-smectite (McCarty *et al.*, 2004) and mica-pyrophyllite-smectite (Drits *et al.*, 2007b).

A variety of mixed layers other than those listed above has been reported over the last decade or so. Interstratification of chorite and expandable layers has, for example, been described both in superficial (Hubert *et al.*, 2009) and deep environments (Lindgreen *et al.*, 2002). Beaufort *et al.* (1997) challenged the commonly accepted description of the saponite-to-chlorite conversion series reporting the existence of chlorite-corrensite, both from XRD data and lattice-fringe images obtained under the transmission electron microscope (TEM). Chlorite layers can also be interstratified with 1:1 phyllosilicates (serpentine – Drits *et al.*, 2001, 2007b; Lindgreen *et al.*, 2002; McCarty *et al.*, 2004). Serpentine layers are usually required to fit the increased breadth of odd order chlorite reflections, compared to even ones, and their relative low intensities compared to the ideal intensity distribution calculated for a periodic chlorite with the correct composition (Moore & Reynolds, 1997; Drits *et al.*, 2001). Kaolinite-smectite (Sakharov & Drits, 1973; Sudo & Shimoda, 1977; Brindley *et al.*, 1983; McCarty *et al.*, 2004; Lanson *et al.*, 2009), serpentine-smectite (Sakharov *et al.*, 2004a), illite-chlorite (Hubert *et al.*, 2011), and illite-kaolinite (Hubert *et al.*, 2011) have also been reported.

3.1.3. Additional complexity of naturally occurring mixed layers

Using a whole-pattern profile-modelling approach, such as the multi-specimen approach, rather than the position of the main diffraction lines to determine the clay paragenesis allowed the uncovering of their frequent polyphasic character. Various mixed layers are indeed frequently reported as coexisting with discrete phases (Drits *et al.*, 1997, 2002a, 2002b, 2004, 2007b; Sakharov *et al.*, 1999a, 1999b; Lindgreen *et al.*, 2002, 2008; Claret *et al.*, 2004; McCarty *et al.*, 2004, 2008, 2009; Aplin *et al.*, 2006; Hubert *et al.*, 2009; Lanson *et al.*, 2009; Ferrage *et al.*, 2011b). Two major factors contribute to this recently revealed mineralogical complexity. The first is the coexistence of mixed layers having similar although distinct compositions or ordering parameters owing to the kinetics of clay mineral (trans)formation processes at low temperature. The coexistence of such mixed layers, whose contributions to the diffracted intensity are closely related, modifies the reflection profiles, and more especially their low-intensity side tails. In particular, mixed layers that do not contribute to any of the well defined maxima can be overlooked completely (Fig. 13). The second factor, which is not necessarily separate from the first, is the frequent presence of mixed layers characterized by broad and poorly defined contributions to the diffracted intensity, especially in the low-angle region (Fig. 13). Careful XRD profile fitting has, for example, allowed us to reveal the frequent presence of randomly interstratified illite-smectite with a high illite content (60–70% illite) in diagenetically altered sediments, soils, *etc.* (Claret *et al.*, 2004; Lanson *et al.*, 2005, 2009; Aplin *et al.*, 2006; Lindgreen *et al.*, 2008; McCarty *et al.*, 2008, 2009; Hubert *et al.*, 2009). The specific diffraction fingerprint, without a significant maximum in the low-angle region (Fig. 13), probably hinders recognition of such highly illitic randomly interstratified mixed layers in natural samples and is probably responsible for the small number of descriptions in works based on the position of diffraction maxima.

In addition, the existence of partial segregation, or partial ordering, has frequently been reported in studies of natural and synthetic mixed layers performed with the multi-specimen method (Drits *et al.*, 1997, 2002b, 2004; Sakharov *et al.*, 1999a, 1999b; Claret *et al.*, 2004; Inoue *et al.*, 2005; McCarty *et al.*, 2008, 2009; Hubert *et al.*, 2009, 2011; Lanson *et al.*, 2009; Ferrage *et al.*, 2011b). The high frequency of natural mixed layers exhibiting junction probabilities different from the usual $S = 0$ and $S = 1$ for MPDO cases clearly demonstrates that XRD profile modelling is the one tool that can provide an accurate structure characterization of mixed layers as peak migration curves are not available for junction probabilities different from these 'ideal' cases.

3.1.4. Recent developments and new insights into the actual structure of mixed layers: structure of elementary layers

Quantitative description of mixed-layer diffraction patterns, which has been used increasingly over the past decade, has not only allowed a more realistic description of mixed layers but has also brought significant new insights into the structure of individual

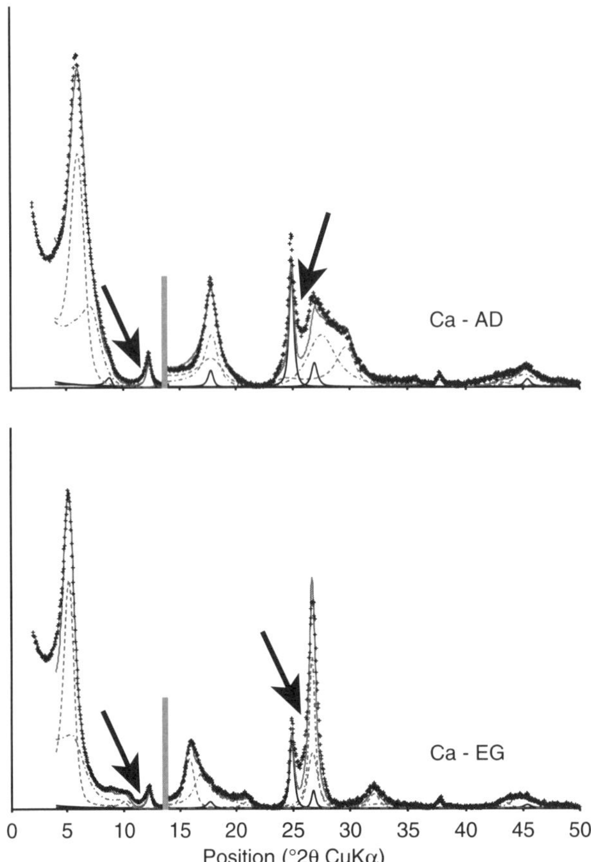

Fig. 13. Comparison between experimental and calculated XRD patterns for the diagenetically altered clayey sample from the Gulf Coast shown in Figure 10 (adapted from Lanson *et al.*, 2009). Unlike the optimum fit shown in Figure 10a, the contribution of the randomly interstratified kaolinite-smectite to the diffracted intensity has not been considered. Readers of the paper version of this chapter may wish to download a colour version of this figure from www.minersoc.org/emu-notes/emu-11/11-4-colour.pdf.

layers. For example, the structure description of interlayer H_2O in expandable layered silicates was improved, despite their intrinsic hydration heterogeneity. It was, in particular, possible to refute the widely used model with three planes of H_2O molecules on each side of the interlayer mid-plane reported by Moore & Reynolds (1997). Instead, a model with a single plane of H_2O molecules on each side of the interlayer mid-plane, which contains charge-compensating cations, was proposed (Ferrage *et al.*, 2005a, 2005b, 2010). Furthermore, it has been shown that the total number of H_2O molecules filling these planes is not constant but rather increases with increasing RH conditions. At the same time, the layer-to-layer distance, but also the distance – perpendicular to the layer plane – from the interlayer mid-plane to the plane of H_2O molecules, increases also to accommodate the larger number of H_2O molecules in the smectite interlayers. Finally, it was shown that thermal motion, commonly described as the Debye-Waller factor, accounted inadequately for the positional disorder of H_2O molecules about these planes. This disorder was better described with a Gaussian function, in agreement with previous models proposed for smectite interlayer H_2O (de la Calle *et al.*, 1977; Ben Brahim *et al.*, 1983; Slade *et al.*, 1985). This configuration allows reproduction of all 00*l* reflections with high precision, with only one new variable parameter (the width of the Gaussian function). In addition, the proposed configuration allows us to match more closely the amount of interlayer water that can be determined independently from water-vapour adsorption/desorption isotherm experiments. Additional Monte Carlo simulations performed in the NVT (Ferrage *et al.*, 2005a) or μVT (Ferrage *et al.*, 2011a) ensembles further supported the proposed model. The combined modelling of X-ray and neutron diffraction data has even allowed us to assess the different sets of potential available in the literature for the clay layer; the *CLAYFF* model proposed by Cygan *et al.* (2004) was shown to account better for the water content and organization compared to the model developed by Skipper *et al.* (1995) and modified by Smith (1998). However, fitting diffraction data for bi-hydrated samples using *CLAYFF* simulations requires us to refine further Lennard-Jones parameters for oxygen atoms of the clay layer (Ferrage *et al.*, 2011a). When combined with the SPC/E water model, this modified version of *CLAYFF* allows matching of experimental water contents and fitting of the complete set of X-ray and neutron-diffraction data obtained on both mono- and bi-hydrated smectites (Fig. 14).

In addition to the distribution of interlayer species, XRD profile modelling has allowed the refinement of structural parameters such as the layer-to-layer distance corresponding to the different layer types for a variety of interlayer cations and a large range of RH values. It was also possible to show that for a given cation and RH value, the basal distance actually scatters about an average distance, and that accurate modelling of experimental data requires that we consider second-order fluctuations of the layer-to-layer distance (Ferrage *et al.*, 2005a, 2005b, 2010). These fluctuations are probably linked to the incomplete, and thus heterogeneous, filling of smectite interlayers (Fig. 15 – Ferrage *et al.*, 2005c). Experience has shown that even for complex mixtures

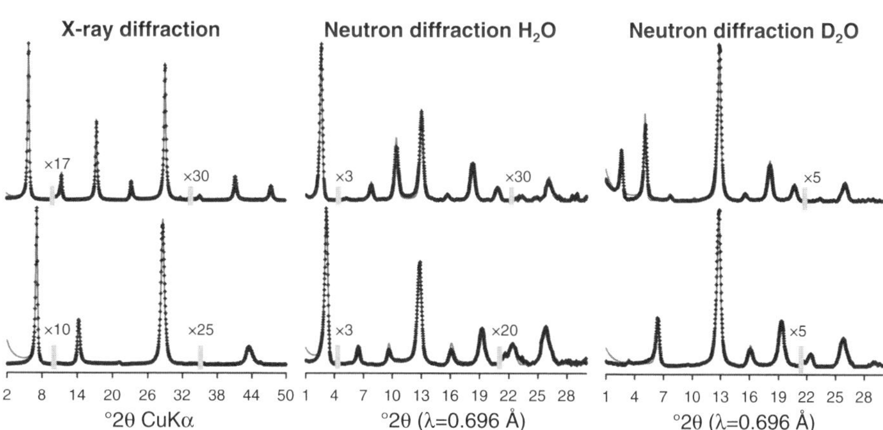

Fig. 14. Comparison between experimental and calculated diffraction patterns for the low-charge saponite shown in Figure 11 (adapted from Ferrage *et al.*, 2011a). The density profiles of interlayer species were derived from Monte-Carlo simulations performed in the Grand Canonical ensemble using a modified version of the CLAYFF model (modified Lennard-Jones parameters for oxygen atoms of the clay layer) and the SPC/E water model. Readers of the paper version of this chapter may wish to download a colour version of this figure from www.minersoc.org/emu-notes/emu-11/11-4-colour.pdf.

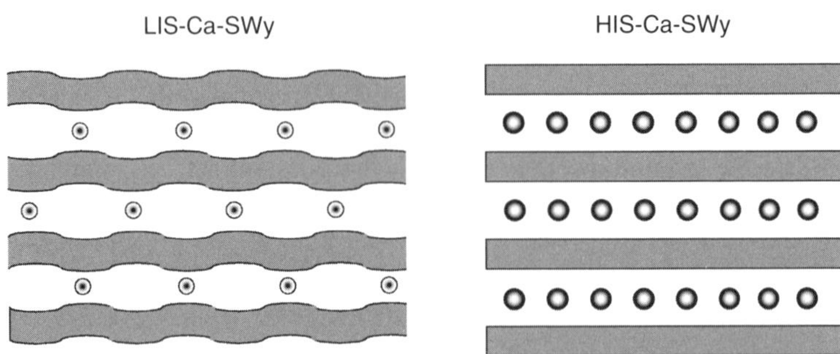

Fig. 15. Schematic representation of the influence of $MeCl^+$ ion pairs on smectite interlayers (adapted from Ferrage *et al.*, 2005c). The incomplete interlayer filling of sample SWy Ca-saturated using low ionic strength solutions (left) leads to significant fluctuations of the layer-to-layer distance because of the local balance between attractive (between layer and interlayer cations) and repulsive (between adjacent layers) forces. For sample SWy Ca-saturated using high ionic strength solutions (right), the more complete filling of the interlayer space by $CaCl^+$ pairs leads both to an increased layer-to-layer distance and to a more homogeneous distribution of interlayer species, thus reducing the fluctuation of the layer-to-layer distance.

of natural mixed layers, the description of experimental data is significantly improved when such fluctuations are considered (Drits *et al.*, 1997, 2002b; Sakharov *et al.*, 1999a; Lindgreen *et al.*, 2000; Drits, 2003), although the parameter quantifying the extent of the fluctuation is not systematically reported.

3.1.5. Recent developments and new insights into the actual structure of mixed layers: intra-crystalline defects

Structure models of mixed layers used for the calculation of XRD profiles most often describe coherent scattering domains (CSDs) limited to a few layers. These models describe well the high-angle region of the experimental diffraction patterns but often fail over the low-angle region ($2\theta < 4$–$6°$Cu$K\alpha$), where the calculated intensity is much greater than the experimental intensity. On the other hand, simulation of small-angle X-ray scattering data obtained on related samples commonly indicate the presence of much thicker CSDs (Pons *et al.*, 1981, 1982).

In an effort to reconcile the results from both methods, Plançon proposed a model for the simulation of experimental XRD patterns in which particles, or megacrystals, are significantly thicker than crystals (or CSDs) used in XRD models (Plançon, 2002). These particles contain layer types identical to those in the usual XRD model. Their relative proportion and their stacking sequences are also identical to those in the usual models. However, adjacent layers may be shifted with respect to each other along the c^* axis according to an adjustable probability. As a result, the apparent CSD size decreases with increasing 2θ angle and XRD patterns calculated according to this new model exhibit a significantly lower scattered intensity over the low-angle region compared to the usual XRD models.

However, this model leads to non-negligible discrepancies between the basal-reflection positions and intensities calculated in the high-angle region compared to usual XRD models. To overcome this problem, an improved model describing the degree of coherency within megacrystals has been proposed (Sakharov, 2005). In contrast to the model proposed by Plançon, this model hypothesizes that CSDs are identical to the usual XRD models. In addition, within megacrystals, these CSDs may be shifted with respect to each other along the c^* axis according to an adjustable probability. This model allows reproduction of diffraction features in both high- and low-angle regions of experimental XRD patterns (McCarty *et al.*, 2009).

3.1.6. Recent developments and new insights into the actual structure of mixed layers: outer surfaces of crystals

None of the algorithms routinely available to simulate XRD patterns of mixed layers can account for the possibility that in natural environments the structure and composition of surface layers of crystals may differ from those of 'core' layers. However, knowledge of the nature of the outer surface layer (OSL) is helpful in understanding better the surface properties of mixed layers and/or to derive constraints on their growth conditions. For example, according to high-resolution TEM, illite crystals may terminate on a kaolinite layer (Tsipursky *et al.*, 1992) whereas kaolinite crystals may have pyrophyllite or smectite layers as surface terminations (Ma & Eggleton, 1999). Kogure *et al.* (2001) also reported crondstedtite crystals exhibiting chlorite OSLs.

Alternative algorithms were thus proposed that allow us to determine the nature of OSLs in mixed layers from the simulation of XRD patterns (Sakharov *et al.*, 1999c, 2004b; Plançon, 2003). In particular, Sakharov *et al.* (2004b) showed that among the usual mixed layers found in natural samples, most significant effects are calculated for those containing elementary chlorite layers. For chlorite samples, the relative intensities of the odd reflections depend not only on the distribution of Fe in the chlorite structure over the 2:1 and 0:1 layers, but also on the nature of the OSLs. For the two samples they investigated, brucite sheets were present on the outer crystal surfaces. However, comparison of the OSL nature determined from XRD profile modelling with that deduced from direct observations using electron or atomic force microscopies remains essential for mixed layers because similar diffraction effects may be obtained by varying the structural and chemical parameters of the mixed layers on the one hand and the OSL nature on the other.

Sakharov *et al.* (2004b) showed also that for periodic structures containing only one layer type, the influence of OSLs may be predicted from simple calculations, and is independent of the scattering power of the OSL. Such a prediction is not possible for mixed layers.

3.2. Natural layered silicates: interstratification of structural fragments differing in terms of their internal structure or their stacking mode

To complement the structural characterization of phyllosilicate expandable interlayers in projection along the c^* axis (see section 3.1.4), simulation of *hkl* reflections, or of *hk* bands, has been used to determine also the x and y coordinates of interlayer H_2O molecules. For example, Ben Brahim *et al.* (1984) performed such a study on bi-hydrated beidellite. Despite the large proportion of random stacking faults (~70%), these authors determined the nature of the layer displacements, when not random, the occurrence probabilities of the different possible displacements, and the atomic coordinates of interlayer species. Similarly, Argüelles *et al.* (2010) refined simultaneously the layer displacement and the atomic coordinates of interlayer H_2O molecules in bi-hydrated vermiculite. The refined model includes the random alternation of $+b/3$ and $-b/3$ layer displacements, thus defining two layer types differing by their interlayer configuration and the position of their interlayer species. This model is fully consistent with that proposed by de la Calle *et al.* (1988) from the qualitative comparison of their data with XRD patterns calculated for a variety of structure models.

The actual layer stacking, and the nature of stacking defects, has also been determined in kaolinite from the calculation of the diffraction effects corresponding to different structure models. In particular, Plançon & Tchoubar (1977) used this approach to reject both rotational and $\pm b/3$ stacking faults as possible defects in kaolinite. Rather, these authors proposed that stacking defects in kaolinite correspond to the different location of the vacant octahedral layer site in successive layers among the three non-equivalent (A, B, or C) sites. Bookin *et al.* (1989) and Drits & Tchoubar (1990) improved this model and showed that kaolinite is composed only of B layers that can actually present two stacking vectors within the *ab* plane: $\mathbf{t}_1$ (-0.369 $\mathbf{a}_1$, -0.024 $\mathbf{b}_1$) and $\mathbf{t}_2$ (-0.352 $\mathbf{a}_1$, $+0.304$ $\mathbf{b}_1$), corresponding to right- and left-handed kaolinite unit cells. Plançon *et al.* (1989) validated this model by fitting the XRD data obtained from kaolinite samples exhibiting a wide range of structural imperfection.

Recently, this model was validated further by the direct imaging, using high-resolution TEM (HRTEM), of the layer structure and stacking in kaolinite (Kogure & Inoue, 2005). In particular, these authors described the occurrence of $\mathbf{t}_1$, $\mathbf{t}_2$ and $\mathbf{t}_0$ ($\sim -a_1/3$, $\sim -b_1/3$) layer displacements and the sole presence of B layers. Kogure *et al.* (2010) observed also $\mathbf{t}_1$ and $\mathbf{t}_2$ layer displacements between successive kaolinite layers. These two well defined stacking vectors occur either as isolated faults (for example a $\mathbf{t}_2$ layer displacement within an ordered sequence characterized by a $\mathbf{t}_1$ vector), or as the alternation of multilayer ordered blocks with either $\mathbf{t}_1$ or $\mathbf{t}_2$ vectors (segregated mixed layer). The combination of diffraction effects arising from these two types of defects allows us to reproduce, at least qualitatively, XRD intensities recorded for the investigated sample. It was also possible, using HRTEM, to reject the interstratification of kaolinite (sequence of B layers only) and of dickite (BC sequence of layers) fragments in samples from a diagenetic kaolinite-to-dickite transition series (Kameda *et al.*, 2008).

Stacking disorder was characterized in other dioctahedral phyllosilicates by coupling direct HRTEM imaging of stacking and calculation of diffraction effects from the observed stacking faults. In both sudoite (Kameda *et al.*, 2007) and pyrophyllite (Kogure *et al.*, 2006b; Kogure & Kameda, 2008), stacking disorder is, as in kaolinite, essentially induced by the coexistence of two possible layer displacements (Fig. 16). Contrastingly, stacking faults in trioctahedral phyllosilicates are related mainly to rotations between successive 2:1 layers (Kogure *et al.*, 2006b; Kogure & Kameda, 2008). The contrast between di- and trioctahedral varieties was tentatively attributed

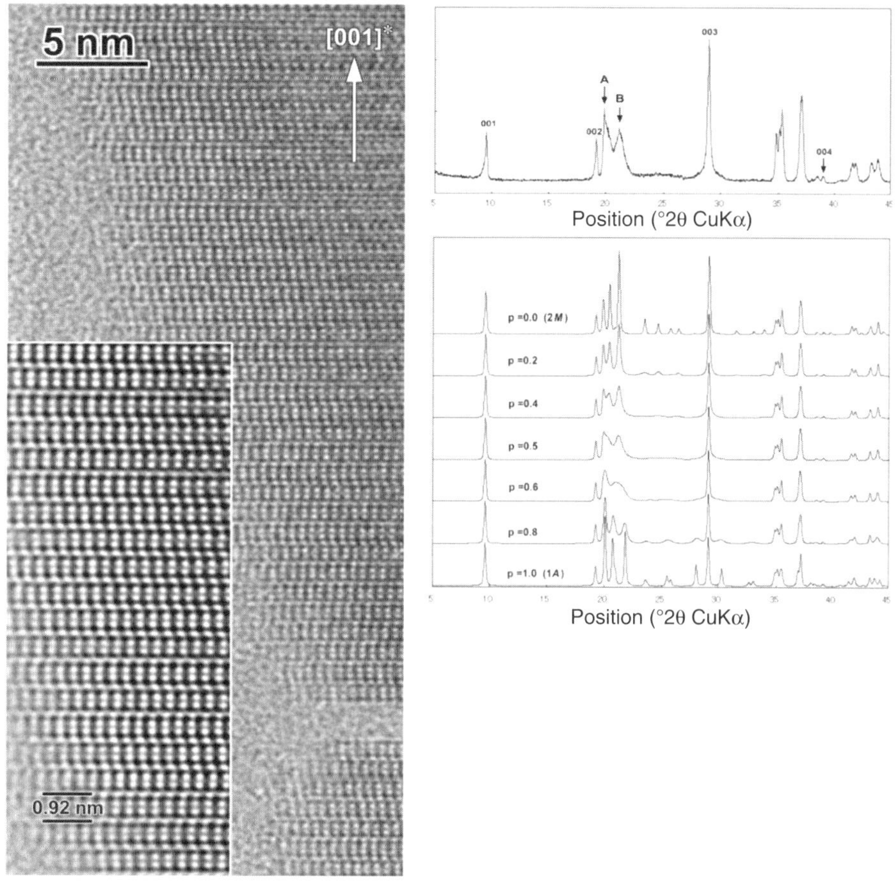

Fig. 16. (left) HRTEM image of the Berozovska pyrophyllite (adapted from Kogure *et al.*, 2006a). The inset at the bottom-left is a filtered and magnified portion of the main image. The contrast for each 2:1 layer is uniform but the direction of shift between adjacent layers is near perfectly disordered. (right) Powder XRD pattern of Nowa pyrophyllite (top) and powder XRD patterns calculated for variable proportions of stacking faults identified on HRTEM images (bottom – adapted from Kogure & Kameda, 2008).

to the corrugation of basal oxygen planes in dioctahedral varieties, although rotational stacking faults were also reported in the dioctahedral mica, celadonite (Kogure *et al.*, 2008).

Reynolds (1993) and McCarty & Reynolds (1995) described the presence of rotational stacking faults in the dioctahedral mica, illite. Simultaneously, these authors determined the internal structure of dioctahedral layered silicates, and more especially the distribution of the layer vacancy between non-equivalent *cis* and *trans* sites. In addition to the modification of the intensity distribution between the different *hkl* reflections, *cis*- and *trans*-vacant dioctahedral micas are characterized by different layer displacements allowing diffraction effects similar to those observed for the interstratification of layers having different layer thickness (see section 2.3.2.). Extending Méring's principle to 3D structures, Drits & McCarty (1996) and Drits *et al.* (1998b) proposed and validated a method, based on peak-migration curves, to obtain a semi-quantitative (and quick) estimate of *cis*- and *trans*-vacant polymorphs in illite and illite-smectite. From systematic calculations of diffraction effects, Drits & Sakharov (2004) demonstrated that the coexistence of *cis*- and *trans*-vacant layers may significantly modify the intensity distribution between the different *hkl* reflections and thus the identification of mica polytypes.

A similar variability of the octahedral cation distribution over *cis* and *trans* sites has been reported in glauconites (Drits *et al.*, 2010), pyrophyllite (Drits *et al.*, 2011), and dioctahedral smectites (Tsipursky & Drits, 1984; Emmerich *et al.*, 2009). A similar fitting of XRD data may be used to unravel this distribution in smectites pending an improvement of their stacking order obtained by the saturation of interlayers with large anhydrous cations (K^+, Cs^+) and subsequent wetting-and-drying cycles (Besson *et al.*, 1983; Tsipursky & Drits, 1984; Cuadros, 2002). It is clear that a comprehensive structural characterization of dioctahedral phyllosilicates should include the description of this distribution, which in turn could provide key information on their (trans)formation mechanisms (Drits & Zviagina, 2009; McCarty *et al.*, 2009).

3.3. Interstratification of other lamellar structures: layered double hydroxides

Layered double hydroxides (LDHs) form a large family of promising materials, natural and synthetic, that can be used as catalysts (Cavani *et al.*, 1991), catalyst supports (Vaccari, 1998), ion exchangers (Ulibarri & Hermosín, 2001), and additives (Leroux & Besse, 2004 – see also Rives, 2001, for a review). These compounds are also referred to as anionic clays or hydrotalcites, from the most common mineral species of this group. Their structure consists of a brucite-like octahedral layer in which some of the M^{2+} cations are replaced isomorphically by M^{3+}. The presence of interlayer anions compensates for the excess positive charge of the layer, whereas H_2O molecules may occupy the remaining interlayer space. As expandable layered silicates, LDHs have the ability to host a variety of interlayer anions and to exchange them dynamically. Similar to expandable layered silicates, the regular interstratification of layers hosting different interlayer anions is not common, but has been described even in naturally occurring varieties. For example, Drits *et al.* (1987) inferred the regular 1:1 interstratification of $Mg_{(3-x)}Al_x(OH)_6$ layers having CO_3^{2-} and SO_4^{2-} anions in their interlayers from the comparison of calculated XRD patterns with those recorded for samples from saline deposits. According to Iyi *et al.* (2007), LDH hydration as a function of RH is also reminiscent of expandable layered silicates: LDHs exhibit a stepwise increase of their layer-to-layer distance upon the incorporation of additional H_2O molecules. Those authors also reported the coexistence of different hydration states within the same crystals as shown by the presence of super-reflections corresponding to the regular alternation of dehydrated and hydrated interlayers. Super-reflections were observed in particular for their higher-charge LDH sample with two large interlayer anions (ClO_4^- and I^-). These observations are consistent with the early works of Brindley & Kikkawa (1980). Newman *et al.* (1998) and Iyi *et al.* (2002) described also the regular interstratification of different hydration states and different interlayer configurations in Mg_2Al LDHs containing a single organic anion (terephtalate and azobenzene derivative, respectively).

Recently, intercalation of LDHs has attracted much interest because this process can modify their electronic, magnetic and optical properties by host-guest effect. Special efforts have been devoted to the simultaneous intercalation of LDHs with different guest species aiming at the synthesis of polyfunctional materials. Over the last decade or so, the development of energy-dispersive XRD on synchrotron sources has been especially useful to investigate the kinetics of intercalation (Fogg *et al.*, 1998a, 1998b; O'Hare *et al.*, 2000). In particular, this technique has allowed us to uncover the formation of 'intermediate' heterostructures corresponding to second staging, or regular 1:1 interstratification, of layers with different interlayer anions, similar to those described in amphiphilic modified clay structures (Ijdo *et al.*, 1996; Ijdo & Pinnavaia, 1998a, 1998b). Such heterostructures, as determined from the presence of a super-reflection characteristic of the regular interstratification (1:1 ratio, $S = 1$ with MPDO), have been reported for a variety of octahedral layer compositions and interlayer anions (Fig. 17a – Fogg *et al.*, 1998a, 1998b; Kayenoshi & Jones, 1998; O'Hare *et al.*, 2000; Pisson *et al.*, 2003; Williams *et al.*, 2003, 2004; Taviot-Guého *et al.*, 2005, 2010; Williams & O'Hare, 2005, 2006; Feng *et al.*, 2006; Zhang *et al.*, 2008; Ragavan *et al.*, 2009). It should be noted that the regular distribution (second staging) of inorganic and organic interlayers can be maintained through subsequent exchange of one of the guest anions as shown by Williams & O'Hare (2006) for the adipate-for-succinate or the adipate-for-tartrate exchange in a Zn_2Cr LDH initially containing Cl^- in addition to the organic anion.

Beside these numerous reports of regularly interstratified mixed layers, additional kinetic studies of the anion-exchange process describe the steady migration of peaks from one end-member to the other during the anion-exchange process (Feng *et al.*, 2006; Taviot-Guého *et al.*, 2010; Fig. 17b), similar to the diffraction behaviour described by Méring for the random interstratification of different layer types (Méring, 1949). Significant variations in the position and breadth of the diffraction peak characteristic of the regular interstratification of two layer types has also been reported for the carbonate-for-tartrate exchange in a Zn_2Cr LDH (Feng, 2006; Fig. 17c). Such variation indicates that the alternation of carbonate and tartrate interlayers departs from the regular 1:1 alternation. Both a carbonate:tartrate ratio other than the 1:1 ratio and junction probabilities other than the $S = 1$ case with MPDO, *i.e.* partial ordering, can account for the observed variation. These non-regular mixed layers have received much less attention than their regular counterparts, however, and stacking sequences have not even been determined.

The alternation, regular or otherwise, of different layer types in LDHs will impact not only the position and profiles of $00l$ basal reflections as described above, but also the position and profiles of non-basal '*hkl*' lines as most LDHs possess a 3D periodicity. Bookin & Drits (1993) reported theoretical calculations performed for all possible LDH structure models, *i.e.* for all possible stacking modes (polytypic fragments) resulting from the different interlayer configurations. The positions and relative intensities

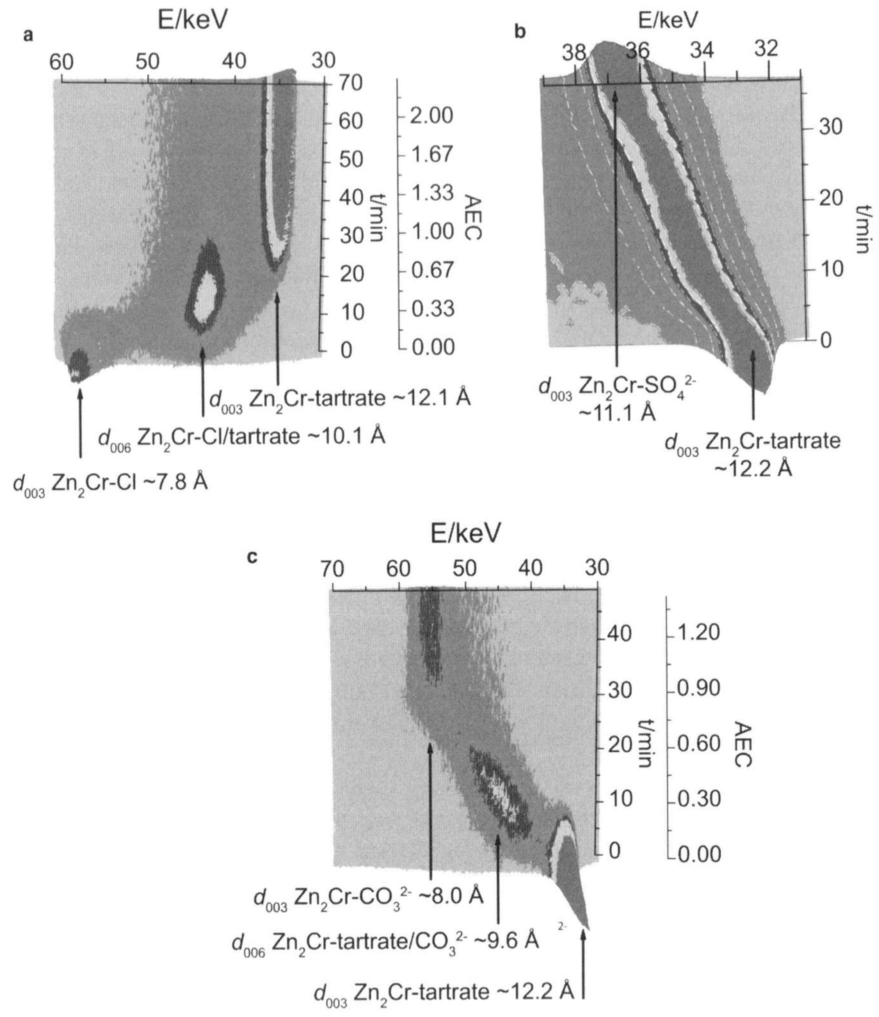

Fig. 17. Dynamic evolution of XRD patterns during the anion exchange of Zn_2Cr LDHs. (***a***) Tartrate-for-chlorine exchange, showing a super-periodicity corresponding to the regular alternation of the two anions in successive interlayers (2nd staging – adapted from Feng, 2006). (***b***) Sulphate-for-tartrate exchange, showing a steady shift of peak position from one end-member to the other, characteristic of the random interstratification of the two types of anions in successive interlayers (adapted from Taviot-Guého *et al.*, 2010). (***c***) Carbonate-for-tartrate exchange (adapted from Feng, 2006). Compared to Figure 17a the peak corresponding to the super-periodicity is shifting as the exchange proceeds, thus suggesting a carbonate:tartrate ratio different from 50:50 in a stacking which is predominantly ordered. For the Zn_2Cr LDH with interlayer tartrate, carbonate, sulphate and chlorine anions the d_{003} values correspond to the layer-to-layer distance because of the 3*R* polytype. For the Zn_2Cr LDH with mixed Cl^-/tartrate or tartrate/CO_3^{2-} anion composition, the d_{006} values correspond to the second order of the basal distance corresponding to the periodic repetition of the two anions in successive interlayers (2nd staging). The AEC scale indicates the amount of tartrate (***a***), SO_4^{2-} (***b***), or CO_3^{2-} (***c***) anions added in solution during these dynamic experiments expressed as a fraction of the anion exchange capacity.

reported for the different LDH polytypes can serve to determine the actual stacking mode, and polytypic variety, of periodic species (Bookin *et al.*, 1993), but can serve also as a basis for the interpretation of diffraction effects arising from the interstratification of different polytypic fragments (Drits & Bookin, 2001). A number of structural studies have actually investigated the interstratification of such structural fragments from the profiles of non-basal reflections, essentially using the recursive description of the layers' stacking and the *DIFFaX* code (Bellotto *et al.*, 1996; Thomas *et al.*, 2004; Radha *et al.*, 2005, 2010; Thomas & Kamath, 2006; Curtius & Ufer, 2007; Radha & Kamath, 2009). These studies consistently show the random interstratification of common rhomboedral, hexagonal, and eventually monoclinic LDH polytype fragments from the migration of the *hkl* reflections corresponding to the predominant polytype towards neighbouring reflections of the minor polytype, which is described as a well defined stacking fault.

3.4. Interstratification of other lamellar structures: layered oxides

3.4.1. Interstratification of commensurate polytype fragments

Interstratification of polytypic fragments has been reported also in layered oxides. For example, in the general context of preparation of heterostructured hybrid materials, Du & O'Hare (2008) reported second staging in α-Co hydroxides from the presence of super-reflections corresponding to the regular interstratification of two layer types. The regular alternation of DDS (dodecyl sulphate) and carbonate layers (27.8 and 8.2 Å, respectively) over extended length scales was confirmed from lattice-fringe images obtained under the TEM.

Interstratification of layers having different layer-to-layer distances has also been reported in $Ni(OH)_2$, which is the positive electrode material of nickel-based alkaline secondary batteries (Ni Cd, Ni metal hybride, $Ni\text{-}H_2$). Two main structures have been reported for this material. $\alpha\text{-}Ni(OH)_2$ is a turbostratic phase exhibiting a basal distance of ~8.4 Å, whereas $\beta\text{-}Ni(OH)_2$ corresponds to the 1*H* polytype and has a ~4.6 Å unit-cell length perpendicular to the layer plane. Using a recursive description of stacking sequences, Ramesh *et al.* (2003) investigated systematically the impact of the random interstratification of these two polytypes on XRD patterns. The influence of such stacking faults on the position and profiles of basal 00*l* reflections was used to report not only their occurrence but also to assess their abundance (relative proportions of the two polytypes – Rajamathi *et al.*, 2000; Tessier *et al.*, 2000a; Ramesh *et al.*, 2003, 2005). The presence of stacking faults in $\beta\text{-}Ni(OH)_2$ not only modifies the position and profile of basal 00*l* reflections, but also impacts non-basal *hkl* reflections as shown for growth and deformation faults (Delmas & Tessier, 1997; Tessier *et al.*, 1999). Within the 1*H* stacking sequences of $\beta\text{-}Ni(OH)_2$ (AbC AbC AbC ...), these two types of faults correspond to the occurrence of twinning (AbC AbC CbA CbA ...) and to a $-a/3$ layer displacement (AbC AbC CaB CaB ...) and induce a significant broadening of 101 and 102 lines compared to other reflections. Tessier *et al.* (1999), Ramesh *et al.* (2003), and Ramesh (2009) calculated systematically the influence of different stacking faults on $\beta\text{-}Ni(OH)_2$ XRD patterns. These calculations were subsequently used to describe, from the modelling of XRD data, the occurrence and abundance of a variety of stacking faults in β-$Ni(OH)_2$ (Delmas & Tessier, 1997; Tessier *et al.*, 1999, 2000a; Guerlou Demourgues *et al.*, 2004; Ramesh *et al.*, 2005, 2006; Ramesh & Kamath, 2008a; Ramesh, 2009). Other structural parameters such as the size of the CSDs and the abundance of vacant layer sites or of isomorphous cationic substitutions, for example, were also determined in some of these studies (Ramesh *et al.*, 2003, 2005, 2006; Ramesh, 2009). In an elegant paper, Ramesh & Kamath (2008b) demonstrated that the quality of the structural characterization is improved significantly, compared to a 'usual' Rietveld refinement, when stacking faults are characterized and quantified, even if the abundance of these faults is limited to a few percent. Finally, clear correlations were determined between the nature and abundance of some of the above defects and the synthesis conditions (Tessier *et al.*, 2000a, 2000b; Guerlou Demourgues *et al.*, 2004; Ramesh *et al.*, 2006) or electrochemical properties of these materials, such as chargeability and electronic conductivity (Tessier *et al.*, 1999; Ramesh & Kamath, 2008a; Ramesh, 2009).

Layered MnO_2 (birnessite or δ-MnO_2) has attracted a sustained interest over the last two decades owing to its ubiquity in superficial oxic environments where its ability to degrade organic pollutants and to fix trace-metal elements has a pivotal influence on the fate of these contaminants, in addition to its potential as a cathode material. As for $Ni(OH)_2$, different stacking sequences have been reported for birnessite octahedral layers, that can also have different symmetry depending on the origin of the layer-charge deficit. Layers with hexagonal symmetry, and small b unit-cell dimensions (2.84–2.85 Å), are characterized by small amounts of Mn^{3+} cations within the octahedral layer, with large layer-vacancy contents being responsible for the layer-charge deficit. On the contrary, when layers are almost devoid of vacancies, the layer-charge deficit arises essentially from the presence of layer Mn^{3+} cations (25–30%). In this case, Jahn-Teller distorted Mn^{3+} octahedra are ordered at room temperature to minimize the steric strains, and their systematic elongation along the a axis leads to an orthogonal layer symmetry with $a > b \times \sqrt{3}$ (Gaillot *et al.*, 2007). Drits *et al.* (2007a) calculated systematically the intensity distribution for all possible polytypes, and relevant associated interlayer configurations, thus providing a sound basis for the identification of these polytypes, but also for the prediction of the diffraction effects arising from their random interstratification. These authors also described the layer symmetry and stacking sequences for most varieties reported in the literature at that time as a function of the synthesis conditions. Owing to this polytypic diversity, 3D-ordered birnessite varieties often appear as mixed layers in which different polytypic fragments are interstratified, as in $Ni(OH)_2$. For example, interstratification of the 1*H* polytype, which is the stable form of birnessite when equilibrated at low pH, with other polytypic variants has been described (Manceau *et al.*, 1997; Lanson *et al.*, 2000, 2002b). By taking into account the interstratification of the different polytypes, it was possible to extract additional structural information from the XRD patterns of these defective structures (Fig. 18). In particular, the number of vacant octahedral sites, the number, location and coordination of interlayer transition metal cations, and the number and location of interlayer H_2O molecules were determined systematically. Interpretation of these fine structure

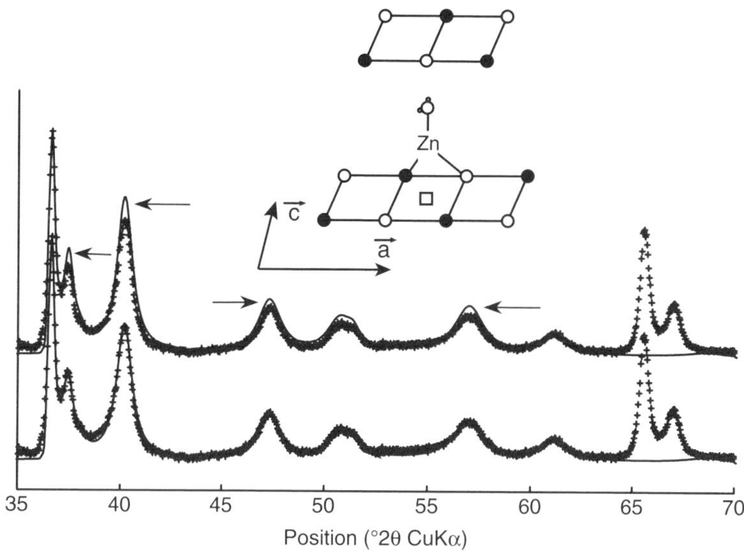

Fig. 18. Comparison between experimental and calculated diffraction patterns for the phyllomanganate birnessite equilibrated at low pH in the presence of Zn (adapted from Lanson *et al.*, 2002b). The agreement between calculated and experimental intensity distributions is significantly improved when both tetrahedral and octahedral coordinations are considered for interlayer Zn cations (0.078:0.040 ratio – bottom) as compared to a unique octahedral coordination (top). The tetrahedral coordination of Zn located above, or below, vacant octahedral layer sites provides additional bonding between adjacent layers shifted by $+a/3$, through the formation of strong H-bonds. Symbols as in Figure 5.

details allowed us to unravel the H-bonding responsible for the respective stability of the different stacking modes (Lanson *et al.*, 2000, 2002b), but also the reaction mechanisms involved in the high to low pH structural transition occurring in birnessite (Lanson *et al.*, 2000) or those leading to the oxidation of Co^{2+} and to its incorporation in the MnO_2 octahedral layer (Manceau *et al.*, 1997). The different polytypes resulting from contrasting synthesis conditions most often include minor amounts of other polytypic fragments randomly interstratified within the main stacking sequence (Gaillot *et al.*, 2003, 2004, 2005). For example, Gaillot *et al.* (2003) were able to refine the structure of a specific $2H$ polytype from a μm-sized monocrystal, but had to consider the contribution from a $2H/3R$ mixed layer to fit the powder XRD pattern recorded for the whole sample. Drits *et al.* (1998a) even reported the complex interstratification of two 4-layer polytypes, in addition to the presence of one of these polytype as a 3D-periodic species. Finally, interstratification of layers having the same stacking vector within the *ab* plane but differing in their layer-to-layer distance was described during the dehydration of birnessite at moderate (100–150°C) temperature (Gaillot *et al.*, 2005).

3.4.2. *Interstratification of incommensurate fragments*

Interstratification may occur between layers each having their own 2D periodicity and irrational a and/or b unit-cell parameters (Drits, 1987, 2003; Organova, 1989;

Fig. 19. Electron diffraction pattern from a Co-asbolane exhibiting reflection networks corresponding to the two incommensurate layers (from Gaillot, 2002).

Makovicky & Hyde, 1992). As a result, the individual structures of these layers can be described with two independent unit cells, and selected area electron diffraction is an especially powerful tool to unravel the interstratification of such incommensurate layers (Fig. 19). For example, the regular alternation, along the axis perpendicular to the layer plane, of octahedral layers having different compositions defines a peculiar group of phyllomanganates whose natural occurrences are reported under the generic name of asbolane (Chukhrov *et al.*, 1982, 1989; Drits, 1985; Manceau *et al.*, 1992; Feng *et al.*, 2001). In these peculiar mixed layers, the MnO_2 layer is continuous whereas the other layer type essentially forms island-like domains sandwiched between the MnO_2 layers to allow for the formation of H-bonds that ensure the overall stacking stability. Regular alternation of incommensurate layers has also been described for other compositions. For example, alternation of brucite and sulphide layers was reported in the minerals valleriite and tochilinite (Evans & Allman, 1968; Organova *et al.*, 1974; Drits, 1987; Organova, 1989). The natural occurrence of regularly alternating chrysotile and hydrotalcite layers was also reported (Drits *et al.*, 1995).

Using the theoretical developments of Plançon (1981), Gaillot *et al.* (2004) revealed a new type of structural disorder in layered structures with the random alternation of partially incommensurate layers. These authors reported the interstratification of layers having hexagonal and orthogonal symmetry, but the same basal distance and the same stacking vectors. In this case, *hk* rods corresponding to the interstratified layer types overlap partially, giving rise to very specific diffraction features. In particular, the breadth of *hkl* reflections corresponding to the mixed layer decreases with increasing values of the Miller index *l*.

3.5. Structural characterization of disordered layered structures

As described in section 2.1.2., the occurrence of random stacking faults leads to diffraction effects strikingly different from those described above. In particular, the resolution of *hkl* reflections is progressively lost when increasing the proportion of such random stacking faults (Fig. 2). In turbostratic compounds, random stacking faults occur systematically between adjacent layers thus inducing 2D crystals, characterized by an asymmetric diffraction profile (*hk* diffraction band) similar to that obtained from a single layer, in addition to 00*l* reflections.

Warren (1941) soon recognized that it was possible to decipher structural information from the profile of *hk* bands, despite the absence of resolved *hkl* lines (Biscoe & Warren, 1942). In their pioneering work, Warren (1941) and Biscoe & Warren (1942) were able, from the analysis of XRD data, to prove the 2D periodicity of carbon blacks and to determine their CSD size within the layer plane. Further information on the actual 2D structure of layers can be retrieved from the modelling of *hk* band profiles. These bands are similar to the structure factor of a single layer and thus comprise the coordinates and occupancies for all atomic sites within the layer. Leoni (2008) reviewed existing methods used to retrieve structural information from such diffraction patterns. The next section will deal with recent results obtained with these different approaches.

3.5.1. Layered silicates

Despite recent studies that challenged the actual turbostratic character of expandable layered silicates (Viani *et al.*, 2002), *hk* band profiles of smectite have been used widely to investigate the 3D organization of their interlayer cations and H_2O molecules (Brindley & Méring, 1951; Méring & Brindley, 1967). Ben Brahim *et al.* (1984) used a similar approach to investigate the positions of cations and H_2O molecules in Na-beidellite, but had to describe also the layer stacking of their sample that was not totally turbostratic. In two articles devoted to the structural characterization of oxidized and reduced nontronite (Fe-rich smectite), Manceau *et al.* (2000a, 2000b) modelled the profiles and relative intensities of the 02,11 and 20,13 bands, to determine the actual occupancy of the different octahedral sites, and in particular, to assess the presence of cations occupying *trans* sites in the essentially *trans*-vacant oxidized species. They showed that all nontronite samples investigated were 100% *trans*-vacant, with a detection limit of ~10% for occupied *trans*-sites. In the companion article devoted to the structural characterization of reduced nontronites (Manceau *et al.*, 2000a), these authors showed the formation of trioctahedral clusters induced by the migration of reduced Fe cations within the octahedral sheet of the smectite layer. Trioctahedral clusters induce in particular a modification of the intensity distribution between 02,11 and 20,13 bands as predicted by Drits *et al.* (1984). They also showed the induced reduction of the CSD size, and the lowering of the original hexagonal layer symmetry hexagonal to an orthogonal one ($b > a \times \sqrt{3}$). The analysis of *hk* band profiles, and especially of their relative intensities, even allowed the presence of Fe cations with tetrahedral coordination within smectite layers to be unravelled and quantified when the Fe content was sufficient (Manceau *et al.*, 2000b; Gates *et al.*, 2002).

3.5.2. Layered oxides

In natural environments, disordered varieties of phyllomanganates (birnessite and its disordered analogue vernadite, or δ-MnO_2) prevail probably as a result of the biogenically mediated oxidation of Mn^{2+}. Modelling studies, similar to those performed on layered silicates, have been carried out recently to determine their layer structure, and, in particular, the origin of the layer-charge deficit (presence of Mn^{3+} cations or of vacant sites within the octahedral layer), and to assess the induced density of reactive sites. From theoretical calculations, Drits *et al.* (2007a) reported the influence of these parameters, but also of the number, location, nature, and coordination of interlayer species on the profile of *hk* bands. In a few reports, the splitting of *hk* bands can be interpreted qualitatively to prove the orthogonal symmetry of the layer, and thus the

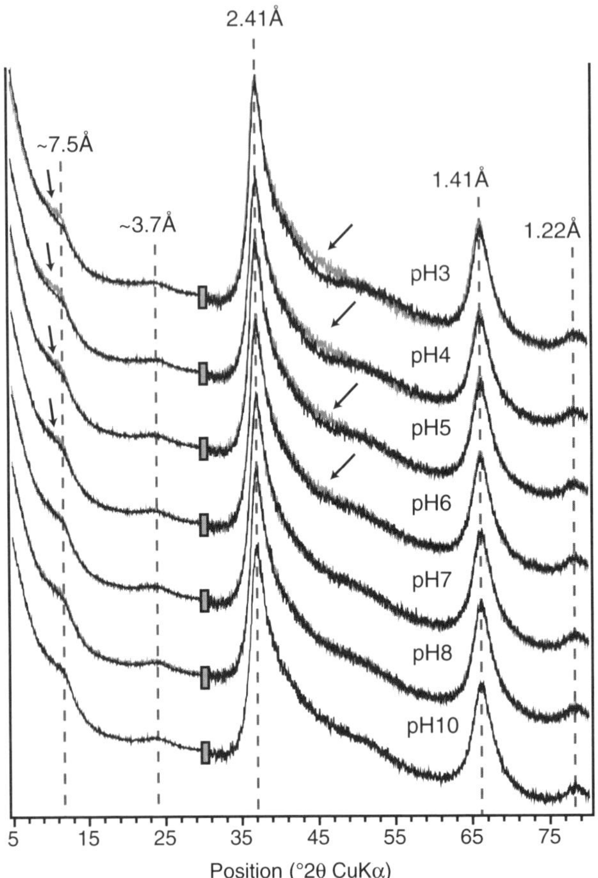

Fig. 20. XRD patterns recorded for the phyllomanganate vernadite (turbostratic birnessite variety) when equilibrated at different pH values (adapted from Grangeon, 2008). The XRD pattern recorded on the pH10 sample is shown systematically as a grey trace to emphasize the differences in the profile shape of the *hk* band at 2.41 Å (arrows). The observed profile modification is linked to the steadily increasing number of interlayer Mn cations located above vacant layer sites when decreasing the pH from 10 (0.075 interlayer Mn cation per layer octahedron) to 3 (0.175 interlayer Mn cation per layer octahedron).

prevailing influence of Mn^{3+} cations within the octahedral sheet (Webb *et al.*, 2005; Zhu *et al.*, 2010). Most often, the layer symmetry is hexagonal, thus indicating the predominant contribution of vacant layer sites to the layer-charge deficit (Fig. 20) (Jurgensen *et al.*, 2004; Villalobos *et al.*, 2006; Grangeon *et al.*, 2008, 2010; Lanson *et al.*, 2008). Depending on the synthesis protocol, or on the living organism responsible for Mn^{2+} oxidation, differences in the profile shape of the decreasing high-angle tail of the 02,11 band reveal a significant scatter in the density of vacant layer sites and thus in the number of reactive sites for the adsorption of di- and trivalent transition metal cations (Villalobos *et al.*, 2006; Grangeon *et al.*, 2008, 2010; Lanson *et al.*, 2008). The size of CSDs, which is also a key factor in the reactivity of these materials from the influence on the relative contribution of edge sites, is also highly variable as a function of the conditions of formation. Finally, the location and coordination of interlayer species was determined, thus allowing a complete structural characterization of these compounds despite their intrinsic defective character. It should be noted that XRD patterns obtained for some of these compounds are actually devoid of 00*l* reflections, despite their undisputable lamellar character, because of the extreme delamination, these lamellar

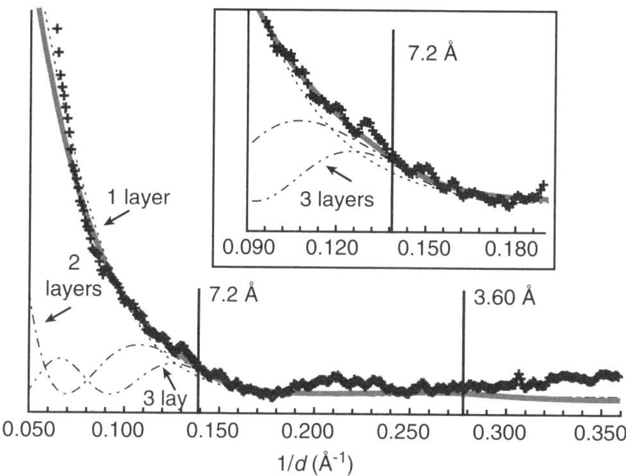

Fig. 21. Comparison between experimental and calculated diffraction patterns (00*l* basal reflections only) for the phyllomanganate vernadite formed in the roots of *Festuca rubra* (adapted from Lanson *et al.*, 2008). 001 and 002 reflections are calculated for crystallites composed of one (dashed line), two (dotted-dashed line) and three (dot-dot-dashed line) parallel layers. The optimal fit (red line) to the data (crosses) is obtained with an assemblage of diffracting crystallites containing 1, 2, and 3 layers in the ratio of 2.0:0.3:0.1. Readers of the paper version of this chapter may wish to download a colour version of this figure from www.minersoc.org/emu-notes/emu-11/11-4-colour.pdf.

compounds occurring essentially as single layers bound together by organic matter (Fig. 21 – Lanson *et al.*, 2008; Grangeon *et al.*, 2010).

4. Conclusions

Modelling of XRD data has been used increasingly over the last few decades to determine the nature, abundance and distribution of structure defects in lamellar compounds. For example, structure models were determined for 2D periodic species that are commonly described as amorphous owing to the absence of resolved *hkl* reflections. Similarly, the nature, abundance and distribution of well defined stacking faults have been uncovered in 3D ordered structures, and the nature of interstratified layer types and their stacking sequences were characterized. Owing to the increased potential of calculation routines, such studies could be carried out successfully even on polyphasic and/or natural samples. In all cases, they provided additional understanding of these complex structures, and the structural parameters determined often proved to be key factors for the comprehension of the material/mineral reactivity. Structural parameters such as layer-stacking sequences in mixed layers were also essential to determine their reaction mechanisms, and, ultimately, to determine thermodynamic data relevant to these systems and possibly the kinetic effects affecting these reactions. As the number of parameters required to describe completely defective structures is increased compared to common structure refinements, additional constraints, either experimental or computational, are extremely beneficial to support proposed structure models. In turn, taking into account structure defects and heterogeneity affecting layered compounds allows us to match more accurately the complementary data. The need to use such a large number of complementary techniques (microscopies, spectroscopies, chemical and thermal analyses, ...) in order to constrain or complement XRD structural studies is emphasized as it is often essential in order to obtain an unambiguous and comprehensive structure model (Drits, 1983, 2003; McCarty *et al.*, 2004; Lindgreen *et al.*, 2008; Drits *et al.*, 2010).

The modelling approach represents the optimum, and at present the sole, quantitative method allowing a thorough structure determination of defective lamellar compounds. This approach has allowed us to reveal the intrinsic complexity of defective lamellar structures and to improve their structural characterization. Lanson *et al.* (2002a) and Ramesh & Kamath (2008b) showed, for example, that structural characterization is significantly enhanced when stacking faults are characterized and quantified, even if the abundance of these faults is limited to a few percent. In addition, the modelling approach has allowed us to unravel specific properties of elementary layers, such as fluctuations of the layer-to-layer distance, or to refine their actual structure. The ability to calculate diffraction effects from multi-component mixed layers has also allowed us to reveal their ubiquitous presence in both natural and synthetic samples, and the frequent contribution to the diffraction intensity of mixed layers lacking a characteristic diffraction fingerprint in the low-angle region. The calculation routines available have refreshed the description of defective structures and a key point when using these tools is thus 'imagination': the search for possible structure models should not be restricted to those that have already been reported in the literature. A wider application of these tools to natural and synthetic samples will undoubtedly allow for new fundamental findings.

The use of these methods remains restricted, however, essentially because of the limited coupling between defective-structure calculation routines and automatic minimization ones. However, when sought, this coupling should not be detrimental to the versatility of the calculation routines. A trial-and-error optimization of the structure model will undoubtedly slow down the characterization, or hamper its generalization to large sample sets, but an (over-)simplified calculation routine will just preclude it. When comprehensive modelling is not performed, direct comparison between experimental XRD patterns and those calculated from the hypothesized structure remains crucial to assess the validity, and possibly the limits, of the proposed structure model.

References

Allegra, G. (1961) A simplified formula for the calculation of the X-ray intensity diffracted by a monodimensionally disordered structure. *Acta Crystallographica*, **14**, 535.

Allegra, G. (1964) The calculation of the intensity of X-rays diffracted by monodimensionally disordered structures. *Acta Crystallographica*, **14**, 579–586.

Aplin, A.C., Matenaar, I.F., McCarty, D.K. & van Der Pluijm, B.A. (2006) Influence of mechanical compaction and clay mineral diagenesis on the microfabric and pore-scale properties of deep-water Gulf of Mexico Mudstones. *Clays and Clay Minerals*, **54**, 500–514.

Argüelles, A., Leoni, M., Blanco, J.A. & Marcos, C. (2010) Semi-ordered crystalline structure of the Santa Olalla vermiculite inferred from X-ray powder diffraction. *American Mineralogist*, **95**, 126–134.

Bailey, S.W., Brindley, G.W., Kodama, H. & Martin, R.T. (1982) Report of the Clay Minerals Society Nomenclature Committee for 1980–1981: Nomenclature for regular interstratifications. *Clays and Clay Minerals*, **30**, 76–78.

Beaufort, D., Baronnet, A., Lanson, B. & Meunier, A. (1997) Corrensite: A single phase or a mixed-layer phyllosilicate in the saponite-to-chlorite conversion series? A case study of Sancerre-Couy deep drill hole (France). *American Mineralogist*, **82**, 109–124.

Bellotto, M., Rebours, B., Clause, O., Lynch, J., Bazin, D. & Elkaïm, E. (1996) A reexamination of hydrotalcite crystal chemistry. *Journal of Physical Chemistry*, **100**, 8527–8534.

Ben Brahim, J., Armagan, G., Besson, G. & Tchoubar, C. (1983) X-ray diffraction studies on the arrangement of water molecules in a smectite. I. Homogeneous two-water-layer Na-beidellite. *Journal of Applied Crystallography*, **16**, 264–269.

Ben Brahim, J., Besson, G. & Tchoubar, C. (1984) Etude des profils des bandes de diffraction X d'une beidellite-Na hydratée à deux couches d'eau. Détermination du mode d'empilement des feuillets et des sites occupés par l'eau. *Journal of Applied Crystallography*, **17**, 179–188.

Bérend, I., Cases, J.M., François, M., Uriot, J.P., Michot, L.J., Masion, A. & Thomas, F. (1995) Mechanism of adsorption and desorption of water vapour by homoionic montmorillonites: 2. The Li^+, Na^+, K^+, Rb^+ and Cs^+ exchanged forms. *Clays and Clay Minerals*, **43**, 324–336.

Bergmann, J. & Kleeberg, R. (1998) Rietveld analysis of disordered layer silicates. In: *Proceedings of the European Powder Diffraction (EPDIC5)* (R. Delhez & E.J. Mittemeijer, editors), p. 300–305.

Besson, G., Glaeser, R. & Tchoubar, C. (1983) Le césium, révélateur de structure des smectites. *Clay Minerals*, **18**, 11–19.

Bethke, C.G., Vergo, N. & Altaner, S.P. (1986) Pathways of smectite illitization. *Clays and Clay Minerals*, **34**, 125–135.

Biscoe, J. & Warren, B.E. (1942) An X-ray study of carbon black. *Journal of Applied Physics*, **13**, 364–371.

Bookin, A.S. & Drits, V.A. (1993) Polytype diversity of the hydrotalcite-like minerals. I. Possible polytypes and their diffraction features. *Clays and Clay Minerals*, **41**, 551–557.

Bookin, A.S., Drits, V.A., Plançon, A. & Tchoubar, C. (1989) Stacking faults in kaolin-group minerals in the light of real structural features. *Clays and Clay Minerals*, **37**, 297–307.

Bookin, A.S., Cherkashin, V.I. & Drits, V.A. (1993) Polytype diversity of the hydrotalcite-like minerals. II. Determination of the polytypes of experimentally studied varieties. *Clays and Clay Minerals*, **41**, 558–564.

Breu, J., Seidl, W., Stoll, A.J., Lange, K.G. & Probst, T.U. (2001) Charge homogeneity in synthetic fluorohectorite. *Chemistry of Materials*, **13**, 4213–4220.

Brindley, G.W. & Kikkawa, S. (1980) Thermal behavior of hydrotalcite and of anion-exchanged forms of hydrotalcite. *Clays and Clay Minerals*, **28**, 87–91.

Brindley, G.W. & Méring, J. (1951) Diffraction des Rayons X par les Structures en Couches Désordonnées. *Acta Crystallographica*, **4**, 441–447.

Brindley, G.W., Suzuki, T. & Thiry, M. (1983) Interstratified kaolinite/smectites from the Paris Basin; Correlations of layer proportions, chemical compositions and other data. *Bulletin de Minéralogie*, **106**, 403–410.

Calarge, L., Lanson, B., Meunier, A. & Formoso, M.L. (2003) The smectitic minerals in a bentonite deposit from Melo (Uruguay). *Clay Minerals*, **38**, 25–34.

Cases, J.M., Bérend, I., François, M., Uriot, J.P., Michot, L.J. & Thomas, F. (1997) Mechanism of adsorption and desorption of water vapour by homoionic montmorillonite: 3. the Mg^{2+}, Ca^{2+}, Sr^{2+} and Ba^{2+} exchanged forms. *Clays and Clay Minerals*, **45**, 8–22.

Cavani, F., Trifirò, F. & Vaccari, A. (1991) Hydrotalcite-type anionic clays: Preparation, properties and applications. *Catalysis Today*, **11**, 173–301.

Cesari, M. & Allegra, G. (1967) The intensity of X-rays diffracted by monodimensionally disordered structures: Case of identical layers and three different translation vectors. *Acta Crystallographica*, **23**, 200–205.

Cesari, M., Morelli, G.L. & Favretto, L. (1965) The determination of the type of stacking in mixed-layer clay minerals. *Acta Crystallographica*, **18**, 189–196.

Chipera, S.J. & Bish, D.L. (2001) Baseline studies of the Clay Minerals Society Source Clays: Powder X-ray diffraction analyses. *Clays and Clay Minerals*, **49**, 398–409.

Chukhrov, F.V., Gorshkov, A.I., Vitovskaya, I.V., Drits, V.A., Sistov, A.V. & Rudnitskaya, Y.S. (1982) Crystallochemical nature of Co-Ni asbolane. *International Geology Review*, **24**, 598–604.

Chukhrov, F.V., Gorshkov, A.I. & Drits, V.A. (1989) *Supergenic manganese hydrous oxides*. Nauka, Moscow, 208 pp.

Claret, F., Sakharov, B.A., Drits, V.A., Velde, B., Meunier, A., Griffault, L. & Lanson, B. (2004) Clay minerals in the Meuse-Haute marne underground laboratory (France): Possible influence of organic matter on clay mineral evolution. *Clays and Clay Minerals*, **52**, 515–532.

Cuadros, J. (2002) Structural insights from the study of Cs-exchanged smectites submitted to wetting-and-drying cycles. *Clay Minerals*, **37**, 473–486.

Curtius, H. & Ufer, K. (2007) Eu incorporation behavior of a Mg-Al-Cl layered double hydroxide. *Clays and Clay Minerals*, **55**, 354–360.

Cygan, R.T., Liang, J.-J. & Kalinichev, A.G. (2004) Molecular models of hydroxide, oxyhydroxide, and clay phases and the development of a general force field. *Journal of Physical Chemistry B*, **108**, 1255–1266.

D'yakonov, Y.S. (1961) Application of Fourier transform methods for the interpretation of mixed-layer mineral diffraction patterns. *Kristallografia*, **6**, 624–625 (in Russian).

D'yakonov, Y.S. (1962) Interpretation of mixed-layer mineral diffraction patterns by direct methods of Fourier transforms (in Russian). In: *X-ray Study of Mineral Raw Materials* (G.A. Sidorenko, editor). Nedra, Moscow, pp. 34–42.

de la Calle, C., Pezerat, H. & Gasperin, M. (1977) Problèmes d'ordre-désordre dans les vermiculites: Structure du minéral calcique hydraté à deux couches. *Journal de Physique*, **C7**, 128–133.

de la Calle, C., Suquet, H. & Pons, C.H. (1988) Stacking order in a 14.30 Å Mg-vermiculite. *Clays and Clay Minerals*, **38**, 481–490.

Delmas, C. & Tessier, C. (1997) Stacking faults in the structure of nickel hydroxide: a rationale of its high electrochemical activity. *Journal of Materials Chemistry*, **7**, 1439–1443.

Drits, V.A. (1983) Some aspects of the study of the real crystal structure of clay minerals. In: *Proceedings of the 5th Meeting of the European Clay Groups (Prague)* (J. Konta, editor), Univerzita Karlova, Prague, pp. 33–42.

Drits, V.A. (1985) Mixed-layer minerals: Diffraction methods and structural features. In: *Proceedings of the International Clay Conference (Denver)* (L.G. Schultz, H. Van Olphen & F.A. Mumpton, editors). The Clay Minerals Society, Boulder, Colorado, USA, pp. 33–45.

Drits, V.A. (1987) *Electron Diffraction and High-resolution Electron Microscopy of Mineral Structures*. Springer Verlag, Berlin, Heidelberg, 304 pp.

Drits, V.A. (1997) Mixed-layer minerals. In: *Modular Aspects of Minerals* (S. Merlino, editor). EMU Notes in Mineralogy, **1**, Eötvös University Press, Budapest, pp. 153–190.

Drits, V.A. (2003) Structural and chemical heterogeneity of layer silicates and clay minerals. *Clay Minerals*, **38**, 403–432.

Drits, V.A. & Bookin, A.S. (2001) Crystal structure and X-ray identification of layered double hydroxides. In: *Layered Double Hydroxides: Present and Future* (V. Rives, editor). Nova Science Publishers, New York, pp. 41–100.

Drits, V.A. & McCarty, D.K. (1996) The nature of diffraction effects from illite and illite-smectite consisting of interstratified *trans*-vacant and *cis*-vacant 2:1 layers: A semi-quantitative technique for determination of layer-type content. *American Mineralogist*, **81**, 852–863.

Drits, V.A. & Plançon, A. (1994) Expert system for structural characterization of phyllosilicates. II. Application to mixed-layer minerals. *Clay Minerals*, **29**, 39–45.
Drits, V.A. & Sakharov, B.A. (1976) *X-ray Structure Analysis of Mixed-layer Minerals*. Nauka, Moscow, 256 pp.
Drits, V.A. & Sakharov, B.A. (2004) Potential problems in the interpretation of powder X-ray diffraction patterns from fine-dispersed $2M_1$ and $3T$ dioctahedral micas. *European Journal of Mineralogy*, **16**, 99–110.
Drits, V.A. & Tchoubar, C. (1990) *X-ray Diffraction by Disordered Lamellar Structures: Theory and Applications to Microdivided Silicates and Carbons*. Springer-Verlag, Berlin, 371 pp.
Drits, V.A. & Zviagina, B.B. (2009) *Trans*-vacant and *cis*-vacant 2:1 layer silicates: Structural features, identification, and occurrence. *Clays and Clay Minerals*, **57**, 405–415.
Drits, V.A., Plançon, A., Sakharov, B.A., Besson, G., Tsipursky, S.J. & Tchoubar, C. (1984) Diffraction effects calculated for structural models of K-saturated montmorillonite containing different types of defects. *Clay Minerals*, **19**, 541–561.
Drits, V.A., Sokolova, T.N., Sokolova, G.V. & Cherkashin, V.I. (1987) New members of the hydrotalcite-manasseite group. *Clays and Clay Minerals*, **35**, 401–417.
Drits, V.A., Varaxina, T.V., Sakharov, B.A. & Plançon, A. (1994) A simple technique for identification of one-dimensional powder X-ray diffraction patterns for mixed-layer illite-smectites and other interstratified minerals. *Clays and Clay Minerals*, **42**, 382–390.
Drits, V.A., Gorshkov, A.I., Mokhov, A.V. & Pokrovskaya, E.V. (1995) New variety of mixed-layer mineral having unusual structure and morphology of crystals. *Lithology and Raw Materials*, **30**, 227–235 (in Russian).
Drits, V.A., Sakharov, B.A., Lindgreen, H. & Salyn, A. (1997) Sequential structure transformation of illite-smectite-vermiculite during diagenesis of Upper Jurassic shales from the North Sea and Denmark. *Clay Minerals*, **32**, 351–371.
Drits, V.A., Lanson, B., Gorshkov, A.I. & Manceau, A. (1998a) Substructure and superstructure of four-layer Ca-exchanged birnessite. *American Mineralogist*, **83**, 97–118.
Drits, V.A., Lindgreen, H., Salyn, A.L., Ylagan, R.F. & McCarty, D.K. (1998b) Semiquantitative determination of trans-vacant and cis-vacant 2:1 layers in illites and illite-smectites by thermal analysis and X-ray diffraction. *American Mineralogist*, **83**, 1188–1198.
Drits, V.A., Ivanovskaya, T.A., Sakharov, B.A., Gor'kova, N.V., Karpova, G.V. & Pokrovskaya, E.V. (2001) Pseudomorphous replacement of globular glauconite by mixed-layer chlorite-berthierine in the outer contact of dike: Evidence from the Lower Riphean Ust'Il'ya Formation, Anabar uplift. *Lithology and Mineral Resources*, **36**, 337–352.
Drits, V.A., Lindgreen, H., Sakharov, B.A., Jakobsen, H.J., Salyn, A.L. & Dainyak, L.G. (2002a) Tobelitization of smectite during oil generation in oil-source shales. Application to North Sea illite-tobelite-smectite-vermiculite. *Clays and Clay Minerals*, **50**, 82–98.
Drits, V.A., Sakharov, B.A., Dainyak, L.G., Salyn, A.L. & Lindgreen, H. (2002b) Structural and chemical heterogeneity of illite-smectites from Upper Jurassic mudstones of East Greenland related to volcanic and weathered parent rocks. *American Mineralogist*, **87**, 1590–1606.
Drits, V.A., Lindgreen, H., Sakharov, B.A., Jakobsen, H.J. & Zviagina, B.B. (2004) The detailed structure and origin of clay minerals at the cretaceous/tertiary boundary, Stevns Klint (Denmark). *Clay Minerals*, **39**, 367–390.
Drits, V.A., Sakharov, B.A., Salyn, A.L. & Lindgreen, H. (2005) Determination of the content and distribution of fixed ammonium in illite-smectite using a modified X-ray diffraction technique: Application to oil source rocks of western Greenland. *American Mineralogist*, **90**, 71–84.
Drits, V.A., Lanson, B. & Gaillot, A.C. (2007a) Birnessite polytype systematics and identification by powder X-ray diffraction. *American Mineralogist*, **92**, 771–788.
Drits, V.A., Lindgreen, H., Sakharov, B.A., Jakobsen, H.J., Fallick, A.E., Salyn, A.L., Dainyak, L.G., Zviagina, B.B. & Barfod, D.N. (2007b) Formation and transformation of mixed-layer minerals by tertiary intrusives in cretaceous mudstones, west Greenland. *Clays and Clay Minerals*, **55**, 260–283.

Drits, V.A., Ivanovskaya, T.A., Sakharov, B.A., Zvyagina, B.B., Derkowski, A., Gor'kova, N.V., Pokrovskaya, E.V., Savichev, A.T. & Zaitseva, T.S. (2010) Nature of the structural and crystal-chemical heterogeneity of the Mg-rich glauconite (Riphean, Anabar uplift). *Lithology and Mineral Resources*, **45**, 555–576.

Drits, V.A., Derkowski, A. & McCarty, D.K. (2011) New insight into the structural transformation of partially dehydroxylated pyrophyllite. *American Mineralogist*, **96**, 153–171.

Du, Y. & O'Hare, D. (2008) Observation of staging during intercalation in layered α-cobalt hydroxides: A synthetic and kinetic study. *Inorganic Chemistry*, **47**, 11839–11846.

Emmerich, K., Wolters, F., Kahr, G. & Lagaly, G. (2009) Clay profiling: The classification of montmorillonites. *Clays and Clay Minerals*, **57**, 104–114.

Evans, H.T. & Allman, R. (1968) The crystal structure and crystal chemistry of valleriite. *Zeitschrift für Kristallographie*, **127**, 73–93.

Feng, Q., Xu, Y.H., Kajiyoshi, K. & Yanagisawa, K. (2001) Hydrothermal soft chemical synthesis of Ni(OH)(2)-birnessite sandwich layered compound and layered LiNi1/3Mn2/3O2. *Chemistry Letters*, **30**, 1036–1037.

Feng, Y. (2006) *Formation and properties of second-stage layered double hydroxide materials*. PhD thesis, University of Clermont-Ferrand, France, 137 pp.

Feng, Y., Williams, G.R., Leroux, F., Taviot-Guého, C. & O'Hare, D. (2006) Selective anion-exchange properties of second-stage layered double hydroxide heterostructures. *Chemistry of Materials*, **18**, 4312–4318.

Ferrage, E. (2004) *Etude expérimentale de l'hydratation des smectites par simulation des raies 00l de diffraction des rayons X. Implications pour l'étude d'une perturbation thermique sur la minéralogie de l'argilite du site Meuse - Haute Marne*. PhD thesis, University of Grenoble, France, 326 pp.

Ferrage, E., Lanson, B., Malikova, N., Plançon, A., Sakharov, B.A. & Drits, V.A. (2005a) New insights on the distribution of interlayer water in bi-hydrated smectite from X-ray diffraction profile modeling of *00l* reflections. *Chemistry of Materials*, **17**, 3499–3512.

Ferrage, E., Lanson, B., Sakharov, B.A. & Drits, V.A. (2005b) Investigation of smectite hydration properties by modeling experimental X-ray diffraction patterns: Part I. Montmorillonite hydration properties. *American Mineralogist*, **90**, 1358–1374.

Ferrage, E., Tournassat, C., Rinnert, E., Charlet, L. & Lanson, B. (2005c) Experimental evidence for Ca-chloride ion pairs in the interlayer of montmorillonite. An XRD profile modeling approach. *Clays and Clay Minerals*, **53**, 348–360.

Ferrage, E., Tournassat, C., Rinnert, E. & Lanson, B. (2005d) Influence of pH on the interlayer cationic composition and hydration state of Ca-montmorillonite: Analytical chemistry, chemical modelling and XRD profile modelling study. *Geochimica et Cosmochimica Acta*, **69**, 2797–2812.

Ferrage, E., Kirk, C.A., Cressey, G. & Cuadros, J. (2007a) Dehydration of Ca-montmorillonite at the crystal scale. Part I: Structure evolution. *American Mineralogist*, **92**, 994–1006.

Ferrage, E., Kirk, C.A., Cressey, G. & Cuadros, J. (2007b) Dehydration of Ca-montmorillonite at the crystal scale. Part 2. Mechanisms and kinetics. *American Mineralogist*, **92**, 1007–1017.

Ferrage, E., Lanson, B., Sakharov, B.A., Geoffroy, N., Jacquot, E. & Drits, V.A. (2007c) Investigation of dioctahedral smectite hydration properties by modeling of X-ray diffraction profiles: Influence of layer charge and charge location. *American Mineralogist*, **92**, 1731–1743.

Ferrage, E., Lanson, B., Michot, L.J. & Robert, J.L. (2010) Hydration Properties and Interlayer Organization of Water and Ions in Synthetic Na-Smectite with Tetrahedral Layer Charge. Part 1. Results from X-ray Diffraction Profile Modeling. *Journal of Physical Chemistry C*, **114**, 4515–4526.

Ferrage, E., Sakharov, B.A., Michot, L.J., Delville, A., Bauer, A., Lanson, B., Grangeon, S., Frapper, G., Jimenez-Ruiz, M. & Cuello, G.J. (2011a) Hydration properties and interlayer organization of water and ions in synthetic Na-smectite with tetrahedral layer charge. Part 2. Towards a precise coupling between molecular simulations and diffraction data. *Journal of Physical Chemistry C*, **115**, 1867–1881.

Ferrage, E., Vidal, O., Mosser Ruck, R., Cathelineau, M. & Cuadros, J. (2011b) A reinvestigation of smectite illitization in experimental conditions: Results from X-ray diffraction and transmission electron microscopy. *American Mineralogist*, **96**, 207–223.

Fogg, A.M., Dunn, J.S. & O'Hare, D. (1998a) Formation of second-stage intermediates in anion-exchange intercalation reactions of the layered double hydroxide $[LiAl_2(OH)_6]Cl.H_2O$ as observed bu time-resolved, *in situ* X-ray diffraction. *Chemistry of Materials*, **10**, 356–360.

Fogg, A.M., Dunn, J.S., Shyu, S.-G., Cary, D.R. & O'Hare, D. (1998b) Selective ion-exchange intercalation of isomeric dicarboxylate anions into the layered double hydroxide $[LiAl_2(OH)_6]Cl.H_2O$. *Chemistry of Materials*, **10**, 351–355.

Gaillot, A.-C. (2002) *Caractérisation structurale de la birnessite: Influence du protocole de synthèse*. PhD thesis, University of Grenoble, France, 392 pp.

Gaillot, A.-C., Flot, D., Drits, V.A., Manceau, A., Burghammer, M. & Lanson, B. (2003) Structure of synthetic K-rich birnessite obtained by high-temperature decomposition of $KMnO_4$. I. Two-layer polytype from 800°C experiment. *Chemistry of Materials*, **15**, 4666–4678.

Gaillot, A.-C., Drits, V.A., Plançon, A. & Lanson, B. (2004) Structure of synthetic K-rich birnessites obtained by high-temperature decomposition of $KMnO_4$. 2. Phase and structural heterogeneities. *Chemistry of Materials*, **16**, 1890–1905.

Gaillot, A.-C., Lanson, B. & Drits, V.A. (2005) Structure of birnessite obtained from decomposition of permanganate under soft hydrothermal conditions. 1. Chemical and structural evolution as a function of temperature. *Chemistry of Materials*, **17**, 2959–2975.

Gaillot, A.-C., Drits, V.A., Manceau, A. & Lanson, B. (2007) Structure of the synthetic K-rich phyllomanganate birnessite obtained by high-temperature decomposition of KMnO4 - Substructures of K-rich birnessite from 1000°C experiment. *Microporous and Mesoporous Materials*, **98**, 267–282.

Gates, W.P., Slade, P.G., Manceau, A. & Lanson, B. (2002) Site occupancies by iron in nontronites. *Clays and Clay Minerals*, **50**, 223–239.

Glaeser, R., Mantine, I. & Méring, J. (1967) Observations sur la beidellite. *Bulletin du Groupe Français des Argiles*, **19**, 125–130.

Grangeon, S. (2008) *Cristallochimie des phyllomanganates nanocristallins désordonnés. Implication pour l'adsorption d'éléments métalliques*. PhD thesis, University of Grenoble, France, 210 pp.

Grangeon, S., Lanson, B., Lanson, M. & Manceau, A. (2008) Crystal structure of Ni-sorbed synthetic vernadite: a powder X-ray diffraction study. *Mineralogical Magazine*, **72**, 1279–1291.

Grangeon, S., Lanson, B., Miyata, N., Tani, Y. & Manceau, A. (2010) Structure of nanocrystalline phyllomanganates produced by freshwater fungi. *American Mineralogist*, **95**, 1608–1616.

Guerlou Demourgues, L., Tessier, C., Bernard, P. & Delmas, C. (2004) Influence of substituted zing on stacking faults in nickel hydroxide. *Journal of Materials Chemistry*, **14**, 2649–2654.

Guggenheim, S., Adams, J.M., Bain, D.C., Bergaya, F., Brigatti, M.F., Drits, V.A., Formoso, M.L.L., Galán, E., Kogure, T. & Stanjek, H. (2006) Summary of recommendations of nomenclature committees relevant to clay mineralogy: report of the Association Internationale pour l'Etude des Argiles (AIPEA) Nomenclature Committee for 2006. *Clay Minerals*, **41**, 863–877.

Guinier, A. (1964) *Théorie et technique de la radiocristallographie*. Dunod, Paris, 740 pp.

Hendricks, S. & Teller, E. (1942) X-ray interference in partially ordered layer lattices. *Journal of Chemical Physics*, **10**, 147–167.

Hower, J. & Mowatt, T.C. (1966) The mineralogy of illites and mixed-layer illite/montmorillonites. *American Mineralogist*, **51**, 825–854.

Hubert, F., Caner, L., Meunier, A. & Lanson, B. (2009) Advances in characterization of soil clay mineralogy using X-ray diffraction: from decomposition to profile fitting. *European Journal of Soil Science*, **60**, 1093–1105.

Hubert, F., Caner, L., Ferrage, E. & Meunier, A. (2011) Refinement of soil clay mineralogy based on the multispecimen XRD profile fitting procedure. *European Journal of Soil Science*, submitted.

Ijdo, W.L. & Pinnavaia, T.J. (1998a) Solid solution formation in amphiphilic organic-inorganic clay heterostructures. *Chemistry of Materials*, **11**, 3227–3231.

Ijdo, W.L. & Pinnavaia, T.J. (1998b) Staging of organic and inorganic gallery cations in layered silicate heterostructures. *Journal of Solid State Chemistry*, **139**, 281–289.

Ijdo, W.L., Lee, T. & Pinnavaia, T.J. (1996) Regularly interstratified layered silicate heterostructures: Precursors to pillared rectorite-like intercalates. *Advanced Materials*, **8**, 79–83.

Inoue, A., Bouchet, A., Velde, B. & Meunier, A. (1989) Convenient technique for estimating smectite layer percentage in randomly interstratified illite/smectite minerals. *Clays and Clay Minerals*, **37**, 227–234.

Inoue, A., Lanson, B., Marques Fernandes, M., Sakharov, B.A., Murakami, T., Meunier, A. & Beaufort, D. (2005) Illite-smectite mixed-layer minerals in the hydrothermal alteration of volcanic rocks: I. One-dimensional XRD structure analysis and characterization of component layers. *Clays and Clay Minerals*, **53**, 423–439.

Iwasaki, T. & Watanabe, T. (1988) Distribution of Ca and Na ions in dioctahedral smectites and interstratified dioctahedral mica/smectites. *Clays and Clay Minerals*, **36**, 73–82.

Iyi, N., Kurashima, K. & Fujita, T. (2002) Orientation of an organic anion and second-staging structure in layered double-hydroxide intercalates. *Chemistry of Materials*, **14**, 583–589.

Iyi, N., Fujii, K., Okatomo, K. & Sasaki, T. (2007) Factors influencing the hydration of layered double hydroxides (LDHs) and the appearance of an intermediate second staging phase. *Applied Clay Science*, **35**, 218–227.

Jagodzinski, H. (1949) Eindimensionale Fehlordnung in Kristallen und ihr Einfluss auf die Röntgeninterferenzen: I. Berechnung des Fehlordnungsgrades aus der Röntgenintensitaten. *Acta Crystallographica*, **2**, 201–207.

Jurgensen, A., Widmeyer, J.R., Gordon, R.A., Bendell Young, L.I., Moore, M.M. & Crozier, E.D. (2004) The structure of the manganese oxide on the sheath of the bacterium *Leptothrix discophora*: An XAFS study. *American Mineralogist*, **89**, 1110–1118.

Kakinoki, J. & Komura, Y. (1952) Intensity of X-ray diffraction by one dimensionally disordered crystal: I. General derivation in cases of the "Reichweit" $S = 0$ and 1. *Journal of the Physical Society of Japan*, **7**, 30–35.

Kakinoki, J. & Komura, Y. (1954a) Intensity of X-ray diffraction by one dimensionally disordered crystal: II. General derivation in the case of the correlation range $S \geq 2$. *Journal of the Physical Society of Japan*, **9**, 169–176.

Kakinoki, J. & Komura, Y. (1954b) Intensity of X-ray diffraction by one dimensionally disordered crystal: III. The close-packed structure. *Journal of the Physical Society of Japan*, **9**, 177–183.

Kakinoki, J. & Komura, Y. (1965) Diffraction by one dimensionally disordered crystal: I. The intensity equation. *Acta Crystallographica*, **19**, 137–147.

Kameda, J., Miyawaki, R., Kitagawa, R. & Kogure, T. (2007) XRD and HRTEM analyses of stacking structures in sudoite, di-trioctahedral chlorite. *American Mineralogist*, **92**, 1586–1592.

Kameda, J., Saruwatari, K., Beaufort, D. & Kogure, T. (2008) Textures and polytypes in vermiform kaolins diagenetically formed in a sandstone reservoir: a FIB-TEM investigation. *European Journal of Mineralogy*, **20**, 199–204.

Karmous, M.S., Ben Rhaiem, H., Robert, J.L., Lanson, B. & Amara, A.B.H. (2009) Charge location effect on the hydration properties of synthetic saponite and hectorite saturated by Na^+, Ca^{2+} cations: XRD investigation. *Applied Clay Science*, **46**, 43–50.

Kayenoshi, M. & Jones, W. (1998) Exchange of interlayer terephtalate anions from a Mg-Al layered double hydroxide: formation of intermediate intersratified phases. *Chemical Physics Letters*, **296**, 183–187.

Kogure, T. & Inoue, A. (2005) Determination of defect structures in kaolin minerals by high-resolution transmission electron microscopy (HRTEM). *American Mineralogist*, **90**, 85–89.

Kogure, T. & Kameda, J. (2008) High-resolution TEM and XRD simulation of stacking disorder in 2:1 phyllosilicates. *Zeitschrift für Kristallographie*, **223**, 69–75.

Kogure, T., Hybler, J. & Durovic, S. (2001) A HRTEM study of cronstedtite: Determination of polytypes and layer polarity in trioctahedral 1:1 phyllosilicates. *Clays and Clay Minerals*, **49**, 310–317.

Kogure, T., Jige, M., Kameda, J., Yamagishi, A., Miyawaki, R. & Kitagawa, R. (2006a) Stacking structures in pyrophyllite revealed by high-resolution transmission electron microscopy (HRTEM). *American Mineralogist*, **91**, 1293–1299.

Kogure, T., Kameda, J., Matsui, T. & Miyawaki, R. (2006b) Stacking structure in disordered talc: Interpretation of its X-ray diffraction pattern by using pattern simulation and high-resolution transmission electron microscopy. *American Mineralogist*, **91**, 1363–1370.

Kogure, T., Kameda, J. & Drits, V.A. (2008) Stacking faults with 180° layer rotation in celadonite, an Fe- and Mg-rich dioctahedral mica. *Clays and Clay Minerals*, **56**, 612–621.

Kogure, T., Elzea-Kogel, J., Johnston, C.T. & Bish, D.L. (2010) Stacking disorder in a sedimentary kaolinite. *Clays and Clay Minerals*, **58**, 62–71.

Lanson, B., Drits, V.A., Silvester, E.J. & Manceau, A. (2000) Structure of H-exchanged hexagonal birnessite and its mechanism of formation from Na-rich monoclinic buserite at low pH. *American Mineralogist*, **85**, 826–838.

Lanson, B., Drits, V.A., Feng, Q. & Manceau, A. (2002a) Structure of synthetic Na-birnessite: Evidence for a triclinic one-layer unit cell. *American Mineralogist*, **87**, 1662–1671.

Lanson, B., Drits, V.A., Gaillot, A.-C., Silvester, E., Plançon, A. & Manceau, A. (2002b) Structure of heavy-metal sorbed birnessite: Part 1. Results from X-ray diffraction. *American Mineralogist*, **87**, 1631–1645.

Lanson, B., Sakharov, B.A., Claret, F. & Drits, V.A. (2005) Diagenetic evolution of clay minerals in Gulf Coast shales: New insights from X-ray diffraction profile modeling. In: *Proceedings of the 42nd Annual Meeting, Clay Minerals Society (Burlington, Vermont, USA)*, p. 69.

Lanson, B., Marcus, M.A., Fakra, S., Panfili, F., Geoffroy, N. & Manceau, A. (2008) Formation of Zn-Ca phyllomanganate nanoparticles in grass roots. *Geochimica et Cosmochimica Acta*, **72**, 2478–2490.

Lanson, B., Sakharov, B.A., Claret, F. & Drits, V.A. (2009) Diagenetic smectite-to-illite transition in clay-rich sediments: a reappraisal of X-ray diffraction results using the multi-specimen method. *American Journal of Science*, **309**, 476–516.

Leoni, M. (2008) Diffraction analysis of layer disorder. *Zeitschift für Kristallographie*, **223**, 561–568.

Leoni, M., Gualtieri, A.F. & Roveri, N. (2004) Simultaneous refinement of structure and microstructure of layered materials. *Journal of Applied Crystallography*, **37**, 166–173.

Leroux, F. & Besse, J.-P. (2004) Layered double hydroxide/polymer nanocomposites. In: *Clay Surfaces: Fundamentals and Applications* (F. Wypych & K.G. Satyanarayana, editors). Elsevier, Amsterdam, pp. 459–495.

Lindgreen, H., Drits, V.A., Sakharov, B.A., Salyn, A.L., Wrang, P. & Dainyak, L.G. (2000) Illite-smectite structural changes during metamorphism in black Cambrian Alum shales from the Baltic area. *American Mineralogist*, **85**, 1223–1238.

Lindgreen, H., Drits, V.A., Sakharov, B.A., Jakobsen, H.J., Salyn, A.L., Dainyak, L.G. & Kroyer, H. (2002) The structure and diagenetic transformation of illite-smectite and chlorite-smectite from North Sea Cretaceous-Tertiary chalk. *Clay Minerals*, **37**, 429–450.

Lindgreen, H., Drits, V.A., Jakobsen, F.C. & Sakharov, B.A. (2008) Clay mineralogy of the Central North Sea Upper Cretaceous-Tertiary Chalk and the formation of clay-rich layers. *Clays and Clay Minerals*, **56**, 693–710.

Ma, C. & Eggleton, R.A. (1999) Surface layer types of kaolinite: A high-resolution transmission electron microscope study. *Clays and Clay Minerals*, **47**, 181–191.

MacEwan, D.M.C. (1956) Fourier transform methods for studying X-ray scattering from lamellar systems: I. A direct method for analysing interstratified mixtures. *Kolloidzeitschrift*, **149**, 96–108.

Makovicky, E. & Hyde, B.G. (1992) Incommensurate two-layer structures with complex crystal chemistry. *Materials Science Forum*, **100–101**, 1–100.

Manceau, A., Gorshkov, A.I. & Drits, V.A. (1992) Structural Chemistry of Mn, Fe, Co, and Ni in Mn hydrous oxides. II. Information from EXAFS spectroscopy, electron and X-ray diffraction. *American Mineralogist*, **77**, 1144–1157.

Manceau, A., Drits, V.A., Silvester, E.J., Bartoli, C. & Lanson, B. (1997) Structural mechanism of Co^{2+} oxidation by the phyllomanganate buserite. *American Mineralogist*, **82**, 1150–1175.

Manceau, A., Drits, V.A., Lanson, B., Chateigner, D., Wu, J., Huo, D., Gates, W.P. & Stucki, J.W. (2000a) Oxidation-reduction mechanism of iron in dioctahedral smectites: II. Crystal chemistry of reduced Garfield nontronite. *American Mineralogist*, **85**, 153–172.

Manceau, A., Lanson, B., Drits, V.A., Chateigner, D., Gates, W.P., Wu, J., Huo, D. & Stucki, J.W. (2000b) Oxidation-reduction mechanism of iron in dioctahedral smectites: I. Crystal chemistry of oxidized reference nontronites. *American Mineralogist*, **85**, 133–152.

McCarty, D.K. & Reynolds, R.C., Jr (1995) Rotationally disordered illite/smectite in Paleozoic K-bentonites. *Clays and Clay Minerals*, **43**, 271–284.

McCarty, D.K., Drits, V.A., Sakharov, B., Zviagina, B.B., Ruffell, A. & Wach, G. (2004) Heterogeneous mixed-layer clays from the Cretaceous greensand, Isle of Wight, southern England. *Clays and Clay Minerals*, **52**, 552–575.

McCarty, D.K., Sakharov, B.A. & Drits, V.A. (2008) Early clay diagenesis in Gulf Coast sediments: New insights from XRD profile modeling. *Clays and Clay Minerals*, **56**, 359–379.

McCarty, D.K., Sakharov, B.A. & Drits, V.A. (2009) New insights into smectite illitization: A zoned K-bentonite revisited. *American Mineralogist*, **94**, 1653–1671.

Méring, J. (1949) L'interférence des rayons-X dans les systèmes à stratification désordonnée. *Acta Crystallographica*, **2**, 371–377.

Méring, J. & Brindley, G.W. (1967) X-ray diffraction band profiles of montmorillonite. Influence of hydration and of the exchangeable cations. *Clays and Clay Minerals*, **15**, 51–60.

Méring, J. & Glaeser, R. (1954) Sur le rôle de la valence des cations échangeables dans la montomorillonite. *Bulletin de la Société Française de Minéralogie et de Cristallographie*, **77**, 519–530.

Meunier, A. (2006) Why are clay minerals small? *Clay Minerals*, **41**, 551–566.

Meunier, A., Lanson, B. & Velde, B. (2004) Composition variation of illite-vermiculite-smectite mixed-layer minerals in a bentonite bed from Charente (France). *Clay Minerals*, **39**, 317–332.

Möller, M.W., Handge, U.A., Kunz, D.A., Lunkenbein, T., Altstädt, V. & Breu, J. (2010a) Tailoring shear-stiff, mica-like nanoplatelets. *ACS Nano*, **4**, 717–724.

Möller, M.W., Hirsemann, D., Haarmann, F., Senker, J. & Breu, J. (2010b) Facile Scalable Synthesis of Rectorites. *Chemistry of Materials*, **22**, 186–196.

Moore, D.M. & Hower, J. (1986) Ordered interstratification of dehydrated and hydrated Na-smectite. *Clays and Clay Minerals*, **34**, 379–384.

Moore, D.M. & Reynolds, R.C., Jr (1997) *X-ray Diffraction and the Identification and Analysis of Clay Minerals*. Oxford University Press, New York, 378 pp.

Murakami, T., Inoue, A., Lanson, B., Meunier, A. & Beaufort, D. (2005) Illite-smectite mixed-layer minerals in the hydrothermal alteration of volcanic rocks: II. One-dimensional HRTEM structure images and formation mechanisms. *Clays and Clay Minerals*, **53**, 440–451.

Newman, S.P., Williams, S.J., Coveney, P.V. & Jones, W. (1998) Interlayer arrangement of hydrated MgAl layered double hydroxides containing guest terephtalate anions: Comparison of simulation and measurement. *Journal of Physical Chemistry B*, **102**, 6710–6719.

O'Hare, D., Evans, J.S.O., Fogg, A.M. & O'Brien, S. (2000) Time-resolved, *in situ* X-ray diffraction studies of intercalation in lamellar hosts. *Polyhedron*, **19**, 297–305.

Organova, N.I. (1989) *Crystal Chemistry of Incommensurate and Modulated Mixed-layer Minerals*. Nauka, Moscow, (in Russian), 140 pp.

Organova, N.I., Drits, V.A. & Dimitrik, A.L. (1974) Selected-area electron diffraction study of a type II "valleriite-like" mineral. *American Mineralogist*, **59**, 190–200.

Oueslati, W., Ben Rhaiem, H., Lanson, B. & Amara, A.B. (2009) Selectivity of Na-montmorillonite in relation with the concentration of bivalent cation (Cu^{2+}, Ca^{2+}, Ni^{2+}) by quantitative analysis of XRD patterns. *Applied Clay Science*, **43**, 224–227.

Perry, E.A., Jr. & Hower, J. (1970) Burial diagenesis in Gulf Coast pelitic sediments. *Clays and Clay Minerals*, **18**, 165–177.

Pevear, D.R. & Schuette, J.F. (1993) Inverting the Newmod X-ray diffraction forward model for clay minerals using genetic algorithms. In: *Computer Applications to X-ray Powder Diffraction Analysis of Clay Minerals* (R.C. Reynolds, Jr & J.R. Walker, editors). CMS workshop lectures, **5**. The Clay Minerals Society, Boulder, Colorado, USA.

Pisson, J., Taviot-Gueho, C., Israeli, Y., Leroux, F., Munsch, P., Itie, J.P., Briois, V., Morel-Desrosiers, N. & Besse, J.P. (2003) Staging of organic and inorganic anions in layered double hydroxides. *Journal of Physical Chemistry B*, **107**, 9243–9248.

Plançon, A. (1981) Diffraction by layer structures containing different kinds of layers and stacking faults. *Journal of Applied Crystallography*, **14**, 300–304.

Plançon, A. (2002) New modeling of X-ray diffraction by disordered lamellar structures, such as phyllosilicates. *American Mineralogist*, **87**, 1672–1677.

Plançon, A. (2003) Modelling X-ray diffraction by lamellar structures composed of electrically charged layers. *Journal of Applied Crystallography*, **36**, 146–153.

Plançon, A. & Drits, V.A. (1994) Expert system for structural characterization of phyllosilicates. I. Description of the expert system. *Clay Minerals*, **29**, 33–38.

Plançon, A. & Drits, V.A. (2000) Phase analysis of clays using an expert system and calculation programs for X-ray diffraction by two- and three-component mixed-layer minerals. *Clays and Clay Minerals*, **48**, 57–62.

Plançon, A. & Roux, J. (2010) Software for the assisted determination of the structural parameters of mixed-layer phyllosilicates. *European Journal of Mineralogy*, **22**, 733–740.

Plançon, A. & Tchoubar, C. (1977) Determination of structural defects in phyllosilicates by X-ray powder diffraction – II. Nature and proportion of defects in natural kaolinites. *Clays and Clay Minerals*, **25**, 436–450.

Plançon, A., Drits, V.A., Sakharov, B.A., Gilan, Z.I. & Ben Brahim, J. (1983) Powder diffraction by layered minerals containing different layers and/or stacking defects. Comparison between Markovian and non-Markovian models. *Journal of Applied Crystallography*, **16**, 62–69.

Plançon, A., Sakharov, B.A. & Drits, V.A. (1984a) Diffraction effects from a homogeneous aggregate of fine crystals with packing effects. *Soviet Physics Crystallography*, **29**, 392–395.

Plançon, A., Sakharov, B.A., Gilan, Z.I. & Drits, V.A. (1984b) Diffraction effects from a homogeneous aggregate of mixed-layer crystals. *Soviet Physics Crystallography*, **29**, 389–392.

Plançon, A., Giese, R.F., Snyder, R., Drits, V.A. & Bookin, A.S. (1989) Stacking faults in the kaolin-group minerals: Defect structures of kaolinite. *Clays and Clay Minerals*, **37**, 203–210.

Pons, C.H., Rousseaux, F. & Tchoubar, D. (1981) Utilisation du rayonnement synchrotron en diffusion aux petits angles pour l'étude du gonflement des smectites: I. Etude du système eau-montmorillonite-Na en fonction de la température. *Clay Minerals*, **16**, 23–42.

Pons, C.H., Rousseaux, F. & Tchoubar, D. (1982) Utilisation du rayonnement synchrotron en diffusion aux petits angles pour l'étude du gonflement des smectites: II. Etude de différents systèmes eau-smectites en fonction de la température. *Clay Minerals*, **17**, 327–338.

Radha, A.V. & Kamath, P.V. (2009) Polytype selection and structural disorder mediated by intercalated sulfate ions among the layered double hydroxides of Zn with Al and Cr. *Crystal Growth and Design*, **9**, 3197–3203.

Radha, A.V., Shivakumara, C. & Kamath, P.V. (2005) DIFFaX simulations of stacking faults in layered double hydroxides (LDHs). *Clays and Clay Minerals*, **53**, 520–527.

Radha, A.V., Antonyraj, C.A., Kamath, P.V. & Kannan, S. (2010) Polytype transformations in the SO_4^{2-} containing layered double hydroxides of Zinc with Aluminum and Chromium: The metal hydroxide layer as a structural synthon. *Zeitschrift Fur Anorganische Und Allgemeine Chemie*, **636**, 2658–2664.

Ragavan, A., Williams, G.R. & O'Hare, D. (2009) A thermodynamically stable layered double hydroxide heterostructure. *Journal of Materials Chemistry*, **19**, 4211–4216.

Rajamathi, M., Kamath, P.V. & Seshadri, R. (2000) Polymorphism in nickel hydroxide: role of interstratification. *Journal of Materials Chemistry*, **10**, 503–506.

Ramesh, T.N. (2009) Crystallite size effects in stacking faulted nickel hydroxide and its electrochemical behaviour. *Materials Chemistry and Physics*, **114**, 618–623.

Ramesh, T.N. & Kamath, P.V. (2008a) The effect of stacking faults on the electrochemical performance of nickel hydroxide electrodes. *Materials Research Bulletin*, **43**, 2827–2832.

Ramesh, T.N. & Kamath, P.V. (2008b) Planar defects in layered hydroxides: Simulation and structure refinement of beta-nickel hydroxide. *Materials Research Bulletin*, **43**, 3227–3233.

Ramesh, T.N., Jayashree, R.S. & Kamath, P.V. (2003) Disorder in layered hydroxides: DIFFaX simulation of the X-ray powder diffraction patterns of nickel hydroxide. *Clays and Clay Minerals*, **51**, 570–576.

Ramesh, T.N., Kamath, P.V. & Shivakumara, C. (2005) Correlation of structural disorder with the reversible discharge capacity of nickel hydroxide electrode. *Journal of the Electrochemical Society*, **152**, A806–A810.

Ramesh, T.N., Kamath, P.V. & Shivakumara, C. (2006) Classification of stacking faults and their stepwise elimination during the disorder ⇒ order transformation of nickel hydroxide. *Acta Crystallographica*, **B62**, 530–536.

Reynolds, R.C., Jr. (1967) Interstratified clay systems: Calculation of the total one-dimensional diffraction function. *American Mineralogist*, **52**, 661–672.

Reynolds, R.C., Jr. (1980) Interstratified clay minerals. In: *Crystal Structures of Clay Minerals and their X-ray Identification* (G.W. Brindley & G. Brown, editors). Monograph **5**, Mineralogical Society, London, pp. 249–359.

Reynolds, R.C., Jr. (1989) Diffraction by small and disordered crystals. In: *Modern Powder Diffraction* (D.L. Bish & J.E. Post, editors). Reviews in Mineralogy, **20**, Mineralogical Society of America, Washington, D.C., pp. 145–181.

Reynolds, R.C., Jr. (1993) Three-dimensional X-ray powder diffraction from disordered illite: Simulation and interpretation of the diffraction patterns. In: *Computer Applications to X-ray Powder Diffraction Analysis of Clay Minerals* (R.C. Reynolds, Jr & J.R. Walker, editors). Clay Minerals Society workshop lectures, **5**. The Clay Minerals Society, Boulder, Colorado, USA, pp. 44–78.

Rietveld, H.M. (1967) Line profiles of neutron-diffraction peaks for structure refinement. *Acta Crystallographica*, **22**, 151–152.

Rives, V. (2001) Layered double hydroxides: Present and future. Nova Science Publishers, New York, 439 pp.

Sakharov, B.A. (2005) Improved model for simulation of experimental XRD patterns for clays including mixed-layer clay minerals. In: *Proceedings of the Clay Minerals Society 42nd Annual Meeting (Burlington, Vermont, USA)*, p. 95.

Sakharov, B.A. & Drits, V.A. (1973) Mixed-layer kaolinite-montmorillonite: A comparison of observed and calculated diffraction patterns. *Clays and Clay Minerals*, **21**, 15–17.

Sakharov, B.A., Naumov, A.S. & Drits, V.A. (1982a) X-ray intensities scattered by layer structures with short-range ordering parameters $S \geq 1$ and $G \geq 1$. *Doklady Akademii Nauk*, **265**, 871–874 (in Russian).

Sakharov, B.A., Naumov, A.S. & Drits, V.A. (1982b) X-ray diffraction by mixed-layer structures with random distribution of stacking faults. *Doklady Akademii Nauk*, **265**, 339–343 (in Russian).

Sakharov, B.A., Lindgreen, H., Salyn, A. & Drits, V.A. (1999a) Determination of illite-smectite structures using multispecimen X-ray diffraction profile fitting. *Clays and Clay Minerals*, **47**, 555–566.

Sakharov, B.A., Lindgreen, H., Salyn, A.L. & Drits, V.A. (1999b) Mixed-layer kaolinite-illite-vermiculite in North Sea shales. *Clay Minerals*, **34**, 333–344.

Sakharov, B.A., Plançon, A. & Drits, V.A. (1999c) Influence of outer surface structure of crystals on X-ray diffraction. In: *Proceedings of the Euroclay conference (Krakow, Poland)*, p. 129.

Sakharov, B.A., Dubinska, E., Bylina, P., Kozubowski, J.A., Kapron, G. & Frontczak & Baniewicz, M. (2004a) Serpentine-smectite interstratified minerals from Lower Silesia (SW Poland). *Clays and Clay Minerals*, **52**, 55–65.

Sakharov, B.A., Plançon, A., Lanson, B. & Drits, V.A. (2004b) Influence of the outer surface layers of crystals on the X-ray diffraction intensity of basal reflections. *Clays and Clay Minerals*, **52**, 680–692.

Sato, T., Watanabe, T. & Otsuka, R. (1992) Effects of layer charge, charge location, and energy change on expansion properties of dioctahedral smectites. *Clays and Clay Minerals*, **40**, 103–113.

Sato, T., Murakami, T. & Watanabe, T. (1996) Change in layer charge of smectites and smectite layers in illite/smectite during diagenetic alteration. *Clays and Clay Minerals*, **44**, 460–469.

Scherrer, P. (1918) Bestimmung der Grösse und der inneren Struktur von Kolloidteilchen mittels Röntgenstrahlen. *Nachrichten von der Gesellschaft der Wissenschaften zu Göttingen, Mathematisch-Physikalische Kl.*, 98–100 (in German).

Skipper, N.T., Chang, F.R.C. & Sposito, G. (1995) Monte Carlo simulation of interlayer molecular structure in swelling clay minerals. 1. Methodology. *Clays and Clay Minerals*, **43**, 285–293.

Slade, P.G., Stone, P.A. & Radoslovitch, E.W. (1985) Interlayer structures of the two-layer hydrates of Na- and Ca-vermiculites. *Clays and Clay Minerals*, **33**, 51–61.

Smith, D.E. (1998) Molecular computer simulations of the swelling properties and interlayer structure of cesium montmorillonite. *Langmuir*, **14**, 5959–5967.

Środoń, J. (1980) Precise identification of illite/smectite interstratifications by X-ray powder diffraction. *Clays and Clay Minerals*, **28**, 401–411.

Środoń, J. (1984) X-ray powder diffraction of illitic materials. *Clays and Clay Minerals*, **32**, 337–349.

Sudo, T. & Shimoda, S. (1977) Interstratified clay minerals – Mode of occurrence and origin. *Minerals Science Engineering*, **9**, 3–24.

Taviot-Guého, C., Leroux, F., Payen, C. & Besse, J.-P. (2005) Cationic ordering and second-staging in copper-chromium and zinc-chromium layered double hydroxides. *Applied Clay Science*, **28**, 111–120.

Taviot-Guého, C., Feng, Y., Faour, A. & Leroux, F. (2010) Intercalation chemistry in a LDH system: anion exchange process and staging phenomenon investigated by means of time-resolved, *in situ* X-ray diffraction. *Dalton Transactions*, **39**, 5994–6005.

Tessier, C., Haumesser, P.H., Bernard, P. & Delmas, C. (1999) The structure of $Ni(OH)_2$: From the ideal material to the electrochemically active one. *Journal of the Electrochemical Society*, **146**, 2059–2067.

Tessier, C., Guerlou Demourgues, L., Faure, C., Basterreix, M., Nabias, G. & Delmas, C. (2000a) Structural and textural evolution of zinc-substituted nickel hydroxide electrode materials upon ageing in KOH and upon redox cycling. *Solid State Ionics*, **133**, 11–23.

Tessier, C., Guerlou Demourgues, L., Faure, C., Demourgues, A. & Delmas, C. (2000b) Structural study of zinc-substituted nickel hydroxides. *Journal of Materials Chemistry*, **10**, 1185–1193.

Thomas, G.S., Rajamathi, M. & Kamath, P.V. (2004) DIFFaX simulations of polytypism and disorder in hydrotalcite. *Clays and Clay Minerals*, **52**, 693–699.

Thomas, G.S. & Kamath, P.V. (2006) Line broadening in the PXRD patterns of layered hydroxides: The relative effects of crystallite size and structural disorder. *Journal of Chemical Sciences*, **118**, 127–133.

Treacy, M.M.J., Newsam, J.M. & Deem, M.W. (1991) A general recursion method for calculating diffracted intensities from crystals containing planar faults. *Proceedings of the Royal Society of London*, **433**, 499–520.

Tsipursky, S.J. & Drits, V.A. (1984) The distribution of octahedral cations in the 2:1 layers of dioctahedral smectites studied by oblique-texture electron diffraction. *Clay Minerals*, **19**, 177–193.

Tsipursky, S.J., Eberl, D.D. & Buseck, P.R. (1992) Unusual tops (bottoms?) of particles of 1M illite from the Silverton caldera (CO). In: *Proceedings of the American Society of Agronomy Annual Meeting*. American Society of Agronomy, Madison, Wisconsin, USA, pp. 381–382.

Tsybulya, S.V., Cherepanova, S.V. & Kryukova, G.N. (2004) Full profile analysis of X-ray diffraction patterns for investigation of nanocrystalline systems. In: *Diffraction Analysis of the Microstructure of Materials* (E.J. Mittemeijer & P. Scardi, editors). Springer Verlag, Berlin, pp. 93–123.

Ufer, K., Roth, G., Kleeberg, R., Stanjek, H., Dohrmann, R. & Bergmann, J. (2004) Description of X-ray powder pattern of turbostratically disordered layer structures with a Rietveld compatible approach. *Zeitschift für Kristallographie*, **219**, 519–527.

Ufer, K., Kleeberg, R., Bergmann, J., Curtius, H. & Dohrmann, R. (2008) Refining real structure parameters of disordered layer structures within the Rietveld method. *Zeitschift für Kristallographie*, **S27**, 151–158.

Ufer, K., Kleeberg, R., Bergmann, J. & Dohrmann, R. (2010) Simultaneous refinement of multi-device and/or multi-specimen XRD data of mixed layered structures. In: *Proceedings of the SEA-CSSJ-CMS trilateral meeting on clays (Sevilla)*, p. 46.

Ulibarri, M.A. & Hermosín, M.C. (2001) Layered double hydroxides in water decontamination. In: *Layered Double Hydroxides: Present and Future* (V. Rives, editor). Nova Science Publishers, New York, pp. 251–284.

Vaccari, A. (1998) Preparation and catalytic properties of cationic and anionic clays. *Catalysis Today*, **41**, 53–71.

Velde, B., Suzuki, T. & Nicot, E. (1986) Pressure-temperature-composition of illite/smectite mixed-layer minerals: Niger delta mudstones and other examples. *Clays and Clay Minerals*, **34**, 435–441.

Viani, A., Gaultieri, A.F. & Artioli, G. (2002) The nature of disorder in montmorillonite by simulation of X-ray powder patterns. *American Mineralogist*, **87**, 966–975.

Villalobos, M., Lanson, B., Manceau, A., Toner, B. & Sposito, G. (2006) Structural model for the biogenic Mn oxide produced by *Pseudomonas putida*. *American Mineralogist*, **91**, 489–502.

Walker, G.F. (1956) The mechanism of dehydration of Mg-vermiculite. *Clays and Clay Minerals*, **5**, 101–115.

Warren, B.E. (1941) X-ray diffraction in random layer lattices. *Physical Review*, **59**, 693–698.

Watanabe, T. (1981) Identification of illite/montmorillonite interstratification by X-ray powder diffraction. *Journal of the Mineralogical Society of Japan*, **15**, 32–41.

Watanabe, T. (1988) The structural model of illite/smectite interstratified mineral and the diagram for their identification. *Clay Science*, **7**, 97–114.

Webb, S.M., Tebo, B.M. & Bargar, J.R. (2005) Structural characterization of biogenic Mn oxides produced in seawater by the marine *Bacillus sp.* strain SG-1. *American Mineralogist*, **90**, 1342–1357.

Williams, G.R. & O'Hare, D. (2005) Factors influencing staging during anion-exchange intercalation into $[LiAl_2(OH)_6]X{\bullet}mH_2O$ ($X = Cl^-$, Br^-, NO_3^-). *Chemistry of Materials*, **17**, 2632–2640.

Williams, G.R. & O'Hare, D. (2006) Towards understanding, control and application of leyered double hydroxide chemistry. *Journal of Materials Chemistry*, **16**, 3065–3074.

Williams, G.R., Norquist, A.J. & O'Hare, D. (2003) The formation of ordered heterostructues during the intercalation of phosphonic acids into a double layered hydroxide. *Chemical Communications*, 1816–1817.

Williams, G.R., Norquist, A.J. & O'Hare, D. (2004) Time-resolved, in situ X-ray diffraction studies of staging during phosphonic acid intercalation into $[LiAl_2(OH)_6]Cl{\cdot}H_2O$. *Chemistry of Materials*, **16**, 975–981.

Zhang, W., He, J. & Guo, C. (2008) Second staging of tartrate and carbonate anions in Mg–Al layered double hydroxide. *Applied Clay Science*, **39**, 166–171.

Zhu, M., Ginder-Vogel, M., Feng, X.H. & Sparks, D.L. (2010) Cation effects on the layer structure of biogenic Mn-oxides. *Environmental Science & Technology*, **44**, 4465–4471.

EMU Notes in Mineralogy, Vol. 11 (2011), Chapter 5, 203–236

Applications of computational atomistic methods to phyllosilicates

C. IGNACIO SAINZ-DÍAZ

Instituto Andaluz de Ciencias de la Tierra (IACT) CSIC-Universidad de Granada, Av. de las Palmeras, 4, 18100-Armilla, Granada, Spain
e-mail: ignacio.sainz@iact.ugr-csic.es

A review of the main computational methods applied to layered silicates and other oxides is described from an atomistic point of view. Every macroscopic phenomenon is the result of a complex junction of many nanoscopic phenomena based on interactions between atoms and molecules. Different methods are presented below in order of theory-level complexity, starting from methods of experimental data analysis including simulated annealing methods. Several classic mechanics forcefields, based on empirical interatomic potentials, are presented. More sophisticated methods based on quantum mechanics are described, applying molecular cluster models and crystal periodic systems. Molecular dynamics simulations are also included. Applications of all these methods to the study of phyllosilicates and layered oxides are reviewed, focusing on structural, crystallographic and spectroscopic properties, reactivity, surface interactions, adsorption of organic molecules and water interactions.

1. Introduction

Phyllosilicates and other layered oxides/hydroxides form an interesting group that is present in many domains of the biosphere and in the interplanetary world. They are distributed in many kinds of soils, on ocean floors and are also suspended in the atmosphere. Their study is interesting from the point of view of geological and environmental phenomena, *e.g.* pollutant diffusion and groundwater quality; and also for industrial and health applications, *e.g.* as catalysts, adsorptive materials, additives in new materials, in pesticides and in drug delivery. One common property of these layered minerals is the ability to adsorb, intercalate and exchange a large variety of ions and molecules into their interlayer space, from small cations to bio-molecules. This property has led some authors to consider these minerals as a pre-biotic inorganic membrane with a relatively selective entrance and output of bioactive molecules. This fact and the catalytic properties have led some to suggest these minerals as a support for the origin of life on Earth.

These layered minerals are very variable because of the possibility of isomorphous cation substitutions in the tetrahedral and octahedral sheets of phyllosilicates and layered double hydroxides. However, the cation substitutions have a high level of disorder between layers and within the same sheet. These minerals also have nano-sized particles with very small crystalline domains and the layers are not ordered relative to each other. All these properties make it difficult to study many of the structural features

DOI: 10.1180/EMU-notes.11.5

experimentally. The behaviour in the interlayer space of phyllosilicates is also not well understood experimentally. How can small variations in the chemical composition change so significantly the swelling properties of phyllosilicates? How can small variations in the nature of the charge defect change the swelling capacity of phyllosilicates? To understand fully the complex structure and properties of these minerals, the theoretical methods of 'computational mineralogy' from an atomistic point of view are useful.

This chapter provides an overview of the philosophy, methods and applications of computational modelling based on an atomistic approach. After an initial definition of 'computational mineralogy' and how it is used in this chapter, different methods with different scale approaches are described with respect to experimental information, from purely experimental data to zero experimental information from *ab initio* methods. Finally, applications of all these methods to the study of phyllosilicates and layered metal oxides at different scales are presented.

2. Computational mineralogy

When we try to study a natural phenomenon, firstly we observe it and describe it in as much detail as possible. Secondly, we ask ourselves why this phenomenon occurs and why it occurs in the way that it does. We usually discover that the phenomenon is a very complex process. Every natural phenomenon is a complex junction of many phenomena that happen at nanoscopic scale.

Computational modelling techniques can explore systems from the macroscopic (km–mm) through mesoscopic (mm–μm) to nanoscopic (nm) scales (Flekkoy & Coveney, 1999).

In recent decades instrumental techniques have improved enormously and now provide close to nanoscale resolutions. However, in some systems with small particle sizes and significant disorder, as in the case of the layered phyllosilicates and metal oxides/hydroxides, it is difficult to interpret some experimental data with respect to the origin of their behaviour at the atomic and molecular level. A useful tool to study these systems at nanoscale and atomistic levels are the theoretical methods of computational chemistry and physics. The field of application of these methods to the different phenomena related to mineralogy is known as 'computational mineralogy'. Computation mineralogy can reproduce and interpret experimental behaviours and also allow us to explore systems in extreme conditions of temperature and pressure, deep ocean conditions, geological periods of time, stratospheric and interplanetary conditions, where experiments require economy of effort.

However, computational mineralogy is not a magic 'black box' where you put your problem, enter some commands and the solution to your problem appears! Computational mineralogy is an inter- and multi-disciplinary subject where different aspects of computing science, physics, chemistry, geology and mathematics meet. The rapid growth in computer processing power and the development of new codes for intensive calculations have allowed extension of the use of theoretical simulations to complex systems like these minerals. There are several excellent comprehensive reviews of the methods that can be used in computational mineralogy (Bleam, 1993; Cygan *et al.*, 2009; Frenkel & Smit, 1996; Schleyer, 1998; Cygan & Kubicki, 2001; Kubicki & Bleam, 2003).

In the present review we focus on modelling related to atoms and molecules. Different levels can be considered in the application of computational mineralogy: advanced methods of data analysis, classical-mechanics methods based on empirical interatomic potentials, and quantum-mechanics methods.

3. Methods of data management for analysis of experiments

Any experimental study can generate a data series. For example, an instrumental analysis detector can give a signal from the interaction of radiation with the sample and a computer transforms this signal into numerical data through a conversion equation. These data have to be analysed and interpreted to obtain results and conclusions. Sometimes these data give only indirect information about the sample or depend on a great number of variables to obtain a direct interpretation. It is necessary, therefore, to apply computational methods to obtain understandable results. The growth in the power of computers has allowed them to be used for instrument control and the mining and management of many data in analysis equipment. This growth has led to the development of many computational methods. These methods are relatively fast and are very specific to each sample or phenomenon, requiring the creation of a specific calculation code. Many of these methods are based on statistical analysis, such as multiple regressions, probabilistic methods, Monte Carlo (MC) methods, simulated annealing, *etc*. These methods have allowed the spectacular development of the application of some experimental techniques, such as X-ray diffraction, neutron diffraction, electron diffraction, *etc*. There are also methods based on databases of empirical information of kinetic constants, equilibrium constants, electronegativity, ionic radii, *etc.*, which allow simulation of the equilibria between species in dissolution, and this is a good tool for speciation in dissolutions. There are many databases of geochemical interest and speciation codes, such as *PHREEQE* (Parkhurst *et al.*, 1980), *MINTEQA2* (Allison *et al.*, 1991), *SXLSQA* (Baes *et al.*, 1990).

Some of these experimental data are related directly to atomic properties. Many methods, used to extract information from these data, are based on Metropolis-MC methods. The MC method is a technique used to sample randomly an infinite number of available configurations of any material. In general, a model of any crystalline mineral can be generated with a structure based on atomic positions applying periodical boundary conditions in 3-D dimensions (Allen & Tildesley, 1987). A series of mathematical functions can be applied to relate this model to one experimental property in such a way that a determined value of this property should correspond to one configuration of this model. Our strategy will consist of trying to find a configuration that reproduces the experimental data starting from a hypothetical initial model. We can define a magnitude, named 'energy' (E), as the difference between the value of a certain property

(h) for one theoretical configuration, h(calc), and the experimental value, h(exp):

$$E = \mathrm{F}[\mathrm{h(exp)} - \mathrm{h(calc)}] \quad (1)$$

Then, starting from an initial random configuration we will search for new configurations that yield a value of this property closer to the experimental value. Hence we will try to minimize E.

The Metropolis-MC method consists of an algorithm for generating randomly a new configuration from a previous one that will yield a change in energy (ΔE); *e.g.* this configuration change can be an atomic displacement or an atomic exchange. When $\Delta E < 0$, leading to a decrease in energy, this change is accepted automatically. On the other hand, if $\Delta E > 0$, the new configuration is not rejected but it can be accepted with a certain probability (P). This P can depend on the energy change, $P = \exp(-\Delta E)$. If ΔE is large, the probability of the new configuration (P) being accepted will be very small. This method avoids disparities during the minimization process. We can also tune this probability including an additional parameter that can be referred to fictitiously as 'temperature', T: $P = \exp(-\Delta E/kT)$. So, more configurations will be accepted for large values of kT and *vice versa*. This procedure is called simulated annealing and avoids the user being trapped in local minima of the configuration space. This process is repeated for a large number of steps, leading to an evolution of the system through 3-D phase space and ensures that the sampling procedure is consistent with thermodynamics.

These methods have been used in layered minerals for several forms of data analysis, *e.g.* X-ray diffraction (XRD) analysis (Viani *et al.*, 2002; Vucelic *et al.*, 1997), nuclear magnetic resonance (NMR) analysis (Dove, 1997), and infrared (IR) spectroscopy (Cuadros *et al.*, 1999). Some applications of these methods to experimental and theoretical data in phyllosilicates will be described below.

4. Atomistic methods based on classic mechanics

All atomistic methods have the hypothesis that every experimental phenomenon is a consequence of many interactions occurring between atoms and molecules. Study of these interactions can therefore give some information about the experimental behaviour.

One possible approach is to apply classical mechanics and to consider the atoms as spheres. These spheres are labelled as different atoms and have a certain spatial arrangement in a molecular, amorphous or crystalline structure according to the known experimental information. They can have different charges and the interactions between atoms are determined by a series of effective interatomic potentials depending on the charges and the distance or orientation between atoms. These potentials are grouped in a forcefield that is an integrated set of analytical functions which describes the potential energy of an assemblage of atoms. There are many kinds of functions but the main functional forms are summarized in Table 1 (Cygan, 2003).

Table 1. Main functions describing interatomic potentials.*

Bond stretching (harmonic)	$E = k_2(r - r_o)^2$
Bond stretching (Morse)	$E = D\{1 - \exp[-\alpha(r - r_o)]\}^2$
Bond-angle bending (three body)	$E = k_2(\theta - \theta_o)^2$
Torsion of dihedral bond angle (four body)	$E = k_\phi[1 + \cos(n\phi - \phi_o)]$
Electrostatic (Coulomb)	$E = q_i q_j / r_{ij}$
Lennard-Jones (non-bonding, van der Waals)	$E = A_{ij} r_{ij}^{-m} - B_{ij} r_{ij}^{-n}$
Buckingham	$E = A_{ij} \exp(-r_{ij}/\rho) - C r_{ij}^{-6}$
Born-Mayer	$E = A_{ij} \exp(-r_i / \rho)$

*The parameters k, D, α, n, m, A, B, ρ and C are constants. The variables r, θ, ϕ, r_o, θ_o and ϕ_o are the bond distance, bond angle and bond torsion angle between atoms and the corresponding nominal equilibrium values; q_i is the charge of the i atom; r_{ij} represents the distance of the covalently non-bonded i and j atoms.

The final total energy term will be the summation of all these potentials for every atom with all atoms of the system:

$$E_{\text{tot}} = \sum (E_{\text{bond}} + E_{\text{angle}} + E_{\text{torsion}} + E_{\text{Coulomb}} + E_{\text{non-bond}} + \cdots) \qquad (2)$$

The non-bonding potentials, Born-Mayer and Buckingham, are used mainly for ionic systems (metal oxides and silicates) describing short-range repulsive and long-term dispersive (the second term of Buckingham one) interactions along with the long-range Coulomb and van der Waals interactions. In the Lennard-Jones potential, the exponent m is usually 12 and n is usually 6 or 9. The summations in the non-bonding potentials are performed over all possible ij interatomic interactions, being limited by a certain cutoff distance saving useless computational time. However, in periodic systems the long-range nature of the electrostatic potential can be non convergent. To overcome this problem, the Ewald summation method is used (Gale, 1997).

Most models consider the atoms as hard spheres, but a shell model can also be used. This shell model is very useful for oxides, minerals with a large oxygen content, and allows the inclusion of the electron polarization effect by coupling within the sphere, by means of a harmonic spring potential, an internal ion core to a shell that has no mass (Fig. 1). This model is implemented in the *GULP* code (Cygan, 2003; Gale, 1997).

All functions used in a forcefield depend heavily on the parameters described in Table 1. The values of these parameters can be obtained by fitting these functions to empirical data (geometric features, vibration frequencies, elastic constants, thermodynamic properties, *etc.*) or quantum mechanical calculations (Lasaga & Gibbs, 1987; Bandura & Kubicki, 2003). The charges of these

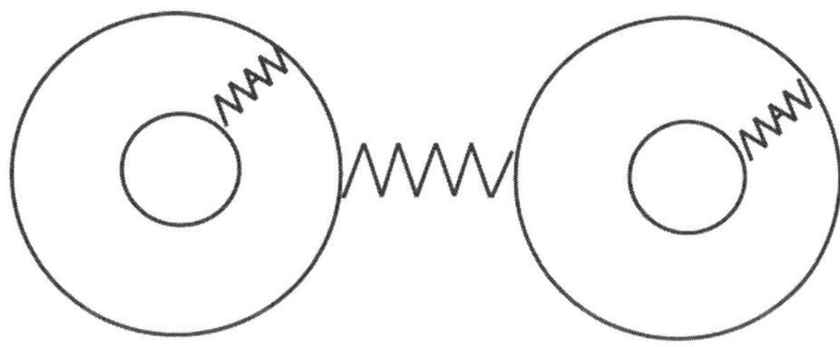

Fig. 1. A scheme of the shell model applied in atoms. The springs represent the intra-atomic and inter-atomic harmonic potentials, the inner sphere is the core and the external surface is the shell.

atoms are another parameter that can be formal charges, charges based on quantum-mechanical calculations (Raghavachari *et al.*, 1998), or empirical charge equilibration, QEq, based on geometry and electronegativities or other methods. A forcefield can be obtained to simulate the properties of a specific system. A good forcefield should have a "high transferability" reproducing properties of different systems without losing accuracy (Dove, 2008).

Many forcefields have been developed for organic molecules, *e.g.* the classical force-fields referred to as MM2, MM3, MM4 (Allinger, 1977; Lii & Allinger, 1989), and for biomacromolecules, *e.g.* the forcefields referred to as AMBER, CHARMM, GROMOS. Some of these forcefields have been adapted for solids, *e.g.* the Consistent Valence Force Field (CVFF) and the augmented version (CVFF_aug) and CFF91. The Universal forcefield (UFF) (Rappé *et al.*, 1992) was parameterized for the complete periodic table and has been used successfully for organics, metals complexes, oxides, zeolites and silicates. DREIDING (Mayo *et al.*, 1990) and COMPASS (Sun, 1998), and many other forcefields have being used for organic and inorganic systems. Some forcefields have been parameterised for phyllosilicates, such as the CFF91 modified with *ab initio* calculations by Teppen *et al.* (1997), the CVFF and CPFF versions of Heinz *et al.* (2005), and CLAYFF (Cygan *et al.*, 2004a). Another strategy is to select the interatomic potentials from different authors in order to refine and obtain a more robust forcefield. Using potentials from metal oxides and framework silicates and a more complete set of three-body terms, the structure and properties of phyllosilicates have been modelled, reproducing the weak interactions of the interlayer space (Sainz-Díaz *et al.*, 2001a; Palin *et al.*, 2001).

These forcefields allow us to calculate the total energy of any system and may be included in a calculation code already or they can be introduced to a code. They have been used mainly for organic molecules, macromolecules (polymers, proteins) and solvated systems. These forcefields can also be applied in codes with periodical boundary conditions and be used for periodical condensed systems, such as liquids, dissolutions and mineral crystal structures. There are many published codes for static energy calculations and with appropriate algorithms for optimization of the geometry of these structures to yield a minimum energy state, *e.g.* *GULP* (Gale, 1997), *Discover* (Accelrys, 2009), *Forcite* (Accelrys, 2009), *LAMMPS* (Plimpton, 1995), *DL_POLY* (Smith *et al.*, 2002), *etc.*

5. Quantum mechanical methods

Even though the modelling methods above describe many physical properties of molecules and solid minerals quite well, they do not take into account the existence of electrons and protons. Therefore, they cannot be used for processes where the electrons may be involved, *e.g.* chemical reactions, electronic structures, metal conductivity and semi-conductivity. In some experimental systems the experimental data on which the empirical potentials are based are not completely accurate. In such cases we have to use more sophisticated methods, based on quantum mechanics, where the real atomic components, electrons and nuclei, are taken into account and the influence of the empirical information is minimal. These methods of electronic-structure calculations require much more computational effort and time than the aforementioned methods and they are more limited to small systems. They have been developed mainly for molecular systems and in more recent decades have been applied to crystalline solids also. The impact and utility of these methods are so important that J.A. Pople and W. Kohn received the Nobel Prize in Chemistry for developing them.

5.1. Theoretical approaches

Quantum-mechanics calculations are often referred to as '*ab initio*' or 'First-Principles' calculations because they do not use empirical data. However, in order to apply this theory, many approximations must be implemented and so some inconsistency in the terminology has arisen. 'Quantum mechanics' is the most correct expression to define this calculation level. It is based on solving the Schrödinger equation:

$$H\Psi = E\Psi \tag{3}$$

where H is the Hamiltonian differential operator, Ψ is the wavefunction dependent on the spatial position of electrons and nuclei, and E is the total energy of the system. This operator has terms of kinetic and potential energy analogous to classical mechanics:

$$\left(-\frac{1}{2}\left(\frac{\mathrm{h}}{2\pi}\right)^2 \sum_i \frac{1}{m_i}\nabla^2 + \sum_{i \neq j} \frac{e_i e_j}{r_{ij}}\right)\Psi = E\Psi \tag{4}$$

where E is the electronic energy for one specific position of the nuclei, h is Planck's constant, m is the mass, ∇^2 is the Laplacian operator, e is the charge of the electrons or nuclei, and r_{ij} is the distance between these particles.

This equation has a solution only for the H atom which has only one electron. However, for systems with more electrons and nuclei, many approximations are necessary to solve this equation. One is the Born-Oppenheimer approximation that consists of decoupling the motions of the nuclei and electrons because of the great difference in mass between them. Another improvement is the Hartree-Fock approximation introducing an antisymmetric determinant of one-electron orbitals, matching the Pauli Exclusion Principle, to define the total wavefunction, treating the electrons individually surrounded by a spherical distribution of the rest of electrons. The molecular orbitals are expressed as linear combinations of atomic orbitals (LCAO). Each atomic orbital (AO) is expressed as Gaussian functions, referred to as 'basis sets'. A greater number of Gaussian functions will lead to a better quality of basis set and more computational effort. The wavefunction orbitals and their coefficients are refined through an iterative process starting with an initial guess of the orbitals until the system reaches a steady result in a self-consistent field (SCF). However, this Hartree-Fock (HF) approximation does not describe the interaction between vicinal electrons well and the total energy is too high as a result. The difference between the true total energy and the Hartree-Fock energy is known as 'correlation energy'. This energy can be estimated by means of significant computational demand by more sophisticated methods: (1) configurational interactions (expressing the wavefunction as a linear combination of Slater determinants) with an enormous computational demand (*e.g.* coupled cluster (CC), configuration interaction (CI), quadratic configuration interaction (QCI)); and (2) perturbative methods (Moller-Plesser method, MP2) (Hehre *et al.*, 1986).

Another important approximation is the Density Functional Theory (DFT) considering the electron density instead of the AO of individual electrons used in HF. Here, the total energy of a system can be expressed in terms of functionals of the electronic charge density. This energy should be exact and the difference from the true total energy is known as the exchange-correlation energy in the scheme of the Khon-Sham approach. All DFT approximations attempt to calculate this energy with some approximations. The local density approximation (LDA) was used for many years and considers that the exchange-correlation potential depends only on the electron density of a uniform electron gas. This approach was improved with generalized gradient approximation (GGA) including the gradient of the charge density. Additional parameterizations for the correlation functional have been developed for LDA (Cerperley-Alder, VWN, PW92, *etc.*) and GGA (PBE, rPBE, BLYP, *etc.*) (Schleyer, 1998).

In order to calculate this exchange-correlation energy, several hybrid methods have been developed, combining Hartree-Fock exchange with DFT exchange-correlation, including parameterizations from empirical thermodynamic data (*e.g.* BLYP, BY3LP, BH and HLYP, *etc.*) (Schleyer, 1998). The DFT methods have been used extensively in recent years due to the computational economy in the scaling with the number of electrons compared with the HF methods and the easy application to the solid-state periodic systems.

Another approximation is the semi-empirical methods, where some empirical input is used to obtain approximate solutions of the Schrödinger equation avoiding the large computational effort of the HF methods. Many semi-empirical methods have been developed, such as, MINDO, ZINDO, AM1, PM3, *etc.* (Pople & Beveridge, 1970), although their use is decreasing due to the success of DFT methods and the increase in computational power available.

Therefore, we can classify the quantum-mechanical methods groups into: (1) Hartree-Fock methods; (2) post-Hartree-Fock methods (correlated and perturbative methods); (3) DFT methods; and (4) semi-empirical methods. In all there is a balance between accuracy and computational cost, and this is related directly to the nature and size of the systems.

There are also embedding methods known as QM/MM, where the whole system is treated by classic mechanical methods (MM) and a small specific zone, where our problem is localized, is treated at quantum-mechanical level (QM). This approach allows us to study large systems but the user must be careful when describing the electron density in the frontiers between the MM and QM zones. These methods, like ONIOM, are useful for solvations, studies in dissolutions, interactions with solvents and proteins, *etc.*

The quantum mechanical methods have been used widely for molecules but in recent decades solids, including crystalline minerals, have also been studied at this level using two main models: molecular clusters and periodical models.

5.2. Molecular cluster models

In crystalline solids, the number of atoms for a minimal size of the model and the periodical boundary conditions sometimes require an enormous computational effort, especially for highly accurate theory levels. In these cases and when the property to be studied is dependent on a local environment, the solid can be modelled as molecular clusters. The local area to be studied is cut and separated from the crystal lattice and the dangling bonds are neutralized with hydrogen atoms (Lasaga, 1992; Bleam, 1993; Sauer *et al.*, 1994). In the case of phyllosilicates, different clusters can be generated (Fig. 2): (1) one unit of silicic, disilicic, or trisilicic acid, or a ring of SiO_2 tetrahedra to represent the tetrahedral sheet; (2) a pair of octahedra with Al^{3+}, Mg^{2+}, Fe^{2+} or Fe^{2+} including the OH groups for modelling the octahedral sheet; or (3) a ring of SiO_2 tetrahedra joined to a pair of octahedra can be selected to model both the tetrahedral and octahedral sheets.

With this approach, the same codes used for calculating molecules can be used for these clusters, such as, *GAUSSIAN* (Frisch *et al.*, 2004), *GAMESS*, *DMOL*, *NEWCHEM*, *HONDO*, *Spartan*, *HyperCHEM*, *etc.* Chemical reactions of silicates have been modelled using cluster models, *e.g.* dissolution reactions (Lasaga, 1992, 1995; Kubicki & Apitz, 1998), spectroscopic properties (Sainz-Díaz *et al.*, 2000), octahedral cation substitutions in phyllosilicates and layered metal oxides/hydroxides

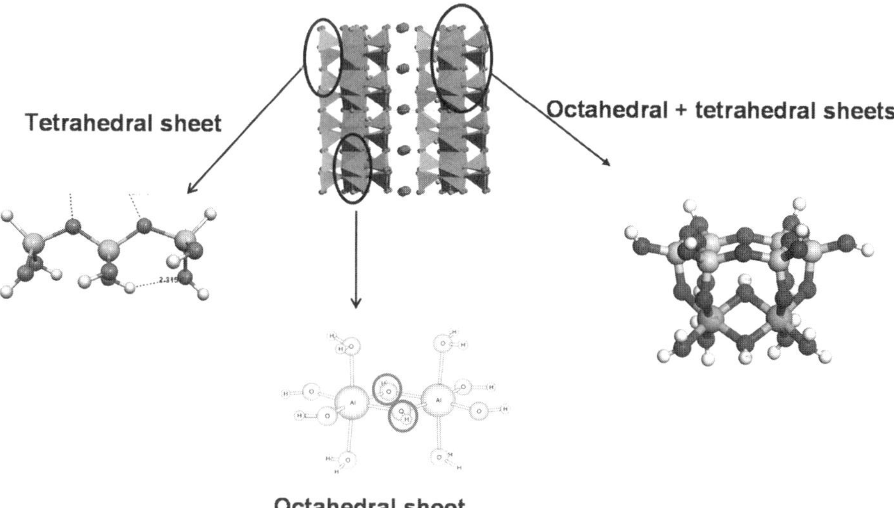

Fig. 2. Possible clusters generated from a phyllosilicate layer. Readers of the paper version of this chapter may wish to download a colour version of this figure from www.minersoc.org/emu-notes/emu-11/11-5-colour.pdf.

(Timón *et al.*, 2003; Chatterjee *et al.*, 2000), adsorption of organics on silicate surface (Kubicki *et al.*, 1999), reactions of free radicals with organics on silicate surfaces (Iuga *et al.*, 2008; 2010), *etc.*

5.3. Periodical models for crystalline solids

The molecular cluster model has a size limitation and some long-range interactions exist in the crystal lattice that affect many properties and are not represented in the cluster model. There are also other properties that depend directly on the crystal lattice structure, *e.g.* crystallographic properties, elastic constants, cation ordering, equation-of-state, surface features, *etc.* Therefore, it is necessary to study directly the crystal lattice that, by translational symmetry, can be considered as a periodic system from the unit cell.

In recent decades, several quantum-mechanical codes have been developed with periodical boundary conditions, applying the Bloch theorem where the wavefunction is expressed by the vectors of the reciprocal space of the Bravais lattice (Payne *et al.*, 1992). Many DFT methods describe the electron density as plane-waves (*e.g.* the codes known as *ABINIT*, *CASTEP*, *VASP*, *CPMD*). Some DFT methods are based on LCAO basis sets (*e.g.* the codes known as *SIESTA*, *ONETEP*) (Soler *et al.*, 2002) improving the scaling up to a larger number of atoms and avoiding calculations of the empty space in surface calculations. Other methods are based on the Hartree-Fock approximation and LCAO (*e.g.* the code known as *CRYSTAL*). Most of these codes use the pseudopotential approach in order to decrease the computational cost. Only the valence electrons are considered for the calculations and the internal core electrons are described by a pseudopotential. Other codes also use all electrons of the periodic system (*e.g.* the code known as *WIEN*) instead of pseudopotentials.

6. Modelling tools

All of the theoretical methods described above can calculate the total energy of one system and are incorporated in codes or packages along with tools for geometry optimization, vibrational analysis, molecular dynamics, *etc.*

6.1. Minimization of geometry

When an atomistic model with its spatial coordinates is introduced into our code, we can calculate the energy of this geometry but we need to know that this system is stable and corresponds to a minimum energy stationary state. The energy of this initial configuration is first determined and then the atomic positions are adjusted to obtain a lower energy configuration using the first derivative of potential energy with respect to the displacement produced. This process is repeated several times until reaching these derivatives (forces) close to zero with defined tolerances (*e.g.* 10^{-n}) for the energy difference and derivatives between successive steps, obtaining the potential energy minimum of the potential energy surface. In a periodical system this minimization can be performed at two levels: (1) at constant volume where the lattice parameters are fixed and only the atom positions are displaced; or (2) at variable cell or variable volume, where the atomic positions and lattice parameters are optimized at the same time.

Several algorithms are used in this optimization procedure, *e.g.* line searches, steepest gradient, conjugate gradient, Newton-Raphson, or Berny methods (Schleyer, 1998). Sometimes it is difficult to obtain the minimum state and some approximations can be applied, *e.g.* imposing constraints in some geometrical features and the optimization can be performed in steps (*e.g.* to optimize a crystal lattice at constant volume and, subsequently, at variable volume). Another risk is that a local minimum energy configuration can be obtained instead of the true global energy minimum. Different optimization methods or several initial configurations should be applied to ensure that the configuration of global energy minimum has been achieved. One procedure to check a minimum energy state is to calculate the second derivative of the potential energy. When this is greater than zero, the configuration is a minimum energy state (global or local), but if this second derivative is negative the configuration is not a minimum energy state but it is a maximum energy state in the potential energy surface of the system.

6.2. Monte Carlo simulations

Sometimes the optimization procedure does not explore all possible configurations in order to find the global minimum. This occurs very often in layered minerals where many local minima can exist, especially at the surfaces of these minerals. Monte Carlo simulations allow us to explore the whole configurational space with small random displacements, following the Metropolis-MC procedure explained above. For each step, the energy can be calculated at classic mechanical or quantum mechanical level, although, commonly, empirical forcefields are used due to the large number of steps used in MC; *e.g.* 1 million steps can be required for equilibration and two million steps for sampling of energetic and structural properties in a phyllosilicate with interlayer cations and water molecules (Skipper, 2003). This method is very dependent on the energy calculation forcefields, equilibration and sampling tuning.

This sort of simulation has been applied when exploring the swelling properties of phyllosilicates with interactions of different interlayer cations and water molecules (Bleam, 1993; Karaborni *et al.*, 1996; Delville, 1993; Skipper *et al.*, 1995), and also for adsorption of organic molecules along the surfaces of layered minerals (Sposito *et al.*, 1999; Sato *et al.*, 2001).

6.3. Molecular-dynamics simulations

Energy minimizations are performed at 0 K, where the thermal vibrations and kinetic energy are ignored, whereas most of the empirical information used for the forcefields is taken at room temperature. More realistic methods are needed. Other atomistic simulations can be performed including the ‘time’ as a variable, by using Newton’s equation of motion; this is the ‘Molecular Dynamic’ (MD) method. Starting from one configuration of atoms, we can calculate the energy of the system and the force on each atom with classic- or quantum-mechanics methods. These forces can be converted to accelerations and with a certain time step we can calculate the velocities and the new positions of the atoms, obtaining a trajectory of the atoms along the simulation time. The time steps are commonly $\sim 1/20^{th}$ of the smallest natural period of vibration and the simulation runs are for time of $1-1000$ ps, depending on the energy calculation method (quantum or classic mechanics). The velocity of the atoms is related directly to the temperature at which the simulation is run. Different thermodynamic ensembles can be selected, keeping constant the number of particles (N), the volume of the system (V), the external pressure (P), the temperature (T), or the energy (E). The typical ensembles used include those known as NVT (N, V and T being constants), NVE, NPT and NPE.

Molecular-dynamic simulations can generate a vast amount of data and the problem is the analysis of these data. With MD we can explore the evolution of the system with time, with temperature and pressure, and avoid finishing at local minima configurations. We can compute conformational analysis, equations of state for a range of temperatures and pressures, phase transitions, distribution functions (pair distribution function, averaged radial distribution function), and analysis of dynamics with a time-correlation function (*e.g.* diffusion constants, vibration frequencies, phonon density of states, thermal conductivity or thermodynamic integration).

The development of the DFT methods to periodical systems has enhanced the development of *ab initio* molecular dynamics (AIMD). These methods can be based on the Born-Oppenheimer approximation, where the motion equations are applied to the nuclei and the electronic structure is calculated associated with the nuclei in the each position along the trajectory. The methods can also be based on the Car-Parrinello approximation, where both the wavefunction and the nuclei are treated as dynamic variables and both can be propagated satisfactorily with Lagrangian equations of motion without the extra effort of converging the wavefunction at each step (Car & Parrinello, 1985).

6.4. Vibration spectra

Vibrational spectroscopy is an experimental technique which allows us to observe a signal which originated from the atomic motions. The theoretical atomistic methods can be used to calculate the frequencies and intensities of the vibration spectra. This is one close connection between experimental and theoretical research.

The vibration frequencies are commonly calculated using the harmonic approximation. The first derivatives of potential energy with respect to atomic motion within the system must be zero (a stationary state) and the second derivatives must be greater than zero (a minimum energy state). A common method of calculating frequencies is by finite displacements. Starting from a minimum energy state, each atom is displaced slightly in two senses for each orthogonal direction ($\pm x$, $\pm y$, $\pm z$) and the forces are calculated for each displacement. Each displacement generates a matrix of forces for the whole system and all displacements generate the Hessian or potential energy second derivative matrix, all the eigenvalues of which should be positive. Solving this matrix, the frequencies and the atomic displacements of each normal vibration mode can be obtained. Hence, we can identify the vibrations and interpret the experimental spectra.

Unfortunately, these frequencies are harmonic and the experimental frequencies also have an anharmonic component. This challenge, and the difficulty in describing the electron correlation and the limited size of the basis set can all contribute to the difference between experimental and theoretical vibration frequencies. Generally, these errors are systematic and a scale factor can be applied to fit these frequencies to experimental values.

In order to overcome the problems with the anharmonicity, the vibrational frequencies can be obtained from molecular dynamics. A velocity autocorrelation function (VACF) is calculated from a certain sampling of atomic coordinates and velocities during the dynamics. By means of the Fourier transform of this VACF, the power spectrum is obtained, yielding the vibrational density of states. If the VACF includes the derivatives of the dipole moment and the polarization, the IR or Raman spectrum can be generated. The implementation of this procedure has become very popular in recent years but sometimes it is difficult to identify the nature of the normal vibration modes.

Sometimes it is not easy to assign one band to any vibration mode experimentally because the resolution of the bands is insufficient to distinguish them or the system is more complex and there are too many bands at the same frequency. In these cases, some approximations are necessary, decomposing the bands in different components with different forms (Gaussian, Lorentzian, *etc.*). The theoretical methods can be very useful in these cases because it is possible to identify each vibration mode separately.

7. Applications in phyllosilicates

7.1. Ordering in cation substitutions in phyllosilicates

This is a good example because it can be studied at the various levels described above, experimental data analysis, empirical force fields, and quantum-mechanical calculations. The different approaches described above are used here to study the same phenomenon with the limitations and the main results given.

One characteristic of phyllosilicates is the great variety of possible isomorphous cation substitutions in the tetrahedral and octahedral sheets. However, it is difficult to observe experimentally how the cations are distributed in these sheets. The substitution of Al^{3+} for Si^{4+} in the tetrahedral sheet can alter the ^{29}Si NMR signal, because the short-range environment of Si changes. However, it is impossible to know from this experimental information alone, how these cations are distributed. Applying reverse Monte Carlo simulations with these experimental data, the Si and Al cation distribution in the tetrahedral sheet of mica was described, observing that there are no AlAl pairs according to Lowenstein's rule and the Al cations are not distributed randomly but are separated from each other along the tetrahedral sheet (Dove, 1997; Herrero & Sanz, 1991). This study was also applied to clintonite, a trioctahedral mica showing Si:Al = 1:1 (Sanz *et al.*, 2003).

In the octahedral sheets of phyllosilicates, there are also cation substitutions, such as Mg^{2+}, Fe^{3+} for Al^{3+} in dioctahedral series, and it is necessary to know their distribution along the octahedral sheet. Infrared, Mössbauer and NMR spectroscopies are especially useful for cation-distribution analysis as they probe local environments and can detect short-range cation relationships. Schroeder & Pruett (1996), using NMR, found that Fe was not randomly located in the octahedral sheet but segregated from Al in kaolinite. From IR studies of celadonites, Slonimskaya *et al.* (1986) found that divalent and trivalent cations alternate in the octahedral sheet and Besson *et al.* (1987) found that octahedral cation distribution was not random but showed a tendency for Al and Fe to be segregated from each other. Drits *et al.* (1997) studied celadonites, glauconites and Fe-illites by IR, Mössbauer, and extended X-ray absorption fine structure (EXAFS) spectroscopies, finding a certain short-range ordering. Using NMR, Schroeder (1993) found in illite/smectite interstratifications that Fe mixes with Al in samples with small Fe contents but that Fe segregates from Al in Fe-rich smectites. These different results make it difficult to extract a definitive conclusion from all these experimental studies. The use of computational methods as an auxiliary tool can help the experimentalist to interpret the experiments.

The properties of the hydroxy groups can change depending on the nature of cations to which they are attached. For example, the frequency of the OH vibration modes can be different for the AlOHAl, AlOHFe, AlOHMg, FeOHFe, FeOHMg or MgOHMg species. When these bands can be detected and assigned, a quantitative analysis of the bands can yield an approximate relative proportion of these species along the octahedral sheet but not the cation distribution. Applying Monte Carlo calculations to these data (Reverse Monte Carlo), it is possible to find cation distributions that reproduce the relative intensities of those species (Cuadros *et al.*, 1999). We can also consider additional experimental data. In NMR, the paramagnetic effect of Fe can alter the intensity of the ^{27}Al NMR signal by inhibiting the response of the Al cations that are surrounding each Fe cation. When the Fe cations are dispersed, the inhibitor effect is greater than when the Fe cations are clustered. This behaviour can give us indirect information about the cation distribution of the octahedral cations. It is possible to apply Reverse Monte Carlo calculations to find cation configurations that reproduce the results of both different experimental techniques. These configurations showed that there is a short-range ordering where the Fe cations tend to form small clusters whereas the Mg cations tend to be highly dispersed (Fig. 3) (Sainz-Díaz *et al.*, 2001b). This procedure is a very useful analysis method to visualize the most probable cation distributions (lowest energy) of these minerals. However, this result does not prove that the effect of Fe is greater in illites than in smectites.

This study is very limited, however, in terms of the accuracy of the experimental data and the homogeneity of chemical composition of the samples which is difficult to find in phyllosilicates. In order to overcome these possible limitations, we can choose an energetic criterion where the cation distribution with the lowest energy will be the most likely to be found in experiments. With this approach, we can generalize this phenomenon to other phyllosilicates with different chemical compositions and also other layered metal oxides/hydroxides.

Firstly, we define an atomistic method to evaluate the energy based on empirical interatomic potentials that can be transferred to any phyllosilicates. The phyllosilicates as metal oxides have a large amount of oxygen atoms and a SHELL model will be good to represent the polarization of these atoms. Using the *GULP* code and applying a combination of empirical potentials obtained for metal oxides, the crystallographic properties of phyllosilicates were reproduced for different chemical compositions (Sainz-Díaz *et al.*, 2001a).

An energy criterion must also be found to evaluate the cation ordering. The energy of a crystal can be represented by the energy related to the inter-cation interactions and a residual energy corresponding to the other interactions of the crystal lattice:

$$H = E_0 + \sum_n (N^n_{A-A}E^n_{A-A} + N^n_{M-M}E^n_{M-M} + N^n_{A-M}E^n_{A-M}) \quad (5)$$

where A and M are two types of cation, n is the type of neighbourhood, N is the number of pairs, and E is the energy of each interaction. In a bi-cation system, we can define a cation-exchange potential:

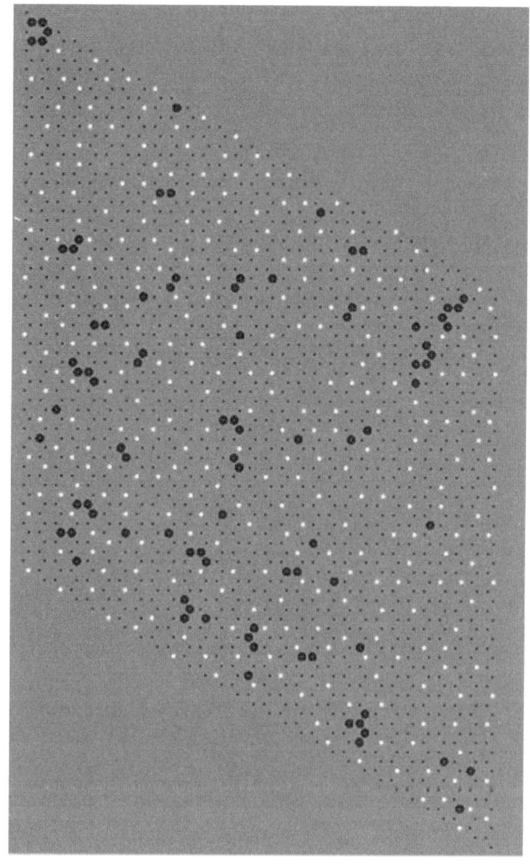

Fig. 3. Octahedral cation distribution fitted to IR and NMR experimental data by reverse Monte Carlo calculations for an illite-smectite interlayered mineral. The octahedral sheet is 625 nm^2 (representing a phyllosilicate layer of 51,250 atoms) and periodical boundary conditions. The Fe and Mg cations are represented by black and yellow balls, respectively. Readers of the paper version of this chapter may wish to download a colour version of this figure from www.minersoc.org/emu-notes/emu-11/11-5-colour.pdf.

$$J_n = E_{A-A} + E_{M-M} - 2E_{A-M} \quad (6)$$

where J is the energy associated with the exchange of two cations (A and M) to form AA and MM pairs instead of AM pairs. Then, the energy of the system can be simplified to:

$$H = E_0 + \sum_n N^n_{A-A}J_n \quad (7)$$

A neighbourhood system can be defined based on the interaction distance where the interactions will decrease with increase in distance. For example, for the octahedral sheet of dioctahedral phyllosilicates, the cation pairs can be defined as first, second, third and fourth neighbours to those placed at 2.7–3.4 Å, 3.5–5.5 Å, 5.6–6.5 Å and 6.6–7.5 Å, respectively (Fig. 4).

In order to calculate the exchange potentials with a statistical value, we have to generate a large number, 50–100, of random configurations of octahedral cations for a certain composition along with some highly ordered configurations. The generation of these configurations and the calculation of the number of cation pairs can be performed with a home-made code or using the *MCclay* code (Sainz-Díaz *et al.*, 2001b). The calculation cell should be big enough to include the J_i neighbours. After calculating the energy of all these configurations and taking into account the number of different kinds of cation pairs, J_i, the exchange potentials can be calculated by multiple regression (Bosenick *et al.*, 2001). With these exchange potentials, Monte Carlo simulations can be performed in big supercells ($16 \times 16 \times 1$) and with a massive number of random cation

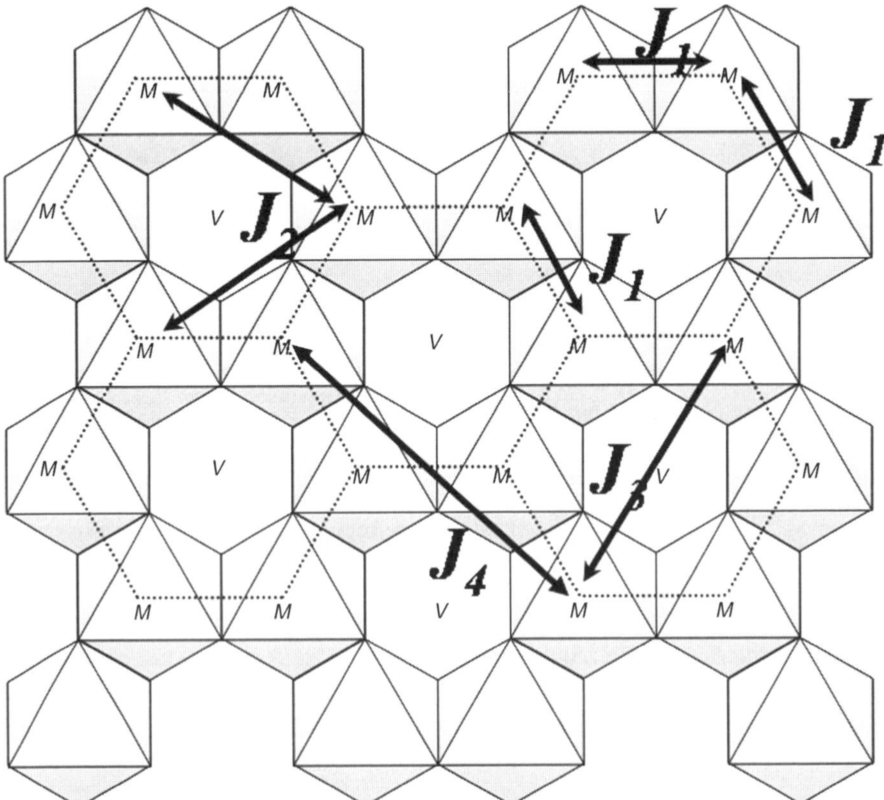

Fig. 4. Neighbourhood of the octahedral cations of dioctahedral phyllosilicates. *M* is the octahedral cation and V is the vacancy. J_i ($i = 1–4$) is the interaction between cation pairs as first, second, third and forth neighbours (reproduced with kind permission from Springer Science from Sainz-Díaz *et al.*, 2003b).

exchanges (10^6-10^8) at different temperatures, obtaining configurations of minimum energy and ordering phase transition in phyllosilicates (Warren *et al.*, 2001).

This study was applied to the octahedral cations for different compositions (Al^{3+}/Mg^{2+}, Al^{3+}/Fe^{3+} and Fe^{3+}/Mg^{2+}) with several interlayer charges and interlayer cations. It was confirmed that the Mg tends to be disperse and mixed with Al along the octahedral sheet forming ordered configurations of low energy (Fig. 5a). The Fe cations follow a similar tendency, though weaker (an ordering phase transition at lower temperature), forming different ordering patterns (Fig. 5b). Phase transitions were observed during the cooling process from the highly disordered to the ordered configurations (Fig. 5c).

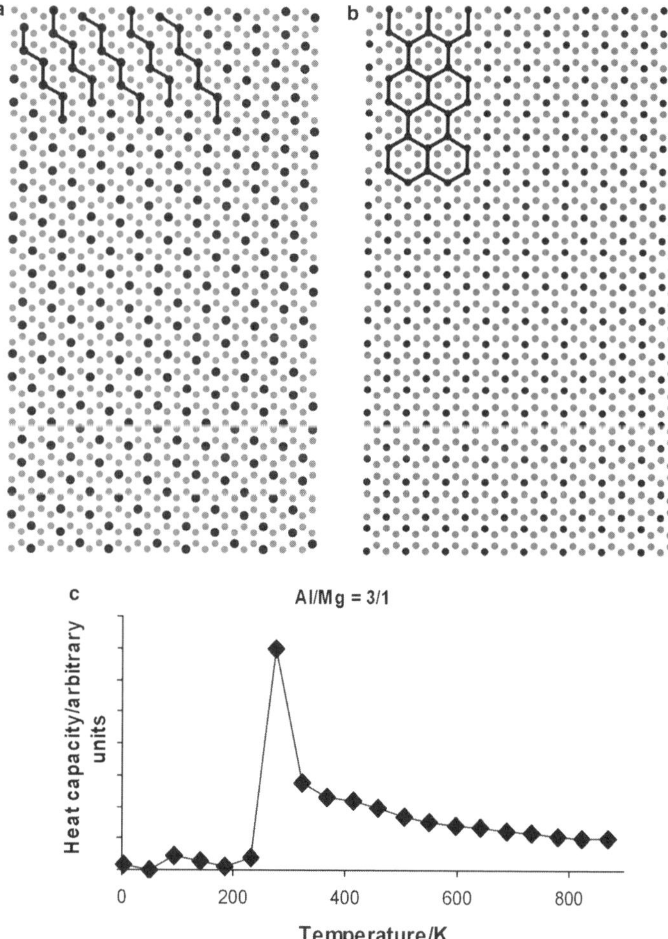

Fig. 5. Snapshots of the low-energy ordered configurations of the octahedral cations of dioctahedral phyllosilicates from Monte Carlo simulations for Al_3Mg (***a***) and Al_3Fe (***b***) compositions (Al atoms are grey). (***c***) Ordering phase transition for an Al_3Mg sample (reproduced with kind permission from Springer Science from Sainz-Díaz *et al.*, 2003b).

From these bi-cationic calculations, we can obtain exchange potentials for tri-cationic systems (Al/Fe/Mg) by means of a mathematical development (Sainz-Díaz *et al.*, 2003a) and also apply Monte Carlo simulations for a larger variety of cation compositions closer to natural samples. The cation ordering is very dependent on the chemical composition. For any composition, Mg tends to be dispersed along the octahedral sheet. However, the presence of Mg facilitates Fe cations forming clusters with ordered patterns (Fig. 6a–c) (Palin *et al.*, 2004). In samples with large Al contents, such as illite and smectite, Mg cations are disperse and Fe cations form small clusters with no long-range ordering (Fig. 6d) matching the cation ordering found experimentally from FT-IR, NMR and reverse Monte Carlo studies (Fig. 3). A similar pattern was found for samples with high tetrahedral charges such as illite. The size of the Fe clusters is no bigger in illite than in smectite, though the proportion of Fe clustered is greater in illite than in smectite. This justifies the experimental observation where the effect of the Fe cations on the ^{27}Al NMR signal is greater in illites than in smectites.

This combination of atomistic calculations with forcefields, cation-exchange potential calculations and Monte Carlo simulations can be applied to other studies of

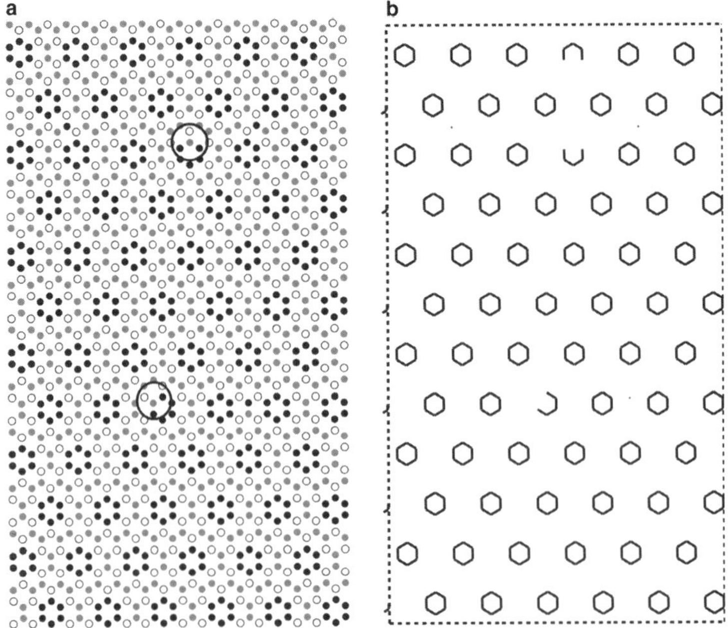

Fig. 6. Snapshots of the most stable configurations of octahedral cations from the Monte Carlo simulations for the relative octahedral compositions: (***a***) Al/Fe/Mg = 1/1/1 (Al, Fe and Mg are in grey, black and white, respectively); (***b***) the same but representing the Fe ordering patterns only; (***c***) Al/Fe/Mg = 3/2/1 (the Fe patterns only); (***d***) Al/Fe/Mg = 4/1/1 (Al, Fe and Mg are in grey, black and white, respectively) (reproduced Sainz-Díaz *et al.*, (2004b) [figs 2, 4 and 8a], with the kind permission of The Clay Minerals Society, publisher of *Clays and Clay Minerals*).

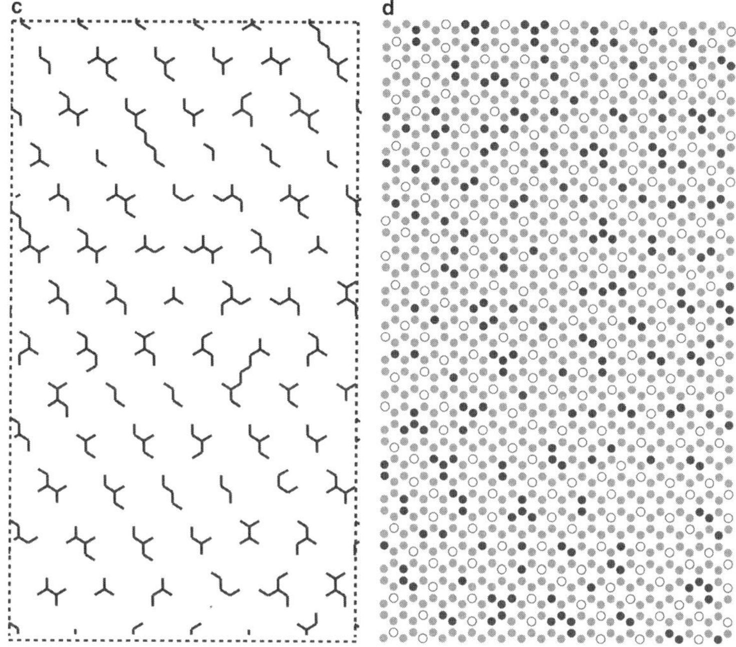

Fig. 6. Continued.

cation ordering in other systems, as in the tetrahedral sheet of phyllosilicates (Palin *et al.*, 2001, 2003; Palin & Dove, 2004).

This cation ordering study can also be performed with First Principles calculations, minimizing the empirical information. The isomorphous cation exchange energy can be deduced by comparing the energy of the cation substitutions as isodesmic reactions in dioctahedral clusters models (Fig. 2) (Timon *et al.*, 2003; Chatterjee *et al.*, 2000):

$$Al_2(OH)_2O_8H_{12} + [Mg_2(OH)_2O_8H_{12}]^{-2} \rightarrow 2[AlMg(OH)_2O_8H_{12}]^{-}$$
$$(\Delta E = -57.1\,\text{kcal/mol})2[AlMg(OH)_2O_8H_{12}]^{-} + Fe_2(OH)_2O_8H_{12} \rightarrow$$
$$[Mg_2(OH)_2O_8H_{12}]^{-2} + 2[AlFe(OH)_2O_8H_{12}]$$
$$(\Delta E = 50.9\,\text{kcal/mol})$$

The first reaction is exothermic showing the great tendency of the Mg^{2+} cations to mix with the Al^{3+} cations in the octahedral sheet as observed above. On the contrary, the second reaction is endothermic showing that Fe tends to be clustered more than mixed with Al or Mg with the presence of Mg^{2+}, according to experiments and the calculations above.

Quantum-mechanical calculations can be applied to cation ordering on periodical crystal models. Because of the large computational cost of these calculations, only a small number of cation configurations was calculated and the energies were compared. The cation distribution has been calculated in the tetrahedral sheet of Mg-vermiculite (Tunega & Lischka, 2003) and in the tetrahedral and octahedral sheets of smectite and illite (Sainz-Díaz *et al.*, 2002; Hernández-Laguna *et al.*, 2006) where the tendency to disperse for octahedral Mg (Fig. 7a) and clustering of Fe (Fig. 7b), and the tendency of tetrahedral Al to be dispersed (Fig. 7c), seen above, are confirmed. Recently, an exhaustive series of intensive DFT calculations of many configurations of octahedral cations in smectites to obtain exchange potentials and applying Monte Carlo simulations were performed for octahedral cation ordering, confirming the results above (Ortega-Castro *et al.*, 2010a).

This study of cation ordering can be extended to layered metal oxides/hydroxides at the different levels described above, where the octahedral cation ordering is very important in the evaluation of their synthesis and reactivity (Vucelic *et al.*, 1997).

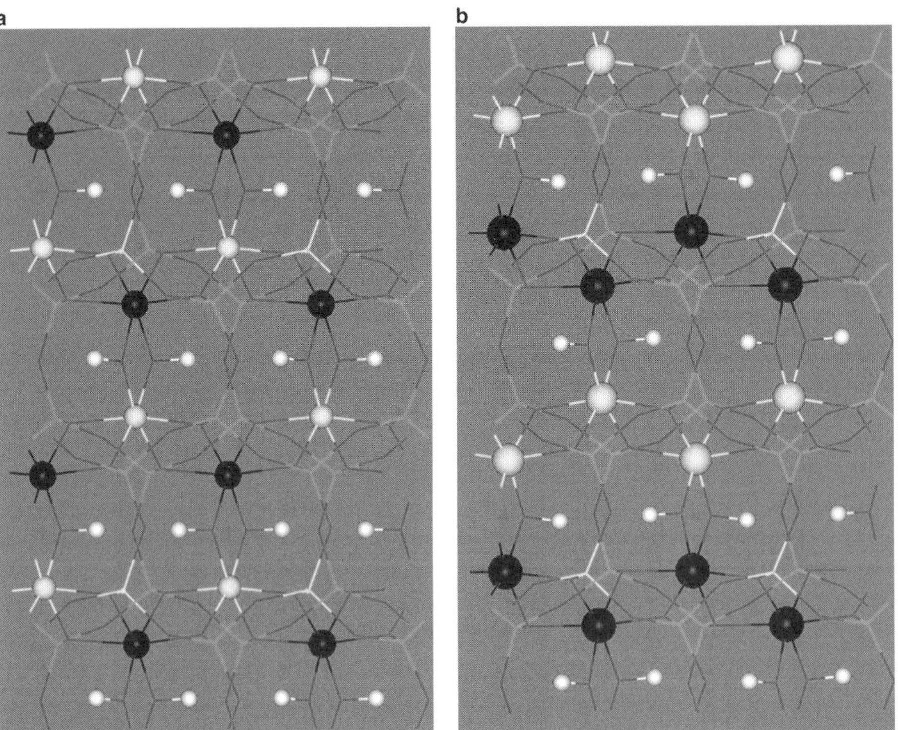

Fig. 7. The most stable octahedral cation configurations for Al_2Mg_2 (***a***), Al_2Fe_2 (***b***) octahedral composition, and Si_7Al tetrahedral composition per unit cell. The atoms Al, Si, O and H are in yellow, pink, red, and white colours, respectively. The Mg (***a***) and Fe (***b***) atoms are in black. In (***c***) the tetrahedral Al and the interlayer cations are highlighted in balls. Readers of the paper version of this chapter may wish to download a colour version of this figure from www.minersoc.org/emu-notes/emu-11/11-5-colour.pdf.

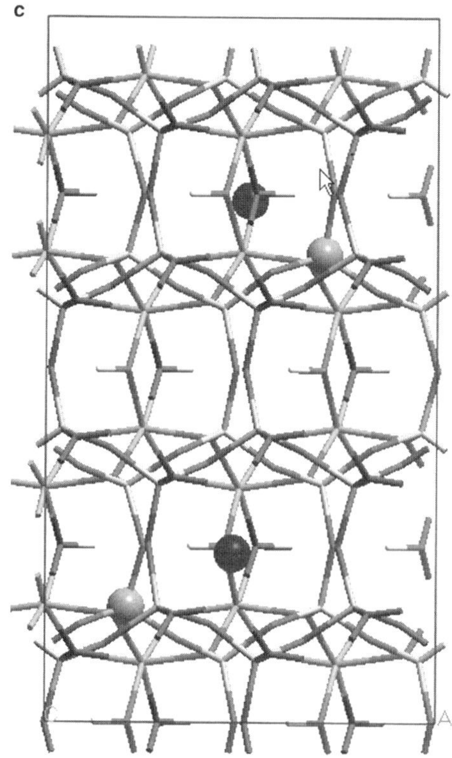

Fig. 7. Continued.

7.2. *Cis*-vacant/*trans*-vacant polymorphism in dioctahedral phyllosilicates

Another kind of polymorphism exists in dioctahedral 2:1 phyllosilicates. In these minerals, one of the three octahedral positions per asymmetric unit is not occupied by a cation, but instead is a vacant site. The hydroxyl groups of the octahedral sheet has two possible dispositions with respect to the vacant site. These OH groups can be on the same side (*cis*-vacant) or on opposite sides (*trans*-vacant) with respect to the vacancy (Fig. 8a,b). The crystal layer has a centre of symmetry in the *trans*-vacant configuration but not in the *cis*-vacant one (Tsipursky & Drits, 1984). The *cis*-vacant/*trans*-vacant proportion can be determined in illite-smectite (I-S) interstratified minerals by XRD and thermal analysis, but with semi-quantitative accuracy only (Drits *et al.*, 1998). In smectite, the octahedral sheet tends to be *cis*-vacant, whereas in illite it is mainly *trans*-vacant. During the illitization process of smectite, the formation of illite layers should lead to an increase in the proportion of *trans*-vacant layers. However, no proportional relationship between the *cis*-/*trans*-vacant site ratio and the rate of smectite/illite transformation was found (Drits, 2003). Besides, the effect of cation substitution on the *cis*-vacant/*trans*-vacant ratio remains poorly understood, as some discrepancies occur in experimental studies. For example, Tsipursky & Drits (1984) found that montmorillonite and Al-rich smectite are *cis*-vacant, whereas McCarty & Reynolds (1995) discovered that the proportion of *cis*-vacant layers increased with tetrahedral Al content. A theoretical approach may be useful to help understand these experiments.

Using an empirical forcefield for phyllosilicates and the SHELL model, the crystallographic properties of different samples of illite, smectite and nontronite with different tetrahedral and octahedral charges and octahedral cation substitution of Mg and Fe were reproduced for *cis*-vacant and *trans*-vacant configuration. The population of *cis*-vacant forms is greater in smectitic samples and the *trans*-vacant forms is greater in illite and nontronite according to experimental results. However, the energy differences between *trans*- and *cis*-configurations are very small (<3 kcal/mol-unit cell) (Sainz-Díaz *et al.*, 2001c). This study was extended using first-principles calculations based on DFT

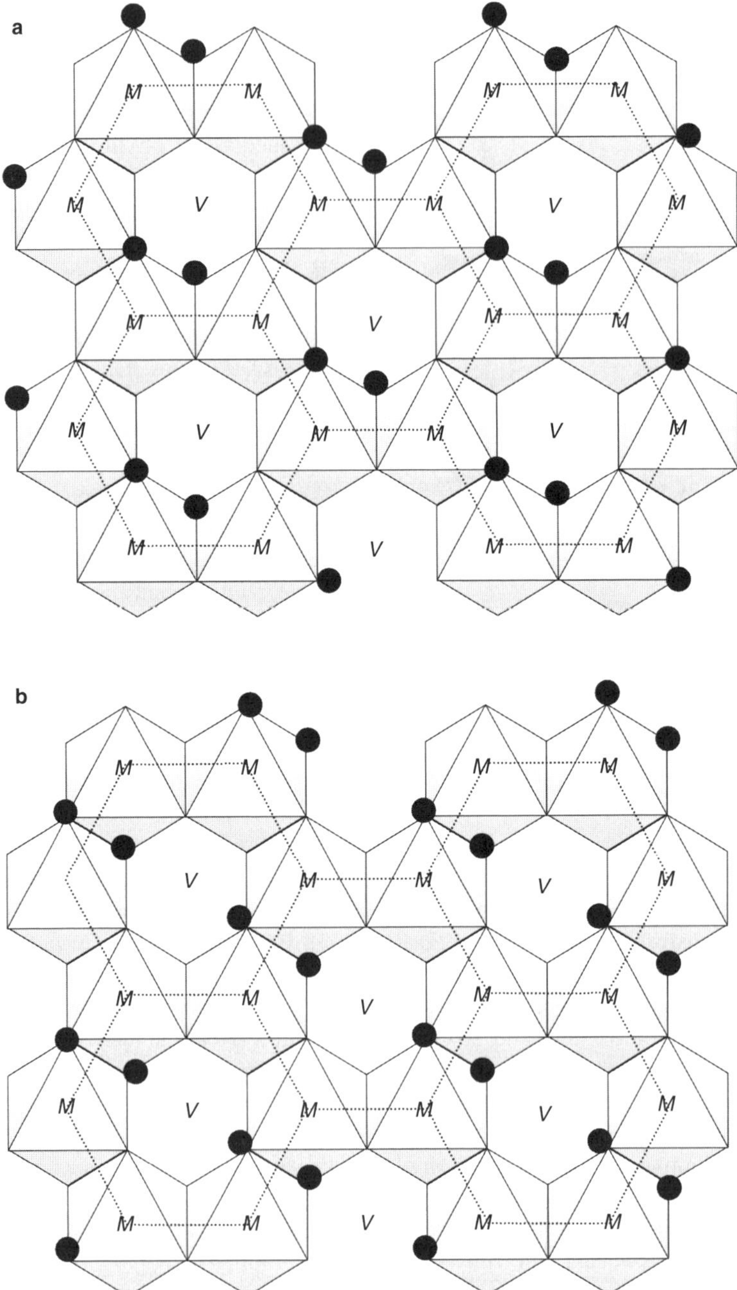

Fig. 8. Scheme of the octahedral sheet for *cis*-vacant (***a***) and *trans*-vacant (***b***) polymorphs of dioctahedral 2:1 phyllosilicates. The black balls represent the OH groups and *M* and *V* are the cation and the vacancy, respectively (from Sainz-Díaz *et al.*, 2005).

approximation for crystalline periodical models (Sainz-Díaz *et al.*, 2005). According to experimental results, in most illite-like compositions, the *trans*-vacant arrangement is more stable than the *cis*-vacant one. However, the energy differences between the *cis*-vacant and *trans*-vacant forms, for a certain cation composition, are smaller than the changes in energy produced by the relative cation distributions in the octahedral sheet. Therefore, the different proportion of these polymorphs in natural samples can be explained mainly by kinetic processes during crystal growth and not just by thermodynamic control.

7.3. Spectroscopic properties of clay minerals

As seen above, IR and Raman spectroscopies are very useful in the study of phyllosilicates and of layered metal oxides/hydroxides (Kubicki, 2001), especially those presenting OH groups, because these techniques can provide information on: (1) the local environment of OH groups; (2) cation substitutions (Ortega-Castro *et al.*, 2009); (3) reactivity (Sainz-Díaz *et al.*, 2004a); and (4) adsorption of pollutants (Sainz-Díaz *et al.*, 2010). Unfortunately the resolution of this vibrational technique is insufficient for these solids and the spectra can have various interpretations. Some deconvolution methods are required to clarify the frequency of a certain band, but the methods are approximate, especially for quantitative analysis. Therefore, theoretical calculations can be useful in the interpretation of these spectra, because it is possible to analyze the group that is responsible for producing one vibration mode with a certain frequency and the changes produced by changing the environment of this group.

Besides the combination of experimental IR spectroscopy and Monte Carlo simulations described above, the spectroscopic properties can be calculated from atomistic methods as this property has its origins in the atomic movements. Empirical forcefields can be useful for calculating frequencies of vibration modes (Braterman & Cygan, 2006), but sometimes these forcefields can be refined very accurately from empirical data for a specific group of compounds. On the other hand, quantum mechanical methods can calculate these vibrations accurately for any system.

Many first-principles calculations on cluster models of silicates have been reported for spectroscopic and reactivity studies (Lasaga, 1995; Sauer *et al.*, 1994; Kubicki & Apitz, 1998; Sainz-Díaz *et al.*, 2000; Chatterjee *et al.*, 2000). These approximations can be useful in the study of the local properties of a solid, like the vibration modes of the OH groups of dioctahedral 2:1 phyllosilicates. As seen above, the octahedral cation substitutions change the vibration frequencies of these hydroxyl groups depending on their local environment and the choice of the right cluster model is critical. A cluster of two edge-sharing octahedra with all dangling bonds protonated contains the most relevant features to study the relative effect of the isomorphous cation substitution in the octahedral sheet at the OH-vibration frequencies (Fig. 2). The variation of the frequency with the cation substitution found experimentally was reproduced with these calculations for δ(OH) (Fig. 9a) and ν(OH) modes (Fig. 9b) (Sainz-Díaz *et al.*, 2000).

However, in ν(OH) mode, the calculated frequencies are systematically greater than the experimental frequency. This is because there is an interaction between the H atom

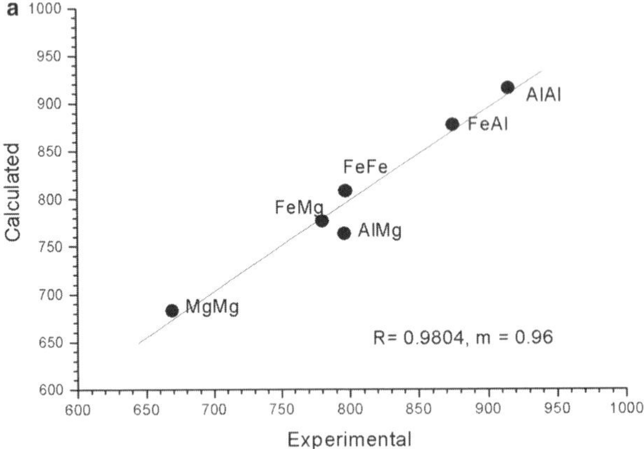

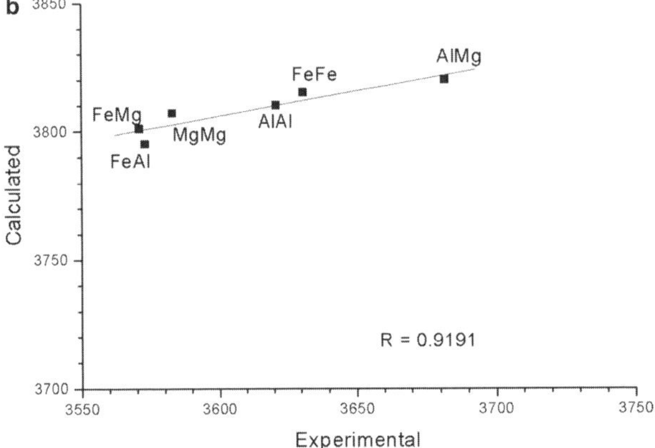

Fig. 9. Relationship between the experimental and calculated effect of the cation substitution on the δ(OH) vibration mode (***a***), and ν(OH) (***b***) (from Timón *et al.*, 2003).

of the OH group with the tetrahedral oxygen atoms which cannot be described following this cluster model. The cluster can be extended by adding a ring of six SiO_4 tetrahedra to the octahedral pair (Fig. 2). The new extended cluster describes the H bonding interactions between the OH groups and the tetrahedral oxygen atoms, decreasing significantly the ν(OH) frequencies (Fig. 10a) (Timón *et al.*, 2003). This decrease was too great, however, because the interlayer cation is not described by this cluster. Therefore, a more realistic periodical crystal lattice model was necessary to describe all interactions for these OH groups and to reproduce the experimental ν(OH) frequencies (Fig. 10b) (Botella *et al.*, 2004).

Using a periodical crystal model, it is possible to study the vibration of each OH group individually, where the frequency changes depending on the relative positions of the tetrahedral Al cation and of the interlayer cation with respect to the hydroxyl group. There are not just one or two ν(AlOHAl) bands, as reported experimentally, but several ν(AlOHAl) bands can exist (Ortega-Castro *et al.*, 2008). This should be taken into account by experimentalists during analysis of spectra with deconvolution methods.

With quantum-mechanical calculations, the IR spectrum of kaolinite has been reproduced including the intensities and the particle-size effect (Balan *et al.*, 2001, 2005). This approach has been applied also for other layered metal oxides/hydroxides (Balan *et al.*, 2006). Molecular dynamics based on empiricial potentials reproduced the vibrational spectrum of brucite (Braterman & Cygan, 2006). By means of *ab initio* molecular dynamics, spectroscopic properties can be reproduced for phyllosilicates (Smrcok *et al.*, 2010).

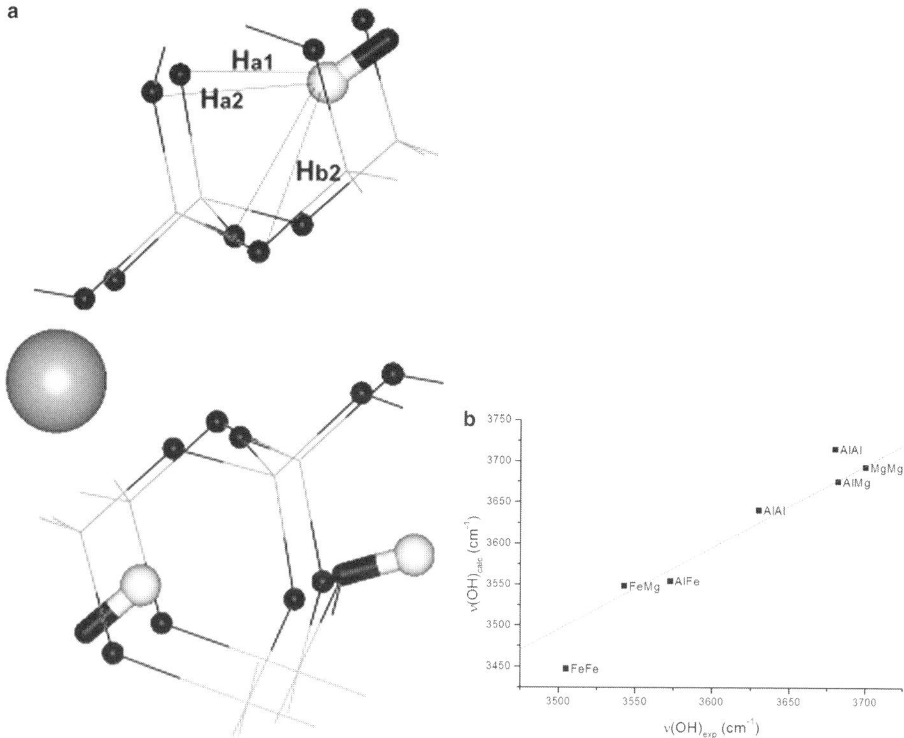

Fig. 10. (***a***) Interactions between the OH group and the surrounding tetrahedral O atoms. (***b***) Relationship between the experimental and calculated effect of the cation substitution on the ν(OH) vibration mode using a periodical crystal model. Reproduced with kind permission from Springer Science from Botella *et al.* (2004). Readers of the paper version of this chapter may wish to download a colour version of this figure from www.minersoc.org/emu-notes/emu-11/11-5-colour.pdf.

7.4. Dehydroxylation-rehydroxylation of phyllosilicates

The hydroxyl groups of phyllosilicates described above can react with each other forming a water molecule and two penta-coordinated Al cations in the octahedral sheet after thermal treatment. This apparently simple reaction happens over a temperature range which is too wide and there are some discrepancies about the mechanism and the possible formation of intermediates. In this case the computational methods can also be useful.

Static first-principles calculations of the reactant, intermediates and product of this reaction on the pyrophyllite layer estimated an activation energy close to the experimental one (Stackhouse *et al.*, 2004). Similar calculations identified the formation of a semi-dehydroxylate intermediate the spectroscopic properties of which reproduced those found experimentally for an unknown intermediate (Fig. 11a,b,c) (Sainz-Díaz *et al.*, 2004a). However, transition states should be identified to calculate accurately the activation energy along the reaction path. Using cluster models (Fig. 12), the

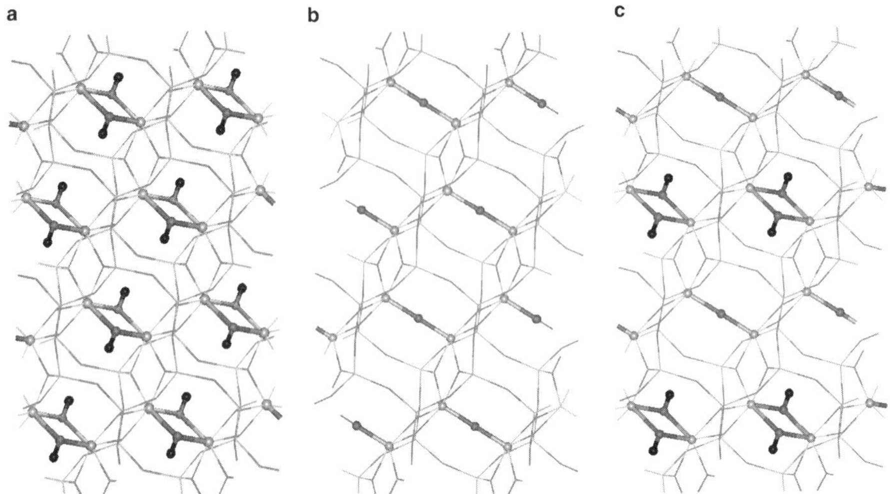

Fig. 11. Crystal structures (views from the 001 plane) of pyrophyllite (***a***), dehydroxylate (***b***) and the semidehydroxylate intermediate (**c**) during the dehydroxylation reaction. The H and O atoms of the OH groups, and the Al atoms are shown as black, dark grey and light grey spheres, respectively.

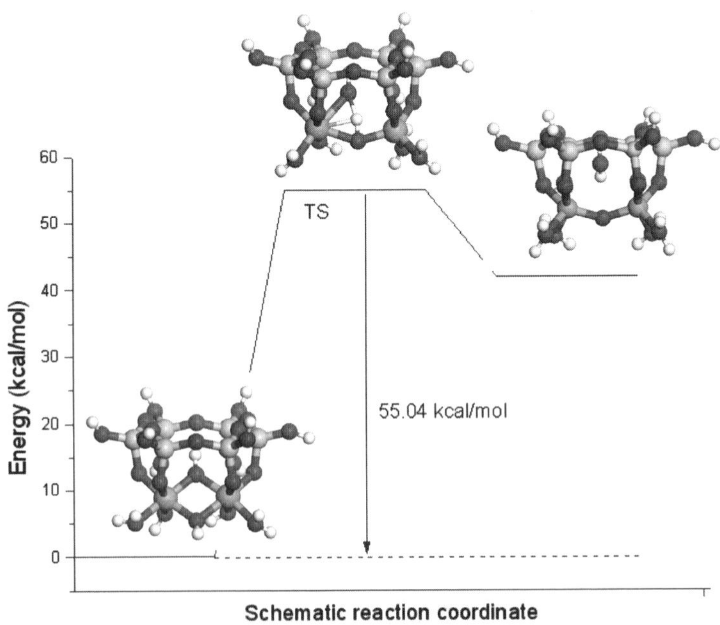

Fig. 12. Dehydroxylation reaction path of pyrophyllite using cluster models. The H, O, Al and Si atoms are in white, black, dark grey and light grey, respectively (reprinted from Molina-Montes *et al.*, 2008, with permission from Elsevier).

transition states were localized and the activation energies were calculated reproducing the variation of reactivity with the Fe^{3+} substitution (Molina-Montes *et al.*, 2008a).

Ab initio molecular dynamics based on the Car-Parrinello method with the *CPMD* code was used to explore this reaction, finding four possible mechanisms and calculating the free energies and localizing the transition states (Molina-Montes *et al.*, 2008b, 2008c). The rehydroxylation reaction of these systems was also found to be highly competitive with respect to the dehydroxylation, justifying the extremely wide range of temperatures found experimentally for the dehydroxylation reaction (Molina-Montes *et al.*, 2010).

7.5. Adsorption phenomena on phyllosilicate surfaces

The adsorption phenomena on the surfaces of phyllosilicates and other layered metal oxides/hydroxides are very important for environmental and industrial purposes. The absorption and swelling properties of phyllosilicates change dramatically with only a small variation in the chemical composition, and the cause of this behaviour is not well understood. Many theoretical studies have been reported, applying empirical mechanics methods in the interaction with water and cations in swelling of the interlayer space reproducing the variation of the interlayer space with the amount of water and the distribution of cations in the interlayer space of clays and hydrotalcites (layered double hydroxides) (Karaborni *et al.*, 1996; Sposito *et al.*, 1999; Cygan *et al.*, 2009; Wang *et al.*, 2003) and also quantum mechanics calculations have been reported for exchange of interlayer cations (Rosso *et al.*, 2001). *Ab initio* calculations have also been applied to study the dissolution reaction of silicates using molecular cluster models (Kubicki *et al.*, 1999; Wallace *et al.*, 2010), and the interaction of organic molecules on the phyllosilicate surfaces (Dios-Cancela *et al.*, 2000; Sainz-Díaz *et al.*, 2010; Muñoz-Santiburcio *et al.*, 2009).

Phyllosilicates also exist under more extreme conditions, *e.g.* deep ocean beds, and experimental study is difficult at these pressure conditions. Monte Carlo and Molecular Dynamics simulations at empirical molecular mechanics level have been applied to study the interaction of methane hydrates on phyllosilicate surfaces (Titiloye & Skipper, 2005; Cygan *et al.*, 2004b; Park & Sposito, 2003). Phyllosilicates are also present in the atmosphere and troposphere, participating in aerosol formation where heterogeneous reactions can occur. Experimental studies at these conditions are also difficult. Free-radical reactions have also been modelled at *ab initio* level in organic molecules adsorbed on silicate surfaces (Iuga *et al.*, 2008, 2010).

7.6. Effect of pressure

Although most phyllosilicates exist close to the Earth's surface, the effect of pressure on these minerals can be important for some industrial applications and for their behaviour under deep sediments and in the subduction zones, where it is not easy to perform experiments. In this case, theoretical modelling research can be very useful.

Wang *et al.* (2004) explored the effect of pressure on the crystallographic properties of phyllosilicates in subduction-zone conditions by using empirical molecular dynamics. Empirical forcefields were used to calculate the elastic constants of serpertine (Auzende *et al.*, 2006) and micas (Collins & Catlow, 1992). Quantum mechanical calculations were applied to calculate the compressibility and the effect of pressure on phyllosilicates (Mookherjee & Stixrude, 2009; Ortega-Castro *et al.*, 2010b).

8. Future perspectives and conclusions

From this chapter we summarize that theoretical modelling techniques can be applied at three main levels: (1) data-analysis methods of experimental data; (2) atomistic calculations based on empirical classic mechanics; and (3) quantum-mechanical calculations. The first level is used extensively by experimentalists as new instrumental techniques are more sophisticated and give large volumes of data which require complex analysis. The other, atomistic computational levels are less accepted among experimentalists, as that approach still seems to be too far away from macroscopic 'reality'. Computational modelling should not be considered as a 'black box' or as an isolated theoretical exercise. Many calculations are necessary to test different computational tools and parameters. Validation of atomistic calculations with accurate experimental results is critical and evaluates the limits on the application of these calculations. Nevertheless, computational modelling is becoming more widely used and collaborative research between theoreticians and experimentalists should help both approaches and the interpretation and analysis of results will be easier.

More progress is necessary in nano-mineralogy, focusing on multi-scale simulation methods combining the macroscopic finite-elements methods, molecular-dynamics simulations, and *ab initio* electronic structure calculations. On the other hand, there are many discrepancies in the adsorption processes with weak interactions, hydrogen bonding and van der Waals interactions, especially in the calculation of the adsorption energies due to the lack of accurate experimental values. Quantum-mechanics methods generally do not describe the weak dispersion interactions well and the empirical methods depend heavily on parameterization from experimental data and the interatomic potentials should be transferable to organic and inorganic solid states simultaneously. Many studies are underway to improve these methods.

Atomistic modelling is especially interesting for phyllosilicates to help interpret experimental behaviour. For example, many calculations have been reported describing the swelling behaviour of phyllosilicates but the origin of the swelling property and its variation with slight chemical changes is still unknown. Hence, the theoretical approach will be able to predict experimental behaviour and help to design specific experiments.

Acknowledgements

The author is grateful to A. Hernández-Laguna for many years of fruitful collaboration in modelling research and for his comments about this paper, to M.T. Dove and to all collaborators, postdoctoral and doctoral students for their collaboration on this subject, and also to the Andalusia regional government (Project RNM-3581) for financial support.

References

Accelrys Software Inc. (2009) *Materials Studio package*, version 4.0. Accelrys Software Inc., San Diego, California, USA.

Allen, M.P. & Tildesley, D.J. (1987) *Computer Simulations of Liquids*. Clarendon Press, Oxford, UK, 408 pp.

Allinger, N.L. (1977) Conformational analysis. 130. MM2. A hydrocarbon force field utilizing V1 and V2 torsional terms. *Journal of the American Chemical Society*, **99**, 8127–8134.

Allison, J.D., Brown, D.S. & Novo-Gradac, K.J. (1991) *MINTEQA2/PRODEFA2, a geochemical assessment model for environmental systems: Version 3.0 User's Manual*. Report EPA/600/3-91/021, United States Environmental Protection Agency, Office of Research and Development, Washington, D.C., USA, 107 pp.

Auzende, A.L., Pellenq, R.J.M., Devouard, B., Baronnet, A. & Grauby, O. (2006) Atomistic calculations of structural and elastic properties of serpentine minerals: the case of lizardite. *Physics and Chemistry of Minerals*, **33**, 266–275.

Baes, C.F., Moyer, B.A., Case, G.N. & Case, F.I. (1990) SXLSQA, A Computer Program for Including Both Complex Formation and Activity Effects in the Interpretation of Solvent Extraction Data. *Separation Science and Technology*, **25**, 1313, 1675–1688.

Balan, E., Saita, A.M., Mauri, F. & Calas, G. (2001) First principles modelling of the infrared spectrum of kaolinite. *American Mineralogist*, **86**, 1321–1330.

Balan, E., Lazzeri, M., Saita, A.M., Allard, T., Fuchs, Y. & Mauri, F. (2005) First-principles study of OH stretching modes in kaolinite, dickite and nacrite. *American Mineralogist*, **90**, 50–60.

Balan, E., Lazzeri, M., Morin, G. & Mauri, F. (2006) First-principles study of the OH stretching modes of gibbsite. *American Mineralogist*, **91**, 115–119.

Bandura, A.V. & Kubicki, J.D. (2003) Derivation of force field parameters for TiO_2-H_2O systems from ab initio calculations. *Journal of Physical Chemistry* B, **107**, 11072–11082.

Besson, G., Drits, V.A., Daynyak, L.G. & Smoliar, B.B. (1987) Analysis of cation distribution in dioctahedral micaceous minerals on the basis of IR spectroscopy data. *Clay Minerals*, **22**, 465–478.

Bleam, W.F. (1993) Atomic theories of phyllosilicates: quantum chemistry, statistical mechanics, electrostatic theory, and crystal chemistry. *Reviews in Geophysics*, **31**, 51–73.

Bosenick, A., Dove, M.T., Myers, E.R., Palin, E., Sainz-Díaz, C.I., Guiton, B., Warren, M.C., Craig, M.S. & Redfern, S.A.T. (2001) Computational methods for the study of energies of cation distributions: applications to cation-ordering phase transitions and solid solutions. *Mineralogical Magazine*, **65**, 197–224.

Botella, V., Timón, V., Hernández-Laguna, A. & Sainz-Díaz, C.I. (2004) Hydrogen bonding and vibrational properties of hydroxy groups in the crystal lattice of dioctahedral clay minerals by means of First Principles calculations. *Physics and Chemistry of Minerals*, **31**, 475–486.

Braterman, P.S. & Cygan, R.T. (2006) Vibrational spectroscopy of brucite: A molecular simulation investigation. *American Mineralogist*, **91**, 1188–1196.

Car, R. & Parrinello, M. (1985) Unified approach for molecular dynamics and density functional theory. *Physics Review Letters*, **55**, 2471–2474.

Chatterjee, A., Iwasaki, T. & Ebina, T. (2000) A novel method to correlate layer charge and the catalytic activity of 2:1 dioctahedral smectite clays in terms of binding the interlayer cation surrounded by monohydrate. *Journal of Physical Chemistry*, **104**, 8216–8223.

Collins, D.R. & Catlow, C. R. A. (1992) Computer simulation of structures and cohesive properties of micas. *American Mineralogist*, **77**, 1172–1181.

Cuadros, J., Sainz-Díaz, C.I., Ramírez, R. & Hernández-Laguna, A. (1999) Analysis of Fe segregation in the octahedral sheet of bentonitic illite-smectite by means of FT-IR, ^{27}Al MAS NMR and reverse Monte Carlo simulations. *American Journal of Science*, **299**, 289–308.

Cygan, R.T. (2003) Molecular models of metal sorption on clay minerals. In: *Molecular Modeling of Clays and Mineral Surfaces* (J.D. Kubicki & W.F. Bleam, editors). Clay Mineral Society workshop lectures, **12,** The Clay Minerals Society, Aurora, Colorado, USA, pp. 144–194.

Cygan, R.T. & Kubicki, J.D. (editors) (2001) *Molecular Modeling Theory: Applications in the Geosciences*. Reviews in Mineralogy and Geochemistry, **42**. The Mineralogical Society, Washington, D.C. and the Geochemical Society, St. Louis, Missouri, USA, 531 pp.

Cygan, R.T., Liang, J.-J. & Kalinichev, A.G. (2004a) Molecular models of hydroxide, oxyhydroxide, and clay phases and the development of a general force field. *Journal of Physical Chemistry* B, **108**, 1255–1266.

Cygan, R.T., Guggenheim, S. & Koster van Groos, A.F. (2004b) Molecular models for the intercalation of methane hydrate complexes in montmorillonite clay. *Journal of Physical Chemistry* B, **108**, 15141–15149.

Cygan, R.T., Greathouse, J.A., Heinz, H. & Kalinichev, A.G. (2009) Molecular models and simulations of layered materials. *Journal of Materials Chemistry*, **19**, 2470–2481.

Delville, A. (1993) Structure and properties of confined liquids: A molecular model of the clay-water Inteface. *Journal of Physical Chemistry*, **97**, 9703–9712.

Dios-Cancela, G., Alfonso-Méndez, L., Huertas, F.J., Romero-Taboada, E., Sainz-Díaz, C.I. & Hernández-Laguna, A. (2000) Adsorption mechanism and structure of the montmorillonite complexes with $(CH_3)_2XO$ (X = C, and S), $(CH_3O)_3PO$, and CH_3-CN molecules. *Journal of Colloid and Interface Science*, **222**, 125–136.

Dove, M.T. (1997) The use of ^{29}Si MAS-NMR and Monte Carlo methods in the study of Al/Si ordering in silicates. *Geoderma*, **80**, 353–368.

Dove, M.T. (2008) An introduction to atomistic simulation methods. In: *Computer Methods in Mineralogy and Geochemistry* (M. Prieto & C. Brime, editors). Seminarios de la Sociedad Española de Mineralogía, **4,** 7–37. Sociedad Española de Mineralogía, Oviedo, Spain.

Drits, V.A. (2003) Structural and chemical heterogeneity of layer silicates and clay minerals. *Clay Minerals*, **38**, 403–432.

Drits, V.A., Dainyak, L.G., Muller, F., Besson, G. & Manceau, A. (1997) Isomorphous cation distribution in celadonites, glauconites and Fe-illites determined by infrared, Mössbauer and EXAFS spectroscopies. *Clay Minerals*, **32**, 153–179.

Drits, V.A., Lindgreen, H., Salyn, A.L., Ylagan, R. & McCarty, D.K. (1998) Semiquantitative determination of trans-vacant and cis-vacant 2:1 layers in illites and illite-smectites by thermal analysis and X-ray diffraction. *American Mineralogist*, **83**, 1188–1198.

Flekkoy, E.G. & Coveney, P. (1999) From molecular dynamics to dissipative particle dynamics. *Physical Review Letters*, **83**, 1775–1778.

Frenkel, D. & Smit, B. (1996) *Understanding Molecular Simulation*. Academic Press, New York, USA, 443 pp.

Frisch, M.J., Trucks, G.W., Schlegel, H.B., Scuseria, G.E., Robb, M.A., Chesseman, J.R., Zarzewki, V.G., Montgomery, J.A., Stratmann, R.E., Burant, J.C., Dapprich, S., Millam, J.M., Daniels, A.D., Kudin, K.N., Strain, M.C., Farkas, O., Tomasi, J., Barone, V., Cossi, M., Cammi, R., Mennucci, B., Pomelli, C., Adamo, C., Clifford, S., Ochterski, J., Petersson, G.A., Ayala, P.Y., Cui, Q., Morokuma, K., Malick, D.K., Rabuck, A.D., Raghavachari, K., Foresman, J.B., Cioslowski, J., Ortiz, J.V., Stefanov, B.B., Liu, G., Liashenko, A., Piskorz, P., Komaromi, I., Gomperts, R., Martin, R.L., Fox, D.J., Keith, T.A., Al-Laham, M.A., Peng, C.Y., Nanayakkara, A., Gonzalez, C., Challacombe, M., Gill, P.M.W., Johnson, B.G., Chen, W., Wong, M.W., Andres, J.L., Head-Gordon, M., Replogle, E.S. & Pople, J.A. (2004) *Gaussian 03 (Revision A.1)*, Gaussian, Inc., Pittsburgh, Pennsylvania, USA.

Gale, J.D. (1997) A computer program for the symmetry-adapted simulations of solids. *Journal of Chemical Society Perkin Transactions*, **93**, 629–637.

Hehre, W.J., Radom, L., Schleyer, P.v.R. & Pople, J.A. (1986) *Ab Initio Molecular Orbital Theory*. John Wiley and Sons, New York, USA, 396 pp.

Heinz, H., Koerner, H., Anderson, K.L., Vaia, R.A. & Farmer, B.L. (2005) Force field for mica-type silicates and dynamics of octadecylammonium chains grafted to montmorillonite. *Chemistry of Materials*, **17**, 5658–5669.

Hernández-Laguna, A., Escamilla-Roa, E., Timón, V., Dove, M.T. & Sainz-Díaz, C.I. (2006) DFT study of the cation arrangements in the octahedral and tetrahedral sheets of dioctahedral 2:1 phyllosilicates. *Physics and Chemistry of Minerals*, **33**, 655–666.

Herrero, C.P. & Sanz, J. (1991) Short-range order of the Si, Al distribution in layer silicates. *Journal of Physical Chemistry of Solids*, **52**, 1129–1135.

Iuga, C., Vivier-Bunge. A., Hernández-Laguna, A. & Sainz-Díaz, C.I. (2008) Quantum Chemistry and computational kinetics of the reaction between OH radicals and formaldehyde adsorbed on small silica aerosol models. *Journal of Physical Chemistry C*, **112**, 4590–4600.

Iuga, C., Sainz-Díaz, C.I. & Vivier-Bunge, A. (2010) On the OH initiated oxidation mechanism of aliphatic aldehydes in the presence of mineral aerosols, *Geochimica et Cosmochimica Acta*, **74**, 3587–3597.

Karaborni, S., Smit, B., Heidug, W., Urai, J. & van Oort, E. (1996) The swelling of clays: molecular simulations of the hydration of montmorillonite. *Science*, **271**, 1102–1104.

Kubicki, J.D. (2001) Interpretation of vibrational spectra using molecular orbital theory calculations. In: *Molecular Modeling Theory: Applications in the Geosciences* (R.T. Cygan & J.D. Kubicki, editors). Reviews in Mineralogy and Geochemistry, **42**, Mineralogical Society of America, Washington, D.C. and the Geochemical Society, St. Louis, Missouri, USA, pp. 459–483.

Kubicki, J.D. & Apitz, S.E. (1998) Molecular cluster models of aluminum oxide and aluminum hydroxide surfaces. *American Mineralogist*, **83**, 1054–1066.

Kubicki, J.D. & Bleam, W.F. (editors) (2003) *Molecular Modeling of Clays and Mineral Surfaces*, CMS Workshop Lectures, **12**, Aurora, Colorado, USA, 229 pp.

Kubicki, J.D., Sykes, S. & Apitz, S.E. (1999) Ab initio calculation of aqueous aluminum and aluminum-carboxylate complex energetics and ^{27}Al NMR Chemical shifts. *Journal of Physical Chemistry*, **103**, 903–915.

Lasaga, A.C. (1992) Ab initio methods in mineral surface reactions. *Reviews of Geophysics*, **30**, 269–303.

Lasaga, A.C. (1995) Fundamental approaches in describing mineral dissolution and precipitation rates. In: *Chemical Weathering Rates of Silicate Minerals* (A.F. White & S.L. Brantley, editors). Reviews in Mineralogy, **31,** 23–86. Mineralogical Society of America, Washington, D.C.

Lasaga, A.C. & Gibbs, G.V. (1987) Application of quantum mechanical potential surfaces to mineral physics calculations. *Physics and Chemistry of Minerals*, **14**, 107–117.

Lii, J.-H. & Allinger, N.L. (1989) Molecular Mechanics. The MM3 Force Field for Hydrocarbons. 3. The van der Waals Potentials and Crystal data for Aliphatic and Aromatic Hydrocarbons. *Journal of American Chemical Society*, **111**, 8576–8582.

Mayo, S.L., Olafson, B.D. & Goddard III, W.A. (1990) DREIDING: a generic force field for molecular simulations. *Journal of Physical Chemistry*, **94**, 8897–8909.

McCarty, D.K. & Reynolds, R.C. (1995) Rotationally disordered illite-smectite in paleozoic K-bentonites. *Clays and Clay Minerals*, **43**, 271–284.

Molina-Montes, E., Timón, V., Hernández-Laguna, A. & Sainz-Díaz, C.I. (2008a) Dehydroxylation mechanisms in Al^{3+}/Fe^{3+} dioctahedral phyllosilicates by quantum mechanical methods with cluster models. *Geochimica et Cosmochimica Acta*, **72**, 3929–3938.

Molina-Montes, E., Donadio, D., Hernández-Laguna, A., Sainz-Díaz, C.I. & Parrinello M. (2008b) DFT research on the dehydroxylation reaction of pyrophyllite 1: First principles molecular dynamics simulations. *Journal of Physical Chemistry* B, **112**, 7051–7060.

Molina-Montes, E., Donadio, D., Hernández-Laguna, A. & Sainz-Díaz, C.I. (2008c) DFT research on the dehydroxylation reaction of pyrophyllite II. Characterization of reactants, intermediates and transition states along the reaction path. *Journal of Physical Chemistry A*, **112**, 6373–6383.

Molina-Montes, E., Donadio, D., Hernández-Laguna, A. & Sainz-Díaz, C.I. (2010) Exploring the rehydroxylation reaction of pyrophyllite by ab initio molecular dynamics. *Journal of Physical Chemistry B*, **114**, 7593–7601.

Mookherjee, M. & Stixrude, L. (2009) Structure and elasticity of serpentine at high-pressure. *Earth and Planetary Science Letters*, **279**, 11–19.

Muñoz-Santiburcio, D., Ortega-Castro, J., Sainz-Díaz, C.I., Huertas, F.J. & Hernández-Laguna, A. (2009) Theoretical study of the adsorption of 2-nitro-1-propanol on smectite surface models. *Journal of Molecular Structure (THEOCHEM)*, **912**, 95–104.

Ortega-Castro, J., Hernández-Haro, N., Hernández-Laguna, A. & Sainz-Díaz, C.I. (2008) DFT calculation of crystallographic properties of dioctahedral 2:1 phyllosilicates. *Clay Minerals*, **43**, 351–361.

Ortega-Castro, J., Hernández-Haro, N., Muñoz-Santiburcio, D., Hernández-Laguna, A. & Sainz-Díaz, C.I. (2009) Crystal structure and hydroxyl group vibrational frequencies of phyllosilicates by DFT methods. *Journal of Molecular Structure (THEOCHEM)*, **912**, 82–87.

Ortega-Castro, J., Hernández-Haro, N., Dove, M.T., Hernández-Laguna, A. & Sainz-Díaz, C.I. (2010a) Density functional theory and Monte Carlo study of octahedral cation ordering of Al/Fe/Mg cations in dioctahedral 2:1 phyllosilicates. *American Mineralogist*, **95**, 209–220.

Ortega-Castro, J., Hernández-Haro, N., Timón, V., Sainz-Díaz, C.I. & Hernández-Laguna, A. (2010b) High-pressure behavior of $2M_1$ muscovite. *American Mineralogist*, **95**, 249–259.

Palin, E.J. & Dove, M.T. (2004) Investigation of Al/Si ordering in tetrahedral phyllosilicate sheets by Monte Carlo simulation. *American Mineralogist*, **89**, 176–184.

Palin, E.J., Dove, M.T., Redfern, S.A.T., Bosenick, A., Sainz-Díaz, C.I. & Warren, M.C. (2001) Computational study of tetrahedral Al-Si ordering in muscovite. *Physics and Chemistry of Minerals*, **28**, 534–544.

Palin, E.J., Dove, M.T., Redfern, S.A.T., Sainz-Díaz, C.I. & Lee, W.T. (2003) Computational study of tetrahedral Al–Si and octahedral Al–Mg ordering in phengite. *Physics and Chemistry of Minerals*, **30**, 293–304.

Palin, E.J., Dove, M.T., Sainz-Díaz, C.I. & Hernández-Laguna, A. (2004) A computational investigation of the Al/Fe/Mg order-disorder behaviour in the dioctahedral sheet of phyllosilicates. *American Mineralogist*, **89**, 164–175.

Park, S.-H. & Sposito, G. (2003) Do montmorillonite surfaces promote methane hydrate formation? Monte Carlo and molecular dynamics simulations. *Journal of Physical Chemistry* B, **107**, 2281–2290.

Parkhurst, D.L., Thorstenson, D.C. & Plummer, L.N. (1980) PHREEQE – A computer program for geochemical calculations. *United States Geological Survey Water Resources Investment*, 80-96. United States Government Printing Office, Washington, D.C.

Payne, M.C., Teter, M.P., Allen, D.C., Arias, T. & Joannopoulus, J.D. (1992) Iterative minimization techniques for ab initio total-energy calculations: molecular dynamics and conjugate gradients. *Reviews of Modern Physics*, **64**, 1045–1097.

Plimpton, S.J. (1995) Fast Parallel Algorithms for Short-Range Molecular Dynamics. *Journal of Computational Physics*, **117**, 1–19.

Pople, J.A. & Beveridge, D.L. (1970) *Approximate Molecular Orbital Theory*. McGraw-Hill, New York, 163 pp.

Raghavachari, K., Foresman, J.B., Cioslowski, J., Ortiz, J.V., Frisch, M.J. & Frisch, A. (1998) *GAUSSIAN 98 User's Reference*. Gaussian Inc., Pittsburgh, Pennsylvania, USA.

Rappé, A.K., Casewit, C.J., Colwell, K.S., Goddard III, W.A. & Skid, W.M. (1992) UFF, a full periodic table force field for molecular mechanics and molecular dynamics simulations. *Journal of American Chemical Society*, **114**, 10024–10035.

Rosso, K.M., Rustad, J.R. & Bylaska, E.J. (2001) The Cs/K exchange in muscovite interlayers: an ab initio treatment. *Clays and Clay Minerals*, **49**, 500–513.

Sanz, J., Herrero, C.P. & Robert, J.-L. (2003) Distribution of Si and Al in clintonites: a combined NMR and Monte Carlo study. *Journal of Physical Chemistry* B, **107**, 8337–8342.

Sainz-Díaz, C.I., Timón, V., Botella, V. & Hernández-Laguna, A. (2000) Isomorphous substitution effect on the vibration frequencies of hydroxyl groups in molecular cluster models of the clay octahedral sheet. *American Mineralogist*, **85**, 1038–1045.

Sainz-Díaz, C.I., Hernández-Laguna, A. & Dove, M.T. (2001a) Modelling of dioctahedral 2:1 phyllosilicates by means of transferable empirical potentials. *Physics and Chemistry of Minerals*, **28**, 130–141.

Sainz-Díaz, C.I., Cuadros, J. & Hernández-Laguna, A. (2001b) Cation distribution in the octahedral sheet of dioctahedral 2:1 phyllosilicates by using inverse Monte Carlo methods. *Physics and Chemistry of Minerals*, **28**, 445–454.

Sainz-Díaz, C.I., Hernández-Laguna, A. & Dove, M.T. (2001c) Theoretical modelling of cis-vacant and trans-vacant configurations in illites and smectites. *Physics and Chemistry of Minerals*, **28**, 322–331.

Sainz-Díaz, C.I., Timón, V., Botella, V., Artacho, E. & Hernández-Laguna, A. (2002) Quantum mechanical calculations of dioctahedral 2:1 phyllosilicates: Effect of octahedral cation distribution in pyrophyllite, illite, and smectite. *American Mineralogist*, **87**, 958–965.

Sainz-Díaz, C.I., Palin, E.J., Dove, M.T. & Hernández-Laguna, A. (2003a) Ordering of Al, Fe and Mg cations in the octahedral sheet of smectites and illites by means of Monte Carlo simulations. *American Mineralogist*, **88**, 1033–1045.

Sainz-Díaz, C.I., Palin, E.J., Hernández-Laguna, A. & Dove, M.T. (2003b) Octahedral cation ordering of dioctahedral 2:1 phyllosilicates by means of Monte Carlo simulations. *Physics and Chemistry of Minerals*, **30**, 382–392.

Sainz-Díaz, C.I., Escamilla-Roa, E. & Hernández-Laguna, A. (2004a) Pyrophyllite dehydroxylation process by first-principles calculations. *American Mineralogist*, **89**, 1092–1100.

Sainz-Díaz, C.I., Palin, E.J., Hernández-Laguna, A. & Dove, M.T. (2004b) Effect of the tetrahedral charge on the order-disorder of the cation distribution in the octahedral sheet of smectites and illites by computational methods. *Clays and Clay Minerals*, **52**, 357–374.

Sainz-Díaz, C.I., Escamilla, E. & Hernández-Laguna, A. (2005) Quantum mechanical calculations of trans-vacant and cis-vacant polymorphism in dioctahedral 2:1 phyllosilicates. *American Mineralogist*, **90**, 1827–1834.

Sainz-Díaz, C.I., Francisco-Márquez, M. & Vivier-Bunge, A. (2010) Molecular structure and spectroscopic properties of polyaromatic heterocycles by first principle calculations: spectroscopic shifts with the adsorption of thiophene on phyllosilicate surface, *Theoretical Chemistry Accounts*, **125**, 83–95.

Sato, H., Yamagishi, A. & Kawamura, K. (2001) Molecular simulation for flexibility of a single clay layer. *Journal of Physical Chemistry* B, **105**, 7990–7997.

Sauer, J., Ugliengo, P., Garrone, E. & Saunders, V.R. (1994) Theoretical study of van der Waals complexes at surface sites in comparison with the experiment. *Chemical Reviews*, **94**, 2095–2160.

Schleyer, P. von R. (1998) *Encyclopedia of Computational Chemistry*, Editor-in-Chief. John Wiley and Sons, Inc., Chichester, UK.

Schroeder, P.A. (1993) A chemical, XRD, and ^{27}Al MAS NMR investigation of Miocene Gulf Coast shales with application to understanding illite-smectite crystal-chemistry. *Clays and Clay Minerals*, **41**, 668–679.

Schroeder, P.A. & Pruett, R.J. (1996) Fe ordering in kaolinite: Insights from ^{29}Si and ^{27}Al MAS NMR spectroscopy. *American Mineralogist*, **81**, 26–38.

Skipper, N.T. (2003) Monte Carlo and molecular dynamics computer simulations of aqueous interlayer fluids in clays. In: *Molecular Modeling of Clays and Mineral Surfaces* (J.D. Kubicki & W.F. Bleam, editors). The Clay Mineral Society Workshop Lectures, **12**, Aurora, Colorado, USA, pp. 101–142.

Skipper, N.T., Chang, F-R.C. & Sposito, G. (1995) Monte Carlo simulation of interlayer molecular structure in swelling clay minerals. 2. Monolayer Hydrates. *Clays and Clay Minerals*, **43**, 294–303.

Slonimskaya, M.V., Besson, G., Dainyak, L.G., Tchoubar, C. & Drits, V.A. (1986) Interpretation of the IR spectra of celadonites and glauconites in the region of OH-stretching frequencies. *Clay Minerals*, **21**, 377–388.

Smith, W., Yong, C.W. & Rodger, P.M. (2002) DL_POLY: Application to molecular simulation. *Molecular Simulations*, **28**, 385-471.

Smrcok, F., Tunega, D., Ramirez-Cuesta, A.J. & Scholtzová, E. (2010) The combined inelastic neutron scattering and solid state DFT study of hydrogen atoms dynamics in a highly ordered kaolinite. *Physics and Chemistry of Minerals*, **37**, 571–579.

Soler, J.M., Artacho, E., Gale, J.D., García, A., Junquera, J., Ordejón, P. & Sánchez-Portal, D. (2002) The SIESTA method for ab initio order-N materials simulation. *Journal of Physics: Condensed Matter*, **14**, 2745–2779.

Sposito, G., Skipper, N.T., Sutton, R., Park, S.-H., Soper, A.K. & Greathouse, J.A. (1999) Surface geochemistry of the clay minerals. *Proceedings of the National Academy of Science USA*, **96**, 3358–3364.

Stackhouse, S., Coveney, P.V. & Benoit, D.M. (2004) Density-Functional-Theory-Based study of the dehydroxylation behaviour of Aluminous dioctahedral 2:1 layer-type clay minerals. *Journal of Physical Chemistry* B, **108**, 9685–9694.

Sun, H. (1998) COMPASS: An ab Initio Force-Field Optimized for Condensed-Phase Applications – Overview with Details on Alkane and Benzene Compounds. *Journal of Physical Chemistry*, **102**, 7338.

Teppen, B.J., Rasmussen, K., Bertsch, P.M., Miller, D.M. & Schäfer, L. (1997) Molecular dynamics modeling of clay minerals. 1. Gibbsite, kaolinite, pyrophyllite, and beidellite. *Journal of Physical Chemistry* B, **101**, 1579–1587.

Timón, V., Sainz-Díaz, C.I., Botella, V. & Hernández-Laguna, A. (2003) Isomorphous cation substitution in dioctahedral 2:1 phyllosilicates by means of ab initio quantum mechanical calculations on clusters. *American Mineralogist*, **88**, 1788–1795.

Titiloye, J.O. & Skipper, N.T. (2005) Monte Carlo and molecular dynamics simulations of methane in potassium montmorillonite clay hydrates at elevated pressures and temperatures. *Journal of Colloid and Interface Science*, **282**, 422–427.

Tsipursky, S.I. & Drits, V.A. (1984) The distribution of octahedral cations in the 2:1 layers of dioctahedral smectites studied by oblique-texture electron diffraction. *Clay Minerals*, **19**, 177–193.

Tunega, D. & Lischka, H. (2003) Effect of the Si /Al ordering on structural parameters and the energetic stabilization of vermiculites – a theoretical study. *Physics and Chemistry of Minerals*, **30**, 517–522.

Viani, A., Gualtieri, A.F. & Artioli, G. (2002) The nature of disorder in montmorillonite by simulation of X-ray powder patterns. *American Mineralogist*, **87**, 966–975.

Vucelic, M., Jones, W. & Moggridge, G.D. (1997) Cation ordering in synthetic layered double hydroxides. *Clays and Clay Minerals*, **45**, 803–813.

Wallace, A.F., Gibbs, G.V. & Dove, P.M. (2010) Influence of ion-associated water on the hydrolysis of Si-O bonded interactions. *Journal of Physical Chemistry* A, **114**, 2534–2542.

Wang, J., Kalinichev, A.G., Amonette, J.E. & Kirkpatrick, R.J. (2003) Interlayer structure and dynamics of Cl-bearing hydrotalcite: far infrared spectroscopy and molecular dynamics modelling. *American Mineralogist*, **88**, 398–409.

Wang, J., Kalinichev, A.G. & Kirkpatrick, R.J. (2004) Molecular modeling of the 10 Å phase at subduction zone conditions. *Earth and Planetary Science Letters*, **222**, 517–527.

Warren, M.C., Dove, M.T., Myers, E.R., Bosenick, A., Palin, E.J., Sainz-Díaz, C.I., Guiton, B.S. & Redfern, S.A.T. (2001) Monte Carlo methods for the study of cation ordering in minerals. *Mineralogical Magazine*, **65**, 221–248.

EMU Notes in Mineralogy, Vol. 11 (2011), Chapter 6, 237–258

The concept of layer charge of smectites and its implications for important smectite-water properties

GEORGE E. CHRISTIDIS

Technical University of Crete, Department of Mineral Resources Engineering, 73100 Chania, Greece
e-mail: christid@mred.tuc.gr

Layer charge is an important intrinsic property of smectites which stems from substitutions in the octahedral and/or tetrahedral sheet or from vacancies in the octahedral sheet. The layer charge is balanced by the interlayer cations which are exchangeable. There are three methods for determining layer charge: the structural formula method (SFM); the alkylammonium method (AMM); and the potassium-saturation method (KSM) which has been calibrated with the SFM. The layer charge affects important physical water-rock properties of smectites. Smectites with small layer charge have smaller cation exchange capacities than their high-charge counterparts. In contrast, during crystalline swelling, the activity of water at which the transition in hydration and dehydration occurs, increases with increasing layer charge of the smectites, suggesting that low-charge smectites have greater swelling capacity. In contrast, data on the influence of layer charge on double layer swelling are inconclusive, although in general low-charge smectites have greater swelling capacity than high-charge smectites. Similarly, the limited existing data suggest that, in general, low-charge smectites form more viscous suspensions than their high-charge counterparts. The influence of layer charge on the double layer swelling and the viscosity is attributed to the formation of quasicrystals, *i.e.* small stacks of smectite layers, which may breakup by hydrodynamic forces under shearing. Layer charge is a property of the unit cell, *i.e.* it refers to the atomic not the macroscopic level and it does not reflect the charge of the smectite particles. Smectite-water properties can be better explained by the concept of fundamental particle charge, *i.e.* the electric charge of the single 10 Å thick smectite particles.

1. Introduction

Smectites are clay minerals with unique physical and chemical properties which are valued by industry. These properties stem from their crystal-chemical characteristics, their very small particle size and the correspondingly large specific surface area and include cation exchange capacity (CEC), swelling and rheological properties, hydration and dehydration, high plasticity, bonding capacity and their ability to react with inorganic and organic reagents (Odom, 1984). Smectites form small crystals, usually <0.5 μm in size, and their crystal shape varies from subhedral lamellae with irregular outlines, to euhedral lamellae with rhombic outlines and lath and ribbon crystallites (Fig. 1). Smectite layers (Fig. 2) display 'turbostratic' stacking without ordering, *i.e.* they are randomly stacked one on the top of the other, like a pile of playing cards (Moore & Reynolds, 1997). When dispersed in water at low concentrations they form

DOI: 10.1180/EMU-notes.11.6

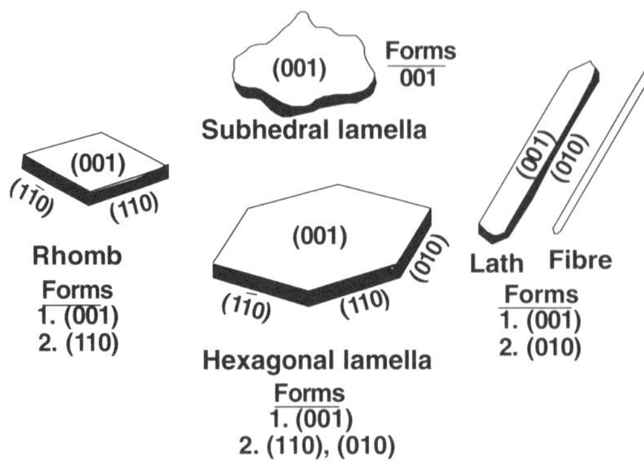

Fig. 1. Idealized shapes of smectite particles (adapted from Güven, 1988). Each particle consists of several layers.

suspensions with Newtonian properties (Rand *et al.*, 1980; Kasperski *et al.*, 1986; Brandenburg & Lagaly, 1988), attributed to hydrodynamic forces, whereas with increasing concentration, interactions between the particles induce a non-Newtonian behaviour (Gillespie, 1960). Some smectites may exhibit Newtonian behaviour even at high suspension concentrations exceeding 6% (Christidis *et al.*, 2006).

Of particular importance is the layer charge of smectites, due to substitutions in the tetrahedral or the octahedral sheet, or to the existence of vacancies in the octahedral sheet, which is associated with CEC (Güven, 1992; Laird, 1999) and the smectite-water properties. Indeed it has been shown that layer charge is related to the colloidal properties of smectites such as swelling (Slade *et al.*, 1991; Laird, 2006, Christidis *et al.*, 2006); charge heterogeneity, which includes both charge magnitude and charge localization (Christidis & Eberl, 2003) is also related to these properties. Moreover, the magnitude of the layer charge is related to the adsorption of organic compounds on the smectite surface (Laird *et al.*, 1992). Evidence for the influence of layer charge on the smectite-water properties is not conclusive, however.

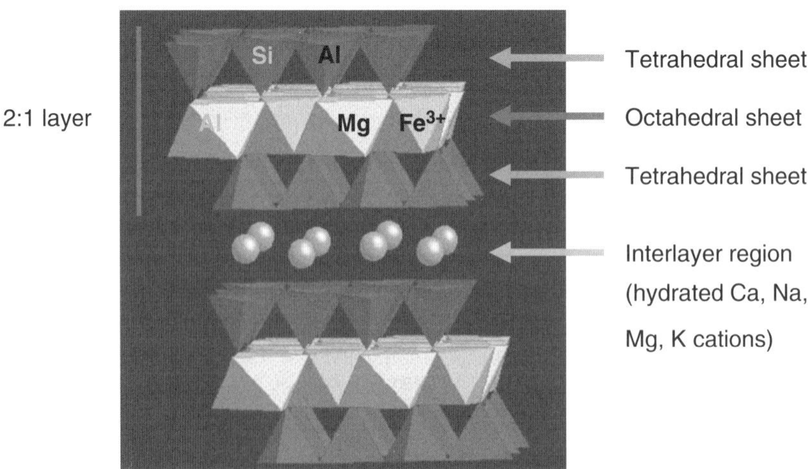

Fig. 2. Idealized structure of dioctahedral smectites, depicting the possible substitutions in the 2:1 layer and the source of layer charge. Readers of the paper version of this chapter may wish to download a colour version of this figure from www.minersoc.org/emu-notes/emu-11/11-6-colour.pdf.

The variation of electrostatic potential on the smectite particle surface, affected by the magnitude and distribution of the layer charge on the silicate layers, is an important factor controlling hydration of the interlamellar space and thus swelling (McEwan & Wilson 1984; Güven, 1992). This may explain, at least partly, the inconclusive evidence for the influence of the layer charge on the smectite-water properties. Nevertheless, it is difficult to determine the charge of smectite particles in suspension, because (1) the smectite particles are not isolated but tend to form quasicrystals or more complex aggregates (see below), (2) isolated smectite particles in real suspensions are the exception rather than the rule, and (3) the charge of the particles may vary considerably within the suspension, because of the extreme heterogeneity which is commonly observed in smectites. In the latter case the average particle charge would be used. These difficulties notwithstanding, there is a need to formulate a model for calculation of the particle charge of smectites, which would be used to examine how the smectite particle charge affects important smectite-water properties such as viscosity and swelling. Therefore the concept of 'fundamental particle charge' of smectites has recently been introduced (Pratikakis *et al.*, 2010). A brief account of this concept, which is not the main topic of this contribution, will be given in the final part of this work.

This contribution consists of four sections. The first section deals with the structure of smectite and the concept of layer charge and the second gives an account of the existing methods used to calculate the layer charge of smectites. The third section shows existing evidence for the influence of layer charge on the interaction of smectite with water (rheological properties, swelling). Finally, the fourth section deals with the concept of fundamental particles (Nadeau *et al.*, 1984a) and gives a brief account of the concept of fundamental-particle charge (Pratikakis *et al.*, 2010).

2. Structural features and layer charge of the smectites

Smectites are phyllosilicates characterized by a T-O-T layer, in which an octahedral sheet (O) is linked with two tetrahedral sheets (T) to form a 2:1 layer (Bailey, 1984; Brigatti *et al.*, 2006) (Fig. 2). Due to ionic substitutions in the tetrahedral sheet (Al^{3+} for Si^{4+}) and the octahedral sheets (Mg^{2+}, Fe^{3+} and Fe^{2+} for Al^{3+} in the dioctahedral smectites or Li^{+} for Mg^{2+} in the trioctahedral smectites) or to the existence of vacancies in the octahedral sheet, an excess charge is created. The existence of vacancies as a source of layer charge is the exception rather than the rule, and is the dominant source of charge deficit in a trioctahedral smectite called stevensite (see below). According to the Nomenclature Committee of AIPEA (Guggenheim *et al.*, 2006) the layer charge of smectites varies from 0.2 to 0.6 electrons per half unit cell (e/h.u.c.). The layer charge is balanced by the interlayer cations which are exchangeable, leading to an important property, the 'cation exchange capacity'. This charge is 'permanent', because it is not affected by the pH of the aqueous media in which the smectites are dispersed.

Except for the permanent charge, clay minerals may bear 'non-permanent charge', which depends on the pH. The non-permanent charge results mainly from uptake or removal of H^+ ions from functional groups at the edges of the clay mineral crystals (aluminols or silanols), according to the pH of the aqueous medium. A schematic representation of the protonation-deprotonation of an edge functional group (aluminol) is shown below:

$$\underset{\text{Positive charge at low pH}}{\text{Al-OH}_2^+} \leftrightarrow \underset{\text{No charge at neutral pH}}{\text{Al-OH}^{\text{o}} + \text{H}^+} \leftrightarrow \underset{\text{Negative charge at high pH}}{\text{Al-OH}^- + \text{H}^+}$$

The non-permanent charge is the main source of charge in kaolin-group minerals and in oxides. In contrast, it is of less importance for smectites.

Smectites are either dioctahedral or trioctahedral. In order to consider the composition of the various smectites it is useful to use as a reference, pyrophyllite [$Al_2Si_4O_{10}(OH)_2$] or talc [$Mg_3Si_4O_{10}(OH)_2$] for the dioctahedral or trioctahedral members, respectively. Substitution of octahedral Mg^{2+} for Al^{3+} in the structure of pyrophyllite yields montmorillonite, the most common smectite, whereas substitution of tetrahedral Si^{4+} by Al^{3+} yields beidellite and substitution of Si by Fe^{3+} yields nontronite. Hence, the layer charge is in the octahedral sheet in montmorillonite and in the tetrahedral sheet in beidellite and nontronite. In nontronite, the main octahedral cation is Fe^{3+}. Another major chemical element present in the octahedral sheet is Fe^{3+}, which does not contribute to layer charge, however. Smectites containing >0.3 Fe^{3+} atoms p.h.u.c. are known as Fe-rich smectites, either Fe-rich montmorillonite or Fe-rich beidellite (Brigatti 1983; Güven, 1988). Volkonskoite is a rare dioctahedral smectite with Cr^{3+} as the main octahedral cation, and with layer charge located mainly in tetrahedral sites (Foord *et al.*, 1987) (see also Brigatti *et al.*, 2011; Guggenheim, 2011, this volume).

Dioctahedral smectites are often characterized as low- and high-charge according to their layer charge. Typical high-charge dioctahedral smectite is Cheto montmorillonite (sample SAz-1 from The Clay Minerals Society repository) and a typical low-charge smectite is Wyoming montmorillonite (SWy-1 or SWy-2, also from The Clay Minerals Society repository). Nevertheless, there is still no acceptable classification of smectites according to the layer charge. Christidis *et al.* (2006), based on the XRD characteristics of K-saturated ethylene glycol-solvated smectites proposed classification of dioctahedral smectites according to layer charge (Fig. 3). Low-charge smectites have layer charge of <0.425 e/h.u.c. and high-charge smectites have layer charge >0.475 e/h.u.c. Smectites with layer charge between 0.425 and 0.475 e/h.u.c. are intermediate-charge smectites. Note that in the proposed classification scheme, the layer-charge boundaries are not placed symmetrically in the layer-charge interval assigned to smectites (0.2–0.6 e/h.u.c.). A different classification of smectites according to the layer charge, which uses the same terminology, however, has been proposed by Emmerich *et al.* (2009). In the latter classification scheme, the layer-charge boundaries are placed symmetrically in the layer-charge interval assigned to smectites. Thus the high-charge smectites have layer charge >0.425 e/h.u.c. and the low-charge smectites have layer charge <0.375 e/h.u.c. However, the reasoning behind the setting of the specific layer-charge boundaries has not been justified. Moreover, most of the

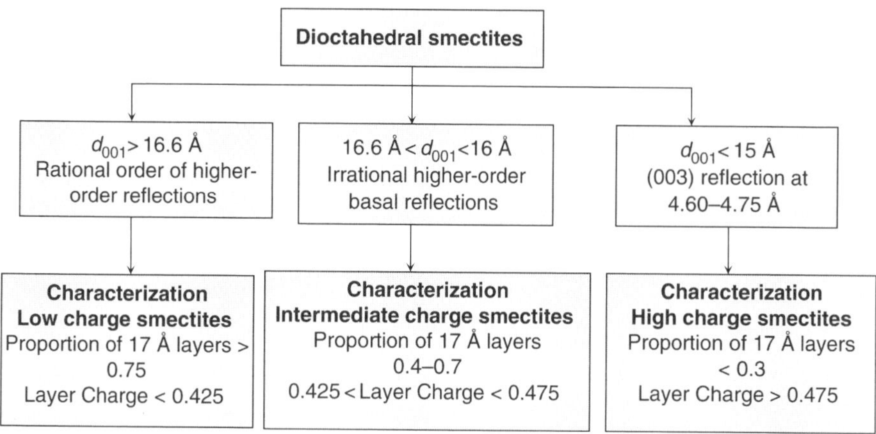

Fig. 3. Classification scheme of dioctahedral smectites according to Christidis *et al.* (2006). Layer charge in e/h.u.c.

Wyoming montmorillonites which are considered as low-charge smectites are, in fact, intermediate-charge smectites (*i.e.* they have layer charge between 0.375 and 0.425 e/h.u.c.) in the classification scheme of Emmerich *et al.* (2009).

Most natural dioctahedral smectites have compositions between montmorillonite and beidellite because pure end members are extremely rare. In natural smectites, beidellites have >50% of layer charge in tetrahedral sites and montmorillonites have >50% of layer charge in octahedral sites. Emmerich *et al.* (2009) proposed a nomenclature for the intermediate members between montmorillonite and beidellite. According to this nomenclature, the end members montmorillonite and beidellite have >90% octahedral and tetrahedral charge, respectively. Those smectites with 10–50% tetrahedral charge are classified as beidellitic montmorillonites, whereas those with 50–90% tetrahedral charge are classified as montmorillonitic beidellites. This nomenclature has not been adopted by AIPEA, however.

According to their octahedral composition and the layer charge, montmorillonites have been divided into Otay, Chambers, Tatatilla and Wyoming types (Schultz, 1969; Newman & Brown, 1987). The Otay, Chambers and Tatatilla montmorillonites are often described as Cheto-type montmorillonites (Güven, 1988) and so we have the well known division of montmorillonites into Wyoming and Cheto types (Grim & Kulbicki, 1961). However, these classification schemes are now considered obsolete and the term 'Al-rich montmorillonites' has replaced all these terms. The aforementioned terms are now used to denote the origin of the smectites. More specifically, the term Wyoming montmorillonite refers to smectites from Wyoming, USA, of Cretaceous age, and does not imply any particular crystal-chemical characteristics.

In trioctahedral smectite, hectorite is derived from talc by substitution of Li^{+} for octahedral Mg^{2+}. In saponite, the layer charge stems mainly from substitution of tetrahedral Si by Al^{3+}. Saponite is different from the other smectites as part of the

Table 1. Classification scheme for natural smectites (adapted from Güven, 1988).

	Dioctahedral smectites		Trioctahedral smectites	
Ratio between tetrahedral (z_t) and octahedral (z_o) charge	**Predominant octahedral cation(s)**	**Smectite species**	**Predominant octahedral cation(s)**	**Smectite species**
$z_o > z_t$	$Al^{3+}(R^{2+})^*$	Montmorillonite	Mg Mg(Li)*	Stevensite Hectorite
$z_o < z_t$	Al^{3+} Fe^{3+} Cr^{3+} V^{3+}	Beidellite Nontronite Volkonskoite Vanadium smectite	Mg Fe^{2+} Zn	Saponite Fe-saponite Sauconite

*Octahedral substitutions

negative tetrahedral charge is balanced by substitution of octahedral Mg^{2+} by trivalent cations, Al^{3+} or Fe^{3+}, *i.e.* the octahedral sheet often bears a positive charge. However, the tetrahedral charge, due to substitution of Si^{4+} by Al^{3+} is much greater and outbalances any possible positive octahedral charge. Stevensite is also a different trioctahedral smectite in the sense that the layer charge stems from a small deficiency in octahedral cations, not from substitutions in the structure (Brindley, 1984). Stevensite can be considered as a phase with a variable number of octahedral vacancies, which yield smectite-like and talc-like domains (Christidis & Mitsis, 2006). Finally, sauconite is a rare trioctahedral smectite with Zn^{2+} as the main octahedral cation and layer charge located mainly in tetrahedral sites. Recently, Christidis & Mitsis (2006) described the first occurrence of Ni-stevensite in nature. A classification scheme for the smectites is shown in Table 1. The classification scheme for layer charge applied to dioctahedral smectites (Christidis *et al.*, 2006) may also be applied to trioctaherdral smectites.

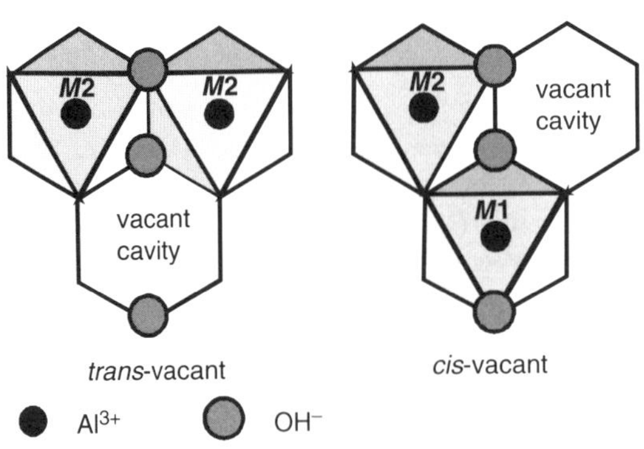

Fig. 4. Octahedral *cis*- and *trans*-configurations in smectites. In *trans*-configuration the hydroxyls are across the octahedral vacant site whereas in *cis*-configuration, the hydroxyls are adjacent to one vacant side of the octahedron.

As well as chemical variations, dioctahedral smectites display variability in terms of octahedral site occupancy. An octahedral vacant site may exist either in the position of *trans*-octahedron (*i.e.* the position usually called *M*1, where OH goups lie on the polyhedron diagonal) or of *cis*-octahedron (*i.e.* the position, usually called *M*2, showing OH groups along the same edge) (Fig. 4). This structural difference seems to control the dehydroxylation temperature of smectites, because *cis*-vacant smectites have greater dehydroxylation temperatures (650–700°C) compared to their *trans*-vacant counterparts (<600°C) (Drits *et al.*, 1998), and dictates the performance of smectites in foundry applications, which are characterized by high temperatures. Fe-rich smectites have *trans*-vacant configuration and therefore are unsuitable for applications in which high temperatures are involved.

3. Methods of determining layer charge

3.1. General characteristics

The experimental methods used to determine total layer charge and charge heterogeneity of smectite include: (1) micro-calorimetry (Talibudeen & Goulding, 1983), in which the heat released during determination of an exchange isotherm is related to different types of exchange sites and hence to charge heterogeneity; (2) measurement of the structural formula (structural formula method, SFM) using chemical or microbeam methods (Weaver & Pollard, 1973; Newman & Brown, 1987; Christidis & Dunham, 1993, 1997; Laird & Fleming, 2007), in which the oxide content of a purified smectite sample is measured and then converted into a structural formula; and (3) by X-ray diffraction (XRD) analysis after saturation with inorganic or organic cations (Tettenhorst & Johns 1966; Čičel & Machajdik, 1981; Stul & Mortier, 1974; Lagaly, 1981, 1994; Olis *et al.*, 1990; Christidis & Eberl, 2003).

The micro-calorimetric method is difficult to apply without specialized equipment and does not give the total layer charge. Rather it provides a qualitative estimate of the layer charge. The SFM, while potentially the most accurate for determining total layer charge, cannot determine layer-charge heterogeneity, and can be affected by the presence of impurities such as silica phases (quartz, opal-CT), other clay minerals, Fe oxides, or X-ray amorphous materials (Si, Al gels) in the clay fraction. These impurities can often be detected and quantified by XRD, or removed by chemical treatments (Jackson, 1985). Laird (1994) has given an account of the shortcomings of the SF method and the influence of impurities. The microbeam techniques for determining structural formulae (Electron probe microanalysis – EPMA; Analytical Transmission Electron Microscopy – AEM-TEM) have certain shortcomings because of the existence of fine-grained impurities such as Fe-, Mn-, Ti- or amorphous oxides in close association with clays, because of analytical problems such as the volatilization of light elements, or because of experimental constraints and errors (*cf.* Velde, 1984; Warren and Ransom, 1992; Christidis & Dunham, 1997; Christidis *et al.*, 2006). Recently it was demonstrated that with careful experimental setup, the microbeam techniques may provide reliable structural formulae and thus layer charge of smectites (Christidis, 2008).

X-ray diffraction, which provides the most convenient set of methods, has been used extensively to determine total layer charge and the charge heterogeneity of smectites. In these methods, the clay fraction is either saturated with an inorganic cation of low hydration energy, preferably potassium, and subsequently treated with ethylene glycol (Tettenhorst & Johns, 1966; Čičel & Machajdik, 1981; Christidis & Eberl, 2003), or is saturated with alkylammonium ions (see Lagaly, 1994, for a review). Both methods assume that smectite is composed of a random interstratification of smectite 2:1 layers with different charges. The alkylammonium method does not recognize end-member smectite layers having distinct layer charges, but instead yields a continuous distribution of layer charges. In contrast, the potassium saturation methods assume the existence of three types of end-member smectite layers, with different swelling capacities. The alkylammonium method is the most widely used technique despite significant shortcomings, which include difficulty in sample preparation and under- or overestimation of permanent layer charge of high- and low-charge smectites (Maes *et al.*, 1979; Laird *et al.*, 1989; also see Laird, 1994, for a review).

3.2. The structural formula method (SFM)

The SFM is based on the calculation of the structural formula from chemical analysis of monomineralic clay fractions or from EPMA. A thorough review of the methods used to calculate the structural formula of a mineral was given by Newman & Brown (1987). The most accurate calculations are obtained when the unit-cell dimensions and density are known, in addition to the chemical composition. In the case of smectites and other clay minerals, it is usually impossible to determine accurately the unit-cell dimensions, the particle density or the structural water content. In this case, the structural formulae are calculated from the chemical composition alone (Newman & Brown, 1987; Laird, 1994). The cations are assigned in tetrahedral, octahedral and exchangeable sites according to Pauling's rules (Pauling, 1960).

In EPMA or in analysis in which the smectite has not been rendered homoionic with an index cation other than Mg, allocation of Mg is usually a difficult task, because some of the Mg may be exchangeable. There is independent evidence that the octahedral occupancy of dioctahedral smectites is essentially very close to the ideal 2 atoms (Schultz, 1969; Vogt & Köster, 1978; Newman & Brown, 1987; Christidis, 2008). Therefore, in EPMA or in non-homoionic dioctahedral smectites it is relatively safe to assume that octahedral occupancy in smectites is 2 and that excess Mg can be assigned to interlayer sites. If the number of octahedral cations is <2 in dioctahedral smectites or <3 in trioctahedral smectite, then vacancies may be present. These vacancies are sources of negative charge equal to the valence of the cation that should be present. Magnesium is considered to be the usual 'missing' cation from octahedral sites. In the case of trioctahedral smectite, the existence of vacancies yields stevensite.

After determination of the structural formula, the calculation of layer charge is possible by subtraction of the tetrahedral and octahedral cation charges from the framework anionic charge (F). The total layer charge is the net anionic charge per formula unit (or per half unit cell) and should be equal but opposite in sign to the interlayer cation charge. The tetrahedral charge is the difference between the theoretical tetrahedral anionic charge (–16.000) and the sum of charges of the tetrahedral cations (Si^{4+} and Al^{3+}). Finally, the octahedral charge is the difference between the theoretical octahedral anionic charge (–6.000) and the sum of charges of the tetrahedral cations (Al^{4+}, Fe^{3+} and Mg^{2+}). Laird (1994) has shown that calculation of total layer charge by the

SF method is not affected by contamination by mineral impurities, but is very sensitive to contamination by index cations such as Ca. Therefore minerals such as calcite or gypsum must be removed before chemical analysis. In contrast, mineral impurities affect calculation of tetrahedral or octahedral charge (Laird, 1994).

3.3. The alkylammonium method (AMM)

The AMM for determining layer charge of clay minerals was developed by Lagaly & Weiss (1969). Since then several refinements have been proposed. Alkylammonium-saturated clays are prepared by exchange of the interlayer cations with straight-chain alkylammonium cations, having 6–18 carbon atoms, according to the following schematic reaction:

$$R\text{-NH}_4^+ + \text{Na-clay} = \text{Na}^+ + R\text{-NH}_4^+\text{-clay} \quad (1)$$

According to the number of carbon atoms (n_c) and the layer charge of the clay, alkylammonium cations may be arranged with monolayer-, bilayer-, pseudotrilayer- or paraffin-type configuration in the interlayer space (Fig. 5). The XRD basal spacings obtained for monolayer, bilayer, pseudotrilayer configurations are 13.6 Å, 17.6 Å and 22 Å. In the paraffin configuration, d spacing increases linearly with n_c.

The AMM assumes that in the monolayer and bilayer configurations the alkylammonium molecules are placed flat, parallel to the silicate layers (Lagaly, 1981, 1994). If the planar area necessary for each alkylammonium cation (A_c) exceeds the average planar area available for each charge site (A_e), the alkylammonium configuration shifts from monolayer to bilayer type. According to Lagaly (1981, 1994) $A_c = 5.7(n_c) + 14$ and $A_e = ab/2\sigma$, where a and b are the unit-cell dimensions ($ab \approx 46.5$ Å^2), σ is the layer charge and n_c is the critical carbon-chain length at which the monolayer to bilayer transition takes place. The layer charge can be calculated from the following equation:

$$\sigma = \frac{\frac{a \times b}{2}}{5.7(n_c) + 14} \approx \frac{23.25}{5.7(n_c) + 14} \quad (2)$$

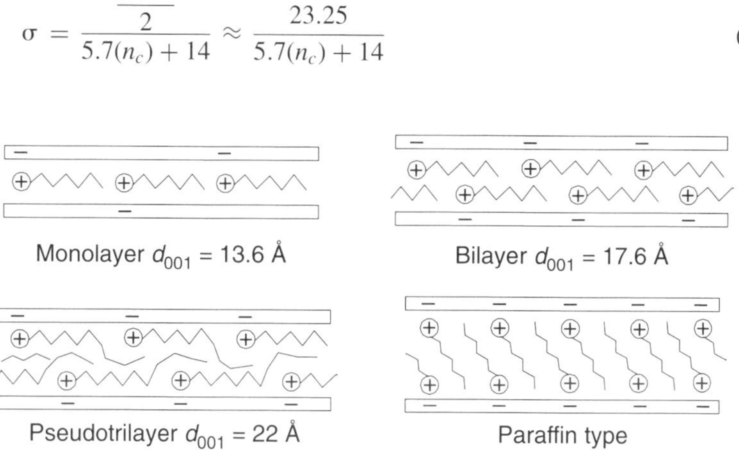

Fig. 5. Schematic configuration of the alkylammonium ions in the interlayer space of smectite.

One of the most intriguing aspects of the AMM is the estimation of the charge heterogeneity in 2:1 minerals. The approach is based on the observation that in most real smectite, the transition from monolayer to bilayer configuration does not occur through a single discrete step, but rather through a series of intermediate steps. This gradual transition has been attributed to random interstratification between lower- and higher-charge layers having monolayer and bilayer configurations, respectively (Lagaly & Weiss, 1969; Lagaly, 1981, 1994). After calculating the proportion of monolayers and bilayers in each sample, the layer charge is subdivided into different charge classes and the charge heterogeneity can thus be estimated. This approach has been challenged, however, as in the bilayer configuration, >45% of the interlayer space may be empty (Laird, 1994). In this case, islands of bilayers of alkylammonium cations may form in the interlayer space surrounded by monolayers, yielding monolayer to bilayer transitions in samples with homogeneous charge distribution (Laird, 1994). Such configuration will underestimate layer charge.

Another source for underestimation of layer charge with the AMM is the influence of the particle size. Those alkyl chains located close to the lateral edges of the clay particles may protrude out of the interlayer space. Hence, the average area occupied by an alkylammonium cation is less than predicted by the AMM. This effect is more pronounced in small particles and empirical corrections have been applied (Lagaly & Weiss, 1971; Stul & Mortier, 1974). In general the AMM provides lower layer charge in smectites compared to the SFM (Maes *et al.*, 1979; Laird *et al.*, 1989; Nikitaki, 2007). The discrepancy may be as much as 40% for high-charge smectites.

3.4. The K-saturation method (KSM)

The KSM was proposed by Christidis & Eberl (2003) based on previous observations (Tettenhorst & Johns, 1966; Čičel & Machajdik, 1981) that saturation of smectites with K yields randomly interstratified layers with spacings of ~17.1 Å, 13.5 Å and 9.98 Å. The *d*-spacing values of the three types of layers are assumed to be related to at least three different values for layer charge: with the lowest charge, 17.1 Å layers absorbing two glycol layers; the intermediately charged, 13.5 Å layers absorbing one glycol layer; the highest charge, 9.98 Å layers being completely collapsed.

Determination of layer charge is feasible with the *LayerCharge* program (Christidis & Eberl, 2003), where XRD traces of K-saturated, ethylene glycol-solvated smectites are compared with simulated XRD traces calculated for three-component interlayering (fully expandable 17.1 Å layers, partially expandable 13.5 Å layers and non-expandable 9.98 Å layers). The program finds the calculated pattern which minimizes the sum of square differences between: (1) the experimental and calculated peak positions for the first six basal reflections or (2) the whole-pattern fit of experimental and calculated intensities normalized to the intensity of the most intense peak. Then it assigns a proportion of 17.1 Å, 13.5 Å and 9.98 Å layers (*i.e.* charge heterogeneity) and calculates the layer charge. The program was calibrated using 28 well characterized smectites from various locations around the world. The layer charge of these smectites available in the literature had been determined using the SFM. Therefore the KSM has been calibrated to the SFM.

Using this approach, Christidis *et al.* (2006) proposed a classification scheme of smectites according to the layer charge based on their XRD characteristics (Fig. 3). The method is not affected by the presence of fine-grained, non-smectitic impurities that may be present in the clay fraction, such as kaolinite, quartz or feldspar. Application of this method has been used to show that 'high-charge smectite' is different from vermiculite and to explain why high-charge smectites convert more quickly to illite after subsequent wetting and drying cycles (Christidis & Eberl, 2003). Moreover, it has been used to interpret smectite:water properties like viscosity and free swelling (Christidis *et al.*, 2006).

4. Important smectite-water properties related to layer charge

Smectites develop important properties related to layer charge when they are dispersed in suspensions. These properties have significant industrial implications. Here we will consider the influence of layer charge on ion exchange, free swelling and viscosity.

4.1. Influence of layer charge on ion exchange

The CEC, which is linked to layer charge, is an important property with direct industrial implications. It involves the exchangeable cations located in the interlayer, which balance the layer charge of smectites. The theoretical aspects of ion exchange are not the topic of this contribution and the reader is encouraged to read more specialized works (*e.g.* Laudelout, 1987; Hellferich, 1992; McBride, 1994 among many others). Although the CEC is not strictly a smectite-water property, ion-exchange reactions are facilitated by the presence of an aqueous medium because the presence of an aqueous medium facilitates diffusion of cations to and from the surface of the exchanger. Moreover, the concentration of the aqueous solution controls ion exchange, even causing reversal of the exchange reaction (Hellferich, 1992).

Layer charge affects ion-exchange reactions both in univalent and bivalent cations (Maes & Cremers, 1977, 1978; Shainberg *et al.*, 1987). The selectivity for K over Ca or Na increases with increasing layer charge in natural smectites (Shainberg *et al.*, 1987) and in smectites with reduced layer charge (Maes & Cremers, 1977, 1978). Nevertheless, contradictory results have been obtained for natural smectites compared to smectites with artificially reduced layer charge in the Ca-for-Na exchange (Maes & Cremers, 1978; Shainberg *et al.*, 1987). The influence of the layer charge of smectites on ion exchange reactions has been attributed to steric effects linked to swelling (Verburg & Baveye, 1994; Laird *et al.*, 1995). Hence, in smectites with higher layer charge, swelling of layers is lower compared to smectites with lower layer charge, thus hindering the accessibility of incoming ions to the exchange sites. In this case the formation and preservation of quasicrystals (see below) is a controlling parameter of ion exchange. High-charge smectites tend to form thicker quasicrystals (Kleijn & Oster, 1982; Laird *et al.*, 1995; Laird, 2006).

The influence of layer charge on ion-exchange reactions has direct implications for industrial practice. Most of the commercial bentonites consist of Ca-smectites, which are not suitable in most industrial applications. The industrial practice involves the so-called alkali or soda activation, which includes addition of sodium carbonate either in dry form or in the form of aqueous spray. The target of this activation process is the production of homoionic Na-smectites. It has been recognized, however, that alkali-activation yields end products with unpredictable properties (Odom, 1984). This unpredictability must be looked for in the incomplete Na-for-Ca exchange in high-charge smectites, due to their greater selectivity for Ca, which is associated with the steric effects linked to swelling, mentioned above (Laird *et al.*, 1995). This means that the distribution of layer charge of smectites is a key parameter for evaluation of the bentonite deposits (Christidis & Huff, 2009).

4.2. Influence of layer charge on swelling and viscosity

Due to the low layer charge, the interlayer cations of smectites are fully hydrated leading to the remarkable property of swelling when hydrated. Swelling may also take place in the presence of certain polar organic compounds such as ethylene glycol or glycerol. These properties are of great importance both for industrial and for mineral-identification purposes. Hence, swelling in the presence of these polar organic solvents has been used as a routine technique for identification of the smectites and their differentiation from non-expandable minerals (MacEwan & Wilson, 1984). The different degree of hydration of cations of low hydration energy such as K^+ has also been used as a probe for estimating the magnitude of layer charge and the charge heterogeneity of smectites using the *LayerCharge* program (see previous section).

The hydration and swelling of smectites has been studied extensively for nearly 60 years. Smectites swell when added to water or polar organic solvents or when they are in contact with a vapour of the solvent having a large vapour pressure. Two main stages have been distinguished, 'crystalline swelling' and 'osmotic or double layer swelling' (Norrish 1954; MacEwan & Wilson, 1984; Güven, 1992) (Fig. 6). The influence of layer charge is different in the two stages of swelling.

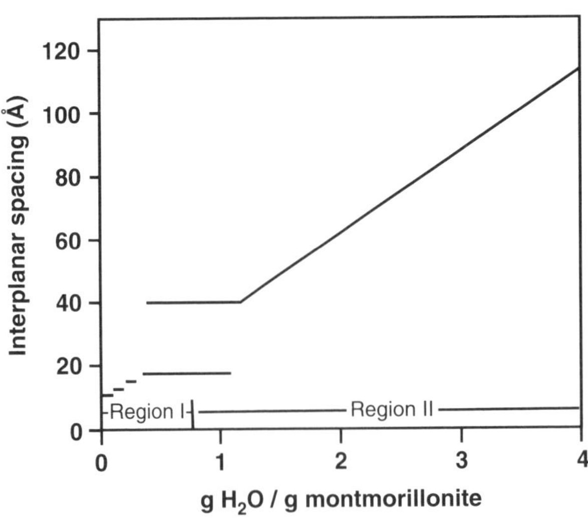

Fig. 6. Schematic representation of crystalline swelling (region I) and double layer (osmotic) swelling (region II) based on the amount of water adsorbed (modified from Norrish, 1954).

During crystalline swelling, the basal spacing of 2:1 phyllosilicates expands or collapses between 10 and 22 Å in a series of steps (10, 12.5, 15, 17.5 and 20 Å) due to the intercalation of 0–4 discrete layers of water molecules (Norrish, 1954). Deviation from the ideal *d* spacing reflects interstratification of layers with different numbers of hydrates and/or different packing arrangement of the interlayer water molecules (Laird, 2006). Although it is generally assumed that anions are excluded from the interlayer because they bear negative charge and are thus repelled by the negative smectite surface, there is experimental evidence for the presence of Cl^- anions in association with the exchangeable cations (Charlet & Tournassat, 2005; Ferrage *et al.*, 2005).

Crystalline swelling is controlled by a balance between attraction and repulsion forces (Norrish, 1954; Kittrick, 1969). Attraction depends on Coulomb effects and on van der Waals forces between adjacent layers (Laird, 2006). Coulomb attraction results between the negative surface charge, due to substitutions in the structure (*i.e.* essentially due to the layer charge), and the positive charge of the interlayer cations. Repulsive forces result from hydration of the interlayer cations and to a lesser degree from the hydration of the surface charge sites. The role of layer charge in swelling is well established from detailed experimental work in a controlled atmosphere (Slade *et al.*, 1991; Laird *et al.*, 1995; Laird, 1999). It has been found that the activity of water at which the transition in hydration and dehydration occurs, increases linearly with increasing layer charge of the smectites (Laird, 2006). The crystalline swelling displays hysteresis loops during progressive hydration and dehydration under controlled atmosphere.

In spite of the distinct role of layer charge in crystalline swelling, the influence on double layer swelling is unclear because experimental evidence is contradictory and the results inconclusive. Low (1980) found increasing water content with increasing layer charge for Na-smectites equilibrated at suction pressures of 0.025–3 atm. In contrast, Christidis *et al.* (2006) reported a decrease in the free swelling volume with increasing layer charge for Na-smectites with intermediate layer charge, with high-charge and low-charge smectites not following the overall trend (Fig. 7b). Moreover,

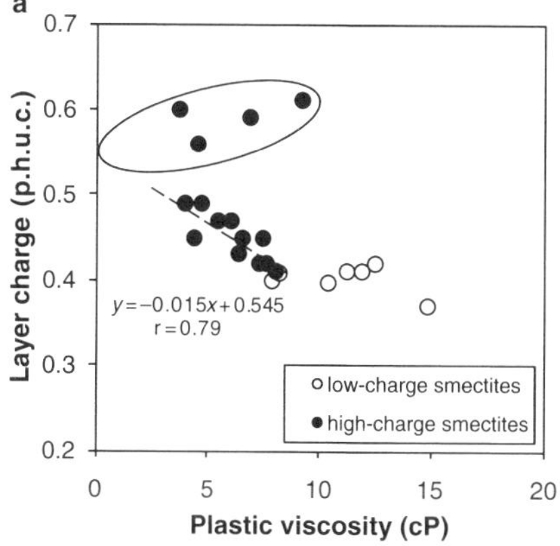

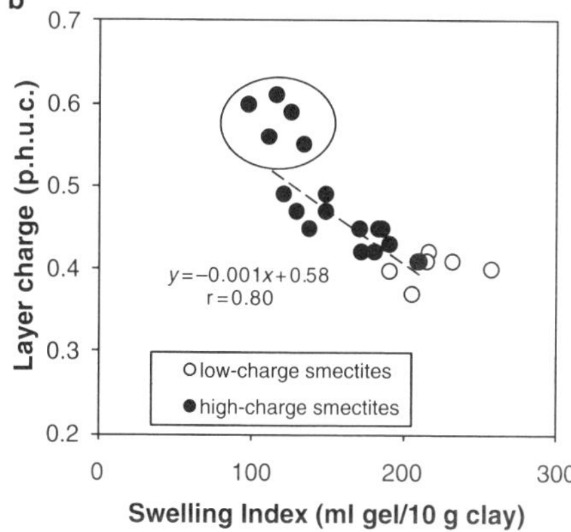

Fig. 7. Relationship between layer charge (e/h.u.c.) and: (**a**) plastic viscosity (cP), (**b**) free swelling volume (ml gel/10 g clay) (after Christidis *et al.*, 2006).

they observed a well expressed negative trend between the free swelling volume and the fraction of non-swelling high-charge layers. Foster (1953) and Viani *et al.* (1983), using different experimental setups (free swelling and interparticle separations at various pressure potentials respectively), did not observe any relationship between layer charge and double layer swelling of smectites. Theoretical calculations based on the fundamental Gouy-Chapman equation in the diffuse double layer theory, which describes the relationship between the surface-charge density, the surface potential, the valence of exchangeable cations and the electrolyte concentration, do not predict significant influence of the layer charge for distances >5 Å from the smectite surface (Laird, 2006). Nevertheless, it has been observed that low-charge smectites form more stable suspensions than their high-charge counterparts. For example, the low-charge Wyoming smectites develop much greater free swelling volumes than other smectites with greater layer charge. Finally, it was well established by the pioneering work of Norrish (1954) and verified by numerous subsequent workers that double layer swelling is controlled by the electrolyte concentration of smectite suspensions.

Smectites are often used in suspensions, which can display Newtonian, Bingham plastic, shear thickening (pseudoplastic) or shear thinning (dilatant) behaviour and may develop yield stress (Fig. 8a) (Luckham & Rossi, 1999). Moreover they can develop time-dependent phenomena such as thixotropic or rheopectic (antithixotropic) behaviour (Brandenburg & Lagaly, 1988; Lagaly, 1989; Luckham & Rossi, 1999) (Fig. 8b). The term 'thixotropy' refers to the ability of a suspension to form a gel upon standing and to become fluid when subjected to shear stress, *i.e.* under stirring or agitation. Due to the presence of electric charge, the colloidal properties of smectite crystallites depend on repulsive forces which cause the electroviscous effect and on attractive interactive forces, which lead to flocculation in addition to Brownian forces and to hydrodynamic forces (Adachi *et al.*, 1998). The available literature on the influence of layer charge on the rheological properties of smectites is very scarce. In their study on high- and low-charge smectites, Brandenburg & Lagaly (1988) observed

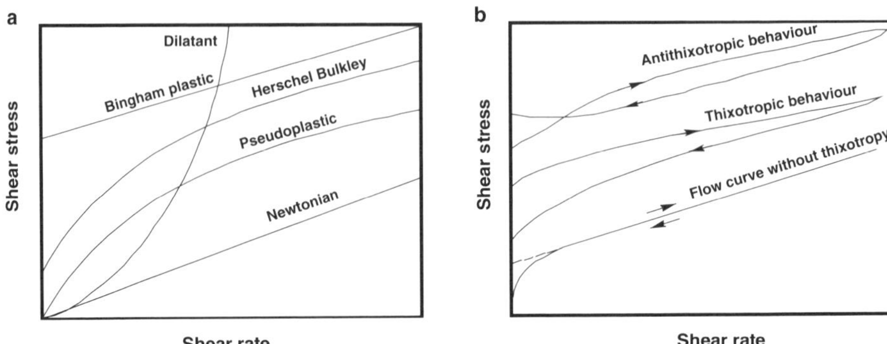

Fig. 8. (**a**) Different types of flow curves observed in clay suspensions. (**b**) Typical flow curves for concentrated thixotropic clay suspensions.

that the low-charge smectite developed much greater viscosity than its high-charge counterpart. Christidis *et al.* (2006) reported a linear decrease in viscosity with increasing layer charge for intermediate-charge smectites (Fig. 7a). However, the low-charge and the high-charge smectites did not follow the same trend.

The influence of layer charge on the double layer swelling and the viscosity must be looked for in the properties of quasicrystals (see below). The magnitude of layer charge controls the electrostatic component of interaction of the smectite particles, which can be important especially in high-charge Ca-smectites (Kleijn & Oster, 1982). Such smectites tend to form quasicrystals even at very low electrolyte concentrations, the thickness of which varies according to the layer charge. Although Ca-smectites can be rendered homoionic *via* Na-exchange, complete Na-for-Ca exchange is not expected in high-charge smectites, due to their greater selectivity for Ca, which is often associated with steric effects linked with swelling (Laird *et al.*, 1995). Even if the Na-for-Ca exchange is complete, the high-charge smectites will tend to form large quasicrystals, which will develop considerably smaller diffuse double layers and thus will bind fewer water molecules compared to deflocculated smectite suspensions. Therefore, the secondary electroviscous effect will not be important in these smectites. The quasicrystals tend to form band-like (ribbon) structures (Brandenburg & Lagaly, 1988; Tombacz & Szekeres, 2004), with weak links between them due to the large floc size, which break upon shearing even at low shear rates. Thus, high-charge smectites are expected to develop inferior colloidal properties approaching the behaviour of vermiculites and this has been observed in practice by Christidis *et al.* (2006).

5. Fundamental particles, quasicrystals, aggregates and the concept of fundamental-particle charge

5.1. Fundamental particles, quasicrystals and aggregates

In the 1980s, Nadeau and coworkers proposed the concept of fundamental particles (Nadeau *et al.*, 1984a). According to this model, clay minerals occur as individual sheet- or lath-shaped particles with thickness approximately integral multiples of 10 Å, which are called 'fundamental particles'. The fundamental particles are individual or free particles that yield single-crystal electron diffraction patterns from the *ab* plane (Nadeau *et al.*, 1984a, 1984b). In the case of pure smectite, each fundamental particle is 10 Å thick, it has variable basal area and carries electric charge balanced by the exchangeable cations. Illite particles are at least 20 Å thick. Coherently diffracting stacks of illite crystals and their interfaces yield an XRD pattern *via* 'interparticle diffraction' (Nadeau *et al.*, 1984b). The concept of fundamental particles was supported by the fact that mixtures of samples consisting of pure smectite particles with samples consisting of R1 mixed-layer illite-smectite particles yielded XRD patterns similar to R0 mixed-layer illite-smectite. In contrast, mixtures of samples consisting of pure smectite particles with samples consisting of illite particles showed no indication of interstratification (Nadeau *et al.* 1984a, 1984b, 1984c). Nadeau *et al.* (1984b)

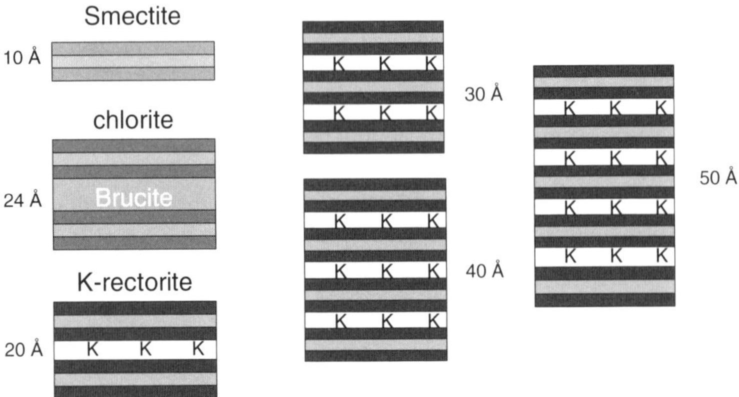

Fig. 9. Clay minerals described by the fundamental particle model. Smectite particles (green) are 10 Å thick, chlorite particles (blue-yellow) are 24 Å thick and illite particles (red-yellow) are 20–50 Å thick (after Nadeau *et al.*, 1984b). Readers of the paper version of this chapter may wish to download a colour version of this figure from www.minersoc.org/emu-notes/emu-11/11-6-colour.pdf.

distinguished three main types of fundamental particles, 10 Å thick smectite particles, 24 Å thick chlorite particles and 20–50 Å thick illite particles (Fig. 9).

The fundamental-particle approach assumes that in the illite and smectite particles only the external surfaces may adsorb water or organic molecules and therefore behave as real smectitic ones (Nadeau *et al.*, 1984a, Środoń *et al.*, 1992). Also, the chlorite particles may adsorb water forming corrensite particles (Nadeau *et al.*, 1984a). Hence in Figure 9, the smectite particles have 100% expandable surfaces, whereas the 20 Å illite particles have 50% expandable (*i.e.* smectitic) surfaces yielding rectorite. The chlorite particles may also have up to 50% expandable surfaces yielding corrensite. In this contribution, only those fundamental particles 10 Å thick with 100% expandable surfaces (*i.e. real* smectite particles) have been considered for the formulation of the fundamental-particle charge model.

In real Na-smectite suspensions, however, particles are not isolated as free fundamental particles but form 'quasicrystals', even in very dilute suspensions (Fig. 10). The term 'quasicrystal' was introduced by Quirk & Aylmore (1971) to describe small stacks made of smectite layers (Fig. 11b). Although the term quasicrystal is often used interchangeably with the term 'tactoid', the two are different because (1) the interlayer separation is much larger in tactoids whereas it is limited to a few well organized water layers in quasicrystals and (2) in quasicrystals the smectite layers are held together by attractive electrostatic forces whereas the tactoids are maintained by long-range repulsive forces due to overlapping double layers of smectite (Güven, 1992). Quasicrystals may break up into smaller ones by hydrodynamic forces and can be almost completely delaminated in dilute suspensions, when saturated with Na or Li. Conversely, they may combine and form larger quasicrystals if their kinetic energy suffices to overcome the double layer repulsion (Laird, 2006). In the dry state, quasicrystals usually combine to form composite particles or aggregates (Fig. 11a).

5.2. The concept of fundamental-particle charge

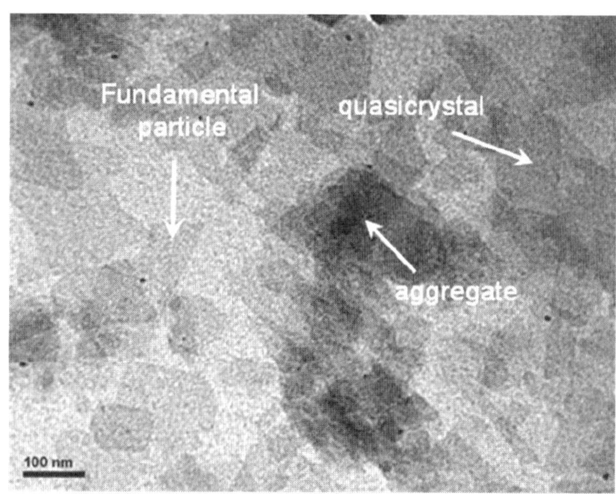

Fig. 10. TEM image of smectite particles precipitated on copper grids from very dilute suspensions (<0.001% w/v).

Although the layer charge is an important property of smectites, which is expected to influence significant smectite:water properties such as crystalline swelling or viscosity (Brandenburg & Lagaly, 1988; Slade *et al.*, 1991; Laird, 2006; Christidis *et al.*, 2006), it is still a property of the unit cell, *i.e.* it is not linked directly to the smectite particles. In contrast, the colloidal properties are macroscopic and refer to particles consisting of an unknown number of unit cells (Fig. 12). The number of the unit cells per particle depends on the smectite particle size. Recently Pratikakis *et al.* (2010) proposed a model for calculation of the charge of fundamental particles of smectites and introduced the concept of 'fundamental particle charge (*Fpc*)'. Moreover, they examined how this charge may affect important properties of smectite, such as viscosity and swelling. The importance of the fundamental particle charge, which also is a macroscopic property, has been evaluated and the results are very promising.

The proposed model of *Fpc* takes into account the basal area and the surface charge density of the smectite particles. The surface charge density of 2:1 clay minerals expressed in Coulomb/m^2 (C/m^2) is (Lagaly 1994):

$$\sigma_0 = 1.602 \times 10^{-19}(x + y)/a \times b \tag{3}$$

where σ_0 is the surface charge density in C/m^2, 1.602×10^{-19} is the elementary charge of one electron in Coulombs, a and b are the dimensions of the unit cell (in m^{-10}, *i.e.* Å)

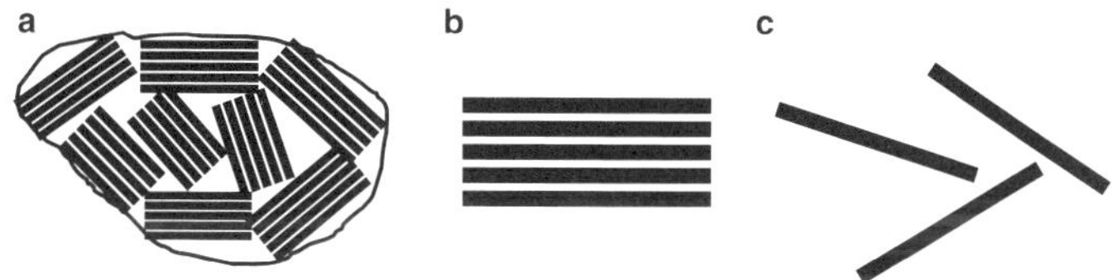

Fig. 11. Schematic representation of smectite: (**a**) particle or aggregate, (**b**) quasicrystal, and (**c**) fundamental particles. The outline of the smectite particle (or aggregate) denotes a smectite grain.

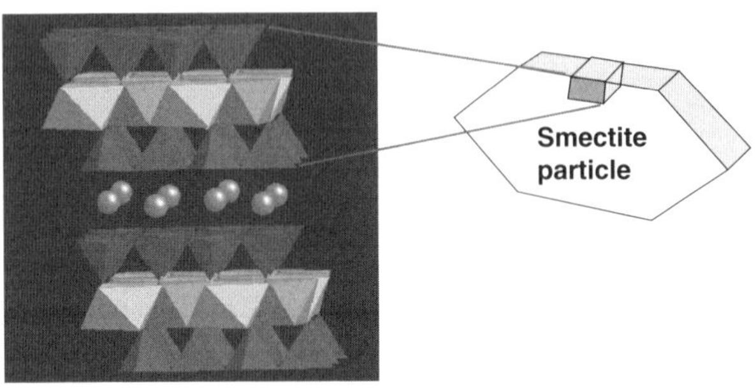

Fig. 12. Schematic relationship between unit cell and a fundamental smectite particle. Readers of the paper version of this chapter may wish to download a colour version of this figure from www.minersoc.org/emu-notes/emu-11/11-6-colour.pdf.

and $(x + y)$ is the charge in e/h.u.c. (y is the tetrahedral charge and x is the octahedral charge).

Smectite particles consist of quasicrystals, which can break up into single layers 1 nm thick with variable basal area, when the exchangeable cation is Na or Li (Fig. 11c). These single layers are the smectite fundamental particles (Nadeau *et al.*, 1984b). Fundamental particles do not form when K, Ca or Mg are exchangeable cations, as smectites saturated with bivalent cations tend to form quasicrystals (Fig. 11b) in aqueous suspensions (Kleijn & Oster, 1982; Güven, 1992; Laird, 2006).

A charge of $1.602 \times 10^{-19}(x + y)$ corresponds to each $a \times b$ area (in Å^2). The fundamental particle carries charge equal to $2 \times (A/\alpha b) \times 1.602 \times 10^{-19}(x + y)$. Consequently, the *Fpc* will be (Pratikakis *et al.*, 2010):

$$Fpc = 2 \times A \times \sigma_0 \tag{4}$$

where A is the is the basal area of the fundamental particle. The *Fpc* is the sum of the layer charges in Coulomb of all the unit-cell charges in the fundamental particle (Pratikakis *et al.*, 2010). It contains additional information about the electric charge of the smectite particles compared to the layer charge because it combines the data of the surface-charge density and the area of each of the two basal surfaces (size of the fundamental particle). The *Fpc* is a new concept for smectites and it is extremely useful in understanding the behaviour of smectite-water systems and interpreting smectite-water properties. This is because smectites with small layer charge may well form particles with large *Fpc* values and *vice versa*.

Acknowledgements

This work was funded by the Greek Secretariat of Research and Development, S&B Industrial Minerals SA, the Fulbright Institute, the Greek Ministry of Education and the Technical University of Crete. I am indebted to P. Makri and A. Pratikakis for their assistance.

References

Adachi, Y., Nakaishi, K. & Tamaki, M. (1998) Viscosity of a dilute suspension of sodium montmorillonite in an electrostatically stable condition. *Journal of Colloid and Interface Science*, **198**, 100–105.

Bailey, S.W. (1984) Structures of layer silicate. In: *Crystal Structures of Clay Minerals and their X-ray Identification* (G.W. Brindley & G. Brown, editors). Monograph **5**, Mineralogical Society, London, pp. 1–113.

Brandenburg, U. & Lagaly, G. (1988) Rheological properties of sodium montmorillonite dispersions. *Applied Clay Science*, **3**, 263–279.

Brigatti, M.F. (1983) Relationships between composition and structure in Fe-rich smectites. *Clay Minerals*, **18**, 177–186.

Brigatti, M.F., Galán, E. & Theng, B.K.G. (2006) Structures and mineralogy of clay minerals. In: *Handbook of Clay Science* (F. Bergaya, B.K.G. Theng & G. Lagaly, editors). Developments in Clay Science, **1**, Elsevier, Amsterdam, pp. 19–86.

Brigatti, M.F., Malferrari, D., Laurora, A. & Elmi, C. (2011) Structure and mineralogy of layer silicates: Recent perspectives and new trends. In: *Layered Mineral Structures and their Application in Advanced Technologies* (M.F. Brigatti & A. Mottana, editors), EMU notes in Mineralogy, **11**, European Mineralogical Union and the Mineralogical Society of Great Britain & Ireland, pp. 1–71.

Brindley, G.W. (1984) Order-disorder in clay mineral structures. In: *Crystal Structures of Clay Minerals and their X-ray Identification* (G.W. Brindley & G. Brown, editors). Monograph **5**, Mineralogical Society, London, pp. 125–189.

Charlet, L. & Tournassat, C. (2005) Fe(II)-Na(I)-Ca(II) cation exchange on montmorillonite in chloride medium: Evidence for preferential clay adsorption of chloride – Metal ion pairs in seawater. *Aquatic Geochemistry*, **11**, 115–137.

Christidis, G.E. (2008) Validity of the structural formula method for layer charge determination of smectites: A re-evaluation of published data. *Applied Clay Science*, **42**, 1–7.

Christidis, G. & Dunham, A.C. (1993) Compositional variations in smectites derived from intermediate volcanic rocks. A case study from Milos Island, Greece. *Clay Minerals*, **28**, 255–273.

Christidis, G. & Dunham, A.C. (1997) Compositional variations in smectites. Part II: Alteration of acidic precursors. A case study from Milos Island, Greece. *Clay Minerals*, **32**, 255–273.

Christidis, G.E. & Eberl, D.D. (2003) Determination of layer charge characteristics of smectites. *Clays and Clay Minerals*, **51**, 644–655.

Christidis, G.E. & Huff, W.D. (2009) Geological aspects and genesis of bentonites. *Elements*, **5**, 93–98.

Christidis, G.E. & Mitsis, I. (2006) A new Ni-rich stevensite from the ophiolite complex of Othrys, central Greece. *Clays and Clay Minerals*, **54**, 653–666.

Christidis, G.E., Blum, A.E. & Eberl, D.D. (2006) Influence of layer charge and charge heterogeneity of smectites on the flow behavior and swelling of bentonites. *Applied Clay Science*, **34**, 125–138.

Čičel, V. & Machajdik, D. (1981) Potassium- and ammonium-treated montmorillonites. I. Interstratified structures with ethylene glycol and water. *Clays and Clay Minerals*, **29**, 40–46.

Drits, V.A., Lindgreen, H., Salyn, A.L., Ylagan, R. & McCarty, D.K. (1998) Semi quantitative determination of *trans*-vacant and *cis*-vacant 2:1 layers in illites and illite-smectites by thermal analysis and X-ray diffraction. *American Mineralogist*, **83**, 1188–1198.

Emmerich, K., Wolters, F., Kahr, G. & Lagaly, G. (2009) Clay profiling: The classification of montmorillonites. *Clays and Clay Minerals*, **57**, 104–114.

Ferrage, E., Tournassat, C., Rinnert, E., Charlet, L. & Lanson, B. (2005) Experimental evidence for Ca-Chloride ion pairs in the interlayer of montmorillonite. An XRD profile modeling approach. *Clays and Clay Minerals*, **53**, 348–360.

Foord, E.E., Starkey, H.C., Taggart Jr, J.E. & Shawe, D.R. (1987) Reassessment of the volkonskoite-chromian smectite nomenclature problem. *Clays and Clay Minerals*, **35**, 139–149.

Foster, M.D. (1953) Geochemical studies of clay minerals: II Relation between ionic substitution and swelling in montmorillonites. *American Mineralogist*, **38**, 994–1006.

Gillespie, T. (1960) An extension of the Goodeve's impulse theory of viscosity to pseudoplastic systems. *Journal of Colloid Science*, **15**, 219–231.

Grim, R.E. & Kulbicki, G. (1961) Montmorillonite: High temperature reactions and classification. *American Mineralogist*, **46**, 1329–1369.

Guggenheim, S. (2011) An overview of order/disorder in hydrous phyllosilicates. In: *Layered Mineral Structures and their Application in Advanced Technologies* (M.F. Brigatti & A. Mottana, editors). EMU Notes in Mineralogy, **11**, European Mineralogical Union and the Mineralogical Society of Great Britain & Ireland, pp. 73–121.

Guggenheim, S., Adams, J.M., Bain, D.C., Bergaya, F., Brigatti, M.F., Drits, V.A., Formoso, M.L.L., Galán, E., Kogure, T. & Stanjek, H. (2006) Summary of recommendations of nomenclature committees relevant to clay mineralogy: Report of the Association Internationale pour l'Etude des Argiles (AIPEA) Nomenclature Committee for 2006. *Clay Minerals*, **41**, 863–877.

Güven, N. (1988) Smectite. In: *Hydrous Phyllosilicates* (S.W. Bailey, editor). Reviews in Mineralogy, **19**. Mineralogical Society of America, Washington, D.C., pp. 497–559.

Güven, N. (1992) Rheological aspects of aqueous smectite suspensions. In: *Clay-Water Interface and its Rheological Implications* (N. Güven & R.M. Pollastro, editors). CMS Workshop Lectures, **4**, The Clay Minerals Society, Aurora, Colorado, USA, pp. 81–125.

Hellferich, F. (1992) *Ion Exchange*. Dover Publications Inc., New York.

Jackson, M.L. (1985) *Soil Chemical Analysis – Advanced Course*, 2nd edition. Published by the author, Madison, Wisconsin, USA, 895 pp.

Kasperski, K.L., Hepler, C.T. & Hepler, L.G., (1986) Viscosities of dilute suspensions of montmorillonite and kaolinite clays. *Canadian Journal of Chemistry*, **64**, 1919–1924.

Kittrick, J.A. (1969) Interlayer forces in montmorillonite and vermiculite. *Soil Science Society of America Journal*, **44**, 667–676.

Kleijn, W.B. & Oster, J.D. (1982) A model of clay swelling and and tactoid formation. *Clays and Clay Minerals*, **30**, 383–390.

Lagaly, G. (1981) Characterization of clays by organic compounds. *Clay Minerals*, **16**, 1–21.

Lagaly, G. (1989) Principles of flow of kaolin and bentonite dispersions. *Applied Clay Science*, **4**, 105–123.

Lagaly, G. (1994) Layer charge determination by alkylammonium ions. In: *Layer Charge Characteristics of 2:1 Silicate Clay Minerals* (A.R. Mermut, editor). CMS Workshop Lectures, **6**. The Clay Minerals Society, Boulder, Colorado, USA, pp. 1–46.

Lagaly, G. & Weiss A. (1969) Determination of the layer charge in mica-type layer silicates. In: *Proceedings of the International Clay Conference, Tokyo, 1* (L. Heller, editor). Israel University Press, Jerusalem, pp. 61–80.

Lagaly, G. & Weiss A. (1971) Anordnung und Orientierung kationischer Tenside auf Silicatoberflachen, IV: Anordnung von n-Alkylammoniumionen bei niedrig geladenen Schichtsilicaten. *Kolloid-Zeitschrift und Zeitschrift für Polymere*, **243**, 48–55.

Laird, D.A. (1994) Evaluation of structural formulae and alkylammonium methods of determining layer charge. In: *Layer Charge Characteristics of 2:1 Silicate Clay Minerals* (A.R. Mermut, editor). CMS Workshop Lectures, **6**. The Clay Minerals Society, Boulder, Colorado, USA, pp. 80–103.

Laird, D.A. (1999) Layer charge influences on the hydration of expandable 2:1 phyllosolicates. *Clays and Clay Minerals*, **47**, 630–636.

Laird, D.A. (2006) Influence of layer charge on swelling of smectites. *Applied Clay Science*, **34**, 74–87.

Laird, D.A. & Fleming, P. (2007) Analysis of layer charge, cation and anion exchange capacities and synthesis of reduced charge clays. In: *Methods of Soil Analysis. Part 5* (A. Ulery & R. Drees, editors). Mineralogical Methods, SSSA Book Series No. **5**. Soil Science Society of America, Madison, Wisconsin, USA.

Laird, D.A., Scott, A.D. & Fenton, T.E. (1989) Evaluation of the alkylammonium method of determining layer charge. *Clays and Clay Minerals*, **37**, 41–46.

Laird, D.A., Barriuso, E., Dowdy, R.H. & Koskinen, W.C. (1992) Adsorption of attrazine in smectites. *Soil Science Society of America Journal*, **56**, 62–67.

Laird, D.A., Shang, C. & Thomson, M.L. (1995) Hysteresis in crystalline swelling of smectites. *Journal of Colloid and Interface Science*, **171**, 240–245.

Laudelout, H. (1987) Cation exchange equilibria in clays. In: *Chemistry of Clays and Clay Minerals* (A.C.D. Newman, editor). Monograph **6**, Mineralogical Society, London, pp. 225–236.

Low, P.F. (1980) The swelling of clay: II. Montmorillonites. *Soil Science Society of America Journal*, **44**, 667–676.

Luckham, P.F. & Rosi, S. (1999) The colloidal and rheological properties of bentonite suspensions. *Advances in Colloid and Interface Science*, **82**, 43–92.

MacEwan, D.A.C. & Wilson, M.J. (1984) Interlayer and intercalation complexes of clay minerals. In: *Crystal Structures of Clay Minerals and their X-ray Identification* (G.W. Brindley & G. Brown, editors). Monograph **5**, Mineralogical Society, London, pp. 197–248.

Maes, A. & Cremers, A. (1977) Charge density effects in ion exchange. Part 1. Heterovalent exchange equilibria. *Faraday Transactions of the Royal Chemical Society*, **73**, 1807–1814.

Maes, A. & Cremers, A. (1978) Charge density effects in ion exchange. Part 2. Homovalent exchange equilibria. *Faraday Transactions of the Royal Chemical Society*, **74**, 1234–1241.

Maes, A., Stul, M.S. & Cremers, A. (1979) Layers charge-cation-exchange capacity relationships in montmorillonite. *Clays and Clay Minerals*, **27**, 387–392.

McBride, M.B. (1994) *Environmental Chemistry of Soils*. Oxford University Press, New York.

Moore, D.M. & Reynolds, R.C., Jr. (1997) *X-ray Diffraction and the Identification and Analysis of Clay Minerals*, 2nd edition, Oxford University Press, New York.

Nadeau, P.H., Wilson, M.J., McHardy, W.J. & Tait, J.M. (1984a) Interstratified clays as fundamental particles. *Science*, **225**, 923–925.

Nadeau, P.H., Wilson, M.J., McHardy, W.J. & Tait, J.M. (1984b) Interparticle diffraction: A new concept for interstratified clays. *Clay Minerals*, **19**, 757–769.

Nadeau, P.H., Tait, J.M., McHardy, W.J. & Wilson, M.J. (1984c) Interstratified XRD characteristics of physical mixtures of elementary clay particles. *Clay Minerals*, **19**, 67–76.

Newman, A.C.D. & Brown, G. (1987) The chemical constitution of clays. In: *Chemistry of Clays and Clay Minerals* (A.C.D Newman, editor). Monograph **6**, Mineralogical Society, London, pp. 1–128.

Nikitaki, O. (2007) *Influence of Layer Charge and Charge Heterogeneity of Smectite Layers in the Synthesis of Organophilic Bentonites.* Unpubl. M.Sc. thesis, Techincal University of Crete, Chania, Greece (in Greek).

Norrish, K. (1954) The swelling of montmorillonite. *Discussions of the Faraday Society*, **18**, 120–134.

Odom, I.E. (1984) Smectite clay minerals: properties and uses. *Philosophical Transactions of the Royal Society of London*, **A311**, 391–409.

Olis, A.C., Malla, P.B. & Douglas, L.A. (1990) The rapid estimation of the layer charges of 2:1 expanding clays from a single alkylammonium ion expansion. *Clay Minerals*, **25**, 39–50.

Pauling, L. (1960) *The Nature of the Chemical Bond.* Cornell University Press, Ithaca, New York.

Pratikakis, A., Christidis, G.E., Villieras, F. & Michot, L. (2010) Fundamental particle charge and its significance. SEA-CSSJ-CMS trilateral meeting on clays, Sevilla, Spain, 8–10 June 2010, book of abstracts, pp. 154–155.

Quirk, J.P. & Aylmore, L.A.G. (1971) Domains and quasi-crystalline regions in clay systems. *Soil Science Society of America Journal*, **35**, 652–654.

Rand, B., Pekenc, R., Goodwin, J.W. & Smith, R.B. (1980) Investigation into the existence of edge-face coagulated structures in Na-montmorillonite suspensions. *Journal of the Chemical Society Faraday Transactions*, **76**, 225–235.

Schultz, L.G. (1969) Lithium and potassium adsorption, dehydroxylation temperature and structural water content of aluminous smectites. *Clays and Clay Minerals*, **17**, 115–149.

Shainberg, I., Alperovich, N.I. & Keren, R. (1987) Charge density and Na-K-Ca exchange on smectites. *Clays and Clay Minerals*, **35**, 68–73.

Slade, P.G., Quirk, J.P. & Norrish, K. (1991) Crystalline swelling of smectite samples in concentrated NaCl solutions in relation to layer charge. *Clays and Clay Minerals*, **39**, 234–238.

Środoń, J., Elsass, F., McHardy, W.J. & Morgan, D.J. (1992) Chemistry of illite-smectite inferred from TEM measurements of fundamental particles. *Clay Minerals*, **27**, 137–158.

Stul, M.S. & Mortier, W.J. (1974) The heterogeneity of the charge density in montmorillonites. *Clays and Clay Minerals*, **22**, 391–396.

Talibudeen, O. & Goulding, K.W.T. (1983) Charge heterogeneity in smectites. *Clays and Clay Minerals*, **31**, 37–42.

Tettenhorst, R. & Johns, W.D. (1966) Interstratification in montmorillonite. *Clays and Clay Minerals*, **15**, 85–93.

Tombacz, E. & Szekeres, M. (2004) Colloidal behaviour of aqueous montmorillonite suspensions: The specific role of pH in the presence of indifferent electrolytes. *Applied Clay Science*, **27**, 75–94.

Verburg, K. & Baveye, P. (1994) Hysteresis in the binary exchange of cations on 2:1 clay minerals: A critical review. *Clays and Clay Minerals*, **42**, 207–220.

Viani, B.E., Low, P.F. & Roth, C.B. (1983) Direct measurement of the relation between interlayer force and interlayer distance in the swelling of montmorillonite. *Journal of Colloid and Interface Science*, **96**, 229–239.

Vogt, K. & Köster, H.M. (1978) Zur Mineralogie Kristallchemie und Geochemie einiger Montmorillonite aus Bentoniten. *Clay Minerals*, **13**, 25–43.

Velde, B. (1984) Electron microprobe analysis of clay minerals. *Clay Minerals*, **19**, 243–247.

Warren, E.A. & Ransom, B. (1992) The influence of analytical error upon the interpretation of chemical variations in clay minerals. *Clay Minerals*, **27**, 193–209.

Weaver, C.E. & Pollard, L.D. (1973) *The Chemistry of Clay Minerals*. Elsevier, Amsterdam, pp. 55–77.

EMU Notes in Mineralogy, Vol. 11 (2011), Chapter 7, 259–284

Intercalation processes of layered minerals

FAÏZA BERGAYA[1,*] and GERHARD LAGALY[2]

[1]*Centre de Recherche sur la Matière Divisée (CRMD) UMR 6619 CNRS, 1b, Rue de la Férollerie F-45071 Orleans CEDEX 02, France*
e-mail: f.bergaya@cnrs-orleans.fr
[2]*Institut für Anorganische Chemie, Christian-Albrechts-Universität zu Kiel, D-24118 Kiel, Germany*

Understanding clay mineral intercalation is the aim of this chapter. Intercalation, which corresponds to a reversible inclusion of different species between two layers, depends on the geometrical, physical and chemical characteristics of each type of clay mineral.

In most phyllosilicates, the interlayer space is occupied by cations which are more or less hydrated. These water molecules which separate two successive layers enable further intercalation reactions by physical adsorption or by chemical grafting of a great variety of species. It is noteworthy that intercalation of inorganic or organic species by ion exchange of these interlayer cations is often the first step in intercalation and is of primary importance in much basic and applied research of the modified clay minerals obtained. This concerns organo-clay minerals (OC), pillared clays (PILC) and clay mineral-polymer nanocomposites (CPN). In non-swelling clay minerals where, generally, the interlayer space is empty, the layers are held by van der Waals interactions or by hydrogen bonds between the stacked layers. In this case, different mechanisms of intercalation can occur and are described.

X-ray diffraction (XRD) and adsorption methods, which induce swelling, and how they are used to confirm intercalation are described below. The meanings of the commonly used terms 'intercalated' and 'exfoliated structures' in CPN literature and the meanings of the confusing terms, 'exfoliation' and 'delamination', are also discussed.

When the reversibility of the intercalation can be controlled, applications become possible. In conclusion, this chapter aims to draw attention to the importance of the geometrical arrangement of the different clay mineral units in predicting their properties. The unique intercalation property of clay minerals allows us to consider possible further development in aid of our environment, our health and our wellbeing.

1. Introduction

Intercalation is an important process and a huge number of papers on layered-minerals properties has led to many possible applications. In this chapter, focus is placed on layered clay minerals although the variety of layered solids extends beyond this already large group. Layered solids include layered minerals such as the phyllosilicates consisting of only tetrahedral silica sheets (kanemite, magadiite and kenyaite), several other compounds such as titanates, niobates, phosphates, manganates, layered zeolites, layered double hydroxides (LDH) and double hydroxyl salts, and also graphite. The attention paid to clay-minerals studies in the existing literature could be extended to these other layered minerals in order to understand their intercalation process. This

*Corresponding author

DOI: 10.1180/EMU-notes.11.7

chapter is not exhaustive in terms of clay intercalation. The fine details of intercalation depend on the physico-chemical characteristics of each type of clay; properties of clay minerals other than swelling, induced by intercalation, such as the filling interstratifications are not discussed here.

1.1. Definitions

Ther term 'intercalation' refers to the reversible inclusion of species (atoms, ions or molecules) between two other species. In biology, one interesting example concerns DNA intercalation compounds (Richards & Rodgers, 2007). Molecules, called ligands in this particular case, of appropriate size can fit between structural elements of the base pairs of the DNA skeleton which consequently presents a local change of structure. These intercalated molecules are used in chemotherapy to inhibit DNA replication in rapidly growing cancer cells.

In materials science, intercalation refers to the insertion of guest molecules between two layers of a host solid (carbon sheets, clay mineral layers, *etc.*). Intercalated K^+ in graphite is one of the best known examples of intercalation compounds. In this case, the K^+ atoms occupy the position between the graphite layers (interlayers). The composition is frequently associated with the stage, with $\frac{1}{1}$, $\frac{1}{2}$ or $\frac{1}{3}$ of the interlayers occupied corresponding to stage 1, 2 or 3 compounds (Boehm *et al.*, 1994). In layered clay minerals, the layers can be in direct contact, held together by van der Waals interactions as in pyrophyllite or talc. In kaolinite, van der Waals forces and hydrogen bonds between the stacked layers co-exist. However, in most clay minerals, as presented by Brigatti *et al.* (2011, this volume) and by Guggenheim (2011, this volume), the interlayer space is not empty but occupied by cations (or anions in LDH) often together with water molecules which separate the two successive layers and enable further intercalation reactions of inorganic or organic species (Christidis, 2011, this volume). In this chapter, the reader will find the term 'clay minerals' as well as 'clay materials' used for the layered clay minerals because they are unique solids both as minerals for mineralogists and acting as raw materials in several sectors (civil engineering, ceramics, environmental protection, *etc.*) These materials can be modified in many ways, limited only by our imagination. Intercalation plays the most important role in modifying clay minerals.

1.2. Chapter outline

Intercalation is often the first step to other large fields of study, *e.g.* organo-clay minerals (OC), pillared clays (PILC), or nanocomposites. In clay mineral-polymer nanocomposites (CPN) the clay minerals are dispersed in the polymer matrix. In the following sections, after a brief structural description of the clay minerals (section 2), attention is paid first to naturally occurring non-swelling clay minerals (section 3). The different mechanisms of intercalation in swelling clay minerals are reviewed with particular mention of co-intercalation of two products from a binary mixture (section 4). Adsorption methods which are used to prove this intercalation are presented in section 5. Some specific topics, derived from intercalation, which have attracted broader attention recently, *e.g.* PILC and CPN, are summarized in sections 6 and 7. The meanings of the terms 'intercalated and exfoliated structures' as often used in CPN literature, and of the confusing terms, 'exfoliation' and 'delamination', are discussed in section 8. Finally, a short survey is given on applications requiring reversible or irreversible intercalation (section 9) with a brief conclusion on this unique intercalation property of clay minerals and its possible further development.

The aim of the following section is to draw attention to the importance of the geometrical arrangement of the different clay mineral units in predicting their properties. The spatial organization of particles of varied size and structure delimits different empty or filled porous spaces. The evolution of this variable geometry under specific physico-chemical and thermodynamic conditions is, in most cases, essential in relation to intercalation properties.

2. Layered clay minerals: variable geometry, porosity and surfaces

The definition of clay minerals, as adopted in the *Handbook of Clay Science* by Bergaya & Lagaly (2006), includes all the layered minerals which obey three criteria previously defined by the joint Nomenclature Committees (NC) of the AIPEA (Association Internationale pour l'Etude des Argiles) and The Clay Minerals Society (Guggenheim & Martin, 1995). The clay minerals are fine-grained minerals included in clay rocks (also defined by the joint NC), they should harden upon heating, and they should have plasticity at appropriate water contents. Phyllosilicates and also natural or synthetic non-phyllosilicates belong to this category if they meet these three criteria.

The layers of the phyllosilicates can be neutral or negatively charged. Only the layered double hydroxides (LDH, also called anionic clays because of their anion exchange property) consist of positively charged layers. Details of the structure, properties and complex reactions with many other compounds are described in the *Handbook of Clay Science* (Bergaya *et al.*, 2006a).

As far as intercalation is concerned, special attention should be paid to the hierarchical structure of the individual layers which are stacked in particles, themselves forming aggregates with different sizes. Figure 1 shows a TEM image of Wyoming montmorillonite at low magnification and a schematic hierarchical dispersion of the aggregates from the micro-scale and nano-scale towards particles and finally isolated layers at the nano-scale. This hierarchical structure induces porosity also at different scales (micro-porosity, meso-porosity and macro-porosity). The main question is in which pores, *i.e.* at which scale, intercalation of guest compounds can occur? At the smallest scale, in the plate-shaped micro-pores between the layers, *i.e.* in the interlayer space, reactions not proceeding in homogeneous media can occur due to the confinement effect. The particles organize themselves in aggregates of different sizes causing interparticle pores. At a larger scale, pores between these aggregates are called inter-aggregate pores.

The porosity with variable geometry due to the variable stacking of the layers and variable arrangement of the particles and of the aggregates leads to two kinds of

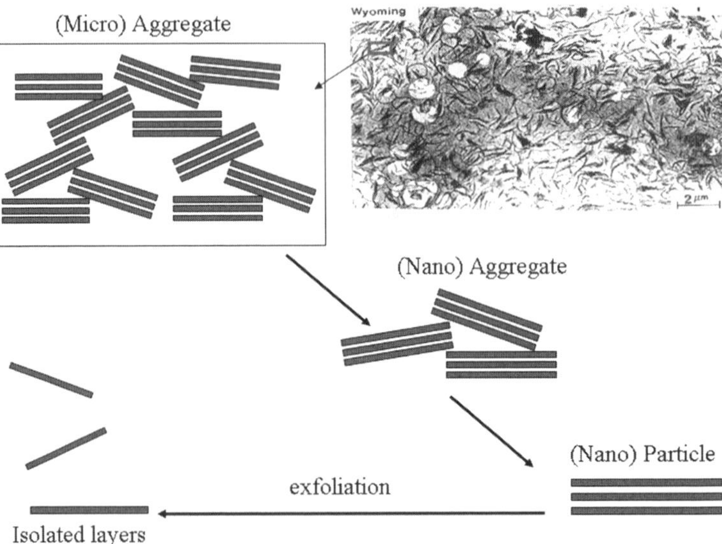

Fig. 1. Hierarchical dispersion of Wyoming Mt. Readers of the paper version of this chapter may wish to download a colour version of this figure from www.minersoc.org/emu-notes/emu-11/11-7-colour.pdf.

surfaces, external and internal. This was proven experimentally for the first time by Bergaya (Annabi-Bergaya, 1978; Annabi-Bergaya *et al.*, 1979) and has an important effect on the accessibility to different molecules, particularly to gases (*e.g.* nitrogen) which are currently used to determine the (easily accessible) specific area of these surfaces.

The properties of the intercalated compounds are completely different if intercalation occurs in the confined space at the nanometer scale or in larger spaces in which the properties of the intercalated compounds are similar to the bulk properties of the compounds in solution.

The classification of clay minerals is presented in Table 1. The clay mineral layers can be stacked regularly or randomly. The layers and the interlayer spaces can be identical with only minor differences in the composition or of different composition and structure. Clay minerals consisting of two or more different layers within the particles are referred to as interstratified minerals.

The TO minerals of the first group (T and O indicate tetrahedral and octahedral sheets, respectively) usually cannot swell. This is because TO minerals have uncharged layers due to the lack of pronounced substitutions and consequently do not contain hydrated cations between the layers. They form particles of many stacked layers which form a regular stacking sequence or which have different degrees of disorder. The layers of kaolinite are held together by strong interactions. This renders the separation (usually called delamination) of the particles very difficult. However, earlier studies and published patents have shown that the complete separation of the layers of TO minerals is possible.

Table 1. Classification of the clay minerals' layer charge and idealized (chemical compositions) formulae of some representative TO and TOT clay minerals.

Charge/formula unit	Dioctahedral species	Trioctahedral species
	Serpentine-kaolin group (non-swelling TO minerals)	
~0	Kaolinite	Serpentine
	$(Si_2)^{IV}(Al_2)^{VI}O_5(OH)_4$	$(Si_2)^{IV}(Mg_3)^{VI}O_5(OH)_4$
	Talc-pyrophyllite group (non-swelling TOT minerals)	
~0	Pyrophyllite	Talc
	$(Si_4)^{IV}(Al_2)^{VI}O_{10}(OH)_2$	$(Si_4)^{IV}(Mg_3)^{VI}O_{10}(OH)_2$
	Smectite group (swelling TOT minerals)	
~0.2–0.6	Montmorillonite	Hectorite
	$(Si_4)^{IV}(Al_{2-y}Mg_y)^{VI}O_{10}(OH)_2, yM^+.nH_2O$	$(Si_4)^{IV}(Mg_{3-y}Li_y)^{VI}O_{10}(OH)_2, yM^+.nH_2O$
	Beidellite	Saponite
	$(Si_{4-x}Al_x)^{IV}(Al_2)^{VI}O_{10}(OH)_2, xM^+.nH_2O$	$(Si_{4-x}Al_x)^{IV}(Mg_3)^{VI}O_{10}(OH)_2, xM^+.nH_2O$
	Vermiculite group (partially swelling TOT minerals)	
~0.6–0.9	Vermiculite	Vermiculite
	$(Si_{4-x}Al_x)^{IV}(Al_{2-y}Mg_y)^{VI}O_{10}(OH)_2, (x+y)K^+$	$(Si_{4-x}Al_x)^{IV}(Mg_{3-y}M_y^{3+})^{VI}O_{10}(OH)_2, (x-y)M^+$
	True (flexible) and Brittle mica group (non-swelling TOT minerals)	
~0.6–1.0	Celadonite	Lepidolite
	$(Si_{4-x}Al_x)^{IV}(Fe_{2-y}Mg_y)^{VI}O_{10}(OH)_2, (x+y)K^+$	$(Si_{4-x}Al_x)^{IV}(Mg_{3-y}Li_y)^{VI}O_{10}(OH)_2, (x+y)K^+$
	Muscovite	Phlogopite
	$(Si_3Al)^{IV}(Al_2)^{VI}O_{10}(OH)_2, K^+$	$(Si_3Al)^{IV}(Mg_3)^{VI}O_{10}(OH)_2, K^+$
~1.8–2.0	Margarite	Clintonite
	$(Si_2Al_2)^{IV}(Al_2)^{VI}O_{10}(OH)_2, Ca^{2+}$	$(Si\,Al_3)^{IV}(Mg_2Al)^{VI}O_{10}(OH)_2, Ca^{2+}$

Among the swelling TOT minerals, particularly in the smectite group, isomorphic substitutions of Al^{3+} (and or Si^{4+}) ions in the octahedral (and/or tetrahedral) sheets by cations of lower charge (Al^{3+} by Mg^{2+} or Si^{4+} by Al^{3+}) generate negative charges. To ensure electroneutrality, these negative charges are counterbalanced by positive charges usually provided by alkaline or alkaline earth cations located in the interlayer space. These interlayer cations can be totally or partially exchanged by other cationic species. The cation exchange capacity (CEC) is usually expressed in equivalents (eq) of cations per gram of the calcined clay mineral or in eq/mol formula unit (Bergaya *et al.*, 2006b). These exchangeable cations are more or less hydrated depending on their physico-chemical parameters (size, charge and polarization). They are of absolute importance in the predicted behaviour of the clay-mineral properties (Annabi-Bergaya *et al.*, 1980a). Among these properties is the swelling of TOT minerals which depends on the type of interlayer cations.

Swelling is usually defined as a macroscopic property when the volume of the clay mineral increases compared to the dry powder, *e.g.* by water uptake. However, at the molecular scale, it is also related to the expansion of the interlayer space by intercalation of different solvents. Thus swelling also depends on the layer charge. The TOT minerals do not swell when the layer charge is very small ($<\sim 0.2$ eq/mol formula unit) or when it is very large (>0.8 eq/mol formula unit). Similarly, the TO minerals do not swell by solvent intercalation because of the almost negligible layer charge.

3. Intercalation in non-swelling minerals

3.1. Kaolinite: a non-swelling natural clay mineral

Kaolinite is the least expensive clay mineral used in the ceramics, paper and rubber markets. It is also the most abundant mineral in kaolin deposits. The kaolin group contains two other polytypes: dickite and nacrite, which are scarcer in the deposits. They differ from kaolinite only by a different type of layer stacking in terms of distribution of the vacant octahedral sites in the successive layers and by the degree of rotation between adjacent layers (see Brigatti *et al.*, 2011, this volume). Halloysite is a kaolinite-like mineral (but with non-planar morphology) in which a monolayer of water is originally intercalated between the layers. By heating at low temperatures ($<100°C$) or under vacuum at room temperature, this water is more or less reversibly lost and the basal spacing ($\sim$1 nm) of the dehydrated halloysite reduces to $\sim$0.7 nm, similar to that of kaolinite.

The kaolinite structure (Figure 2) presents two different TO basal surfaces. The tetrahedral T sheet is composed of siloxane groups and the octahedral O sheet contains mostly hydroxyl groups. The hydrogen bonds between these inner-surface hydroxyl groups and the oxygen atoms of the adjacent layer, together with the van de Waals interaction, reduce the swelling ability of the particles. The inner-hydroxyl groups of the octahedral sheet, which are located at the borderline between the octahedral and tetrahedral sheets, as shown in Figure 2, cannot form hydrogen bonds but can interact with

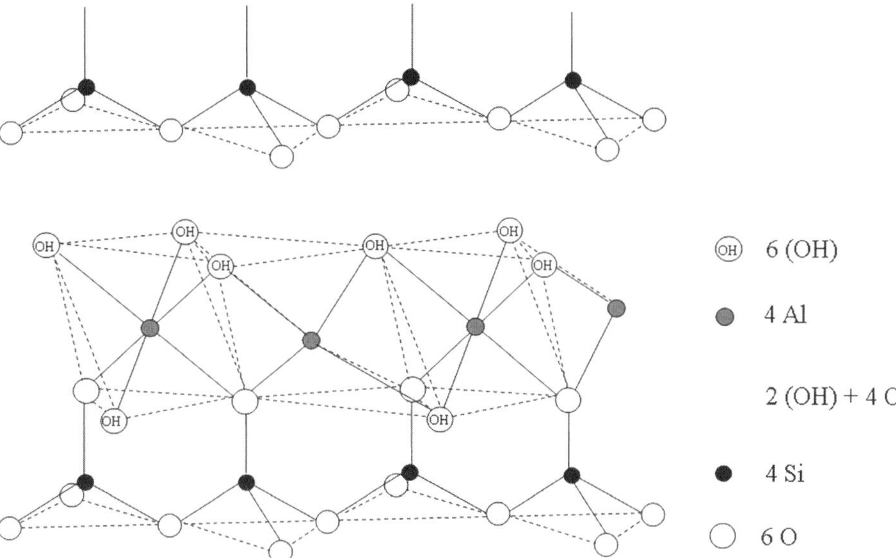

Fig. 2. Kaolinite (TO) structure.

species intercalated between the layers through the ditrigonal holes of the T sheets. Until the 1960s kaolinite was considered to be a non-swelling mineral.

3.1.1. Direct intercalation in kaolinite

A limited number of types of guest molecules can be intercalated into kaolinite. Despite the strength of the interlayer hydrogen bonding, unexpected swelling of the kaolinite occurs with some inorganic salts and polar organic compounds. As hydrogen bonds have an important part in the interaction between the layers, only molecules such as urea which can break the hydrogen bonds are intercalated as guests. The conventional intercalation compounds are molecules with large dipole moment such as dimethyl sulphoxide (DMSO) (Cruz *et al.*, 1973), or molecules containing acceptor groups such as hydrazine or donor groups for hydrogen bonds such as dimethyl formamide (DMF). However, historically, the first two examples of intercalation reactions, published simultaneously and independent of each other, were the reaction of kaolinite with potassium or ammonium acetate (Wada, 1961) as well as with urea (Weiss, 1961). Reaction with potassium acetate increases the basal spacing from 0.7 nm to 1.1 nm; the reason for this intercalation is still unknown. The intercalated molecules are generally easily removed by water.

A critical analysis of the intercalation behaviour of kaolinite was presented by Gardolinsky (2005). This author reviewed the mechanism of intercalation in kaolinite and the numerous factors affecting the intercalation process, including: (1) particle size and degree of crystallinity of the kaolinite, (2) the amount of water or other polar solvent added, (3) factors related to the nature of the intercalated molecules, and (4) external conditions such as temperature or pH of the medium.

The process corresponding to the separation, at the nanoscale, of the individual layers in the particle by the intercalated molecules was referred to as "delamination" (Gardolinsky & Lagaly, 2005). In ceramics technology, delamination has often been reported as a mechanical process which leads to partial amorphization of the kaolinite. It is interesting to note that the term 'exfoliation process' which will be discussed in section 6.8, is also used for the decomposition of large aggregates of kaolinite into smaller particles. All these studies on the intercalation of this originally non-swelling kaolinite continue to be of interest. For example, in a recent study, Sun *et al.* (2011) showed that fast intercalation of DMSO occurs by grinding a mixture of DMSO/kaolinite, followed by heating at 120°C. This technique reduces the reaction period from several days by the traditional direct intercalation, to hours. Mechano-chemical activation played a crucial role in the exfoliation of the kaolinite layers by DMSO.

3.1.2. Indirect intercalation in kaolinite by displacement reactions

The molecules intercalated directly into kaolinite, mentioned above, can also be displaced by many other guest compounds which cannot be intercalated directly. In this way, many other intercalation compounds of kaolinite can be prepared as described by Lagaly *et al.* (2006). Some of these indirectly intercalated compounds obtained from the intercalated precursors are unstable, as the initial raw kaolinite is recovered when dried, but other intercalation compounds remain stable. However, all these indirectly intercalated compounds de-intercalate by washing with water or with any appropriate solvent.

3.2. Micas: natural non-swelling minerals

Natural micas usually do not swell because the interlayer cations compensating the layer charges are unhydrated K^+ ions (see Table 1). Potassium ions in micas fit the hexagonal holes of the tetrahedral surface sheets perfectly and impede any expansion of the interlayer (or interparticle) spaces. In agreement with this experimental evidence, a Monte Carlo simulation study of water confined between particles of micas (Delville, 1993) confirmed the influence of the location of these K^+ ions on the non-swelling properties of the mica-water system. It was claimed by Weiss *et al.* (1956), nevertheless, that micas can swell but that it takes years to do so.

However, commercial highly-charged (type) synthetic micas with Na^+ as exchangeable cations (Na-fluorotetrasilicic mica – Na-TSM) showed pronounced swelling in particular solvents such as ethanol. This intercalation is easily reversible in a dry atmosphere. Neutral compounds such as natural phospholipids which have hydrophobic dual alkyl chains and a hydrophilic head also show a fast intercalation of the lipid dual chains (bilayers) in a chloroform solution of Na-TSM. In this case, cation exchange does not seem to be the driving mechanism as the amount of Na^+ ions released is very small (Kanzaki *et al.*, 1997). These intercalated bi-layers are very stable. They form a wall surrounding the biological cell which acts as selective membrane important in biological activity. These immobilized bi-layers serve as supports for unstable medicines.

4. Intercalation in swelling clay minerals

The TOT clay minerals (see Table 1), particularly the smectites, show pronounced intercalation. Montmorillonite (Mt) (Fig. 3) is taken as an example for illustrating the different mechanisms which can occur in the interlayer space. All Mt are highly disordered, often showing turbostratic disorder, *i.e.* the layers are rotated towards one another in a particle. Montmorillonite is able to adsorb various ionic and neutral inorganic or organic compounds. The intercalation is achieved easily by dispersing the Mt in aqueous or organic solutions of the guest compound. Some guest molecules can also be intercalated by solid-state reactions.

4.1. Solvation by ion dipole and hydrogen-bonding interactions

Hydration or solvation by any polar solvent leads to intercalated water (or intercalated solvent) which shows different properties from those of aqueous solutions (Fripiat *et al.*, 1984). Hydration of clay minerals has been studied widely but is yet not fully understood although experimental tools and modelling have progressed significantly since the famous paper of Norrish (1954) on the swelling of Mt. Norrish established that Mt in water swells in discrete steps, intercalating 1, 2, 3 and 4 water layers. The corresponding basal spacings are 1.24 nm, 1.5, 1.8 and 2.1 nm, respectively. With further addition of water, the basal reflections disappear indicating the complete splitting of the particles into individual layers. The implied mechanism of the intercalated solvent has been demonstrated by studying the interaction of Mt with methanol which is easier to study by spectroscopic techniques because this polar molecule contains only one

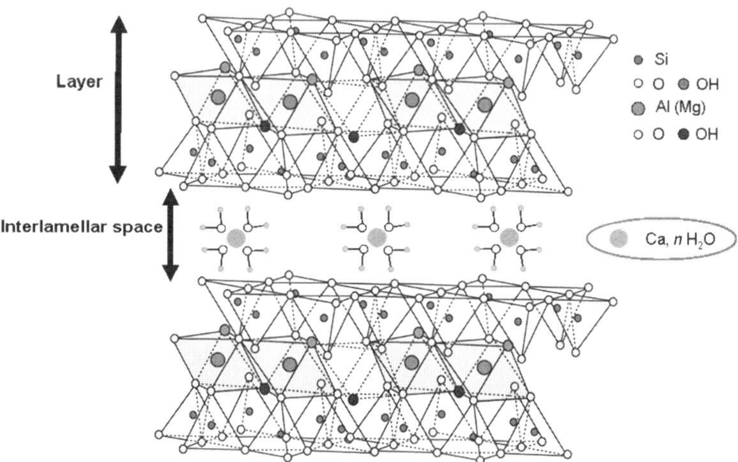

Fig. 3. Ca^{2+}-montmorillonite (TOT) structure with different types of surfaces: external (basal and lateral) surfaces *vs.* internal basal surfaces in the confined interlayer space. Readers of the paper version of this chapter may wish to download a colour version of this figure from www.minersoc.org/emu-notes/emu-11/11-7-colour.pdf.

OH group (Annabi-Bergaya *et al.*, 1980b). The first alcohol molecules are bound by ion-dipole interaction with the interlayer cations. Further methanol molecules are adsorbed by hydrogen bonds to the first adsorbed molecules. Hydrogen bonds between the intercalated molecules and the oxygen atoms of the internal basal plane surfaces are assumed to occur but have never been proven. Other polar solvents can also be intercalated by hydrogen bonds to the first water molecules linked directly and strongly to the cations.

4.2. Ion exchange

A large number of inorganic, organic and organometallic cationic species can intercalate in the interlayer space of the TOT clay minerals by cation exchange. The ion-exchange reactions and measurements have been presented extensively by Bergaya *et al.* (2006b). Two examples of cation exchange reactions are presented below.

4.2.1. Exchange selectivity between inorganic cations

Ion exchange corresponds to a chemical equilibrium which takes place in the interlayer space. This exchange is defined by a preference (exchange selectivity) for some cations compared to others. It is expected that inorganic cations which have a larger radius or greater charge, and therefore strong polarizing power, will exchange more easily than cations presenting the opposite properties (small radius or charge and less polarizing power). However, this selectivity rule has some exceptions. The exchange selectivity in clay minerals presented in different publications seems to fluctuate depending on the experimental conditions. Nevertheless, the cations may be classified into three groups: (1) the particular status of the Li^+ and Na^+ monovalent cations; (2) the other remaining monovalent cations with greater radius, and finally (3) the bivalent and trivalent cations. When an exchange reaction occurs between cations of different groups, a hysteresis effect is observed but it is not yet completely understood. In relation to intercalation, the first group can intercalate a large amount of water in comparison with the other groups. The degree of hydration allows the clay minerals to swell until particles almost fully exfoliate into smaller particles with the presence of some individual layers (Benna *et al.*, 2001).

4.2.2. Organo-clay minerals (OC)

Special attention is given here to the ion-exchange which occurs with large cationic surfactants (Lagaly *et al.*, 2006) such as primary, secondary, tertiary and quaternary alkyl ammonium ions. Unlike three-dimensional systems which have a rigid, non-deformable porous framework, the bi-dimensional montmorillonite can adjust its interlayer space to accommodate the intercalation of the guest surfactant (He *et al.*, 2006). The cation exchange is usually accompanied by the intercalation of ion pairs, *i.e.* the cationic surfactant molecules are accompanied by their counter ions. The additional surfactant cations not compensating layer charges can usually be removed by washing whereas the surfactant cations compensating the layer charges are tightly fixed in the interlayer space and at the external surfaces. These cation exchange reaction and intercalation processes easily extend the interlayer distance along the c axis leading to different structures of the organo-clay minerals (OC). Such organo-clay minerals have been studied for many years (*cf.* Theng, 1974) and produced industrially ('Bentones' is the commercial name) for different purposes. The intercalated structures of the OC (Figure 4) were described for the first time by Lagaly (1986) and Lagaly *et al.* (2006). The intercalation in monolayers (~1.4 nm), bilayers (~1.8 nm), pseudo-trimolecular layers (~2.2 nm) or paraffin-type layers arrangement (>2.2 nm) depends heavily on the nature of the clay mineral and of the surfactant. However, the surfactant/CEC ratio is also another important factor that has been taken into account more recently (Mandalia & Bergaya, 2006). Vaia *et al.* (1994) suggested that the long-chain compounds intercalated in monolayers, bilayers or paraffin-type layers can either be in a solid-like state or assume a disordered liquid-like arrangement, the latter being favoured for the short-chain alkyl ammoniums or for low clay mineral charge. The long-chain compounds are usually intercalated from solution but can also be intercalated by solid state reactions (Ogawa *et al.*, 1989).

The cohesion between the organo-layers can be described thermodynamically by the cleavage energy associated with Coulomb interactions. High cleavage energy impedes the exfoliation of the layers in a polymer matrix; it depends on the CEC and on the surfactant characteristics, *i.e.* the head group and the chain length (Fu & Heinz, 2010). Using a molecular dynamics simulation, Fu & Heinz (2010) showed that the cleavage energy indicates local minima for loosely packed flat layers of surfactant, and local maxima for the flat densely packed structures.

4.3. Protonation

Although ion exchange is the most common method of intercalation in smectites, other mechanisms are also used, *e.g.* protonation. The reason is that polarization of intercalated water molecules by the cations (M^{n+}) increases the probability of dissociating the protons. This water coordinated to the cations is more acidic than bulk water. For instance, ammonia and amines are protonated in the interlayer space as shown in the

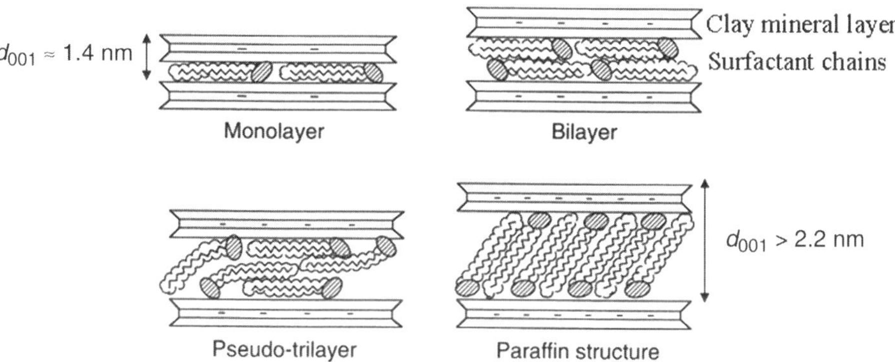

Fig. 4. Organo-clay minerals (modified from Lagaly, 1986).

following equilibria reactions:

$$[M^{n+} - OH_2 + H_2O] = [M^{n+} - OH^- + H_3O^+]$$
$$[M^{n+} - OH^- + H_3O^+ + NH_3] \leftrightarrows [M^{n+} - OH^- + NH_4^+ + H_2O]$$

Some organic proton donors are also a potential source of protons on drying or heating. This deprotonation leads to some important and complex consequences. Migration of the protons occurs after the exchange reaction with the exchangeable cation; stable H^+-Mt is very difficult to obtain. The interlayer protons which migrate into the layer interact with the hydroxyl groups. This process, known as 'autotransformation', is followed by changes in the composition (Komadel & Madejová, 2006) and is of technical importance in the acid-activation of clay minerals (Harvey & Lagaly, 2006). The protons can also reduce the structural Fe^{3+} ions of the smectites causing a drastic change in the layer. According to Heller Kallai (2001), few studies have addressed the question of deprotonation by inorganic proton acceptors. Heller Kallai (2001) presented a critical review of the protonation-deprotonation reactions in smectites.

4.4. Complexation

In the particular case where exchangeable cations are transition metal ions, intercalation can proceed by the formation of complexes. Some examples are copper complexes with water and with ethylene diamine or diethylene triamine. This reaction is the basis of one of the best established methods for CEC measurements (Bergaya & Vayer, 1997; Amman *et al.*, 2005; Bergaya *et al.*, 2006b). Many further examples can be found in Lagaly *et al.* (2006).

4.5. Electron transfer and redox reactions

Iron in the smectite layers is more difficult to reduce than iron in hydr(oxides) but it can lead to redox reactions by charge transfer between the intercalated molecules or ions as reducing agents (including bacteria) and the iron ions in the octahedral sheet (Stucki, 2006; Anastácio *et al.*, 2008). The structural reduced iron ions have a pronounced depressing effect on water retention, thus on swelling. However, the understanding of the mechanism involved causing this alteration is still not fully understood (Stucki, 2006). Similarly, the mechanism of reduction/reoxidation in smectites is partially understood. Although real progress has been made during recent decades, the specific pathway by which the electrons reach the iron in the octahedral sheet, inside the layer, is still unknown. The reader is referred to Stucki (2006) for further information about the redox processes of iron on clay minerals and for a list of the remaining unanswered questions and challenges.

A well studied example is the electron transfer between benzidine and the structural Fe^{3+} ions (Solomon *et al.*, 1968; Furukawa & Brindley, 1973). This mechanism is rather uncommon. Another very interesting case is the reaction of copper Mt with benzene vapour. At certain conditions, single-electron transfer of the benzene molecules to the

Cu^{2+} ions leads to the loss of aromaticity of the benzene molecules (Pinnavaia & Mortland 1971; Pinnavaia *et al.*, 1974); in this case, however, the oxidizing agent would be the interlayer cations rather than a species in the layers as in the preceding example.

An electron-transfer reaction can also occur between an electron donor and excited dye molecules or metal complex molecules adsorbed by clay minerals leading to photochemical reactions (Shichi & Takagi, 2000).

Redox reactions can occur between the reduced structural iron in phyllosilicates and redox-sensitive metals such as Cr^{6+} (Gan *et al.*, 1996). The extent of reduction is greater in Na^+ smectite than in K^+ smectite, indicating that the reaction occurs at basal surfaces (Shen & Stucki, 1994). The redox reactions are generally affected by the Lewis acid-base reaction on the clay minerals' surfaces. A high basicity of the basal oxygen atoms can be provided by substitutions in tetrahedral sheets (Bleam, 1990).

4.6. Grafting

Grafting reactions, *i.e.* attachment of organic groups by covalent bonds, can occur with all the pendant aluminol or silanol groups of the clay mineral edges and of the clay minerals basal surfaces including the internal surfaces of the TOT minerals. In TO minerals, grafting of organic compounds can occur with the aluminol groups of the internal basal surface (Tunney & Detellier, 1993). Interlayer kaolinite grafting was described exhaustively by Gardolinsky in his doctoral thesis (2005).

The advantage of grafting compared to the intercalation mechanism is that it allows an irreversible fixation of organic units on the layers. Grafting reactions at the internal surfaces of kaolinite are only made possible by a preceding swelling step, *e.g.* by intercalation of DMSO (Tonle *et al.*, 2007). The subsequent reaction with aminopropyl triethoxysilane (APTES) yielded a new robust nanohybrid material.

Preliminary grafted organo-kaolinites can be used as intermediates for the preparation of further intercalation compounds to obtain functional nanocomposites based on TO clay minerals. As an example, Letaief & Detellier, (2009) showed that covalent bonds were formed between grafted triethanolamine molecules and the aluminol groups of the internal octahedral sheets of a urea-kaolinite. This grafted kaolinite was quaternarized by iodomethane leading to hybrid organo-inorganic polycations which were further mixed with sodium polyacrylate.

4.7. Co-intercalation of two compounds

Co-intercalation is more common than assumed. When a guest compound is intercalated from solution (often in water), solvent molecules penetrate into the interlayer space together with the guest molecules. This process corresponds to an adsorption from binary solutions when the two end members are liquids, *e.g.* water and ethanol. It is, therefore, difficult to determine correctly the amount of intercalated guest molecules (Lagaly *et al.*, 2006). Co-intercalation processes can cause synergy effects. For instance, corresponding to the base-pairing in DNA, the adsorption of thymine and uracil was enhanced in the presence of adenine already adsorbed in the montmorillonite (Lagaly, 1984; Samii & Lagaly, 1987).

A well known example is the reaction of TOT clay minerals with long-chain alkylammonium ions from aqueous solution. At higher concentrations of added alkylammonium salts, the exchange of the alkylammonium ions is accompanied by the intercalation of ion pairs (alkylammonium ions + the anions) as stated above, and also of some water. In these cases, most of the intercalated ion pairs can be removed by washing while the water is desorbed by drying.

Competitive ion-exchange intercalation when ammonia is added to montmorillonite together with a pillaring solution also has a large influence on the homogeneity of distribution of the intercalated species in PILC (see section 6) and on their acid strength (Figueras *et al.*, 1990).

Co-intercalation of two species is also used to study the most convenient precursor for CPN. For instance, the results of the intercalation of a mixture of stearic acid and octadecylamine (both having the same side chain) have been compared to the results obtained when the two intercalates were added separately in successive steps (Capcova *et al.*, 2006). The structure of the intercalation compounds are compared as a function of the methods of preparation. Stearic acid itself does not intercalate into Na^+-montmorillonite but co-intercalation with octadecylamine led to the formation of octadecyl ammonium stearate, which was successfully intercalated into the interlayer space of montmorillonite.

5. Adsorption and intercalation

As stressed before, surface modification can be performed by physical adsorption or by chemical grafting. Clay minerals can adsorb a great variety of inorganic or organic species by physisorption, by chemisorption or by ion exchange on all the available surfaces. Adsorption in the interlayer space corresponds to intercalation of the species. The advantage of the physical adsorption is that the clay mineral structure is not altered but the disadvantage is that the molecules are energetically weakly adsorbed. The deintercalation is called desorption. Physical and chemical adsorptions are controlled by thermodynamic factors and measured by isotherms of adsorption the shape of which indicates the type of adsorption (Gregg & Sing, 1982; Rouquerol *et al.*, 1994).

Adsorption should not be confused with absorption; see reports from the literature as demonstrated by methylene blue adsorption (Kipling & Wilson, 1960) or methylene blue absorption (Hang & Brindley, 1970). Absorption is the capacity of a liquid to be retained by a solid without any reference to fixation on the surface (adsorption) and without any mention of intercalation. The solid can swell macroscopically by absorption with retention in the macropores, including adsorption on the surface. It seems that in the current literature these two terms are no longer confused. However, some authors also use 'sorption' instead of adsorption (Johnston *et al.*, 2002) but this word is less precise; it could be used as a general term for both absorption and adsorption.

X-ray diffraction (XRD) is the most useful method to establish a diagnosis of the intercalation in clay minerals. A huge literature exists on this technique. Nevertheless, many researchers continue to develop it (Lanson, 2011, this volume) because interpretation is not easy in some particular situations, *e.g.* for interstratified samples. Usually, in simple cases, the distance between the stacked layers along the *c* axis, is deduced from the *d* values of the *00l* reflections (basal reflections) in the powder XRD patterns (if the layers are stacked as in the clay minerals). If the layers and interlayer spaces in the particles are identical, the basal reflections form an integral series and the basal spacing is $d_{001} = 2d_{002} = \ldots nd_{00n}$. Thus, the thickness of the interlayer space is the basal spacing minus the thickness of the layer (0.71 nm for kaolinite, 0.95 nm for smectites). However, note that the preparation of the sample is of fundamental importance in detecting the basal reflections indicative of the intercalation state. The intensity of the basal reflections is increased when oriented samples (with a high degree of parallel orientation of the particles) are prepared, *e.g.* by slow evaporation of aqueous clay mineral dispersion.

Other advanced techniques to define intercalation processes in layered materials are presented in more detail by Mottana & Aldega (2011, this volume).

Adsorption of molecules or cations which induce swelling such as ethylene glycol monoethyl ether (EGME), glycerol, or methylene blue cations (MB) is often used to determine the total (including internal and external) specific surface area of swelling clay minerals. This determination is based on the calculated intercalated amount and on the XRD results. Adsorption of MB is one of the oldest known methods (Johnson, 1957). Methylene blue is still used today as a rough method (spot method) to evaluate the swelling clay minerals content in soils. In laboratories, MB adsorption is currently used to determine both the specific surface area and the CEC values (Hang & Brindley, 1970; Lagaly *et al.*, 2006). However, determination of the CEC by MB is not very reliable. In fact, interpretation of the results are influenced not only by the experimental conditions but also by the nature of the exchangeable cations and by the charge density (Michot & Villieras, 2006).

6. From intercalated clay minerals to pillared clay minerals (PILC)

Intercalation of polynuclear hydroxy metal cations and metal cluster cations in smectites yields pillared clay minerals with pore sizes that can be made larger than those of conventional zeolite catalysts. These compounds often show interesting catalytic properties (Pinnavaia, 1983). Clay minerals have been known since the beginning of the last century for their catalytic properties (Bergaya, 1990). In the 1960s, synthetic zeolites were favoured by those in industry because of their greater activity and selectivity for cracking compared to the clay catalysts used in the Houdry process (Houdry *et al.*, 1938). Research started on PILC as catalysts, after the 'oil shock' of 1973. For cracking of heavy petroleum, zeolites are not ideal as they have limited numbers of micropores and are rapidly 'poisoned' both by coke formation and by the metals present in heavy petroleum. The situation has changed recently because research on zeolites with increasing pores has progressed too. The amount of research on PILC is increasing, however, because the application of these porous materials is not limited to catalysis – they can also be used as adsorbents, *e.g.* in the area of environmental protection and pollution control (Zeng & Liu, 2005).

The idea was to intercalate a limited number of sturdy pillars between the layers to avoid the collapse of the interlayer space which normally occurs after heating hydrated Na^+-clay minerals. The intercalation occurs by cation exchange of the interlayer cation by an inorganic species. This was successful with the famous aluminium pillar known as the Keggin ion or Al_{13} because it contains one tetrahedral Al surrounded by twelve octahedral Al. These polycations $(Al_{13}O_4OH_{24}H_2O_{12})^{7+}$ obtained by hydrolysis of Al salts by a base, not only act as 'provisory pillars' maintaining an expansion of ~1 nm between the basal surfaces of two successive layers but also allow some empty porosity between the pillars in the interlayer space (Figure 5). This opening up of the internal surface is proved by increased specific surface area values obtained after pillaring compared to the initial specific surface area of the pristine clay mineral measured by liquid nitrogen adsorption at −196°C. This result is one of the first tests that must be applied to confirm the success of the intercalation process. However, these intercalated compounds are not very stable and should be heated to at least relatively high

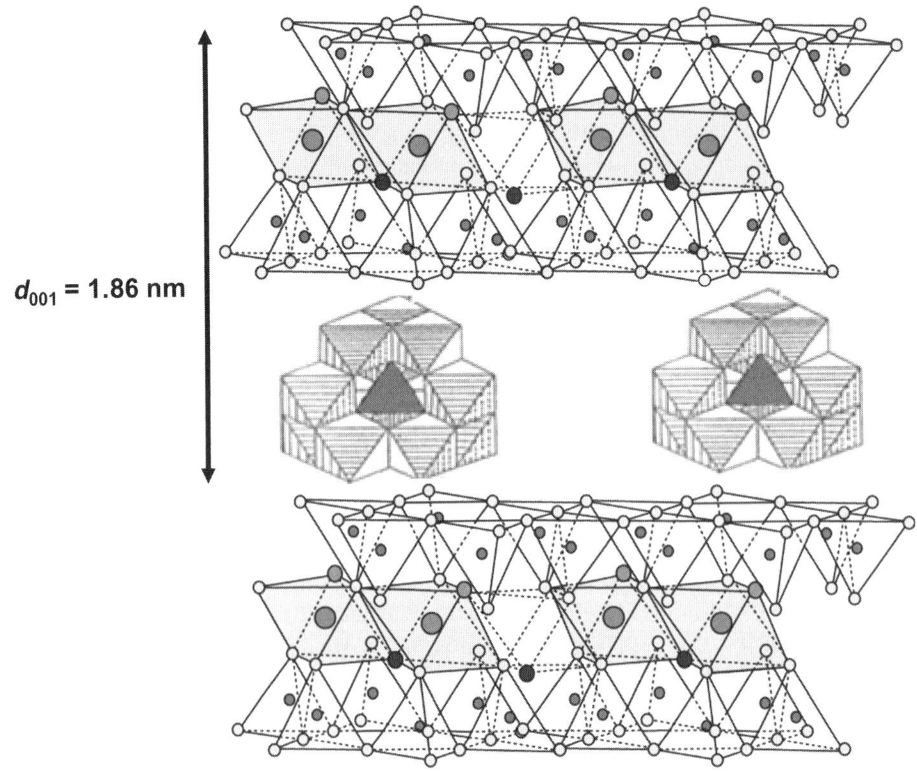

Fig. 5. Two-dimensional microporous Al_{13}-pillared Mt. Readers of the paper version of this chapter may wish to download a colour version of this figure from www.minersoc.org/emu-notes/emu-11/11-7-colour.pdf.

temperature, $>400°C$, to be considered as pillared materials corresponding to a permanent porous structure. This second step is often reported in PILC literature as a 'calcination step' no matter what the temperature is. In other contexts, however, 'calcination' of clay minerals usually means heating at the temperature corresponding to full dehydroxylation, *i.e.* when all the structural hydroxyl groups are removed by water evaporation (*i.e.* at $>900°C$). The pillared clay mineral contains very stable oxy-hydroxides of aluminium, with smaller charge (4+ instead of 7+) and with a very slight decrease in size, which act as 'true pillars' maintaining the distance between the layers at the nanometer scale. The research has been extended to cations other than Al^{3+} which can lead to intercalation of bulkier species. The intercalation was verified by XRD and researchers are always looking for interlayer distances >1 nm. Another aim of PILC research is to obtain more acidic properties than with Al-PILC. The subject of pillared clays has been explored in depth; in addition to the large number of journal articles and reviews, a book was published recently (Gil *et al.*, 2010). There is scope for almost unlimited investigation of PILC because: (1) the nature of the pillaring species was extended from the first known and most studied Keggin cation to the polycations offered by the chemistry of all the elements in the periodic table that can be hydrolysed (Baes & Messmer, 1976), not all of which have been studied so far (Bergaya *et al.*, 2006c). (2) Besides this already large number of polycations, there are several combinations of mixed pillaring agents added simultaneously or successively or formed *in situ* in the interlayer space and which can show a synergetic effect. Other types of interactions with the pillars can be obtained by simple impregnation by cationic species or with pillars doped after intercalation. (3) All members of the large family of clay minerals with different properties can also be used for pillaring.

It is interesting to note that pillared clays can be further treated with surfactants leading to inorgano-organo pillared clays (OPILC) which act as more suitable precursors than OC for particular CPN with apolar polymers (Bergaya *et al.*, 2005). The interaction between organo cations and pillared clays (Mishael *et al.*, 1999) showed improved properties, *e.g.* for adsorption of herbicides (Li *et al.*, 2009).

The aim of this chapter is not to present in detail the structural properties and applications of PILC but note that, as far as intercalation is concerned, no universal explanation of the origin of the bonding between the pillars and the basal surface of the layers is yet available. Although structural models have been proposed for particular clay minerals, such as fluoro hectorite (Pinnavaia *et al.*, 1985), beidellite (Plée *et al.*, 1985) and saponite (Bergaoui *et al.*, 1995), implying the formation of covalent bonds between pillars and the clay mineral layers, the question as to why the pillars can be intercalated permanently in all types of clay minerals and with all of the intercalant species studied remains unanswered.

7. Clay mineral polymer nanocomposites (CPN)

Intercalation of clay minerals by polymers is an old subject (Blumstein, 1965; Theng, 1974). The reason for adding clays as fillers to the polymers was only economic at the beginning. Clay minerals are used now as a low-cost charge to replace others which are more expensive. The new benefits of these fillers added to polymers were discovered more recently, in the last 20 years. In fact, when very small amounts (usually ~5% or less) of clay minerals are added to the polymers, using different methods of preparation (Bergaya & Lagaly, 2007), some of which are presented below, the clay mineral particles are not inert but interact with these polymers, and the CPN properties are improved. Sometimes, even new properties are created compared to the pristine polymer. The question is how this interaction occurs and whether there is intercalation of part of the polymer in the confined interlayer space, or if these clay-mineral layers are dispersed homogeneously or exfoliated (see discussion in section 8) in the major (>~95%) polymer matrix.

7.1. Intercalation from solvents

Among the different methods of preparing CPN, intercalation can occur from solution either directly, by addition of the liquid polymer or of a monomer which is adsorbed and then polymerized in the interlayer space (*in situ* intercalative polymerization). If the polymer is solid, it must be soluble in adequate aqueous or organic solvent. Organo-clay minerals are often used instead of unmodified clay minerals because the intercalated surfactant reduces the surface energy of the clay mineral particles and improves the wetting by the polymer, allowing better intercalation with a greater expansion of the interlayer space than with the unmodified clay mineral. In addition, the presence of some functional groups on the OC can initiate the polymerization of the intercalated monomers (Bergaya *et al.*, 2011a).

7.2. Melt intercalation

Intercalation of modified and unmodified clay minerals by polymers has attracted the attention of many scientists and those in industry in many countries because of Japanese results with the synthesis of Nylon-6-montmorillonite nanocomposite (Fukushima & Inagaki, 1987) and also because of the American discovery of a solvent-free intercalation method (Vaia *et al.*, 1993; Vaia, 1995). The two most commonly used methods cited above were supplanted by a new method known as 'melt intercalation'. This allows the direct intercalation of melted polymer, just above its softening point, in clay minerals (or organo-clay minerals) by simple mixing of the two solid components without any solvent. The degree of intercalation increases with time in the molten state, until a critical content of intercalated polymer is reached corresponding to a d_{001} value of ~4 nm. It is interesting to note that this value corresponds to the maximum basal spacing of ordered water-intercalated pseudo-layers.

In all these methods of preparation, hydrogen bonding is the first mechanism implied between the different polymer groups and the siloxane groups of the clay minerals or *via* the water molecules directly coordinated to the interlayer cations. More details of other mechanisms which can occur with polymers as well as other methods of CPN preparation and processes which influence these mechanisms are presented by Bergaya & Lagaly (2007).

The compatibility between the polymer and the clay mineral is not only obtained by organic modification but also by other modifications such as pillaring or plasma treatments. Finally, the polymer interfacial interaction with the clay-mineral surfaces (both internal and external) is of primary importance in obtaining intercalated and what are known as delaminated or exfoliated structures. The performance of the resultant nanocomposites is heavily dependent on these structures (Bergaya & Lagaly, 2007; Annabi-Bergaya, 2008). Distinction between these possible structures needs to be discussed in more detail.

8. Intercalated *vs.* exfoliated structures

These two terms, used in current CPN literature, are very confusing for the interpretation of the structures described. They are used in different senses by several authors and we summarize them here.

8.1. Intercalated structures

The definition of 'intercalation' phenomena in clay minerals seems to be clearly understood, at least in the following cases: (1) intercalation arises from adsorption of small molecules; (2) intercalation occurs with hydration of clay minerals by between one and four adsorbed pseudo-water layers; and (3) the intercalation of surfactants from monolayers to paraffin-type layers. As long as the interaction between the adjacent layers is clearly established and these layers are stacked in an orderly way, the structure obtained is referred to as an 'intercalated structure'.

Confusion appears when the interlayer distance between the layers increases to $>\sim 4$ nm, and especially when in a particle, the separated layers lose their parallel stacking. So, in the literature, the terms 'delaminated' or 'exfoliated' are used in different ways. An attempt is made here to clarify the two terms, particularly for use in the nanocomposite literature.

'Delamination' of clay minerals is the term used to define the separation process between the planar faces of two adjacent lamellae or layers, involving cohesive interlayer forces. The degree of separation of the layers of the host compound depends on the amount intercalated and on the nature of the (molecules or ions) guests. 'Exfoliation' is taken to mean mechanical separation processes. Originally, the term 'exfoliation' was generally used by those in industry (Jasmund & Lagaly, 1993), but as long as the layers remain oriented along the *c* axis, the structure of particles delaminated with surfactant or polymers is known as an intercalated structure.

8.2. Exfoliated structures

Similarly to what is known as exfoliation at the scale of the aggregates which 'decomposed' into smaller particles, when orientation is lost between the layers, the structure obtained must be defined as an 'exfoliated structure' (although this structure is sometimes called delaminated in some articles based on the fact that the 001 basal reflection disappears due to the high disorder in the layer sequence).

Noting that there appeared to be no clear distinction between delamination and exfoliation in many studies, Bergaya *et al.* (2011b) proposed to use the term exfoliation when referring to the particular stage when the delaminated units (including isolated layers or stackings of a few layers) are dispersed isotropically in a solvent or polymer matrix. The layers eventually became completely independent of one another, losing crystallographic orientation along the *c* axis. Each unit is then freely oriented in space, independently of the others.

Some additional comments are that: (1) delamination and exfoliation actually refer to different stages of the same process; exfoliation is considered to be the ultimate state of delamination; (2) expressions such as 'intercalated or exfoliated nanocomposites' are incorrect. Instead 'clay polymer nanocomposites contain exfoliated layers and/or intercalated particles' should be used; and finally, (3) 'exfoliate' is not a transitive verb as usually reported in the literature. Saying that particles (and or layers) are exfoliated is incorrect, but one can say that particles exfoliate.

It is very important to note that the particles which exfoliate are also composed of intercalated layers leading to the simultaneous presence of intercalated and exfoliated structures as often mentioned in the CPN literature. Depending on the degree of exfoliation, in mixed exfoliated and intercalated structures, the number of elementary layers in a particle showing intercalated structure is at least equal to two. Full exfoliation of all the layers in the polymer matrix without any intercalation domain is rather difficult to reach even in very dilute media.

9. Applications of intercalation/deintercalation processes

Modified clay minerals are important in many fields, from catalysis to medicine. Intercalation of organic surfactants in clay minerals to yield OC finds many applications. The famous commercial 'Bentones' are mostly used due to their rheological properties (*e.g.* thickening and thixotropic effects in paints).

Preparation of OC is also an important preliminary step in developing advanced materials such as CPN for many practical applications.

Intercalation compounds based on layered materials and particularly on clay minerals have received considerable attention as green and sustainable materials. Research on procedures for intercalation are developing quickly in laboratories with the aim of reaching controlled intercalation leading to better control of the properties. For example, intercalated layer by layer (LbL) adsorption in thin-films production is under development (Miao *et al.*, 2010) for sensors and optical applications.

The reversibility of the intercalation process is very important. In some applications, one needs irreversible retention in order to capture definitively the pollutants or other undesired products. In other applications, slow, controlled release is required for fertilizers in agriculture, for example (Mishael *et al.*, 2001).

One of the most interesting topics will be the slow deintercalation of intercalated (bio)molecules for controlling slow release. This can be used in very different areas, from pesticide release in soils to drug release in medical applications (Dong & Feng,

2005). The use of modified clay minerals, intercalated by active molecules, in therapeutic products is increasing considerably (Viseras *et al.*, 2007). Most cardiovascular and anti-inflammatory agents are negatively charged biomolecules and can be intercalated in LDH by ion exchange allowing a controlled release (Nalawade *et al.*, 2009).

10. Conclusion

The swelling of clay minerals is an exceptionally interesting property. This large family of raw materials, abundant throughout the world, is a mine of resources for developed countries but also for developing countries. The intercalation compounds of clay minerals are of prime importance for academic and applied research. In academic research, intercalation in swelling clay minerals can be used as a probe for their identification.

All the characteristics of intercalation of a particular guest depend on each species of clay minerals. From this point of view, this chapter is far from giving a complete survey. As presented briefly above, the intercalation of different components (water, surfactants, oligomer pillars, polymers) is a source of important questions which have attracted the attention of researchers and industrialists for many years, *e.g.* the OC (organo-clay minerals or bentones) in the 1950s (Jordan, 1949a,b), the PILC (pillared clays) in the 1970s and the CPN (clay polymer nanocomposites) in the 1990s. Though some of the problems are not yet understood for all these subjects, research is in active progress (Michot *et al.*, 2002; Lagaly *et al.*, 2006; Bergaya *et al.* 2006c; Carrado & Bergaya, 2007) and the commercial developments are far from being saturated. According to the increasing international awareness of the need for 'green' development, all 'bio-subjects' have recently become very fashionable. Certainly, intercalation of medical substances in clay minerals for drug targeting will provide successes in the near future. The door is just half-open for future research in health-care studies combined with materials science.

Acknowledgements

The authors are grateful to Prof. Jean-François Lambert for his useful comments on the manuscript.

References

Amman, L., Bergaya, F. & Lagaly, G. (2005) Determination of the cation exchange capacity (CEC) of clays with copper complexes – revisited. *Clay Minerals*, **40**, 441–453.

Anastácio, A.S., Aouad, A., Sellin, P., Fabris, J.D., Bergaya, F. & Stucki, J.W. (2008) Characterization of a redox-modified clay mineral with respect to its suitability as a barrier in radioactive waste confinement. *Applied Clay Science*, **39**, 172–179.

Annabi-Bergaya, F. (1978) Organisation de molécules polaires adsorbées par la montmorillonite. PhD thesis, Orléans, France.

Annabi-Bergaya, F. (2008) Layered clay minerals. Basic research and innovative composites applications. *Microporous & Mesoporous Materials*, **107**, 141–148.

Annabi-Bergaya, F., Cruz, M.I., Gatineau L. & Fripiat, J.J. (1979) Adsorption of alcohols by smectites. I. Distinction between internal and external surfaces. *Clay Minerals*, **14**, 249–258.

Annabi-Bergaya, F., Cruz, M.I., Gatineau, L. & Fripiat, J.J. (1980a) Adsorption of alcohols by smectites. II. Role of the exchangeable cations. *Clay Minerals*, **15**, 219–223.

Annabi-Bergaya, F., Cruz, M.I., Gatineau, L. & Fripiat, J.J. (1980b) Adsorption of alcohols by smectites. III Nature of the bonds. *Clay Minerals*, **15**, 225–237.

Baes, C.F. Jr. & Mesmer, R.E. (1976) *The Hydrolysis of Cations*. Wiley, New York, 489 pp.

Benna, M., Kbir-Ariguib, N., Clinard, C. & Bergaya, F. (2001) Card-to-house microstructure of purified sodium-montmorillonite gels suspensions evidenced by filtration properties at different pH. *Colloid and Polymer Science*, **117**, 204–210.

Bergaoui, L., Lambert, J.-F., Franck, R., Suquet, H. & Robert, J.-L. (1995) Al-pillared saponites. Part 3. Effect of parent clay layer charge on the intercalation-pillaring mechanism and structural properties. *Journal of Chemical Society, Faraday Transactions*, **91**, 2229–2239.

Bergaya, F. (1990) Argiles à piliers. In: *Matériaux Argileux. Structures, Propriétés et Applications* (A. Decarreau, editor). Société Française de Minéralogie et de Cristallographie (article in French), pp. 513–537.

Bergaya, F. & Lagaly, G. (2006) General introduction: Clays, Clay minerals and Clay Science. In: *Handbook of Clay Science* (F. Bergaya, B.K.G. Theng & G. Lagaly, editors). Developments in Clay Science, Vol. 1. Elsevier, Amsterdam, pp. 1–18.

Bergaya, F. & Lagaly, G. (2007) Clay mineral properties responsible for clay-based polymer nanocomposites (CPN) performance. In: *Clay-Based Polymer Nanocomposites (CPN)*. CMS Workshop Lectures, **15**. The Clay Mincrals Socicty, Chantilly, Virginia, USA, pp. 61–97.

Bergaya, F. & Vayer, M. (1997) CEC of clays: Measurement by adsorption of a copper ethylenediamine complex. *Applied Clay Science*, **12**, 275–280.

Bergaya, F., Mandalia, T. & Amigouet, P. (2005) Nanocomposite based on PE and novel modified raw clays. *Colloid and Polymer Science*, **283**, 773–782.

Bergaya, F., Theng, B.K.G. & Lagaly, G. (editors) (2006a) *Handbook of Clay Science*. Developments in Clay Science, Vol. 1. Elsevier, Amsterdam, 1224 pp.

Bergaya, F., Lagaly, G. & Vayer, M. (2006b) Cation and anion exchange. In: *Handbook of Clay Science* (F. Bergaya, B.K.G. Theng & G. Lagaly, editors). Developments in Clay Science, Vol. 1. Elsevier, Amsterdam, pp. 979–1001.

Bergaya, F., Mandalia, T. & Aouad, A. (2006c) Pillared clays and clay minerals. In: *Handbook of Clay Science* (F. Bergaya, B.K.G. Theng & G. Lagaly, editors). Developments in Clay Science, Vol. 1. Elsevier, Amsterdam, pp. 393–421.

Bergaya, F., Jaber, M. & Lambert, J.F. (2011a) Organophilic clay minerals. In: *Rubber Clay Nanocomposites* (M. Galimberti, editor). Wiley, New York (in press).

Bergaya, F., Jaber, M. & Lambert, J.F. (2011b) Clays and Clay Minerals. In: *Rubber Clay Nanocomposites* (M. Galimberti, editor), Chapter 1. Wiley (in press).

Bleam, W.F. (1990) Electrostatic potential at the basal (001) surface of talc and pyrophyllite as related to tetrahedral sheet distortions. *Clays and Clay Minerals*, **38**, 522–526.

Blumstein, A. (1965) Polymerisation of adsorbed monolayers. II Thermal degradation of the inserted polymers. *Journal of Polymer Science Part A General papers*, **3**, 7PA, 2665–2673.

Boehm, H.P., Setton, R. & Stumpp, E. (1994) Nomenclature and terminology of graphite intercalation compounds. *Pure and Applied Chemistry*, **66**, 1893–1904.

Brigatti, M.F., Malferrari, D., Laurora, A. & Elmi, C. (2011) Structure and mineralogy of layer silicates: recent perspectives and new trends. In: *Layered Mineral Structures and their Application in Advanced Technologies* (M.F. Brigatti & A. Mottana, editors). EMU notes in Mineralogy, **11**. European Mineralogical Union and the Mineralogical Society of Great Britain & Ireland, pp. 1–71.

Capcova, P.C., Pospisil, M., Valaskova, M., Merinska, D., Trochva, M., Zedlakova, Z., Weiss, Z. & Simonik, J. (2006) Structure of montmorillonite cointercalated with stearic acid and octadecylamine: Modeling, diffraction, IR spectroscopy. *Journal of Colloid and Interface Science*, **300**, 264–269.

Carrado, K.A. & Bergaya, F. (editors) (2007) *Clay-Based Polymer Nanocomposites (CPN)*. CMS Workshop Lectures, **15**. The Clay Minerals Society, Chantilly, Virginia, USA.

Christidis, G.E. (2011) The concept of layer charge of smectites and its implications on important smectite-water properties. In: *Layered Mineral Structures and their Application in Advanced Technologies* (M.F. Brigatti & A. Mottana, editors). EMU notes in Mineralogy, **11**. European Mineralogical Union and the Mineralogical Society of Great Britain & Ireland, London, pp. 237–258.

Cruz, M., Jacobs, H. & Fripiat, J.J. (1973) Interlayer bonding in kaolin minerals. *Proceedings of the International Clay Conference*, Madrid, 25–46.

Delville, A. (1993) Structure and properties of confined liquids: a molecular model of the clay-water interface. *Journal of Physical Chemistry*, **97**, 9703–9712.

Dong, Y. & Feng, S.S. (2005) Poly(D,L-lactide-co-glycolide)/montmorillonite nanoparticles for oral delivery anticancer drugs. *Biomaterials*, **26**, 6068–6076.

Figueras, F., Klapyta, Z., Massiani, P., Mountassir, Z., Tichit, D. & Fajula, F. (1990) Use of competitive ion exchange for intercalation of montmorillonite with hydroxyl-aluminium species. *Clays and Clay Minerals*, **38**, 257–264.

Fripiat, J.J., Levitz, P. & Letellier, M. (1984) Interactions of water with clay surfaces. *Philosophical Transaction of the Royal Society, London*, A, **311**, 287–299.

Fu, Y.-T. & Heinz, H. (2010) Cleavage energy of alkyl ammonium modified montmorillonite and relation to exfoliation in nanocomposites: Influence of cation density, head group structure, and chain length. *Chemistry of Materials*, **22**, 1595–1605.

Fukushima, Y. & Inagaki, S. (1987) Synthesis of an intercalated compound of montmorillonite and 6-polyamide. *Journal of Inclusion Phenomena*, **5**, 473–482.

Furukawa, T. & Brindley, G.W. (1973) Adsorption and oxidation of benzidine and aniline by montmorillonite and hectorite. *Clays and Clay Minerals*, **21**, 279–288.

Gan, H., Bailmey, G.W. & Yu, Y.S. (1996) Morphology of Lead (II) and Chromium (III) reaction products on phyllosilicates surfaces as determined by Atomic Force Microscopy. *Clays and Clay Minerals*, **44**, 734–743.

Gardolinsky, J.E.F.C. (2005) *Interlayer Grafting and Delamination of Kaolinite*. PhD thesis, University of Kiel, Germany.

Gardolinsky, J.E.F.C. & Lagaly, G. (2005) Grafted organic derivatives of kaolinites II. Intercalation of primary n-alkylamines and delamination. *Clay Minerals*, **40**, 547–556.

Gil, A., Korili, S.A., Trujillano, R. & Vicente, M.A. (editors) (2010) *Pillared Clays and Related Catalysts*. Springer, 550 pp.

Gregg, S.J. & Sing, K.S.W. (1982) *Adsorption, Surface Area and Porosity*, second edition. Academic Press, New York, 303 pp.

Guggenheim, S. (2011) An overview of order/disorder in hydrous phyllosilicates. In: *Layered Mineral Structures and their Application in Advanced Technologies* (M.F. Brigatti & A. Mottana, editors). EMU Notes in Mineralogy, **11**. European Mineralogical Union and the Mineralogical Society of Great Britain & Ireland, London, pp. 73–121.

Guggenheim, S. & Martin, R.T. (1995) Definition of Clay and Clay Mineral joint report of the AIPEA nomenclature and CMS nomenclature committees. *Clays and Clay Minerals*, **43**, 255–256 and *Clay Minerals*, **30**, 257–259.

Hang, P.T. & Brindley, G.W. (1970) Methylene blue absorption by clays and clay minerals, determination of surface areas and cation exchange capacities (clay-organic Studies). *Clays and Clay Minerals*, **18**, 203–212.

Harvey, C.C. & Lagaly, G. (2006) Conventional applications, In: *Handbook of Clay Science* (F. Bergaya, B.K.G. Theng & G. Lagaly, editors). Developments in Clay Science, Vol. 1. Elsevier, Amsterdam, pp. 501–540.

He, H., Frost, R.L., Bostrom, T., Yuan, P., Duong, L., Yang, D., Xi, Y. & Kloprogge, J.T. (2006) Changes in the morphology of organoclays with $HDTMA^+$ surfactant loading. *Applied Clay Science*, **31**, 262–271.

Heller-Kallai, L. (2001) Protonation–deprotonation of dioctahedral smectites. *Applied Clay Science*, **20**, 27–38.

Houdry, E., Burt, W.F., Pew, A.E. & Peters, W.A. Jr. (1938) Catalytic processing by the Houdry process. *National Petroleum News*, R570–R580.

Jasmund, K. & Lagaly, G. (1993) *Tonminerale und Tone – Struktur, Eigenschaften, Anwendungen und Einsatz in Industrie und Technik.* Steinkopff Verlag, Darmstadt (article in German) also cited by Gardolinsky (2005).

Johnson, C.E. Jr. (1957) Methylene blue adsorption and surface area measurements. *131st National Meeting of the American Chemical Society*, 7–12.

Johnston, C.T., Sheng, G., Teppen, B.J., Boyd, S.A. & De Oliveira, M.F. (2002) Spectroscopic study of dinitrophenol herbicide sorption on smectite. *Environmental Science & Technology*, **36**, 5067–5074.

Jordan, J.W. (1949a) Organophilic bentonites. I. Swelling in organic liquids. *Journal of Physical Chemistry*, **53**, 294–306.

Jordan, J.W. (1949b) Alteration of the properties of bentonite by reaction with amines. *Mineralogical Magazine*, **28**, 598–605.

Kanzaki, Y., Hayashi, M., Minami, C., Inoue, Y., Kogure, M., Watanabe, Y. & Tanaka, T. (1997) Intercalation and bilayer formation of phospholipids in layered synthetic mica. 2. Solvent effect of the intercalation reaction of natural and reduced-type phosphatidylcholines. *Langmuir*, **13**, 3674–3680.

Kipling, J.J. & Wilson, R.B. (1960) Adsorption of methylene blue in the determination of surface areas. *Journal of Applied Chemistry* (London), **10**, 109–113.

Komadel, P. & Madejová, J. (2006) Acid activation of clay minerals. In: *Handbook of Clay Science* (F. Bergaya, B.K.G. Theng & G. Lagaly, editors). Developments in Clay Science, Vol. 1. Elsevier, Amsterdam, pp. 263–287.

Lagaly, G. (1984) Clay-organic interactions. *Philosophical Transactions of the Royal Society of London*, **311**, 315–332.

Lagaly, G. (1986) Interaction of alkylamines with different types of layered compounds. *Solid State Ionics*, **22**, 43–51.

Lagaly, G., Ogawa, M. & Dekany, I. (2006) Clay-organic interaction. In: *Handbook of Clay Science* (F. Bergaya, B.K.G. Theng & G. Lagaly, editors). Developments in Clay Science, Vol. 1. Elsevier, Amsterdam, pp. 309–377.

Lanson, B. (2011) Modelling of X-ray diffraction profiles: Investigation of defective lamellar structure crystal chemistry. In: *Layered Mineral Structures and their Application in Advanced Technologies* (M.F. Brigatti & A. Mottana, editors). EMU notes in Mineralogy, **11**. European Mineralogical Union and the Mineralogical Society of Great Britain & Ireland, London, pp. 151–201.

Letaief, S. & Detellier, C. (2009) Clay-polymer nanocomposite material from the delamination of kaolinite in the presence of sodium polyacrylate. *Langmuir*, **25**, 10975–10979.

Li, J., Li, Y. & Lu, J. (2009) Adsorption of herbicides 2,4-D and acetochlor on inorganic–organic bentonites. *Applied Clay Science*, **46**, 314–318.

Mandalia, T. & Bergaya, F. (2006) Organo clay mineral–melted polyolefin nanocomposites. Effect of surfactant/CEC ratio. *Journal of Physics and Chemistry of Solids*, **67**, 836–845.

Miao, S.D., Bergaya, F. & Schoonheydt, R. (2010) Ultrathin films of clay-protein composites. *Philosophical Magazine*, **90**, 2529–2541.

Michot, L.J. & Villieras, F. (2006) Surface area and porosity. In: *Handbook of Clay Science* (F. Bergaya, B.K.G. Theng & G. Lagaly, editors). Developments in Clay Science, Vol. 1. Elsevier, Amsterdam, pp. 965–1001.

Michot, L.J., Villieras, F., François, M., Bihannic, I., Pelletier, M. & Cases, J.F. (2002) Water organisation at the solid aqueous solution interface. *Compte Rendus Geosciences*, **334**, 611–631.

Mishael, N.Y., Rytwo, G., Nir, S., Crespin, M., Annabi-Bergaya, F. & Van Damme, H. (1999) Interactions of monovalent organic cations with pillared clays. *Journal of Colloid and Interface Science*, **209**, 123–128.

Mishael, N.Y., Nir, S., Rubin, B., Polubesova, T., Bergaya, F., Van Damme, H. & Lagaly, G. (2001) Clay-based formulations of metolachlor with reduced leaching. *Applied Clay Science*, **18**, 265–275.

Mottana, A. & Aldega, L. (2011) Layered mineral structures and their application in advanced techniques. In: *Layered Mineral Structures and their Application in Advanced Technologies* (M.F. Brigatti &

A. Mottana, editors). EMU notes in Mineralogy, **11**. European Mineralogical Union and the Mineralogical Society of Great Britain & Ireland, London, pp. 285–312.

Nalawade, P., Aware, B., Kadam, V.J. & Hirlekar R.S. (2009) Layered double hydroxides. A review. *Journal of Scientific and Industrial Research*, **68**, 267–272.

Norrish, K. (1954) The swelling of montmorillonite. *Discussions of the Faraday Society*, **18**, 120–134.

Ogawa, M., Kuroda, K. & Kato, C. (1989) Preparation of montmorillonite-organic intercalation compounds by solid-solid state reactions. *Chemical Letters*, **18**, 1659–1662.

Plée, D., Borg, F., Gatineau, L. & Fripiat, J.J. (1985) High-resolution solid-state ^{27}Al and ^{29}Si nuclear magnetic resonance study of pillared clays. *Journal of American Chemical Society*, **107**, 2362–2369.

Pinnavaia, T.J. (1983) Intercalated clay catalysts. *Science*, **220**, 365–371.

Pinnavaia, T.J. & Mortland, M.M. (1971) Interlamellar metal complexes of layer silicates. I. Copper(II)-arene complexes on montmorillonite. *The Journal of Physical Chemistry*, **75**, 3957–3962.

Pinnavaia, T.J., Hall, P.L., Cady, S.S. & Mortland, M.M. (1974) Aromatic radical cation formation on the intracrystalline surfaces of transition metal layer lattice silicates. *The Journal of Physical Chemistry*, **78**, 994–999.

Pinnavaia, T.J., Landau, S.D., Tzou, M.S., Johnson, I.D. & Lipsicas, M. (1985) Layer cross linking in pillared clays. *Journal of American Chemical Society*, **107**, 7222–7224.

Richards, A.D. & Rodgers, A. (2007) Synthetic metallomolecules as agents for the control of DNA structure, *Chemical Society Reviews*, **36**, 471–483.

Rouquerol, J., Rodriguez-Reinoso, F., Sing, K.S.W. & Unger, K.K. (editors) (1994) *Characterization of Porous Solids III*. Studies in Surface Science and catalysis, **87**. Elsevier, Amsterdam, 802 pp.

Samii, M.M. & Lagaly, G. (1987) Adsorption of nuclein bases on smectites In: *Proceedings of the International Clay Conference, Denver, 1985* (L.G. Schultz, H. van Olphen & F.A. Mumpton, editors). The Clay Mineral Society, Bloomington, Indiana, USA, pp. 363–369.

Shen, S. & Stucki, J.W. (1994) Effects of iron oxidation state on the fate and behaviour of potassium in soils. In: *Soil Testing: Prospect for Improving Nutrients Recommendations* (J.L. Havlin & J.S. Jacobsen, editors). SSSA Special Publication **40**, Soil Science Society of America, Madison, Wisconsin, USA, pp. 173–185.

Shichi, T. & Takagi, K. (2000) Clay minerals as photochemical reaction fields. *Journal of Photochemistry and Photobiology C: Photochemistry Reviews*, **1**, 113–130.

Solomon, D.H., Loft, B.C. & Swift, J.C. (1968) Reactions catalysed by minerals – IV. The mechanism of the benzidine-blue reaction on silicate minerals. *Clay Minerals*, **7**, 389–397.

Stucki, J.W. (2006) Properties and behavior of iron in clay minerals. In: *Handbook of Clay Science* (F. Bergaya, B.K.G. Theng & G. Lagaly, editors). Developments in Clay Science, Vol. **1**. Elsevier, Amsterdam, pp. 429–482.

Sun, D., Li, B., Li, Y., Yu, C., Zhang, B. & Fei, H. (2011) Characterization of exfoliated/delamination kaolinite. *Materials Research Bulletin*, **46**, 101–104.

Theng, B.K.G. (1974) *The Chemistry of Clay-Organic Reactions*. Adam Hilger, London, 343 pp.

Tonle, I.K., Diaco, T., Ngameni, E. & Detellier, C. (2007) Nanohybrid kaolinite-based materials obtained from the interlayer grafting of 3-aminopropyl triethoxysilane and their potential used as electrochemical sensors. *Chemistry of Materials*, **19**, 6629–6636.

Tunney, J.J. and Detellier, C. (1993) Interlamellar covalent grafting of organic units on kaolinite. *Chemistry of Materials*, **5**, 747–748.

Vaia, R.A. (1995) *Polymer Intercalation in Mica-Type Layered Silicates*. PhD thesis, Cornell University, New York, USA.

Vaia, R.A., Ishii, H. & Giannelis, E.P. (1993) Synthesis and properties two-dimensional nanostructures by direct intercalation of polymer melts in layered silicates. *Chemistry of Materials*, **5**, 1694–1696.

Vaia, R.A., Teukolsky, R.K. & Giannelis, E.P. (1994) Interlayer structure and molecular environment of alkylammonium layered silicates. *Chemistry of Materials*, **6**, 1017–1022.

Viseras, C., Aguzzi, C., Cerezo, P. & Lopez-Galindo, A. (2007) Uses of clay minerals in semisolid health care and therapeutic products. *Applied Clay Science*, **3**, 37–50.

Wada, K. (1961) Lattice expansion of kaolin minerals by treatment with potassium acetate. *American Mineralogist*, **46**, 78–91.

Weiss, A. (1961) Eine Schichteinschlussverbindung von Kaolinit mit Harnstoff. *Angewandte Chemie*, **73**, 736–737, (in German) also cited by Gardolinsky (2005) and by Lagaly *et al.* (2006).

Weiss, A., Mehler, A. & Hofmann, U. (1956) Kationenaustausch und innerkristallines Quellungsvermögen bei den Mineralen der Glimmergruppe. *Zeitschrift für Naturforschung*, B, **11**, 435–438.

Zeng, X.Q. & Liu, W.P. (2005) Adsorption of Direct Green B on mixed hydroxy-Fe-Al pillared montmorillonite with large basal spacing. *Journal of Environmental Sciences*, **17**, 1159–1162.

EMU Notes in Mineralogy, Vol. 11 (2011), Chapter 8, 285–312

Advanced techniques to define intercalation processes

ANNIBALE MOTTANA and LUCA ALDEGA

Dipartimento di Scienze Geologiche, Università degli Studi Roma Tre,
Largo S. Leonardo Murialdo 1, 00146 Roma, Italy
e-mail: mottana@uniroma3.it

Intercalation is the inclusion or reversible insertion of a guest chemical species (atom, ion, molecule) in a virtually unchanged host-crystal structure. Any type of layer-structured material may give rise to intercalated compounds, the guest species being artificially inserted or naturally included between the host sheets without loss of their planarity. Layer silicates, in particular, may be considered intercalated structures where interlayer guest species and complexes are inserted between the silicate layers. The most common guest species is H_2O, which is generally present under natural conditions in intercalated layer silicates such as smectites, vermiculite and halloysite.

Past research focused attention on the swelling/shrinking behaviour of intercalated compounds with respect to H_2O, and also on the non-stoichiometric, heterogeneous complexes formed from organic liquids such as ethylene glycol and glycerol. The unique combination of layer-silicate features (small crystal size, large surface area) and the small concentrations required to effect a change in the matrix, both coupled with the advanced characterization techniques available, have generated much interest. This interest extends to the special field of nanocomposites, and of graphene, which is also an intercalated layered structure. In general, any guest material inserted into an interlayer space causes a modification in the structure, with spacing-size changes in a particular crystallographic direction (*d* value). First, a brief introduction on conventional and synchroton-based X-ray techniques used to define crystal size and thickness is given. Then, the peak-broadening approach by conventional X-ray diffraction (XRD) techniques, such as the Scherrer method is presented. Further on, the crystallinity measurements and the Bertaut-Warren-Averbach (BWA) method used in the *MudMaster* program are described. A short summary is presented of the grazing-incidence diffraction (GIXRD) technique. Finally, additional and complementary information from X-ray absorption spectrometry (XAS), such as short-range order, and detailed local information on atomic positions by angle-resolved X-ray absorption near-edge stucture (AXANES), polarized extended X-ray absorption fine structure (P-EXAFS), and near-edge extended absorption fine structure (NEXAFS) spectroscopies are analysed and discussed. Examples of the applications of these methods to clay minerals, micas and graphene are given.

1. Introduction

Most intercalated layered structures occur naturally in the form of fine- to very fine-grained particles that are difficult to characterize properly by standard methods. However, advanced techniques using X-rays – those derived from a synchrotron radiation source – have been used to understand the essential components of intercalated structures at the local atomic range, thus making the understanding of the physical and chemical properties of related materials possible. To achieve this result, a model

DOI: 10.1180/EMU-notes.11.8

based on larger particles was needed, and micas – which are intrinsically intercalated and layered structures – were appropriate materials for study. Consequently, most of the following examples use micas, but then we show applications to clays, oxi-hydroxides and, ultimately, to graphene. Graphene is an extreme case of a two-dimensionally extended layered structure, with outstanding physical and chemical properties that are now being explored and improved.

Intercalation compounds are a subgroup of the general class of highly anisotropic, layered solids with structures of fairly rigid two-dimensional (2D) layers of atoms linked by intraplanar binding forces that dominate over interplanar binding forces. The interlayer (*I*) may be empty or contain intercalated atoms, ions and molecules (known as 'guests'), which often but not always are in a 2D arrangement that is controlled by the 'host' structure. Insertion of atoms in a different structural arrangement may result in complex structures which develop over several layers (quasi-2D materials). Structural expansion in a direction perpendicular to the plane of the host sheets is a positive indication of intercalation and, sometimes, gives useful information on the type of intercalated material. Some intercalated species are weakly bonded to the host layered framework and can be exchanged readily; by contrast, others are so strongly bonded that they are able to modify the periodicity of the host framework itself, giving rise to commensurate-incommensurate transitions (Sinha, 1980; Bak, 1982).

The prototype of intercalated layered structures is graphite, in which intercalation was discovered as early as 1841 by Carl Schaffhäutl (Dresselhaus & Dresselhaus, 2002, p. 3). Most basic concepts about intercalation and the finest technological application are from studies on graphite, and most nomenclature and terminology are derived from IUPAC-approved rules (Boehm *et al.*, 1994). The conceptual gap from graphite to graphene is not large; however, it took a long time to achieve advancement in terms of the structure because special preparation methods were required. Currently graphene represents one of the major frontiers of solid-state science (Geim & Novoselov, 2007; Castro Neto *et al.*, 2009).

Among rock-forming minerals, common intercalated structures involve phyllosilicates, which may form regularly stacked structures with or without materials in the interlayer where chemical species of any type may be inserted. As a result, many phyllosilicates are non-stoichiometric intercalate compounds showing stacking variability ('polytypes'). Phyllosilicates of very small grain size, such as smectites among the clay minerals, represent typical cases; the nano-sized varieties of smectites can form films and suspensions with interesting technological applications which have not yet been explored fully.

For many years, powder X-ray diffraction (P-XRD) was used to characterize intercalated materials and this technique is still a very effective probe of 2D structures. Crystal symmetry, *d* values and unit-cell constants may be determined easily. Peak profile analysis may give information on crystallite size and on inter- *vs.* intra-planar atom correlations. Recent development of computer programs has made P-XRD the routine method for clay-sized minerals.

X-ray absorption spectroscopy (XAS) is a less widespread probing method, but it has contributed enormously to our understanding of intercalation features and processes at the atomic level. The XAS is atom-specific because it probes the structural short-range order of an atom selectively and independently of all other atoms present in the compound. By contrast, XRD is structure-specific, as it determines the atom's symmetric long-range order over the entire framework, which usually consists of more than two atom types.

XAS has the intrinsic property of being anisotropic, and the signal measured depends on the X-ray beam polarization direction relative to the crystal axes of the material. This property is particularly important for intercalated structures, because: (1) it involves the 2D development of layered structures, with the incident beam polarizated either parallel or orthogonal to the layers, and (2) it is sensitive to the atom absorption K-edge, because, for 1s transitions, the final local empty p states reflect highly directional bonds. Thus, XAS, when properly angle-resolved, may be suitable for studying not only the guest atom oxidation state and its local coordination, but also its distribution over planes. Site distortion may be investigated also, *i.e.* how the ligands in the host structure deviate from their expected symmetry positions owing to the inequality of the bond lengths and angles they actually form with the probed guest atom.

The aim of this chapter is to note recent results on quasi-2D intercalated solids, mainly phyllosilicates (the most common naturally occurring 2D layered structures) and graphene, which is an intriguing novelty. We show the relative contribution derived from a series of XRD and XAS techniques, integrated with some applications for several compounds. Whereas all XAS methods require a synchrotron-based radiation source to be effective, XRD analyses require a conventional, inexpensive source. Consequently, the present trend in material science is to use a synchrotron source only for complex compounds or for very minute particles (nano-particles).

2. X-ray diffraction

2.1. Introduction

Because an intercalated structure forms from a host crystal structure with a guest chemical species sandwiched between the host layers, phyllosilicates may be described as consisting of three essential components: (1) the phyllosilicate layer itself (1:1 or 2:1 layers), having an overall negative charge owing to ionic substitutions and which constitutes the host layers; (2) the exchangeable cations which compensate the overall negative charge of the layer and which occur between the layers; and (3) neutral molecules, which, if present, are sandwiched between the host layers.

New intercalated structures form when the interlayer components are replaced by foreign materials. Naturally occurring monovalent and divalent cations (*e.g.* K^+, Ca^{2+}) located in the interlayer of phyllosilicates may be replaced by a variety of simple or complex inorganic or organic cations. The most common interlayer neutral molecule is H_2O, which, under natural conditions, is present between the layers of many 2:1 phyllosilicates (*e.g.* smectite, vermiculite). A water molecule may be replaced by glycols, alcohols, amines and other neutral molecules. In 1:1 phyllosilicates, only neutral molecules may occur between the host layers, because these phyllosilicates do not have a net negative charge on the layer. In some 1:1 silicates (*e.g.* halloysite) water molecules are present under natural conditions.

In the early literature on interlayer complexes in phyllosilicates, the focus was on the swelling-shrinking behaviour and the presence of water, and on the complexes that the host phyllosilicate forms with organic liquids such as ethylene glycol and glycerol. Many studies have been performed on smectite hydration states, showing that the basal spacing of smectite is dependent on the different exchange cation (Iwasaki & Watanabe, 1988; Cuadros, 1997) and on the magnitude and location of the layer charge (octahedral *vs.* tetrahedral: see Sato *et al.*, 1992; Ferrage *et al.*, 2007). Different hydration states may correspond to 0, 1, 2 or 3 planes of water molecules present in smectite interlayers. Intercalation of H_2O produces a stepwise hydration depending on relative humidity, which can be measured easily by XRD *via* a change in basal spacing. In general, any material (*e.g.* cations, organic molecules, polymers) naturally included in or artificially inserted into an interlayer causes a major modification in the structure of phyllosilicates, with *d* value changes in a particular crystallographic direction.

Conventional and synchroton-based X-ray techniques allow for the precise description of the intercalation process and provide representative data, because each measurement is the average of millions of small particles. In the next sections, we provide a brief introduction of these techniques, with special attention to those that describe crystal size and thickness, and considering mainly phyllosilicates, clay minerals and films.

2.2. XRD measurement of crystal thickness

Particle-size studies are important for many research fields, from geology to mineralogy, chemistry, engineering, and medicine. Such studies may yield information about sediment provenance, transport mechanism, degree of weathering and/or metamorphism, and their influence on the physical and chemical properties of a compound (*e.g.* rheology, solubility, cation exchange capacity (CEC), surface area).

Particles with diameter of >3 μm are distinguished optically by light microscopy. Minerals of smaller size, such as clay minerals and nano-size crystals, cannot be measured by a petrographic microscope. Laser scattering methods can be used for samples having particles ranging from 0.02 to 2000 μm in size, but they provide only average particle size values, and size cannot be measured along a crystallographic direction. In addition, the particle measured by laser techniques may be composed of several different minerals and/or by aggregates of individual crystals. Measurements by electron and atomic force microscopy are tedious, expensive, and time consuming. The data obtained involve very small sample volumes; they may yield an accurate mean size for one sample, but the data are often too few to determine an accurate distribution of sizes.

Methods based on XRD are more efficient than most other techniques, because XRD provides data averaged over relatively large volumes from many billions of individual crystals. The crystallite size measured by XRD is often related to the average size of the individual crystals of the investigated sample, rather than to the aggregate size of the crystals. Therefore, the average size of a particular mineral, in a mixture of minerals, can be measured by XRD. In addition, the size of that mineral along definite crystallographic directions can be determined. A shortcoming of the XRD technique is that the size that is measured is the X-ray coherent scattering domain size ('crystallite size'), which may or may not be equal to the crystal size.

The 'crystallite size' is defined as equal to

$$(N - 1)d_{hkl} \tag{1}$$

where N is the number of *hkl* planes responsible for a reflection and d_{hkl} is the distance between a set of *hkl* planes. This expression considers the fact that layers and *hkl* planes are not the same, because layers have a thickness and planes do not. Therefore, it is always useful to check the measured XRD crystallite thickness by another method.

Information on crystal size may be extracted from the *00l* XRD peak profiles of layered minerals. Before performing this analysis, layered minerals must be oriented to enhance their *00l* reflections. The most commonly used method to obtain oriented samples is the glass slide method. In this method, either <0.2 or <2 μm grain-size fractions, previously separated from the rock, are sedimented onto a glass microscope slide and then dried in an oven or in air. In our opinion, an ultrasonic probe should be used for a few seconds to completely disperse the clay before making the slide. Moreover, the clay layer should be sufficiently thick to ensure 'infinite thickness' and avoid the broad XRD band from the underlying glass.

Clay-mineral separation is achieved by a variety of mechanical and chemical procedures. The procedure depends on the degree of consolidation of the rock and the presence of floculation or cementing substances (*e.g.* salts, sulphates, oxides, organic matter). For detailed information on clay-mineral separation and other techniques for preparing oriented samples for XRD investigation, we suggest Moore & Reynolds (1997) or Środoń (2006).

The first approach at crystal-size determination dates back to Scherrer (1918), who proposed the following formula based on the observation that XRD peaks are broadened regularly as a function of decreasing crystallite size. After corrections for instrument effects on peak broadening have been made, the crystal size is calculated as a function of peak width [specified as the intensity measured for full width at half maximum (FWHM) of the peak], peak position, and wavelength:

$$\beta_{sh} = K\lambda / T\cos\theta \tag{2}$$

where β_{sh} is the FWHM of the *00l* peak, in radians; K is a dimensionless shape factor; λ is the X-ray wavelength, in nanometres; T is the mean thickness of the coherent scattering domains, and θ is the Bragg angle. The Scherrer formula is applicable only in the particular case where coherent scattering domains have a single thickness. However, minerals exhibit many distributions of thicknesses and β_{sh} depends on the mean thickness of the coherent scattering domains and the law of the thickness distribution (Drits *et al.*, 1997).

The assumption that 'the narrower the diffraction peak the larger the crystal size' has been applied extensively for the last 50 years in geology and metamorphic petrology to evaluate the grade of metapelitic rocks (*e.g.* Kübler, 1967; Árkai *et al.*, 1995; Mählmann, 2001; Aldega *et al.*, 2007; Corrado *et al.*, 2010). Regional studies on clay mineralogy relating to very low-grade metamorphism led to the development of XRD-based clay mineral 'crystallinity' techniques to monitor the transition from diagenesis to the epizone. This term 'crystallinity' is commonly applied to illite, although 'crystallinity' can also be determined on other phyllosilicates (*e.g.* chlorite, kaolinite). The term 'crystallinity' is qualitative and depends on the type of order, the dimensional nature of the periodicity present, and the technique involved in its measurement. For phyllosilicates, which are low-symmetry materials with strongly anisotropic structures, various chemical compositions and with random or ordered interstratifications of various structures, 'crystallinity' is ambiguous. Guggenheim *et al.* (2002) suggested that the use of a 'crystallinity' index should avoided and instead be referred to the name of the author who originally described the parameter (*e.g.* Kübler index for illite).

Currently, the most common parameter used to determine grade in metapelitic sequences is the Kübler index (Kübler, 1967), which measures the FWHM of the first basal 1.0 nm P-XRD peak of dioctahedral illite-muscovite on the <2 μm size-fraction of air-dried clay samples by using Cu$K\alpha$ radiation. The Kübler index is a complex function of the illite-smectite layer ratio in mixed-layer clay minerals, the d value of smectite swelling layers, the ratio between discrete illite and mixed-layer illite-smectite, and the crystal-size distribution of the two clay minerals. The Kübler index measurement can be made more rigorous if a modified Scherrer equation is applied to the 001 peak of K^+-exchanged and heated samples (see Drits *et al.*, 1997 for theory). Under such conditions, the mixed layering effect on peak broadening is avoided and the 1.0 nm peak width becomes exclusively a function of crystal thickness distribution; thus, the mean crystal thickness of the mixture can be measured.

An alternative approach to the Kübler index is the measurement of the crystallite thickness distribution by the Bertaut-Warren-Averbach (BWA) method, which can be applied to clay minerals having a periodic structure along the c axis (Drits *et al.*, 1998). A detailed analysis of crystallite size is possible because the intensity function for XRD peaks can be represented as a Fourier series. Fourier analysis is a mathematical method used to determine the collection of sine waves, differing in frequency and amplitude, that is necessary to approximate a function. The Fourier analysis of the interference function for clay minerals yields Fourier coefficients for the various crystallite thicknesses that can then be analysed according to the BWA theory to yield crystallite-thickness distributions. These distributions are found by calculating the second derivative of a plot of the Fourier coefficients *vs.* the number of layers in the crystallites.

The BWA method has been codified by the *Mudmaster* computer program that extracts information on crystal thickness distributions from the peak shape (Eberl *et al.*, 1996). This information is contained in the interference function (Φ), which can be calculated from the measured diffraction intensity (I) by subtracting the background (bg) and by dividing by the Lorentz-polarization factor (Lp) and by the layer scattering intensity (G^2):

$$I = \mathrm{Lp}G^2\Phi + bg \tag{3}$$

thus,

$$\Phi = (I - \text{bg})/\text{Lp}G^2 \tag{4}$$

For a given sample only I is measured. The remaining factors are approximated to extract Φ, or they can be calculated independently (Moore & Reynolds, 1997). Additional corrections to experimental XRD data can be made by deconvolving the effects of instrumental broadening, of $K\alpha_2$ radiation, and of strain. These calculations are performed by the *MudMaster* computer program (Eberl *et al.*, 1996).

Phyllosilicates and non-swelling clay minerals can be studied by this technique in their natural state. For mixed-layered minerals with a swelling component or other swelling minerals, the swelling effect must be eliminated before the analysis. This is done in two ways. The first technique consists of heating K-saturated samples and then analysing them in dry air. In this case, the *00l* reflections yield thickness information about the stacks of crystals [*e.g.* for mixed-layer illite-smectite (hereafter I-S) it gives the thickness of the MacEwan crystallites, see Reynolds, 1980]. A second technique involves intercalating a polymer, polyvinylpyrrolidone (PVP-10), into the mixed-layer clay minerals to make the crystals swell until they no longer diffract coherently by interparticle diffraction (Eberl *et al.*, 1998a). In this case, the BWA method gives the thicknesses of the individual crystals (the 'fundamental particles' of Nadeau *et al.*, 1984).

This technique has been calibrated on illite crystals that form the 'building blocks' of mixed-layer I-S crystals (Eberl *et al.*, 1998a), but it can be used for other clay minerals also (*e.g.* Šucha *et al.*, 1999; Simić and Uhlík, 2006). The BWA approach cannot detect monolayers of smectite particles because their effect on the XRD peak width is lost in the background. Therefore, this technique is limited to measuring the thickness of fundamental illite particles >2 nm thick. Thus, the PVP-method gives accurate mean thicknesses for mixed-layer I-S that are composed of $<\sim 50\%$ expandable layers.

The upper limit for thickness determinations depends significantly on the accuracy of the diffractometer used to measure instrumental broadening. In general, using narrow slits and a modern X-ray system allows one to ignore instrument-related broadening to a mean thickness of $\sim 25-30$ nm for illite (Kotarba & Środoń, 2000; Eberl *et al.*, 2003). Reflections for non-periodic systems (*e.g.* basal reflections of random mixed-layer structures as well as *hkl* reflections of minerals containing stacking faults) cannot be analysed by the BWA approach.

The BWA technique was applied to track the changes in I-S from the Carpathian foredeep (Kotarba and Środoń, 2000), to study the weathering processes which affected smectite and mixed layer I-S (Šucha *et al.*, 2001), to measure the crystallite size changes of pyrophyllite during grinding (Uhlík *et al.*, 2000), and to reveal the contribution of diagenetic and detrital illite crystals to the shape of crystal-size distributions (Aldega & Eberl, 2005). In addition, the shapes of crystallite thicknesses (lognormal, asymptotic and steady-state) measured by the BWA method can be related to crystal-growth mechanisms (Eberl *et al.*, 1998b) for low-grade and hydrothermal illites (Bove *et al.*, 2002; Brime & Eberl, 2002).

2.3. Grazing incidence X-ray diffraction

Standard XRD techniques are of limited value for very dilute suspensions of clays on glass slides or for nanoscale films grown on a susbstrate. In these cases, the clay films are too thin to provide diffraction tracings with accurate relative intensity using conventional XRD geometry. These limitations are enhanced at high diffraction angles where the X-ray beam penetrates to levels deeper than the film thickness, thus producing patterns dominated by a low signal-to-noise ratio as a result of glass scattering.

Grazing incidence diffraction (GIXRD) and synchrotron sources permit considerable enhancement of the intensity from thin films relative to the substrate, and excellent depth sensitivity for thin films with smooth surfaces and high crystallinity (Marra *et al.*, 1979). Such high intensity enables characterization of surface chemistry at sensitivities approached by few other methodologies. Specific trace-element concentrations of near 10^8 atoms cm^{-2} can be determined even in the presence of other more concentrated impurities (Waychunas, 2002). Therefore, GIXRD is useful for the characterization of extremely small surface concentrations of geochemically important elements.

X-ray penetration depth in GIXRD can be tuned by varying the incidence angle of the incoming beam. By employing an angle of incidence, φ, which is smaller than the critical angle (φ_c), the penetration of the X-ray beam can be limited to a few nanometres (~5.0 nm), allowing surface-sensitive measurement. For larger values of φ, above the critical angle, the beam penetrates through the bulk of the film into the substrate, and the bulk scattering is measured.

GIXRD geometry is based on the concept that the refractive index for most materials is slightly <1.0 for X-ray energies. Thus, total external reflection occurs from a surface if $\varphi<\varphi_c$. At these low angles (typical φ values are 0.2° to 0.6° for Cu*K*α radiation: Huang, 1990), the total externally reflected wave penetrates only a few nanometres into the sample, giving extremely good surface sensitivity. Maintaining a low incidence angle throughout the scan increases the path length through the sample resulting in diffraction patterns with greater relative peak intensities and lower substrate scatter. Because the path lengh increases when the grazing incidence angle is used, the diffracting volume increases proportionally, therefore enhancing the signal strength.

When the specimen is oriented to satisfy Bragg conditions in the plane of the sample (Fig. 1a, b), a strong diffracted beam occurs at the same grazing angle as the incident beam. In such an arrangement, Bragg planes normal to the sample surface are probed. For GIXRD, the incident and diffracted beams are made nearly parallel by means of a narrow slit or a Göbel mirror on the incident beam and Soller slits on the diffracted beam. During the experiment, the sample surface and the X-ray source remain stationary throughout the detector scan (Fig. 1c). GIXRD is a non-focusing geometric arrangement that uses parallel, rather than divergent X-rays diffracted from suitably oriented crystallographic planes.

GIXRD has limitations. The method is only useful with smooth surfaces and at grazing angles most of the incident X-ray beam is wasted. Assuming an incident angle of 0.1° and a surface dimension of 1 cm, only a 2 μm slice of the X-ray beam falls on the surface; the remaining portion either hits the side of the sample or passes over it. Thus, a focusing geometry may or may not help: if the incident beam is converging to a focus at the sample, the beam contains a range of incident directions and not all of them are below the critical angle. Therefore, the need is for a beam that is intrinsically collimated and intense such as that provided by synchrotron sources. However, even with synchrotron radiation, photons are lost in the grazing incidence geometry. For example, at SSRL (SPEAR II) with a 2 mm vertical height beam, 98% of the beam is discarded (Waychunas, 2002). For this reason, many experiments described as being in the grazing incidence geometry actually use incident angles somewhat larger than the critical angle to increase the number of incident photons.

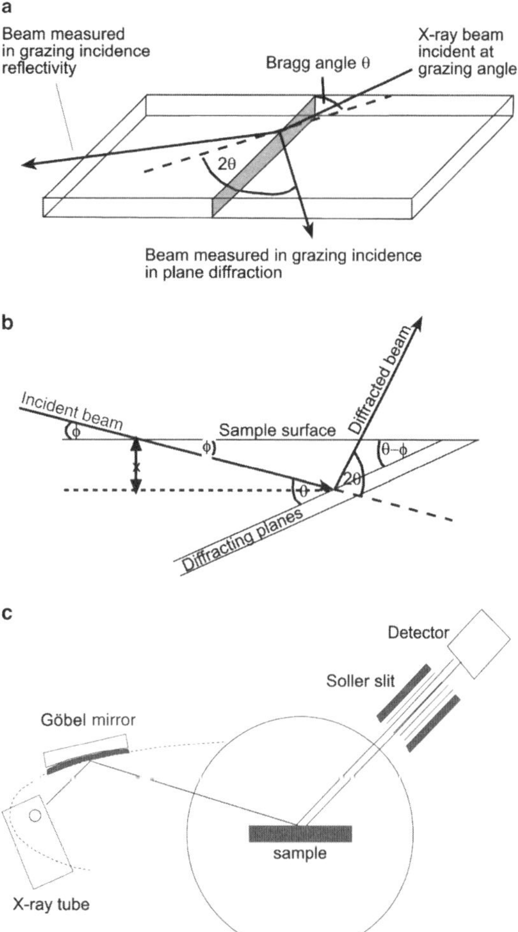

Fig. 1. GIXRD geometry showing the incident, diffracted and specularly reflected beam (***a***); path of a diffracted X-ray beam in GIXRD (after Tanner *et al.*, 2003, redrawn and modified) (***b***); example of GIXRD experimental setup (after Ozalas & Hajek, 1996, redrawn and modified) (***c***).

GIXRD is generally applied in material-science industries to characterize crystalline surface phases of thin films associated with magnetic tapes and superconductive and semicondutive materials (Toney & Brennan, 1989; Rhan *et al.*, 1993). GIXRD has also been used to characterize surface textures and corrosion features of metals (Larsen *et al.*, 1989; Szpunar *et al.*, 1993) and to study biodeterioration and weathering effects on stone buildings, pigments and mixtures (Herrera & Videla, 2009).

Other applications include measuring the thickness of an amorphous surface layer by recording the integral intensity of GIXRD as a function of the incidence angle (*i.e.* as with usual X-ray specular reflectivity measurements, but with the detector in the diffracted beam direction). The latter was very effective for the measurement of amorphization effects in ion implantation (Golovin *et al.*, 1985; Rugel *et al.*, 1993). Recently, Briscoe *et al.* (2007) showed that nanofilm structures on a mica substrate can be described by conventional XRD sources.

GIXRD has not been applied extensively to phyllosilicates and clays, suggesting that this technique is in its infancy for these minerals and further studies are needed.

3. X-ray absorption spectroscopy

The XAS spectrum of an atom theoretically extends from the absorption edge E_0 [eV] to infinity. However, prominent features, *i.e.* those with sufficient signal-to-noise ratio, cannot be recorded above a few hundred eV or, at most, one thousand eV beyond the absorption edge.

Conventionally, a *K*-edge XAS spectrum is divided into three parts, namely the pre-edge (PE) region, the near-edge region (XANES, also called NEXAFS when very light atoms or the *L*-edge spectra are studied), and the extended (EXAFS) region. These three regions provide independent and complementary information on the physical-chemical environment of the atom probed. The potential of EXAFS, *i.e.* the XAS technique that has been applied most often to intercalated structures, has been reviewed extensively (Gates, 2006). Therefore, emphasis here is on the other two regions and techniques and their applications on layered structures, and the information here is updated from previous reviews (*e.g.* Mottana, 2004 and references therein).

The three energy regions of the experimental absorption spectrum are recorded differently, and then analysed by different mathematical formalisms. These analyses are independently useful to extract chemical and structural information. Consequently, there are three techniques (or types) of XAS spectroscopy applied to solid compounds. All these techniques are described by the acronym XAFS (where F stands for fine), indicating the detailed data and analysis of a significant portion of the full absorption spectrum arising from the probed atom located in a crystal structure and subjected to a crystal field. The outline of this section includes (1) the best practice used to most efficiently record the data, (2) the theoretical basis, (3) the results obtained on a series of intercalated mineral structures and, finally, (4) the technologically relevant potential uses of the results. Very important preliminary steps prior to analysis, such as sample preparation (*e.g.* thickness, homogeneity, preferred orientation, *etc.*), instrumental setting (type of source, beam optimization, monochromator alignment, energy calibration, *etc.*) and spectrum normalization (background fitting, signal to noise ratio, *etc.*) must be considered to avoid systematic and accidental errors that could be interpreted as real physical effects instead of artifact.

3.1. Preparing the sample

Most layered materials of technological interest are fine-grained in size and platy in habit; their preparation and placement in the sample holder should exploit these properties, and thus the crystals should lie flat so that they are set parallel to the holder and tightly packed. The incident polarized X-ray beam is oriented with its electrical vector, ε, parallel to the sample surface. The amplitude of the scattered photoelectron wave depends on the angle between the incident beam and the scattering pair of inter-atomic vectors that lie on a lattice plane of the layered structure (Heald & Stern, 1977). In such a setting, the amplitude of the scattered photoelectron wave is maximized.

Powder mounts are normally assumed to be randomly oriented and to produce isotropic representative spectra. This is true for materials that do not exhibit cleavage, but layered minerals have a pronounced cleavage. The results show that nano-sized layer silicates have a crystallite preferred orientation that strongly affects the shape of the XAS by primarily changing peak intensity, which is a significant component of the information. A theoretical study of the orientation effect suggests that the average, isotropic information can be attained by rotating the sample to an angle θ_m such as $\sin^2\theta_m = \frac{1}{3}$, *i.e.* $\theta_m = 35.26^\circ$ around an axis at right angles to both the beam direction and its polarization axis (Pettifer *et al.*, 1992). This value is analogous to the theoretical 'magic angle' spinning value of 54.7° in Mössbauer and nuclear magnetic resonance (NMR) analyses (Pettifer *et al.*, 1992).

Furthermore, powders obtained by grinding require optical examination and by XRD, because grinding may induce damage into solids by making part of the sample amorphous. Only settling from liquid for fixed time (Stokes law) assures a fairly homogeneous grain size and, in particular, undamaged layered mineral samples, thus allowing reproducible experimental XAFS spectra (Manceau *et al.*, 1998).

The orientation dependence of XAS spectra was originally studied only for minerals with perfect cleavage such as halides (Brown *et al.*, 1977). More recently, a practical application was developed for self-supporting clay-mineral thin films (Manceau *et al.*, 1998). For these films, a three-dimensional system of co-ordinates to record spectra in a standard setting was also proposed (Manceau *et al.*, 1999), and applied to several clay-size layer silicates (Manceau *et al.*, 2000; Manceau & Schlegel, 2001).

Single-crystal blades and cleavage flakes lying flat on the sample holder are also suitable for investigation. They are optically oriented with their $\boldsymbol{a} \cong \boldsymbol{b} \perp Z$, where Z is an axis orthogonal to the synchrotron radiation beam forward direction and to the flake surface, *i.e.* ~[001] (Fig. 2). The incident beam is initially collimated to impinge at a right angle ($\theta = 0^\circ$). Then the blade is rotated and the angle θ is stepwise increased to ~80°; this is the maximum rotation allowed by a conventional detection system. Rotation allows scanning atoms and bonds under different angles, although always on a single plane. This plane may be later rotated by 90°, thus exploring the entire specimen geometrical system. The best correspondence of the XAS spectrum for the same material when ground to an

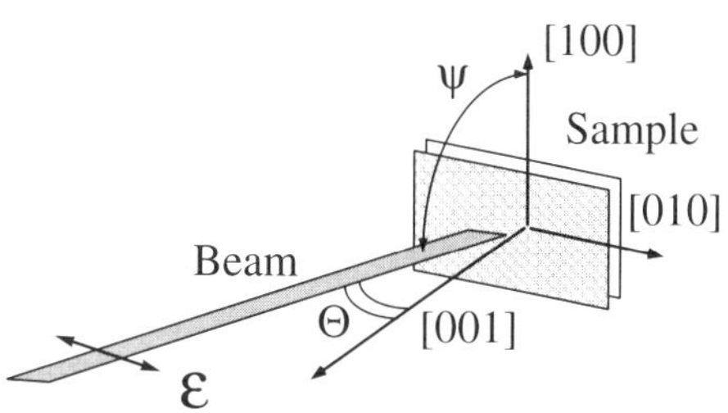

Fig. 2. Instrumental design to record angle-dependent spectra of layer-structured minerals by exploiting the polarized character of the synchrotron radiation beam. ε = SR polarization plane; θ = horizontal rotation angle; ψ = fixed 90° angle between the horizontal beam and the surface of the sample (redrawn and modified after Cibin *et al.*, 2010, fig. 2). The set up can be inverted so as to interchange θ and ψ.

homogeneous fine-grained powder is obtained for $\theta = 35-40^\circ$, *i.e.* the value which corresponds approximately to the magic angle, as expected.

3.2. Choosing the strategy for recording the experimental spectrum

If the energy region to be studied is the XANES (which usually includes the PE also), the signal-to-noise ratio is medium to strong, and thus measurements may be performed in the fluorescence (FY) or total electron yield (TEY) modes. For such regions of the absorption spectrum, any type of flat surface, including cleavage planes of single crystals set onto a metal holder, is appropriate. In contrast, signals in the EXAFS region are rather weak, and measurements are usually performed in transmission mode (TM): powders are compressed into a metal-framed window in sufficient quantity to ensure ~80–90% transmission. However, for many technological applications, powders cannot be analysed without some specific treatments (*e.g.* dispersed in liquids to obtain delaminated particles reacting with suitable chemicals, sprayed or deposited on membranes, or smeared on tape, *etc.*). When using cleaved flakes, possible problems arise from the roughness of the probed surface. This problem is unimportant in most XAS measurements in the hard X-ray energy range (>4 KeV), but it becomes significant in the soft and ultra-soft X-ray ranges, *i.e.* in NEXAFS studies (Kasrai *et al.*, 1998). Observing the surface under the microscope is desirable, followed by re-preparation of the sample if it is rough.

3.3. Recording and optimizing the experimental spectrum

Spectra are normally recorded and processed at room temperature according to standardized procedures (Mottana, 2004). Moving the monochromator by 0.3–0.5 eV steps is a common approach, when operating in TEY mode. TEY strongly reduces self-absorption, if any, because only a very thin surface layer contributes to the signal (~5 nm; Kasrai *et al.*, 1996). Moreover, TEY allows better comparison of spectra taken at θ rotation angles which are very different, especially because TEY does not introduce amplitude deformations as a function of θ.

The full spectrum (*i.e.* PE + XANES + EXAFS) is customarily collected during a single run, which includes several replications, averaging the results, and then producing a convolute final spectrum. This spectrum has an enhanced signal-to-noise ratio and is devoid of artifacts. Thus, collection of data is normally determined such that for each run the three energy regions are optimized independently: two rapid recordings in large eV steps at the beginning and at the end (from −100/200 eV to −20/30 eV, *i.e.* before the expected edge energy [below PE], and from +50/60 eV to +1000/2000 eV after the edge energy *i.e.* in the EXAFS), and an intermediate slow recording in short steps at long counting time [PE + XANES]. Step size and counting time may be different depending on the energy range being scanned, but the resolution is always high ($>10^{-4}$).

The regions recorded at the two ends of the run establish the average absorptions (backgrounds) before (<PE) and after the edge (EXAFS), respectively. They are

fitted by linear or spline functions, their intensity difference determining the edge jump $\Delta\mu_0(E)$. The recorded spectrum is then background-subtracted using standard Victoreen polynomials or other spline functions and normalized to ~50–60 eV above the threshold. The energy position of the absorption threshold, E_0, is determined as the maximum of the first derivative of the pattern, and those of the following absorption features as the minima of the second derivative. Accuracy is ± 0.2–0.3 eV for E_0 and for the large features closest to E_0, and ± 0.5–1 eV for the broad features at higher energy. Intensities (in arbitrary units, a.u.) are accurate to better than 3 rel.% for the strongest features and to <10 rel.% for the weak and broad features.

3.4. Interpreting the normalized spectrum

The theory of XAFS angular dependence derived by Brouder (1990) shows, for the simple case of dipole contributions to dichroism, *i.e.* for two-dimensional structures, that $\sigma^D(\varepsilon)$, the absorption general formula as a function of energy, is expressed as:

$$\sigma^D(\varepsilon) = \sigma^D(0,0) - (1/\sqrt{2})(3\sin^2\theta - 1)\sigma^D(2,0) \qquad (5)$$

where θ is the angle between the rotation symmetry axis and the incident beam. This formula is decomposed into two components (Cibin *et al.*, 2006): parallel (in-plane, $\sigma_\parallel$) and orthogonal (out-of-plane, $\sigma_\perp$) to the main reference atomic plane, respectively:

$$\sigma^D(\varepsilon) = \sigma_\parallel \cos^2\theta + \sigma_\perp \sin^2\theta \qquad (6)$$

Assigning $\theta = 0°$ corresponds to the case where the electric field vector ε is directed along the layered structure *a-b* plane and $\theta = 90°$ is normal to it, with the polarization vector pointing out of that plane. Cibin *et al.* (2006) also developed an algorithm that extracts the angle-dependent dipolar contributions to the XANES spectrum, and decomposes them into two partial patterns, which are solved for the in-plane ($\sigma_\parallel$) contributions and the out-of-plane ($\sigma_\perp$) contributions, respectively. This procedure is accurate for intensity to better than 3% and simplifies considerably the interpretation of multiple scattering (MS) interactions and pathways.

Pleochroism owing to rotation induces energy shifts by ≤ 5 eV for some peaks and variations in their intensity by $\leq 50\%$ (Fig. 3). Peaks may disappear and others may appear. These changes gradually affect all regions of the XAS spectrum, but they may also occur abruptly, although this is rare. What occurs is related to changes in the MS path selection by the photoelectron owing to the changing geometrical relationships with respect to sample: the probability for some MS pathways is weakened, while the probability increases for others. Moreover, the probability of observing scattering paths that are out of the plane of polarization increases, such as the intensity of the pre-edge and main absorption lines in non-centrosymmetric structures. In such cases, comparing theoretical simulations with experimental data can be helpful to correctly interpret the observed spectral changes (*e.g.* Wu *et al.*, 1996).

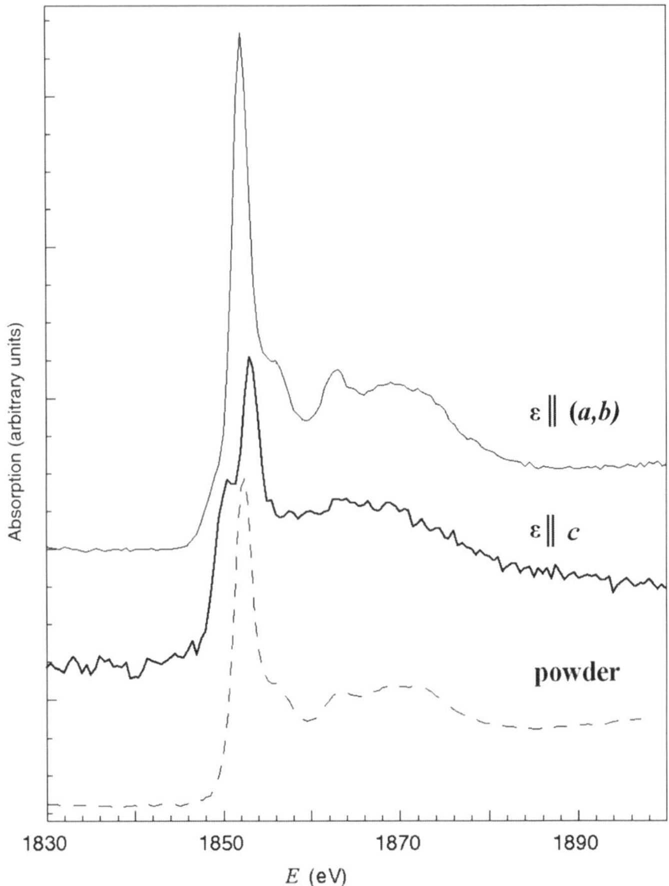

Fig. 3. Changes in the Si *K*-edge XANES spectrum of a muscovite crystal flake resulting from changing its orientation against the horizontally polarized SR beam from orthogonal to almost vertical. The lowest trace is the Si *K*-edge XANES spectrum of the same muscovite recorded on a fine-grained powder mount (modified after Mottana, 2004, p. 506, fig. 12).

3.5. Analysis and extraction of information from the spectral regions

3.5.1. Pre-edge

PE-XAFS records and characterizes the features arising in the XAS spectrum at energies just below the absorption edge ($-15 < E_0 < 0$ [eV]). These features are related to electronic transitions that occur in the atom from core states to empty bound states *i.e.* s → p or p → d (de Groot, 2001; Wu *et al.*, 2002). These transitions are dipole-allowed, but others, such as s → d, are dipole-forbidden; they occur only in the case of a distortion of the next-neighbour environment of the atom. Therefore, PE transitions not only depend on the available number of unoccupied electron sites, but also on symmetry-enhanced mixing between core and empty excited states having p-like character

(Calas & Petiau, 1983) that are induced by site distortions affecting the studied material (*e.g.* strain). For a transition metal, PE features provide accurate information on oxidation state as well as on site coordination and distortion of the absorber atom (Westre *et al.*, 1997; Wilke *et al.*, 2001). Although PE features were believed to occur only in the spectra of atoms of the transition series, they occur in the spectra of other atoms also, if there are empty states available where a core hole electron may leap (Cibin *et al.*, 2003).

PE analysis requires us to extract the configuration of the PE structure by separating the small peaks from the steeply rising main edge contribution (Fig. 4). This is accomplished by fitting the edge rise by an arctangent or spline curve and subtracting this from the low-energy part of the experimental spectrum. The residuals produce small features, and they constitute the PE. They are then fitted with Gaussian curves to determine energy, intensity and FWHM peak positions. Energy has a first-order relationship with the oxidation state of the absorber atom: the higher the oxidation state the more the centroid positions of PE features shift to higher energies (Petit *et al.*, 2001). Intensity is related to transition probabilities, and thus to site coordination (Wilke *et al.*, 2001).

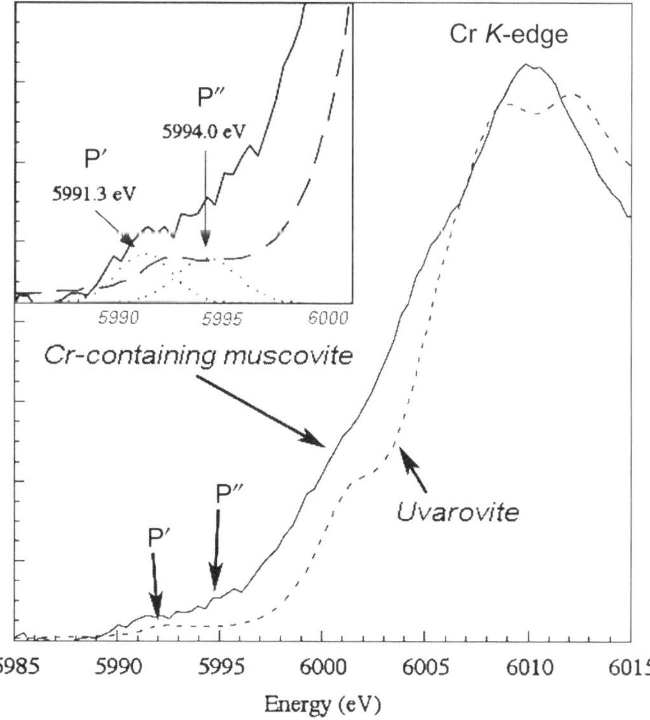

Fig. 4. Extraction of the PE from the Cr *K*-edge XANES spectrum of a chromian muscovite with reference to a uvarovite standard, both set at the "magic angle" (modified after Brigatti *et al.*, 2000, Fig. 9). Two barely recognizable components P′ and P″ (inset) can be extracted by linearizing the residuals obtained by subtracting an arctangent function from the XANES lower limb. Absorption (on the left) is in arbitrary units.

Symmetrical octahedral sites display extremely weak PE features or no PE features at all; however, if the octahedral site distorts increasingly, then the mixing of 4p orbitals and the 3d orbitals increases and PE features appear and increase gradually in intensity. Tetrahedral sites have perfectly cubic symmetry, but no inversion centre. Consequently, a transition atom shows spectra with intense PE features. Small sites, such as tetrahedral sites, attract small cations in a high oxidation state, and have asymmetrical distributions of their outer electrons, thus resulting in an even greater PE feature intensity. FWHM peak positions provide information on accuracy and reliability.

PE features may occasionally be observed when probing an absorber that is not a transition atom. Such anomalous features indicate the presence of $1s \rightarrow d$ transitions that are forbidden in a regular octahedron or cube, but become allowed because of the non-centrosymmetric orbital character of the atom, if located in a deformed 1st shell cage (Cibin *et al.*, 2003). Alternatively, such a feature may be related to extra-transitions towards 3p empty states mixed with the empty density states of the non-transition atom for particularly large clusters extending to the 4th or 5th shell (Wu *et al.*, 1996).

Most applications of the PE features intensity ratios refer to the quantitative determination of the oxidation-state ratio of the atom, which is usually combined with that of coordination and distortion. The most common application is for Fe (*e.g.* Petit *et al.*, 2001; Wilke *et al.*, 2001). The energy difference between Fe^{2+} and Fe^{3+} (based on the average centroid position) is small but readily measurable (1.4 to 2.0 ± 0.1 eV), although their intensities may differ greatly. Fe^{3+} in 4-fold coordination has a PE that is three orders of magnitude more intense than Fe^{3+} in 6-fold coordination.

The Fe pre-edge region undergoes angle-dependent variations, which are small but may become significant for advanced studies. Cibin *et al.* (2008) studied the Fe-PE behaviour in a number of phengite micas and found that all their PE features moved somewhat between two extreme mutually normal settings (Fig. 5). By contrast, the Fe-PE of pyrope garnet showed no variations, in agreement with the isotropic character of the Fe site in this mineral.

Among transition metals other than Fe, Cr was investigated. Cr^{6+} is of interest because it is mobile, highly soluble and poisonous, whereas Cr^{3+} is innocuous and stable. Brigatti *et al.* (2000, 2001) studied the uptake by ferroan smectites of the Cr^{6+} present in aqueous solutions and showed that it is readily adsorbed in the octahedral sheet by transforming to the reduced Cr^{3+} form, and occupying the *M*2 site, just as it does in chromian muscovite.

3.5.2. *XANES*

For a given atom and structure, XANES spectroscopy is able to quantitatively determine four important physical and chemical properties: (1) the local, *i.e.* short-range, nearest-neighbour geometry of the atom site in the crystal structure; (2) the medium-range order, *i.e.* next-nearest neighbour of the atom site in the crystal structure (*i.e.* bond lengths and angles formed with its next-neighbours' outer shells to ~1 nm away); (3) the 'effective charge' (*i.e.* the time-averaged number of valence electrons of the atom, with an accurate evaluation of the coordination); and (4) the electronic structure (*i.e.* charge distribution around the atom).

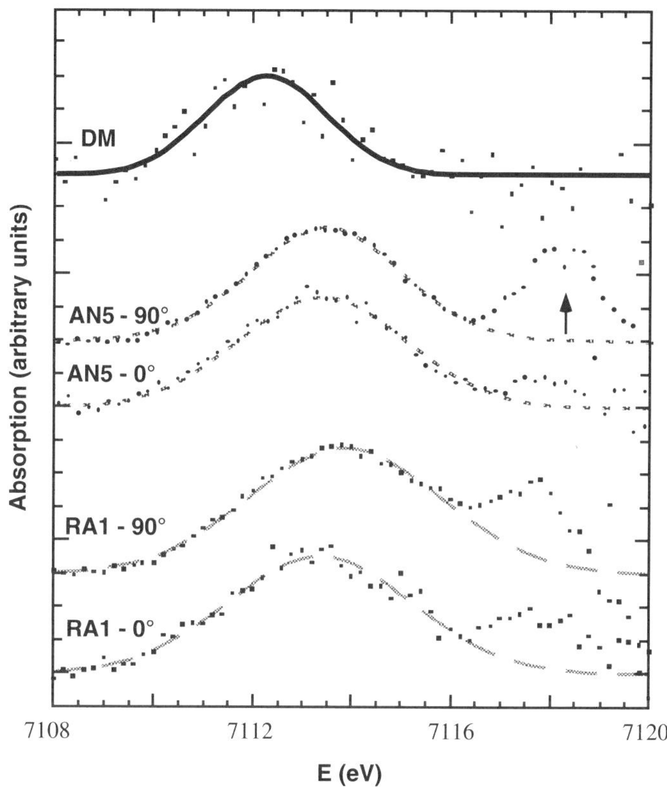

Fig. 5. Orientation dependence of 'chemical' and 'structural' shifts in the pre-edges of Fe-bearing white micas ('phengites') at the Fe *K*-edge: they show up as minor variations in intensity and/or energy, and also as new features outside the known PE variation range. They are probably related to hybridization with distant coordination shells (redrawn and modified after Cibin *et al.*, 2008, fig. 5). DM is garnet, which shows no orientation effects because of its cubic symmetry.

XANES produces the best data when two sub-regions of the spectrum are analysed separately: (1) the Full Multiple Scattering (FMS) sub-region, which extends from the absorption threshold (E_0) to ~30 eV beyond E_0 (typically the deep minimum following the edge-top), and (2) the Intermediate Multiple Scattering (IMS) sub-region, which extends from the minimum to 50–60 eV from the threshold, and merges with the EXAFS oscillations at higher energies.

The FMS sub-region contains a small number of fine features (4 to 6 'peaks', usually partly overlapped), which are related to inelastic scattering of the photoelectron at low kinetic energy. Essentially, the fine features arise from MS interactions with the atoms located in the 1st nearest neighbour's shell around the absorber. Among them, there is the edge-top or 'white-line', *i.e.* the most intense feature of the entire spectrum, which arises from the superposition of many MS contributions and of contributions from the atom electronic properties. Thus, the FMS is particularly difficult to unravel. Over a relatively short energy range, overlapping and superposition of MS contributions of different origins affect the intensity of the experimental features, and a careful deconvolution is required. Nevertheless, some important aspects are clear and are stressed here.

(1) The FMS overall spectral shape is similar for isostructural compounds. Their FMS have the same numbers of features, with limited differences in energy position and intensity. This characteristic helps to identify compounds that are too fine-grained to be identified either optically or by P-XRD. This result is important for compounds that appear to be amorphous because they do not exhibit long-range-order but only a short-range structure. Typical cases are the allophanes, ferrihydrites, and oxy- and hydroxy-manganates, *i.e.* highly disordered layered structures suitable for intercalated components.

(2) The FMS (and the entire XANES spectrum also) undergoes a positive shift in energy as a function of the ionization states of the absorber. The greater the 'chemical' shift, the greater is the formal valence. The chemical shift is determined either precisely, by locating the energy of the inflection point of the arctangent fitting the upward limb of FMS sub-region, or, approximately, by assuming the location at the maximum of the 'white-line'. The shift occurs for transition elements (*e.g.* Fe, V, Mn, Cr, *etc.*), which have several oxidation states coexisting in many layered structures (*e.g.* Fe^{2+} and Fe^{3+} in the phlogopite-ferriphlogopite micas), and for non-transition atoms that show multiple valence states (*e.g.* S, P, As). This edge shift duplicates and amplifies the PE shift, and the shift can be measured with much greater accuracy than the PE shift owing to the much greater intensity of the main edge. However, measurement of this 'chemical' shift is never quite as straightforward as it would appear, because of interference with another shift arising from how near neighbours connect with sites nearby ('structural' shift). Indeed, type, size and symmetry of the nearest-neighbour cage around the absorber atom contribute to modify its absorption energy. The ionic *vs.* covalent amount of the bonds connecting to the absorber atom may also modify shift and make interpretation impossible.

(3) The FMS may also provide information about the average bond length between the absorber and the surrounding ligands, although with a much greater error (~0.03 Å) than single-crystal XRD. In solid compounds, this information is averaged by the simplified relationships $E_r - E_b = const/d^2_{(A-L)}$ (Natoli, 1984), where *const* is an empirical value depending on the structure, E_r and E_b are the energies of the resonance feature and of the electron bound state, respectively, and $d_{(A-L)}$ is the absorber to ligand distance. Thus, the greater the energy of a resonance peak with respect to the 'white-line' energy, the longer the MS pathways of the photoelectron, *i.e.* the distance from the absorber to the relevant coordination shell. Natoli's rule has been shown to be valid only for small variations of d (<~10%, *i.e.* ~0.015–0.025 nm) and for isostructural compounds. The rule does not hold where there are changes in coordination *e.g.* for Fe^{3+} bonds in 4- *vs.* 6-fold coordinated sites as in the phlogopite–ferriphlogopite series (Tombolini *et al.*, 2002).

'Chemical' and 'structural' shifts are rarely used for intercalated layered structures. Successful cases include micas, for Al, which shifts by 2 eV from 4- to 6-fold coordination (Mottana *et al.*, 1997); smectites and chlorites, where, on changing valence from 3+ to 2+, the entire FMS of the Fe *K*-edge spectrum shifts by as much as 7 eV (Brigatti *et al.*, 2000, 2001).

The IMS sub-region features are usually of a limited number ($n < 4$), and probe the MS contributions arriving at the absorber atom from ligands at an intermediate range of distances (to the 4th or 6th shell: Wu *et al.*, 1996; Cabaret *et al.*, 1996). These MS contributions are sharper and stronger than EXAFS oscillations, and therefore they can be recorded accurately. They are helpful for interpreting the spectra in the EXAFS region while contributing to them by using the same algorithms (Bugaev *et al.*, 2001). In practice, the IMS sub-region blends data derived from the local geometry and the overall structure. Thus, the IMS sub-region is affected by atomic interactions that are not as long-range as those probed by XRD, but neither as local and short-range as EXAFS spectra. The information on the atom distribution and properties involves medium-range distances that may extend to 6–8 unit-cells, *i.e.* over a distance <0.4–0.6 nm from the absorber (Bugaev *et al.*, 2001).

3.5.3 EXAFS

Gates (2006) recently reviewed the EXAFS region. Thus, only a few aspects need to be repeated here; in fact, most involve warnings about methods and evaluations of the quality of the results.

EXAFS algorithms calculate local structure properties such as bond length and coordination number in a quantitative manner, and may also be able to provide insight into the absorber nearest neighbours (Debye Weller factors). However, the error in the bond length determined, depending on the phase shift, is usually large (~20%) and even >50% where the coordination number is large. The calculated value for coordination determined, with its decimal places, is often nonsense. EXAFS, obviously, is nearly useless for atoms in very low concentration, when the signal-to-noise ratio may become so small that extraction of reliable data is impossible. Indeed, EXAFS features are, in general, poorly resolved broad undulations that dampen out at high energies. The lower the intensity, the more uncertain is the information extracted.

When the qualities of XANES and EXAFS are compared, XANES requires excellent energy resolution because its features are located in a short energy window around the edge, and they are intense. By contrast, EXAFS requires features with excellent signal-to-noise ratios to obtain reliable atomic distances and coordination numbers. When combined, the two techniques clarify the structure around the probed atom from the very local 1st coordination shell to the 4th–5th coordination shells, *i.e.* to ~6 nm distant, thus bridging with XRD data, which provide interatomic values to ~10 nm. Moreover, both XANES and EXAFS provide directly crystal-chemical information that XRD does not provide. The oxidation state of each atom species present in the compound is better determined by PE and XANES, whereas the coordination is better approximated by EXAFS.

3.6. Conclusions

Theoretical considerations and experimental data indicate that the spectra of layered structures can give extremely sensitive information on both the geometry of the sites and on the chemical environment around the site where the absorbing atom is located, extending as far as the 4th–5th coordination shell. However, to achieve this result, polarization effects must be consideed, particularly when analysing samples with an unknown orientation. Sample orientation has a major impact on the spectral shapes, thus possibly providing false interpretations, if ignored.

A two-component deconvolution is adequate to fit experimental angle-resolved XANES spectra of layered structures, thus simplifying data interpretation. EXAFS patterns show variations among samples that tend to be minor when compared to the large variations seen in the XANES spectra. From EXAFS, useful, often fairly good numerical 2D data for interatomic distances and site coordination can be obtained, but better data, including 3D information, can be obtained from XANES. Obtaining quality data from XANES always requires time-consuming experimental work, followed by careful, sophisticated deconvolution and considerable thought.

4. Practical examples

4.1. The structure of the interlayer of phyllosilicates (*e.g.* clays, illites, micas)

The best examples of using XANES to understand the structure of quasi-2D interlayered compounds involve trioctahedral micas at the potassium *K*-edge, where K^+ is the intercalated 'guest' cation and the 'host' modules are composed of 2 T + 1 O (T = tetrahedral, O = octahedral) sheets. The experimental data and interpretations of Cibin *et al.* (2005, 2006, 2008, 2010) and Brigatti *et al.* (2008) to resolve the interlayer structure document this example exhaustively.

The project started with an interpretion of the XANES spectra of K in trioctahedral micas by recording the standard XAFS spectra on randomly oriented powder samples (Cibin *et al.*, 2005). This first attempt resulted only in confirmation of the relationships between tetrahedral rotation and the Fe^{2+}-Mg_{-1} and $Fe^{3+}Al_{-1}$ exchange vectors known from XRD, *i.e.* in a successful but rather minor result that added little to either the potential of XANES spectroscopy or to the mica structure and crystal-chemical relationships.

To better focus on the peculiarities of layered structures, the research plan moved to study the same micas using cleavage flakes oriented at different angles, *i.e.* to exploit the dependence of XANES features upon orientation, a property known before being theoretically investigated by Brouder (1990) and Pettifer *et al.* (1992), but not often addressed because it is time-consuming. The experimental XANES spectra of micas show significant variation in intensity and also in the energy position of their features as a function of the angle at which the synchrotron beam impinges on the flake (*cf.* Fig. 2). Fitting and decomposing the experimental spectra obtained at six different angles produced two end-member patterns that correspond to the full in-plane absorption component ($\sigma_{\parallel}$, $\theta = 0°$) and to the full out-of-plane normal component ($\sigma_{\perp}$, $\theta = 90°$), respectively.

The former pattern (Fig. 6a) was interpreted as depending on the arrangement of the atoms located in the mica interlayer (mostly K) and in the facing T sheets (O, Si/Al). The latter pattern (Fig. 6b) involves the substitution that occurs in the O sheet (mainly $FeMg_{-1}$), which distorts the atom arrangement of nearby T sheets. Therefore, studying potassium, the 'guest' atom intercalated into the mica 'host' layered structure, provides information not only on interlayer cation ordering over the I plane, but also on how interlayer ordering modifies the arrangement of the atoms in the nearby T and O sheets of the 2:1 module, up to several nm away. Moreover, these patterns give indirect information on the oxidation states of Fe in the O and T sheets, and also on the OH-F substitution at the O4 anion site (Cibin *et al.*, 2006, 2010).

A further analysis of micas (Cibin *et al.*, 2008) showed that decomposition of the AXANES spectra is able to show the effects of collinearity among the interlayer K atoms acting as the absorber, if they are not interrupted by other atoms or by OH, as often occurs in natural phyllosilicates (illite, *etc.*). In addition, the analysis determined the number of independent octahedral sites and their *trans*- *vs.* *cis*-orientation, thus clarifying the extent of the mosaic layered structure of these chemically heterogeneous and structurally defect-rich materials.

This wealth of XAFS information, combined with single-crystal structural data, allowed the assignment of each observed absorption feature to a precise structural pathway followed by the photoelectron ejected by K. This electron either collides with the atoms in the T sheets facing the interlayer, or it crosses them to collide with the O sheet. Thus, the electron also probes the atom substitutions in O for both cations (Al *vs.* Fe, mainly) and anions (OH *vs.* F), as well as the oxidation states of

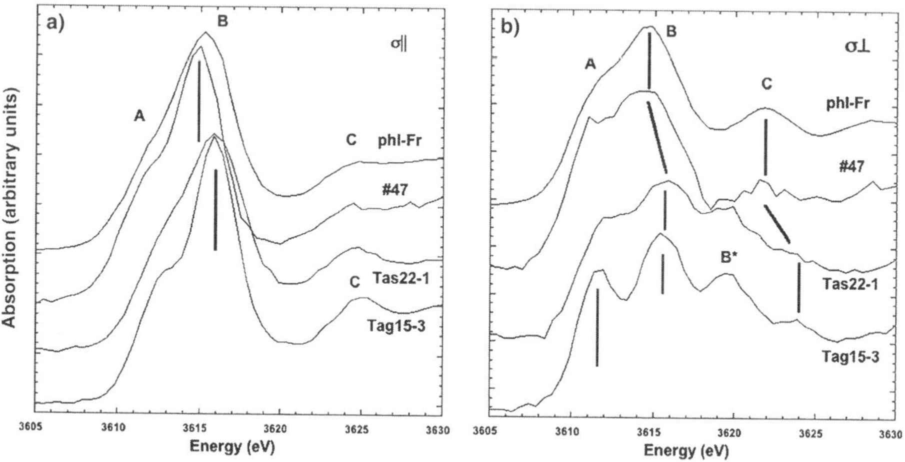

Fig. 6. Orientation dependence of the FMS sub-regions of K *K*-edge XANES spectra for four trioctahedral micas, deconvoluted for their in-plane (***a***: $\sigma_{\|}$, $\theta = 0°$) and out-of-plane (***b***: $\sigma\perp$, $\theta = 90°$) components (redrawn and modified after Cibin *et al.*, 2006, figs 9 and 10). Vertical bars underline the energy changes in the absorption spectrum of the intercalated K^+ 'guest' ion owing to its changing arrangement in the interlayer so as to match the compositional and structural differences of the 'host' 2:1 mica layer modules.

Fe (Brigatti *et al.*, 2008). In conclusion, a quantitative, full crystal-chemical model of the quasi-2D layered structure of micas can be disclosed from measurements of bond distances and angles, and of the energies required at the individual structure sites. In particular, the XANES features observed in the FMS sub-region at energies greater than the 'white-line', and those features occurring in the IMS sub-region, which are generally ignored in the standard evaluation of XANES spectra, produce the most reliable quantitative information (Brigatti *et al.*, 2008).

The potential of IMS features supports the method (Bugaev *et al.*, 2001) which uses these higher-energy XANES features together with the EXAFS features to extract data on distances, coordination, and oxidation states; using the EXAFS features only produces unreliable data. Furthermore, the AXANES method sheds new light on calibrating the binding forces of atoms in the structure that is the scientific pre-requisite for technological application of any material.

A final note: several micas used in these studies were analysed by Tombolini *et al.* (2002) and they determined independently the size of the T sites occupied by Si, Al, and Fe^{3+}. They started from the average $<T{-}O>$ distance of 0.168 nm measured by single-crystal XRD refinement on micas of intermediate T-sheet composition and used the edge-top energy of the FMS sub-region as the reference bound state. They determined, from equations derived from Natoli's rule as applied to the IMS features, the sizes of the individual T sites. Thus, the T site centred by Al has a $<T{-}O>$ distance 0.174 nm, and the site centred by Fe^{3+} is 0.186 nm (the $<T{-}O>$ distance in a Si-centred tetrahedron was assumed to be 0.164 nm). Although these determinations are affected by an error that is one order of magnitude greater than would result from an XRD measurement (0.003 *vs.* 0.0003 nm), they are significant in that the sizes of cations in the same T site could be determined, rather than averaging them to give the site mean size, as XRD refinement results do. This involves a greater understanding of solid solutions, and allows us to plan the synthesis of the most appropriate analogues of layered silicates used for technological applications.

4.2. Graphene and its derivative structures

Currently, graphene is a major research subject for solid state scientists; indeed, studying its physical properties and those of the derived intercalate structures may produce as yet imperfectly known, but certainly highly profitable applications. The structure of graphene is a flat monolayer of carbon atoms tightly packed into a 2D honeycomb pattern, with the C–C bonds of 0.142 nm (Geim & Novoselov, 2007, p. 183). Stacking of graphene monolayers produces the well known 3D structure of graphite, whereas its wrapping to form sphere-like particles produces 0D fullerenes, and by rolling, 1D carbon nanotubes are formed. Initially it was believed that graphene could not exist in the free state, because strictly 2D crystals are thermodynamically unstable. However, monolayers were eventually found as an integral part of larger 3D structures, or as epitaxial-growth layers on crystals with matching surface structures. More recently, self-sustaining graphene wafers were stripped from bulk graphite masses by micromechanical cleavage (using scotch tape or drawing method: Novoselov *et al.*, 2004), and from other substrates. By contrast, producing graphene by chemical exfoliation of bulk graphite that had previously been intercalated with various substances did not succeed. Instead, new 3D materials were produced with large molecules intercalated between planes made of C atom meshes (Dresselhaus & Dresselhaus, 2002).

The bulk properties of graphite are well known (Wallace, 1947), but those of graphene are still under study, and exciting new aspects often occur. Most characterization involves high-resolution transmission electron microscopy (HRTEM), X-ray photoelectron spectroscopy (XPS) and Raman spectroscopy, but a very significant contribution has been made by Near-Edge X-ray Absorption Fine Spectroscopy (NEXAFS), *i.e.* the equivalent of XANES using soft X-rays (<800 eV). Carbon, being a very light atom, requires very soft X-rays. Parallel studies by XAFS using hard X-rays, in such cases, simply support the NEXAFS results in that they produce information on the metal or silicon substrate where graphene was epitaxially grown.

NEXAFS spectra on various graphenes set at the 'magic angle' confirm the close affinity to graphite, and its dissimilarity to diamond (Fig. 7). Moreover, angle-dependent NEXAFS show a very significant structural dependence of the signal from orientation. The observed spectral features originate from transitions of C 1s electrons to states of π- and σ-symmetry. The C 1s $\rightarrow \pi^*$ transition signal is intense for an incident beam electric vector ε parallel to the surface normal, and the C 1s $\rightarrow \sigma^*$ transition signal is intense for vector ε perpendicular to the surface normal (Fig. 8).

5. Conclusions

Theoretical considerations and experimental evidence indicate that the spectra of layered structures can give extremely sensitive information on the geometry of the sites and on the chemical environment around the site where the absorbing atom is located to the 4th–5th coordination shell. However, to obtain this result, polarization effects must be considered, particularly when analysing samples with unknown orientation, because orientation has a major impact on the spectral shapes leading to false interpretations if ignored.

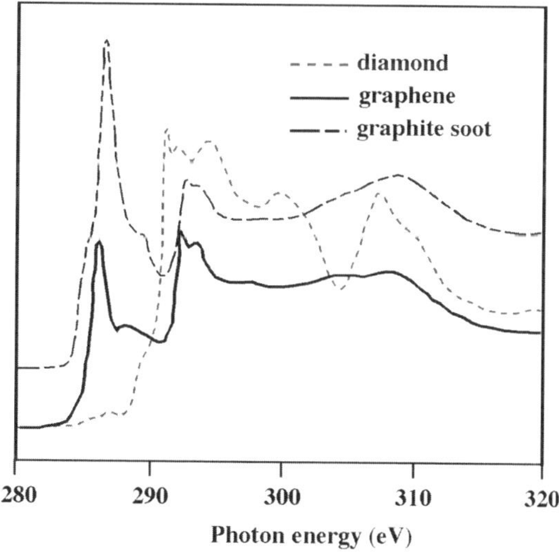

Fig. 7. The NEXAFS spectra of the allotropic forms of carbon in different structural arrangements: graphene (2D) and graphite (3D) have sp^2 hybrid bonds in a planar trigonal coordination; they all show the prominent edge feature at 288 eV. Diamond (3D) has sp^3 hybrid bonds; the edge is shifted to 291 eV and a pre-edge appears due to the distortion intrinsic in the 4-fold C arrangement (redrawn and modified after Li & Hirose, 2009, p. 136, fig. 1). Absorption (vertical axis) is in arbitrary units.

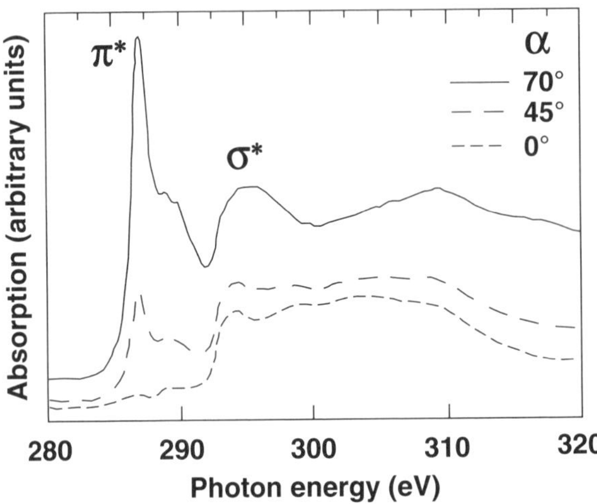

Fig. 8. Orientation-dependent spectra of graphene, showing the strong enhancement of the out-of-plane π* bond intensity with respect to the in-plane σ* bond one on increasing the α angle between the sample surface and the incident beam ε electric vector (redrawn and modified after Vollmer *et al.*, 2009, p. 3, fig. 3).

A two-component deconvolution is adequate to fit experimental angle-resolved XANES spectra of layered structures, thus simplifying data interpretation. EXAFS patterns show variations among samples that tend to be limited in extent when compared to the large variations observed in the XANES spectra. From EXAFS, fairly good numerical 2D data on interatomic distances and site coordination can be obtained, but even better data, which include both 2D and 3D information, can be obtained from XANES, although after a much more time-consuming experimental preparation and deconvolution analysis.

The application of modern spectroscopic and diffraction techniques to any type of layered structure, from the simplest 1D layer of graphene-based soot to the complex structures of chlorites and multiple layered polytypes, which are often multiple quasi-2Ds, creates new ways to understand their physical properties, and to enhance the preparation of synthetic analogues designed for unique technological applications.

Acknowledgements

The Italian Ministry for Education and Research is thanked for PRIN 2008 ('Polarized Spectroscopy (XAFS and FTIR) of micas of petrological significance and their *in situ* transformation products'). A grant-in-aid of LLP-ERASMUS made possible the organization of the school at which this short presentation was used as didactic support. Stephen Guggenheim kindly reviewed the chapter and suggested helpful improvements.

References

Aldega, L. & Eberl, D.D. (2005) Detrital illite crystals identified from crystallite thickness measurements in siliciclastic sediments. *American Mineralogist*, **90**, 1587–1596.

Aldega, L., Botti, F. & Corrado, S. (2007) Clay mineral assemblages and vitrinite reflectance in the Laga Basin (Central Apennines, Italy): What do they record? *Clays and Clay Minerals*, **55**, 504–518.

Árkai, P., Sassi, F.P. & Sassi, R. (1995) Simultaneous measurements of chlorite and illite crystallinity: a more reliable tool for monitoring low- to very low-grade metamorphism in metapelites. A case study from the Southern Alps (NE Italy). *European Journal of Mineralogy*, **7**, 1115–1128.

Bak, P. (1982) Commensurate phases, incommensurate phases and the devil's staircase. *Reports on Progress in Physics*, **45**, 587–629.

Boehm, H.-P., Setton, R. & Stumpp. E. (1994) Nomenclature and terminology of graphite intercalation compounds. *Pure and Applied Chemistry*, **66**, 1893–1901.

Bove, D.J., Eberl, D.D., McCarty, D.K. & Meeker, G.P. (2002) Characterization and modeling of illite crystal particles and growth mechanisms in a zoned hydrothermal deposit, Lake City, Colorado. *American Mineralogist*, **87**, 1546–1556.

Brigatti, M.F., Lugli, C., Cibin, G., Marcelli, A., Giuli, G., Paris, E., Mottana, A. & Wu, Z.Y. (2000) Reduction and sorption of chromium by Fe(II)-bearing phyllosilicates: chemical treatments and X-ray absorption spectroscopy (XAS) studies. *Clays and Clay Minerals*, **48**, 272–281.

Brigatti, M.F., Galli, E., Medici, L., Poppi, L., Cibin, G., Marcelli, A. & Mottana, A. (2001) Chromium-containing muscovite: crystal chemistry and XANES spectroscopy. *European Journal of Mineralogy*, **13**, 377–389.

Brigatti, M.F., Malferrari, D., Poppi, M., Mottana, A., Cibin, G., Marcelli, A. & Cinque, G. (2008) Interlayer potassium and its surrounding in micas: Crystal chemical modeling and XANES spectroscopy. *American Mineralogist*, **93**, 821–835.

Brime, C. & Eberl, D.D. (2002) Growth mechanisms of low grade illites based on shapes of crystal thickness distributions. *Schweizerische Mineralogische und Petrographische Mitteilungen*, **82**, 203–209.

Briscoe, W.H., Chen, M., Dunlop, I.E., Klein, J., Penfold, J. & Jacobs, R.M.J. (2007) Applying grazing incidence X-ray reflectometry (XRR) to characterising nanofilms on mica. *Journal of Colloid and Interface Science*, **306**, 459–463.

Brouder, C. (1990) Angular dependence of X-ray absorption spectra. *Journal of Physics: Condensed Matter*, **2**, 701–738.

Brown, G.S., Eisenberger, P. & Schmidt, P. (1977) Extended X-ray absorption fine-structure studies of oriented single crystals. *Solid State Communications*, **24**, 201–203.

Bugaev, L.A., Sokolenko, A.P., Dmitrienko, H.V. & Flank, A.-M. (2001) Fourier filtration of XANES as a source of quantitative information of interatomic distances and coordination numbers in crystalline minerals and amorphous compounds. *Physical Review B*, **65**, 024105-1-8.

Cabaret, D., Sainctavit, Ph., Ildefonse, P. & Flank, A.-M. (1996) Full multiple-scattering calculations on silicates and oxides at the Al K edge. *Journal of Physics: Condensed Matter*, **8**, 3691–3704.

Calas, G. & Petiau, J. (1983) Coordination of iron in oxide glasses through high-resolution K-edge spectra: information from the pre-edge. *Solid State Communications*, **48**, 625–629.

Castro Neto, A.H., Guinea, F., Peres, N.M.R., Novoselov, K.S. & Geim, A.K. (2009) The electronic properties of graphene. *Reviews of Modern Physics*, **81**, 109–162.

Cibin, G., Marcelli, A., Mottana, A., Giuli, G., Della Ventura, G., Romano, C., Paris, E., Cardelli, A. & Tombolini, F. (2003) Magnesium X-ray absorption near-edge structure (XANES) spectroscopy on synthetic model compounds and minerals. *Frascati Physics Series (19th International Conference on X-ray and Inner-Shell Processes)*, **32**, 95–150.

Cibin, G., Mottana, A., Marcelli, A. & Brigatti, M.F. (2005) Potassium coordination in trioctahedral micas investigated by *K*-edge XANES spectroscopy. *Mineralogy and Petrology*, **85**, 67–87.

Cibin, G., Mottana, A., Marcelli, A. & Brigatti, M.F. (2006) Angular dependence of potassium *K*-edge XANES spectra of trioctahedral micas: Significance for the determination of the local structure of the interlayer site. *American Mineralogist*, **91**, 1150–1162.

Cibin, G., Cinque, G., Marcelli, A., Mottana, A. & Sassi, R. (2008) The octahedral sheet of metamorphic $2M_1$-phengites: A combined EMPA and AXANES study. *American Mineralogist*, **93**, 414–425.

Cibin, G., Mottana, A., Marcelli, A., Cinque, G., Xu, W., Wu, Z. & Brigatti, M.F. (2010) The interlayer structure of trioctahedral lithian micas: An AXANES spectroscopy study at the potassium *K*-edge. *American Mineralogist*, **95**, 1084–1094.

Corrado, S., Invernizzi, C., Aldega, L., D'Errico, M., Di Leo, P., Mazzoli, S. & Zattin, M. (2010) Testing the validity of organic and inorganic thermal indicators in different tectonic settings from continental subduction to collision: the case history of the Calabria-Lucania border (southern Apennines, Italy). *Journal of the Geological Society*, **167**, 985–999.

Cuadros, J. (1997) Interlayer cation effects on the hydration state of smectite. *American Journal of Science*, **297**, 829–841.

De Groot, F. (2001) High-resolution X-ray emission and X-ray absorption spectroscopy. *Chemical Reviews*, **101**, 1779–1808.

Dresselhaus, M.S. & Dresselhaus, G. (2002) Intercalation compounds of graphite. *Advances in Physics*, **51**, 1–187.

Drits, V.A., Środoń, J. & Eberl, D.D. (1997) XRD measurement of mean crystallite thickness of illite and illite/smectite: reappraisal of the Kübler index and the Scherrer equation. *Clays and Clay Minerals*, **45**, 461–475.

Drits, V.A., Eberl, D.D. & Środoń, J. (1998) XRD measurement of mean thickness, thickness distribution and strain for illite and illite-smectite crystallites by the Bertaut-Warren-Averbach technique. *Clays and Clay Minerals*, **46**, 38–50.

Eberl, D.D., Drits, V.A., Środoń, J. & Nüesch, R. (1996) MudMaster: A computer program for calculating crystallite size distributions and strain from the shapes of X-ray diffraction peaks. *U.S. Geological Survey Open File Report 96–171*, 44 p.

Eberl, D.D., Nüesch, R., Šucha, V. & Tsipursky, S. (1998a) Measurement of fundamental illite particle thicknesses by X-ray diffraction using PVP-10 intercalation. *Clays and Clay Minerals*, **46**, 89–97.

Eberl, D.D., Drits, V.A. & Środoń, J. (1998b) Deducing crystal growth mechanisms for minerals from the shapes of crystal size distributions. *American Journal of Science*, **298**, 499–533.

Eberl, D.D., Środoń, J. & Drits, V.A. (2003) Comment on "Evaluation of X-ray diffraction methods for determining the crystal growth mechanisms of clay minerals in mudstones, shales and slates," by L.N. Warr and D.R. Peacor. *Schweizerische Mineralogische und Petrographische Mitteilungen*, **83**, 349–358.

Ferrage, E., Lanson, B., Sakharov, B.A., Geoffroy, N., Jacquot, E. & Drits, V.A. (2007) Investigation of dioctahedral smectite hydration properties by modeling of X-ray diffraction profiles: Influence of layer charge and charge location. *American Mineralogist*, **92**, 1731–1743.

Gates, W.P. (2006) X-ray absorption spectroscopy. In: *Handbook of Clay Science* (F. Bergaya, B.K.G. Theng & G. Lagaly, editors). Developments in Clay Science, **1**, Elsevier, Amsterdam, pp. 789–829.

Geim, A.K. & Novoselov, K.S. (2007) The rise of graphene. *Nature Materials*, **6**, 183–191.

Golovin, A.L., Imamov, R.M., & Kondrashkina, E.A (1985) Potentialities of new X-ray diffraction methods in structural studies of ion-implanted silicon layers. *Physica Status Solidi A*, **88**, 505–514.

Guggenheim, S., Bain, D.C., Bergaya, F., Brigatti, M.F., Drits, V.A., Eberl, D.D., Formoso, M.L.L., Galán, E., Merriman, R.J., Peacor, D.R., Stanjek, H. & Watanabe, T. (2002) Report of the Association Internationale pour l'Etude des Argiles (AIPEA) Nomenclature Committee for 2001: Order, disorder and crystallinity in phyllosilicates and the use of the 'Crystallinity Index'. *Clays and Clay Minerals*, **50**, 406–409.

Heald, S.M. & Stern, E.A. (1977) Anisotropic X-ray absorption in layered compounds. *Physical Review B*, **16**, 5549–5559.

Herrera, L.K. & Videla, H.A. (2009) Surface analysis and materials characterization for the study of biodeterioration and weathering effects on cultural property. *International Biodeterioration & Biodegradation*, **63**, 813–822.

Huang, T.C. (1990) Surface and ultra-thin film characterization by grazing-incidence asymmetric Bragg diffraction. *Advances in X-ray Analysis*, **33**, 91–99.

Iwasaki, T. & Watanabe, T. (1988) Distribution of Ca and Na ions in dioctahedral smectites and interstratified dioctahedral mica/smectites. *Clays and Clay Minerals*, **36**, 73–82.

Kasrai, M., Lennard, W.N., Brunner, R.W., Bancroft, G.M., Bardwell, J.A. & Tan, K.H. (1996) Sampling depths of total electron yield and fluorescence measurements in Si L- and K-edge absorption spectroscopy. *Applied Surface Science*, **99**, 303–312.

Kasrai, M., Fleet, M.E., Muthupari, S., Li, D. & Bancroft, G.M. (1998) Surface modification study of borate materials from B K-edge X-ray absorption spectroscopy. *Physics and Chemistry of Minerals*, **25**, 268–272.

Kotarba, M. & Środoń, J. (2000) Diagenetic evolution of crystallite thickness distribution of illitic material in Carpathian flysch shales studied by Bertaut-Warren-Averbach XRD method (MudMaster computer program). *Clay Minerals*, **35**, 387–395.

Kübler, B. (1967) La cristallinité de l'illite e les zones tout à faires supéeures du métamorphisme. Pp. 105–121 in: *Etages Tectoniques*. Colloque de Neuchatel 1996, France.

Larsen, R.A., McNulty, T.E., Goehner, R.E. & Crystal, K.R. (1989) X-ray diffraction studies of polycrystalline thin films using glancing angle diffractometry. *Advances in X-ray Analysis*, **32**, 311–321.

Li, Y. & Hirose, A. (2009) XAS-assisted materials design and characterization for growth of carbon nanostructures directly on transition metals. *Canadian Light Source 2009 activity report*, 136–137.

Mählmann, R.F. (2001) Correlation of very low grade data to calibrate a thermal maturity model in a nappe tectonic setting, a case study from the Alps. *Tectonophysics*, **334**, 1–33.

Manceau, A. & Schlegel, M. (2001) Texture effect on polarized EXAFS amplitude. *Physics and Chemistry of Minerals*, **28**, 52–56.

Manceau, A., Chateigner, D. & Gates, W.P. (1998) Polarized EXAFS, distance-valence least-squares modeling (DVLS) and quantitative texture analysis approaches to the structural refinement of Garfield nontronite. *Physics and Chemistry of Minerals*, **25**, 347–365.

Manceau, A., Schlegel, M., Chateigner, D., Lanson, B., Bartoli, C. & Gates, W.P. (1999) Application of polarized EXAFS to fine-grained layered minerals. In: *Synchrotron X-ray Methods in Clay Science* (D. Schulze, P. Bertsch & J. Stucki, editors). CMS Workshop Lectures, **9**, The Clay Minerals Society, Boulder, Colorado, USA, pp. 68–114.

Manceau, A., Drits, V.A., Lanson, B., Chateigner, D., Wu, J., Huo, D., Gates, W.P. & Stucki, J.W. (2000) Oxidation-reduction mechanism of iron in dioctahedral smectites: II. Crystal chemistry of reduced Garfield nontronite. *American Mineralogist*, **85**, 153–172.

Marra, J.M.C., Eisenberger, P. & Cho, A.Y. (1979) X-ray total-external-reflection–Bragg diffraction: A structural study of the GaAs-Al interface. *Journal of Applied Physics*, **50**, 6927–6933.

Moore, D.M. & Reynolds, R.C. Jr. (1997) *X-ray Diffraction and the Identification and Analysis of Clay Minerals*. Oxford University Press, New York, 378 pp.

Mottana, A. (2004) X-ray absorption spectroscopy in mineralogy: Theory and experiment in the XANES region. In: *Spectroscopic Methods in Mineralogy* (A. Beran & E. Libowitzky, editors). EMU Notes in Mineralogy, **6**, Eötvös University Press, Budapest, pp. 465–552.

Mottana, A., Robert, J.L., Marcelli, A., Giuli, G., Della Ventura, A., Paris E. & Wu, Z.Y. (1997) Octahedral versus tetrahedral coordination of Al in synthetic micas determined by XANES. *American Mineralogist*, **82**, 497–502.

Nadeau, P.H., Tait, J.M., McHardy, W.J. & Wilson, M.J. (1984) Interstratified XRD characteristics of physical mixtures of elementary clay particles. *Clay Minerals*, **19**, 67–76.

Natoli, C.R. (1984) Distance dependence of continuum and bound state of excitonic resonance in x-ray absorption near-edge structure (XANES). In: *EXAFS and Near-edge Structure III* (K.O. Hodgson, B. Hedman & J.E. Penner-Hahn, editors). Springer, Berlin, pp. 38–42.

Novoselov, K.S., Geim, A.K., Morozov, S.V., Jiang, D., Zhang, Y., Dubonos, S.V., Grigorieva, I.V. & Firsov, A.A. (2004) Electric field effect in atomically thin carbon films. *Science*, **306**, 666–669.

Petit, P.E., Farges, F., Wilke, M. & Solé, V.A. (2001) Determination of the iron oxidation state in Earth materials using XANES pre-edge information. *Journal of Synchrotron Radiation*, **8**, 952–954.

Pettifer, R.F., Brouder, C., Benfatto, M., Natoli, C.R., Hermes, C. & Ruiz López, M.F. (1992) Magic-angle theorem in powder X-ray-absorption spectroscopy. *Physical Review B*, **42**, 37–42.

Reynolds, R.C., Jr. (1980) Interstratified clay minerals. In: *Crystal Structures of Clay Minerals and their X-ray Identification* (G.W. Brindley & G. Brown, editors), Monograph **5**, Mineralogical Society, London, pp. 249–303.

Rhan, H., Pietsch, U., Rugel, S., Metzger, H. & Peisl, J. (1993) Investigations of semiconductor superlattices by depth-sensitive X-ray methods. *Journal of Applied Physics*, **74**, 146–152.

Rugel, S., Wallner, G., Metzger, H. & Peisl, J. (1993) Grazing-incidence X-ray diffraction on ion-implanted silicon. *Journal of Applied Crystallography*, **26**, 34–40.

Sato, T., Watanabe, T. & Otsuka, R. (1992) Effects of layer charge, charge location, and energy change on expansion properties of dioctahedral smectites. *Clays and Clay Minerals*, **40**, 103–113.

Scherrer, P. (1918) Bestimmung der Grösse und der inneren Struktur von Kolloidteilchen mittels Röntgenstrahlen. *Göttinger Nachricthen*, **26**, 98–100.

Simić, V. & Uhlík, P. (2006) Crystallite size distribution of clay minerals from selected Serbian clay deposits. *Annales Géologique de la Péninsule Balkanique*, **67**, 109–116.

Sinha, S.K. (editor) (1980) *Ordering in Two Dimensions*. Elsevier North Holland, New York.

Środoń, J. (2006) Identification and quantitative analysis of clay minerals. In: *Handbook of Clay Science* (F. Bergaya, B.K.G. Theng & G. Lagaly, editors). Developments in Clay Science, **1**, Elsevier, Amsterdam, pp. 765–787.

Šucha, V., Kraus, I., Šamajová, E. & Puškelová, L. (1999) Crystallite size distribution of kaolin minerals. *Periodico di Mineralogia*, **68**, 81–92.

Šucha, V., Środoń, J., Clauer, N., Elsass, F., Eberl, D.D., Kraus, I. & Madejová, J. (2001) Weathering of smectite and illite-smectite in central-European climatic conditions. *Clay Minerals*, **36**, 403–419.

Szpunar, J.A., Ahlroos, S. & Tavernier, Ph. (1993) Method of measurement and analysis of texture in thin films. *Journal of Material Science*, **28**, 2366–2376.

Tombolini, F., Marcelli, A., Mottana, A., Cibin, G., Brigatti, M.F. & Giuli, G. (2002) Crystal-chemical study by XANES of trioctahedral micas: The most characteristic layer silicates. *International Journal of Modern Physics, B*, **16,** 1673–1679.

Toney, M.E. & Brennan, S. (1989) Structural depth profiling of iron oxide thin films using grazing incidence asymmetric Bragg X-ray diffraction. *Journal of Applied Physics*, **65**, 4763–4768.

Uhlík, P., Šucha, V., Eberl, D.D., Puškelová, L. & Čaplovičova, M. (2000) Evolution of pyrophyllite particle sizes during dry grinding. *Clay Minerals*, **35**, 423–432.

Vollmer, A., Feng, X.L., Wang, X., Zhi, L.J., Mullen, K., Koch, N. & Rabe, J.P. (2009) Electronic and structural properties of graphene-based transparent and conductive thin film electrodes. *Applied Physics A*, **94**, 1–4.

Wallace, P.R. (1947) The band theory of graphite. *Physical Review*, **71**, 622–634.

Waychunas, G.A. (2002) Grazing-incidence X-ray absorption and emission spectroscopy. In: *Applications of Synchrotron Radiation in Low-temperature Geochemistry and Environmental Science* (P.A. Fenter, M.L. Rivers, N.C. Sturchio & S.R. Sutton, editors). Reviews in Mineralogy and Geochemistry, **49**. Mineralogical Society of America, Chantilly, Virginia, and the Geochemical Society, St. Louis, Missouri, pp. 267–315.

Westre, T.E., Kennepohl, P., DeWitt, J.G., Hedman, B., Hodgson, K.O. & Solomon, E.I. (1997) A multiplet analysis of Fe *K*-edge 1*s*-3*d* pre-edge features of iron complexes. *Journal of the American Chemical Society*, **119**, 6297–6314.

Wilke, M., Farges, F., Petit, P.-E., Brown, G.E. Jr. & Martin, F. (2001) Oxidation state and coordination of Fe in minerals: An Fe *K*-XANES spectroscopic study. *American Mineralogist*, **86**, 714–730.

Wu, Z.Y., Marcelli, A., Mottana, A., Giuli, G., Paris, E. & Seifert, F. (1996) Effects of higher-coordination shells in garnets detected by XAS at the Al K edge. *Physical Review B*, **54**, 2976–2979.

Wu, Z.Y., Natoli, C.R., Marcelli, A., Paris, E., Bianconi, A. & Saini, N.L. (2002) Unified interpretation of pre-edge X-ray absorption fine structures in 3*d* transition metal compounds. In: *X-ray and Inner-Shell Processes*. 19th International Conference on X-ray and Inner-Shell Processes (A. Bianconi, A. Marcelli & N.L. Saini, editors). AIP Conference Proceedings, **652**, American Institute of Physics, Mineola, New York, pp. 497–506.

EMU Notes in Mineralogy, Vol. 11 (2011), Chapter 9, 313–334

Interaction of organic molecules with layer silicates, oxides and hydroxides and related surface-nano-characterization techniques

GIOVANNI VALDRÈ*, DANIELE MORO and GIANFRANCO ULIAN

Dipartimento di Scienze della Terra e Geologico-Ambientali, Università di Bologna, Piazza di Porta San Donato 1, Bologna I-40126, Italy
e-mail: giovanni.valdre@unibo.it

Knowledge of the surface properties of layered minerals is of great importance to understand both fundamental and applied technological issues, such as, for example, liquid–surface interactions, microfluidity, friction or tribology and biomolecule self-assembly and adhesion.

Recent developments in Scanning Probe Microscopy (SPM) have widened the spectrum of possible investigations that can be performed at a nanometric level at the surfaces of minerals. They range from physical properties such as surface potential and electric field topological determination to chemical and spectroscopic analysis in air, in liquid or in a gaseous environment.

After a brief introduction to new technological developments in SPM, we present recent achievements in the characterization and application of nanomorphology, surface potential and cleavage patterns of layer silicates, in particular chlorite.

Two general research directions will be presented: interaction of organic molecules with layer silicates and synthetic substrates, and mineral hydrophilicity/phobicity and friction/adhesion issues. SPM is used to assess the force-curve, force-volume, adhesion and surface potential characteristics of layer silicates by working in Electric Force Microscopy (static and dynamic EFM) and in Kelvin probe modes of operation. For instance, EFM allows us to measure the thickness of silicate layers and, from frequency, amplitude, phase modulation and Kelvin analysis, to derive the electrostatic force experienced by the probe. We can relate these measurements directly to the electrostatic force gradient at the mineral surface.

Transverse dynamic force microscopy, also known as shear force microscopy is introduced here and examples of the investigation of attractive, adhesive and shear forces of water on layer silicates will be presented. The study of water in confined geometries is very important because it can provide simple models for fluid/mineral interactions.

The ability to control the binding of biological and organic molecules to a crystal surface is fundamental, especially for biotechnology, catalysis, molecular microarrays, biosensors and environmental sciences. For instance, recent studies have shown that DNA molecules have different binding affinities and assume different conformations when adsorbed to different layer silicate surfaces. On certain crystals the electrostatic surface potential anisotropy is able to order and stretch the DNA filament and induce a natural change in its conformation. The active stretching of DNA on extensive layer silicates is a clear indication of the basic and technological potential carried by these minerals when used as substrates for biomolecules. Other examples including amino acids, proteins, nucleotides, nucleic acids and cells are discussed here. Finally, a comparison between experimental data and simulation is presented and discussed.

*Corresponding author

DOI: 10.1180/EMU-notes.11.9

1. Introduction

The physico-chemical surface properties of phyllosilicates and mineral oxides can drive mineral–environment surface interactions. A great variety of phenomena with important geological, environmental, (bio-)technological and economical spin-offs are influenced by the effects of the local surface properties, at the micro and nanometric scale.

Mineral-surface properties may control important interaction processes, such as coagulation, aggregation, sedimentation, filtration, catalysis, and ionic transport in porous media. Mineral floatability is strongly related to the particles' surface charge, as well as mineral structure and composition. This charge can be permanent or constant, resulting, for instance, from isomorphous substitution, or variable, due for example, to protonation and deprotonation of functional groups (Montes *et al.*, 2007; Alvarez-Silva *et al.*, 2010). In addition, the electrical properties of mineral surfaces affect reactions with charged species such as metal ions, biomolecules and cells.

Mineral oxide particles have applications, for example, in many paint and paper formulations as pigments and fillers. Their stability is of importance for the processing properties and the quality of the final products, and knowledge of the interactions between such particles is necessary to prevent them from flocculating or, at least, to control their flocculation.

Phyllosilicates typically present a characteristic surface nanomorphology. Processes of nucleation, growth, and alteration of the nanostructures occur when minerals are exposed to humid air and the environment. Nanomorphology and nanostructures often have physico-chemical properties that are very different from those of the underlying mineral substrates. The overall effective reactivity can be greatly affected, regulating the rate of particle solubility as well as the surface sorption of inorganic and organic components. For instance, cell adhesion and subsequent biofilm formation can be directed by the nanostructure properties rather than by those of the primary particle (Kendall *et al.*, 2008).

The structure and properties of water adsorbed onto clay surfaces are influenced significantly by the mineral surface. The differences in the structural charge, cation occupancies and distributions affect the dynamics and structure of surface water.

In geochemistry and environmental science, water–mineral interactions are important factors controlling processes such as surface ion adsorption and ion exchange, which are crucial for the mobility of contaminants in surface and groundwater systems, weathering, soil development, soil moisture behaviour, water composition and quality, and removal and sequestration of atmospheric carbon dioxide.

The chemical behaviour of mineral–water interfaces is central to aqueous reactivity in natural waters, soil evolution, and atmospheric chemistry and is of direct relevance for maintaining the integrity of waste repositories and remediating environmental pollutants (Yanina & Rosso, 2008).

Furthermore, clay minerals are potentially catalytic surfaces for mineral-mediated biomolecule interactions, such as RNA polymerization and vesicle formation. These interactions can be driven by particle size, shape, composition, charge and surface area (Hanczyc *et al.*, 2006; Kwon *et al.*, 2006; Valdrè, 2007).

To analyse in detail these physico-chemical properties and the local interactions at mineral surfaces, techniques with a high spatial resolution are needed. One of the most powerful and promising techniques is Scanning Probe Microscopy (SPM).

Recent developments in SPM have widened the spectrum of possible investigations that can be performed at a nanometric level at the surfaces of minerals. These range from physical properties such as surface potential and electric field topological determination to chemical and spectroscopic analysis in air, in liquid or in a gaseous environment. Furthermore, new technological solutions make it possible to probe single-molecule interactions at the mineral surfaces with previously unattainable force sensitivity (<pN).

The new technological developments behind SPM are given below. These range from methodologies for the nanomorphological characterization of mineral surfaces to the measurement of nanoconfined surface potentials and the mechanical characterization of the surface nanostructures. Recent achievements in the field of the characterization and application of nanomorphology, surface potential and cleavage patterns of minerals, in particular the behaviour of some layer silicates, oxides, hydroxides and synthetic compounds are presented. Investigations of the interaction of organic molecules with these mineral surfaces are reviewed with respect to single-molecule imaging and the correlation between mineral nanomorphology, surface properties and molecule absorption, self-assembling and nanopatterning. Finally, some computational studies on the modelling of mineral surfaces and the analysis of their interactions with organic molecules, by means of *ab initio* and molecular-dynamics methods, are presented and discussed.

2. Nanoscale surface-imaging techniques of layer silicates, oxides and hydroxides

2.1. Nanotopography and cleavage patterns

Mineral surfaces can be spatially resolved by means of atomic force microscopy (AFM), a SPM methodology which can attain atomic resolution. In AFM there are two main modes of operation: amplitude modulation (AM-AFM) and frequency modulation (FM-AFM). In amplitude modulation the AFM probe is mechanically excited at the cantilever fundamental frequency by a piezoelectric element and the oscillatory motion is detected by the optical system (Lee & Jhe, 2006). AM-AFM uses a large amplitude of oscillation (dozens of nm) to probe the sample surface. The oscillation amplitude is used as the feedback parameter to image the sample topography. This can be achieved by means of a feedback loop using a lock-in amplifier.

By mapping the phase shift between the excitation signal and the response of the cantilever during the scan, phase imaging may provide further information beyond simple topographical mapping to detect, for instance, variations in composition, adhesion, friction and viscoelasticity. Applications include qualitative differentiation among regions of high and low surface adhesion or hardness, identification of contaminants and mapping of different components in composite materials.

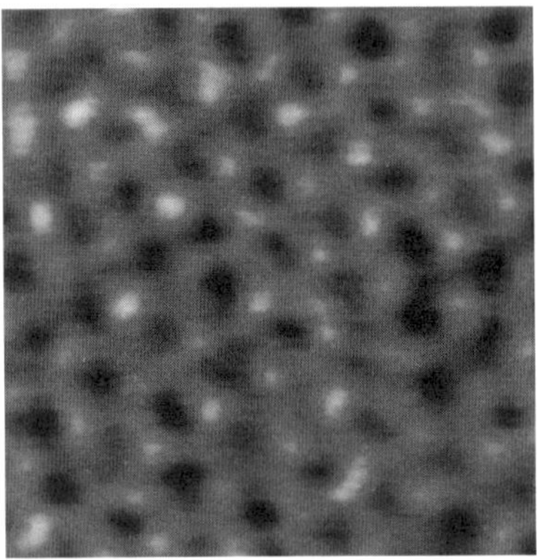

Fig. 1. Atomically resolved image of the cleaved (001) surface of a muscovite taken in water by FM-AFM. Scan size: 4 nm × 4 nm (modified from Fukuma *et al.*, 2005). Readers of the paper version of this chapter may wish to download a colour version of this figure from www.minersoc.org/emu-notes/emu-11/11-9-colour.pdf.

In frequency modulation mode (FM-AFM), an increase in sensitivity is achieved over AM-AFM (Giessibl, 2003). The signal used to produce the image comes from direct measurement of the resonant frequency of the cantilever, which is modified by the tip–surface interaction. In FM-AFM the cantilever is kept oscillating at its current resonant frequency, achieved by means of a phase-locked loop (PLL). The spatial dependence of the frequency shift induced in the cantilever motion by the tip–sample interaction is used as the source of topographical contrast.

Sub-atomic resolution has been achieved by FM-AFM (Giessibl *et al.*, 2000). It is also a promising technique in the imaging of organic systems, for example, single- and double-stranded DNA. Recently, high resolution was also achieved in liquid both of minerals and biomolecules. For instance, Figure 1 shows an atomically resolved image of the surface of muscovite in water obtained by FM-AFM (Fukuma, 2009).

By detecting and measuring precisely the short-range chemical forces between the tip and surface atoms using FM-AFM, Sugimoto *et al.* (2007) achieved the chemical identification of individual surface atoms in ultrahigh vacuum at room temperature (Fig. 2a, b).

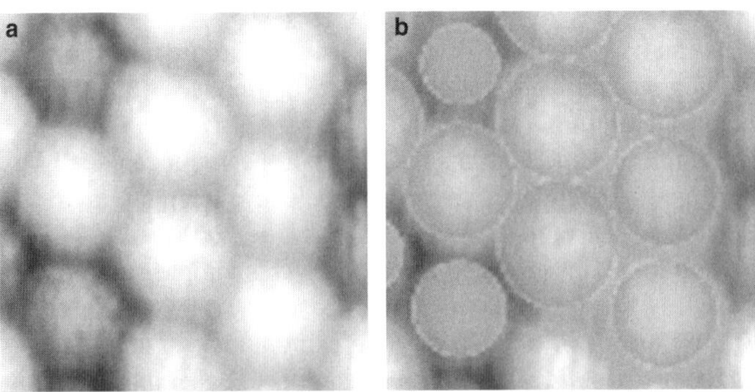

Fig. 2. (**a**) AFM topographic image of Si, Sn and Pb atoms on a Si(111) substrate. (**b**) Single-atom chemical identification of the area shown in image (**a**). Red, blue and green atoms correspond to Sn, Pb and Si, respectively. Scan size: 2 nm × 2 nm (modified from Sugimoto *et al.*, 2007). Readers of the paper version of this chapter may wish to download a colour version of this figure from www.minersoc.org/emu-notes/emu-11/11-9-colour.pdf.

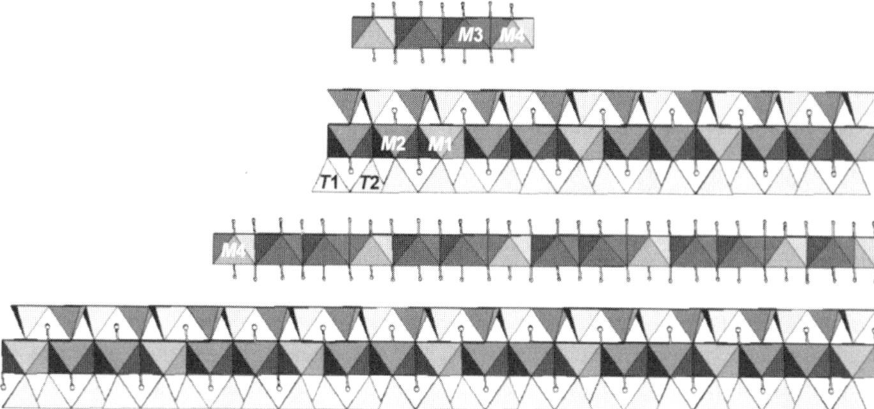

Fig. 3. Schematic structural interpretation of the cleaved surface of a clinochlore (modified from Valdrè *et al.*, 2009). Readers of the paper version of this chapter may wish to download a colour version of this figure from www.minersoc.org/emu-notes/emu-11/11-9-colour.pdf.

As an example of layer-silicate nanomorphology, the surface cleavage pattern of a clinochlore is presented here. Clinochlore is a layer silicate where a brucite-like, positively charged octahedral sheet, ~0.5 nm thick, is sandwiched between two negatively charged TOT layers by means of weak electrostatic forces (Fig. 3). The heterogeneous layer structure of clinochlore results in a mixed surface of brucite-like and talc-like areas after cleavage (Fig. 4a). Using AFM the surface step heights can be measured accurately with a sub-nanometre resolution. Figure 4b shows the profile of a single brucite-like sheet. As expected, the phase-shift signal reveals the presence of two different materials,

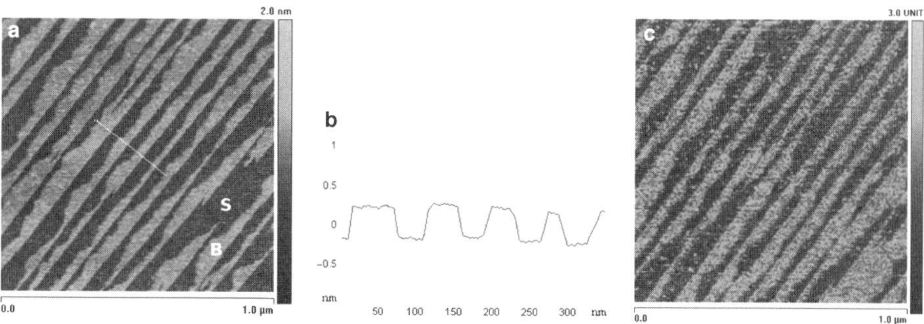

Fig. 4. (**a**) AFM image of the surface nanomorphology of a clinochlore. After cleavage, the clinochlore presented simultaneously zones of brucite-like sheets (B) and zones of siloxane (S) with lateral size ranging from a few nanometer to microns. (**b**) Profile section along the horizontal line in image (**a**) showing the thickness of a single brucite-like sheet. (**c**) AFM phase image of the same zone as in (**a**). Readers of the paper version of this chapter may wish to download a colour version of this figure from www.minersoc.org/emu-notes/emu-11/11-9-colour.pdf.

independent of the height measurements of brucite-like and TOT layers, which are related to the crystal chemistry (Fig. 4c).

2.2. Surface potential and related properties

In SPM the surface potential can be measured using three main methods: static electric force microscopy (EFM), dynamic EFM and Kelvin probe force microscopy (KPFM). The three methods differ in terms of type of signal detected, sensitivity and ease of operation.

In static EFM we measure the cantilever deflection caused by the electrostatic force acting between the probe and the specimen surface. The effective electrostatic force depends on the voltage difference between the probe and the specimen and also on both tip and cantilever geometry (shape and size). For this reason, the effect of both tip and cantilever on the total force gradient must be taken into account for a thorough understanding of the experimental data.

The electrostatic interaction between a conductive tip and sample is given by

$$F = \frac{1}{2}\frac{dC(z)}{dz}V^2$$

where $C(z)$ is the system capacitance, z is the tip–sample distance, and V the tip–sample electric potential difference. At equilibrium, the electrostatic force is balanced by the reacting elastic force of the cantilever, and its measured deflection is directly related to the electrostatic force acting on the cantilever system.

In static EFM, by means of a Force Volume experiment, a bi-dimensional matrix of force curves is acquired, giving us the force field above the sample. The probe is brought into contact with the sample in a controlled way by means of the piezoelectric scanner and then retracted to a chosen height. The cantilever deflection, ΔL, measured point by point during the approach-withdrawal cycle is related to the electrostatic force, F_{E}, by means of the elastic constant of the cantilever, k_L:

$$F_{\mathrm{E}} = \mathrm{k}_L \Delta L$$

A numerical simulation of the same experiment can relate the probe's deflection to the mineral-surface potential (Valdrè & Moro, 2008a, 2008b). The real geometry of the cantilever and tip can be reconstructed by computer-aided design (CAD). The electro-mechanical interaction between the probe and the mineral surface can be studied by 3D finite element analysis (FEA). The coupling between the probe and the mineral sample is modelled with the numerical formulation of the Maxwell stress tensor, which, ignoring electrostriction, is given by:

$$\vec{\vec{T}} = \varepsilon \vec{E}\vec{E} - \frac{1}{2}\varepsilon \vec{E} \cdot \vec{E}\delta$$

where $\vec{E}$ is the electric field, ε is the dielectric tensor and δ the Kroneker delta.

Finite element analysis allows us to determine the mineral-surface potential and further to discriminate the contribution of the various probe areas (tip-apex, tip sides, cantilever) to the measured force, to calculate the spatial resolution of the probe and the force sensitivity.

In dynamic mode EFM, a piezo-mechanically driven conductive cantilever oscillates above the sample surface. In this case, frequency, amplitude and phase detection are used to study the tip–surface interactions. By using the small-amplitude approximation, the interactions can be mapped point by point, eliminating the ambiguities which occur if large-amplitude techniques are used. At small amplitudes the resonance frequency, ω_0, is directly related to the interaction stiffness, k_{ts}, which is the negative value of the force gradient:

$$k_{ts} = k_L \left(\frac{\omega_0^2}{\omega_{0f}^2} - 1 \right)$$

where ω_{0f} is the cantilever free resonance frequency and k_L the elastic constant.

Measuring the frequency shift, we can obtain ω_0 and then the force gradient. The electrostatic force can be associated to the sample surface potential by a numerical simulation of the probe–sample interaction.

In the configuration of a material with fixed charges at its surface, sandwiched between a conductive plane substrate and the EFM tip, the total electrostatic force can be expressed as:

$$F = \frac{1}{2}\frac{dC(z)}{dz}V^2 + \frac{Q_S Q_t}{4\pi\varepsilon_r\varepsilon_0 z^2}$$

where $C(z)$ is the tip-substrate capacitance, V is the tip to substrate voltage, Q_S is the static charge, Q_t is the total charge induced in the tip, ε_0 is the permittivity of free space and ε_r describes the dielectric constant of the materials between the conductive plane substrate and the EFM tip.

In Kelvin probe force microscopy (KPFM) the tip is biased directly by $V_{tip} = V_{dc} + V_{ac}\cos(\omega t)$ to a metallic sample or substrate (Nonnenmacher *et al.*, 1991). In this case, the capacitive force, F_{cap} between the tip and the surface is

$$F_{cap}(z) = \frac{1}{2}(V_{tip} - V_s)^2 \frac{\partial C(z)}{\partial z}$$

where V_s is the surface potential or contact potential difference (between tip and sample), and $C(z)$ is the tip–surface capacitance, which depends on the tip size and geometry, surface topography and tip–surface separation, z. The first harmonic of the force $F_{1\omega}^{cap}$ is related to the difference between the constant component of the tip bias, V_{dc}, and the surface potential V_s by

$$F_{1\omega}^{cap}(z) = \frac{\partial C(z)}{\partial z}(V_{dc} - V_s)V_{ac}.$$

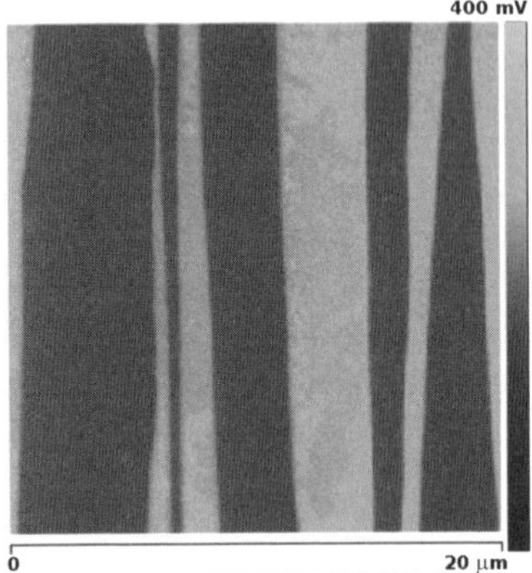

Fig. 5. 20 μm × 20 μm dual-pass KPFM image of the surface potential of μm-long brucite-like and mica-like stripes on a clinochlore surface. The measurement was performed by means of the 'classic' dual pass Kelvin probe force mode. Readers of the paper version of this chapter may wish to download a colour version of this figure from www.minersoc.org/emu-notes/emu-11/11-9-colour.pdf.

Fig. 6. 1 μm × 1 μm image showing the quantitative determination of the surface potential of a highly fragmented clinochlore surface by means of the single-pass Kelvin probe force method. 25 nm-wide surface potential features are clearly resolved. Readers of the paper version of this chapter may wish to download a colour version of this figure from www.minersoc.org/emu-notes/emu-11/11-9-colour.pdf.

By adjusting V_{dc} to nullify the force, we obtain, point by point, a surface potential map of V_s.

If static charges, Q_s, are present on the surface, a supplementary Coulomb force arises between the static charges and the ac charge induced on the tip, which is CV_{ac} (Nyffenegger *et al.*, 1997; Girard, 2001). In this case, $F^{cap}_{1\omega}$ can be written as:

$$F^{cap}_{1\omega}(z) = \left[\frac{\partial C(z)}{\partial z}(V_{dc} - V_s) - \frac{Q_s C(z)}{4\pi\varepsilon_r\varepsilon_0 z^2}\right] V_{ac}.$$

The sign and position of charge Q_s can, in principle, be obtained, but the measurement depends heavily on the particular tip to sample configuration and is not as simple as for voltages; however, an indication of the related surface potential can be obtained.

Typically, Kelvin Probe measurements are achieved by a two-pass procedure. First, the topographic profile is recorded, then the tip is kept at an arbitrary chosen height above the sample and moved following the topographic profile, to minimize interferences of the sample morphology with the electric potential signal.

Figure 5 shows the surface potential distribution measured by two-pass KPFM on the cleaved surface of a clinochlore. The potential measurement was performed by keeping the probe at 30 nm above the mineral surface.

To improve this method in terms of sensitivity and spatial resolution, single-pass procedures have been developed (Ziegler *et al.*, 2007). In this innovative method the probe measures simultaneously the morphology and the surface potential while shortening of the scanning times. The tip to sample distance is minimized, thus maximizing the spatial resolution. To avoid the coupling of the topographical to the surface potential signal, the Kelvin feedback works at a higher flexural mode. Figure 6 presents a single-pass KPFM image of clinochlore revealing highly resolved surface potential features $<$25 nm in size.

Another single-pass method is the multifrequency electric force microscopy (MF-EFM), from which qualitative information on the surface potential of the sample can be obtained (Stark *et al.*, 2007).

A method to investigate the surface potential of a mineral in liquid using the AFM signal phase shift is under development (Valdrè *et al.*, 2011). The phase lag between the excitation signal and the response of the cantilever during a scan can be related to the surface potential (Baumann & Stark, 2010). By monitoring the phase contrast on a mineral surface in a liquid environment, we can obtain information on the surface potential variation. The liquid environment can be controlled easily by means of a fluid cell; therefore we can change the pH of the solution while measuring the phase lag induced by the surface potential variation. Figure 7 shows the surface of a clinochlore imaged at room temperature (25°C) in a water solution at pH 5.5 and 9.0, respectively. The phase differences between the brucite-like and the talc-like layers are related to changes in the surface potential. These experiments can contribute to the determination of the point of zero charge or pH-related variation of surface potential at high spatial resolution.

2.3. Nanomechanical properties

The Pulsed Force Mode (PFM^{TM}, WITec GmbH, Germany) is a new SPM-based technique measuring the sample topography without shear lateral forces at a relatively high scanning speed (Rosa-Zeiser *et al.*, 1997). Besides topography, it is possible to make a simultaneous acquisition of properties such as superficial adhesion, charge distribution and viscoelastic material behaviour.

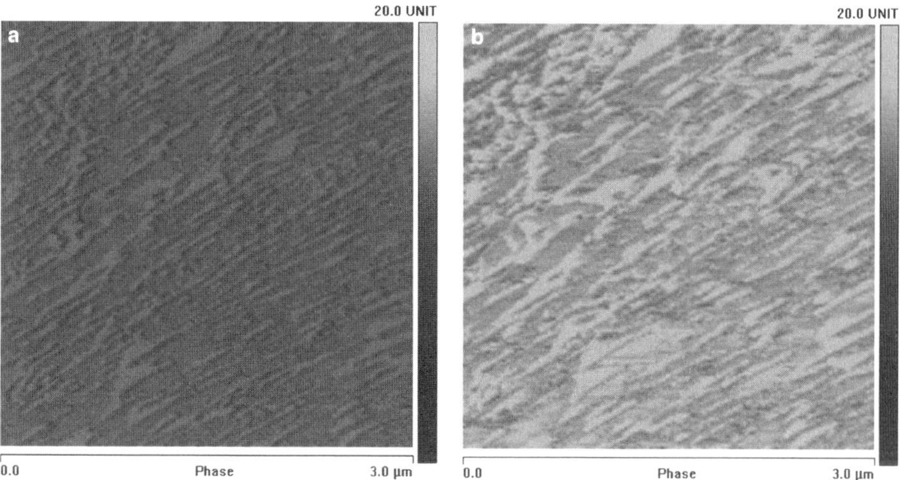

Fig. 7. (**a**) AFM phase contrast at the surface of a clinochlore, acquired in liquid at pH 5.5. (**b**) Phase image of the same zone after changing the solution pH to 9.0. Readers of the paper version of this chapter may wish to download a colour version of this figure from www.minersoc.org/emu-notes/emu-11/11-9-colour.pdf.

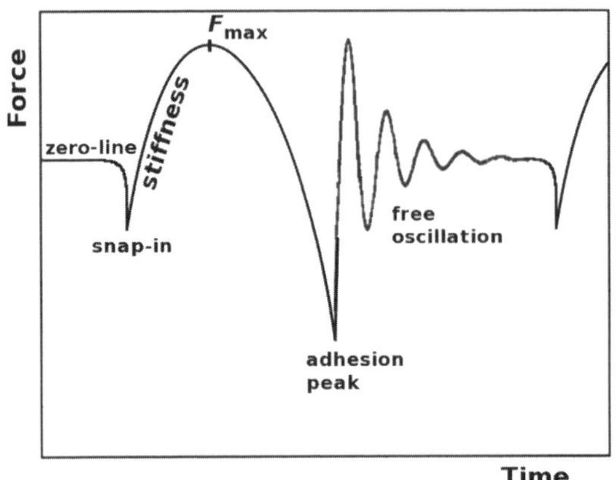

Fig. 8. The force trend *vs.* time of an SPM probe operating in pulsed force mode is shown as a continuous line (modified from Krotil *et al.*, 1999).

In PFM, a function generator modulates the *z* offset (*i.e.* sample to SPM tip distance) with a sinusoidal signal of frequency from 100 Hz to 5 kHz and amplitude in the range 10–500 nm.

Usually, the frequency modulation is chosen to be $\frac{1}{5}$ or less of the cantilever resonant frequency. The amplitude is adjusted after each contact between the surface and the tip during one cycle. The force trend is shown in Figure 8 as force *vs.* time during the cycle.

Initially the tip is kept above the sample surface (baseline) and, moving towards the sample, the tip comes into contact with it (snap-in point). Then the contact is increased until the repulsive force reaches the maximum value (F_{max}). As the piezo-element pushes away the tip, the repulsive force decreases and the force signal changes sign (from repulsive to attractive), because of the adhesive interaction between tip and sample. Finally the tip completely loses the contact and the consequent oscillation damps to the basal line; then a new cycle starts.

The maximum tip–sample force, F_{max}, is the base of feedback control and in PFM the user can control the maximum tip–sample repulsive force, a very important feature when analysing soft materials.

To measure the adhesive forces, the maximum negative force value is recorded when the tip parts from the surface (adhesion peak).

Elastic properties are obtained by measuring the force trends before and after F_{max}. The force difference between a point in the repulsive part of the force signal and the maximum force is closely connected to the sample local stiffness. Finally, long-range forces such as electrostatic forces, can be mapped using the surface potential distribution recording of the zero-line fluctuations. This technique is used primarily as a property mapping method with a trigger force of a few nanonewtons or more.

Another method to measure mechanical properties is the PeakForce Quantitative Nanomechanical Mapping (PeakForce™ QNM™, Bruker AXS GmbH, Germany) which allows the user to acquire high-resolution images at high scanning speed and to measure specific material properties (Pittenger *et al.*, 2010). The feedback circuit controls the maximum peak force on the tip which can range down to the pico-newton.

The modulation frequency is ~2 kHz and the modulated tip to specimen position is measured as a time function. To analyse the sample characteristics it is necessary to transform the time function curves into separation function curves (Fig. 9).

A wide variety of properties can be obtained from the curve analysis, such as elastic modulus, tip–sample adhesion, energy dissipation and the maximum deformation on the sample. Here we report only on the elastic modulus.

To determine the elastic modulus of a material, the bold line in the force–distance curve is fitted using the Derjaguin-Muller-Toporov (DMT) model according to the following equation (Pittenger *et al.*, 2010):

$$F - F_{\mathrm{adh}} = \frac{4}{3}E'\sqrt{R(d-d_0)^3}$$

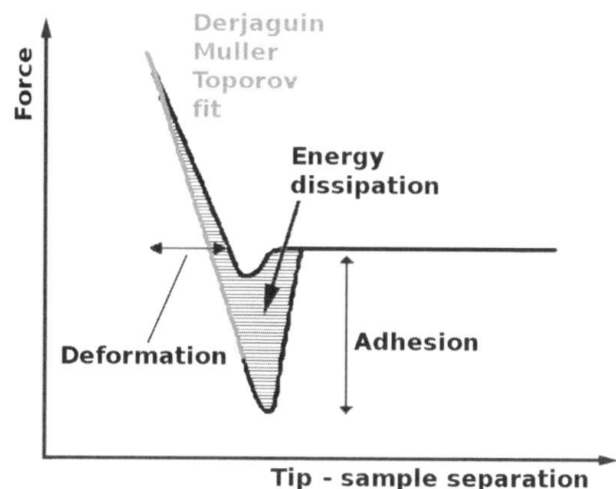

Fig. 9. Example of a SPM force curve, force *vs.* tip–sample separation, with the indication of the mechanical parameters that can be calculated (modified from Pittenger *et al.*, 2010). Readers of the paper version of this chapter may wish to download a colour version of this figure from www.minersoc.org/emu-notes/emu-11/11-9-colour.pdf.

with $F - F_{\mathrm{adh}}$ = force on the cantilever relative to the adhesion force, R = tip radius of curvature and $d - d_0$ = sample deformation.

The reduced modulus, E', can be calculated by data fitting. If the Poisson ratio is known, the material's elastic modulus, E_S, can obtained from E' by the following relation:

$$E' = \left[\frac{1-\nu_s^2}{E_s} + \frac{1-\nu_{\mathrm{tip}}^2}{E_{\mathrm{tip}}}\right]^{-1}$$

By considering the tip elastic modulus $E_{\mathrm{tip}} \gg E_s$, the previous equation can be approximated to:

$$E' = \left[\frac{1-\nu_s^2}{E_s}\right]^{-1}$$

The Poisson ratio generally ranges between ~0.2 and 0.5 (perfectly incompressible) giving a difference between the reduced modulus and the sample modulus of between 4 and 25%. Generally, it is possible to determine the elastic modulus in a range from 700 kPa to 70 GPa. Figure 10 reports a calibration of the SPM-based elastic modulus measurement using some standard materials of known elastic modulus (Pittenger *et al.*, 2010).

Another SPM-based method to measure visco-elastic properties at the surface of minerals is the Transverse Dynamic Force Microscopy (TDFM), also known as shear force microscopy, where a probe is mounted vertically and perpendicular to the sample

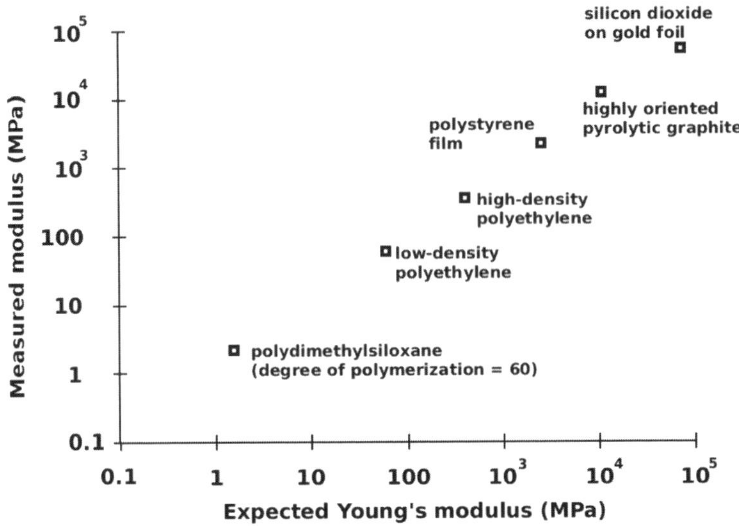

Fig. 10. Experimental calibration of the AFM Young's modulus measurement. AFM experiments (ordinate) by the *PeakForce* QNM method for different types of materials are in agreement with the expected Young's Modulus (from standards) (modified from Pittenger *et al.*, 2010).

surface and set in horizontal oscillation using a dither piezo (Antognozzi *et al.*, 2001, 2003). The oscillation amplitude and the corresponding phase signal are measured quantitatively with a specific optical detection system (Antognozzi *et al.*, 2008). This technique is of great importance for the study of the shear properties of liquids confined to their molecular dimensions. Nanoconfined liquids can reveal properties completely different from those of the bulk. This behaviour has a fundamental consequence in phenomena such as the swelling of clays, viscosity and lubrication. Furthermore, ion adsorption/desorption from the mineral surface can also be investigated.

By TDFM a known shear strain can be imposed on a confined liquid and the corresponding stress response is evaluated. If the confined liquid acts as a visco-elastic material the resulting stress has an in-phase and an out-of-phase component with respect to the strain.

Figure 11a shows the oscillation amplitude and the phase lag of the tip approaching a freshly cleaved muscovite sample in pure water. The curves present four steps in both amplitude and phase, with a mean periodicity of 2.54 Å. Figure 11b shows the elastic (in-phase) and viscous (out-of-phase) forces calculated from the previous data. Note that the dissipative component dominates for tip–sample separations of >1.6 nm, the elastic force increases rapidly with decreasing separation. A different visco-elastic behaviour of the confined liquid is related to each molecular layer and is also dependent on the number of water layers. TDFM can be used to investigate the role of surface-adsorbed cations on the visco-elastic response of nanoconfined water on muscovite and other layered minerals.

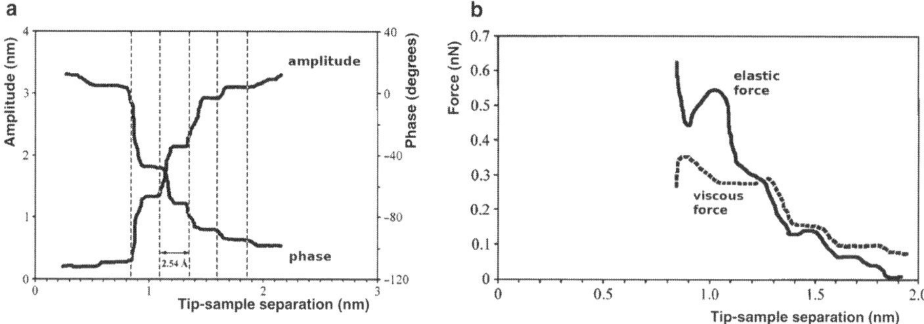

Fig. 11. (**a**) Amplitude and phase approach curves on a freshly cleaved muscovite surface in pure water (continuous lines). The steps highlighted in the curves are due to the molecular layering of water in the confined space between the probe and the muscovite surface. (**b**) Viscous and elastic forces trends calculated from the approach curves shown in (**a**). The dissipative viscous force dominates over the elastic component at the start of the interaction, but becomes less significant when the tip–sample separation is <1.5 nm (modified from Antognozzi *et al.*, 2001).

3. Interaction of organic molecules with mineral and synthetic substrates

3.1. Layer silicates

In recent years, many researchers have studied the interactions that arise between a mineral phase (adsorbent) and biomolecules such as amino acids, proteins, lipids and nucleic acids. Silicates are among the most investigated minerals with a large variety of different analytical methods. In this work we report on scanning probe microscopy (SPM) and related methodologies.

Muscovite, biotite, brucite, chlorite and vermiculite was studied and compared in terms of surface affinity, self-assembly and nanopatterning of adsorbed DNA molecules (Valdrè *et al.*, 2004). Differences in the surface structure and chemistry of these layered silicate minerals were identified as responsible for a remarkable variety of DNA adsorption mechanisms. For instance, the investigations on di- and trioctahedral micas revealed different deposition patterns in terms of the amount of molecules adsorbed and their configurations.

In later studies, the clinochlore surface was studied as a substrate for nucleotide and DNA adsorption (Antognozzi *et al.*, 2006; Valdrè, 2007). Systematic AFM analysis of nucleotide deposition on clinochlore showed the presence of both agglomerates, ordered structures and alignments of nucleotides (Fig. 12). In particular, the edges and ledges of the brucite-like $Mg(OH)_2$ sheet provide preferential adhesion and filament-forming sites of nucleotides (Fig. 13).

Regarding DNA adsorption, the clinochlore surface showed areas with high DNA affinity with ~50% coverage (onto the positively charged brucite-like $Mg(OH)_2$ sheet) and areas where the DNA molecules were almost completely absent (on negatively charged siloxane zones). As shown in Figure 14, single DNA molecules bridge

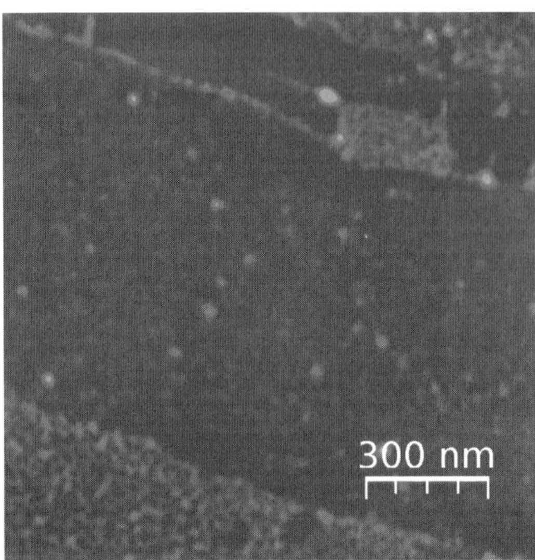

Fig. 12. AFM image of a clinochlore region where nucleotides line-up (modified from Valdrè, 2007). Readers of the paper version of this chapter may wish to download a colour version of this figure from www.minersoc.org/emu-notes/emu-11/11-9-colour.pdf.

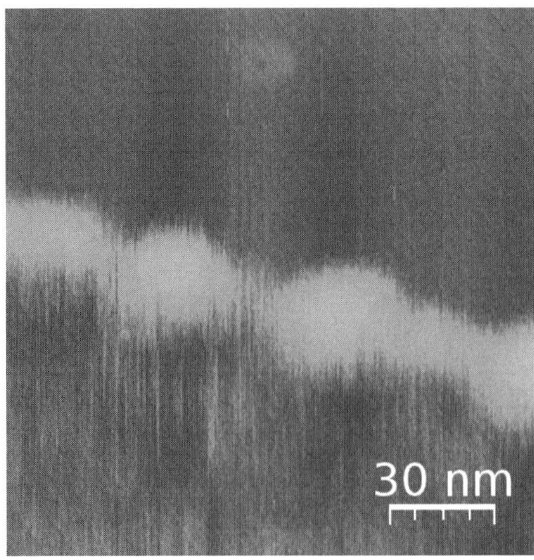

Fig. 13. High-magnification image showing a detail of the line-up of nucleotides along the edge of a brucite-like sheet (modified from Valdrè, 2007). Readers of the paper version of this chapter may wish to download a colour version of this figure from www.minersoc.org/emu-notes/emu-11/11-9-colour.pdf.

between two adjacent $Mg(OH)_2$ brucite-like parallel stripes. The DNA molecules appear stretched (linear conformation) by comparison to the DNA which is not bridged and is relaxed in entangled and bundled conformations. As in the case of nucleotides, DNA tends to agglomerate and order at the edges of the $Mg(OH)_2$ sheets.

Chlorite could also be an ideal model platform to understand the mechanisms regulating the adsorption of molecules at surfaces. Some molecules, such as surfactants, exhibit a hydrophobic chain and a positively charged hydrophilic head group. The 'binary' chlorite surface could be used to interrogate the competition between the hydrophobic and electrostatic forces, as it presents positively charged hydrophobic areas (brucite-like) side by side with negatively charged hydrophilic areas (talc-like), thus improving the understanding of how exactly these two forces balance each other and determine the structure and morphology of the adsorbed layer.

Various layer silicates were also considered as support for bioadsorption of red blood cells (Valdrè & Fabbrizioli, 2006) comparing their fixing behaviour to that of commercial mica (muscovite) without the use of any fixative and binding chemical agent. Figure 15a, for example, shows an AFM image of the morphology of red blood cells after adsorption onto a biotite surface. Vermiculite reveals similar adhesion properties, but with a greater surface adsorption density (Fig. 15b). On the contrary, red blood cells deposited directly on muscovite under the same experimental conditions are dragged away by the SPM tip. On biotite and vermiculite, the cell adhesion is sufficient to allow the observation of membrane channels at the nanoscale (Fig. 15c).

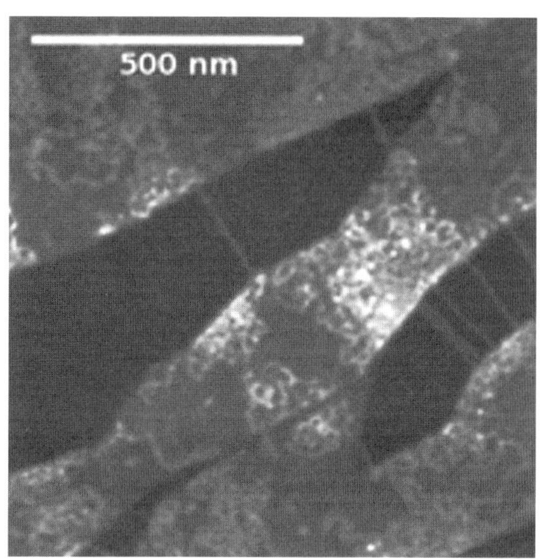

Fig. 14. AFM image of DNA adsorbed on the surface of a cleaved clinochlore crystal. The DNA molecules are confined to the brucite-like regions. The bright height contrast is due to DNA filaments that accumulate and align along the brucite-like edges. Many single filaments are in a stretched conformation between two adjacent brucite-like sheets above the siloxane area. The average DNA length is ~1.5 μm (modified from Antognozzi *et al.*, 2006). Readers of the paper version of this chapter may wish to download a colour version of this figure from www.minersoc.org/emu-notes/emu-11/11-9-colour.pdf.

Molecular-resolution imaging of biological samples in liquid such as proteins adsorbed on muscovite can be obtained using frequency modulation atomic force microscope (FM-AFM). The frequency noise of FM-AFM in liquid can be reduced considerably by the reduction of the noise-equivalent deflection of an optical-beam deflection sensor (Yamada *et al.*, 2009).

Recently, researchers have used muscovite as a substrate for mono and bilayer lipid

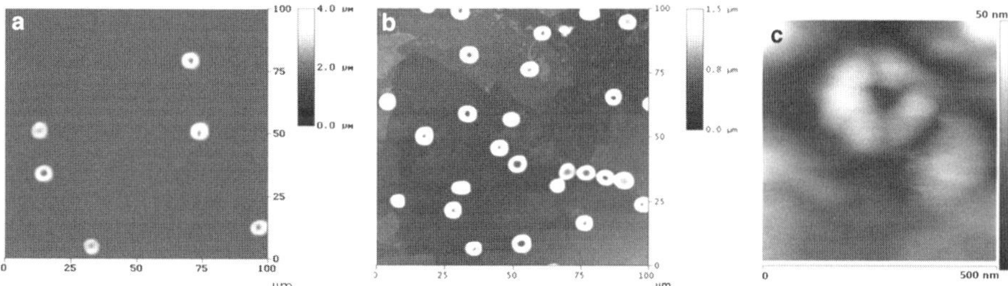

Fig. 15. (**a**) AFM image showing the biotite surface after deposition of red blood cells in air. The cells are perfectly attached to the surface and the preservation of cell morphology is clearly observed. (**b**) AFM image of a vermiculite surface after deposition of red blood cells. Well immobilized cells are observed with a surface density greater than that of biotite. (**c**) Membrane channel of a red blood cell fixed on a biotite surface imaged by AFM. Readers of the paper version of this chapter may wish to download a colour version of this figure from www.minersoc.org/emu-notes/emu-11/11-9-colour.pdf.

systems (Valdrè *et al.*, 1998; Alessandrini *et al.*, 2007). The surfaces of these systems were investigated by AFM; subsequent studies describe the advancing and receding contact angle of polar molecules, namely water, formamide and di-iodomethane (Jurak & Chibowski, 2010). Thermodynamic analysis of the excess area and excess free energy of the lipid mixture indicated that lateral phase separation takes place for the components, which was indeed confirmed by SPM.

Clay minerals, such as montmorillonite, kaolinite, *etc.* were used to study adsorption of nucleic acid bases on their surfaces at different pH (Benetoli *et al.*, 2008). The main finding was the weaker adsorption of uracil and thymine on clays which raises the question of the possibility of a genetic code based on purines only. Fourier-transform infrared spectroscopy at pH 2 showed that the interaction of adenine, cytosine, thymine and uracil with the clays occurs through positively charged, protonated groups.

A similar study was carried out by another group (Cai *et al.*, 2006), where adsorption isotherms of DNA on the soil colloids and minerals examined conformed to the Langmuir equation. The amount of DNA adsorbed followed the order: montmorillonite $\gg$ fine inorganic clay $>$ fine organic clay $>$ kaolinite $>$ coarse inorganic clay $>$ coarse organic clay. The results, pH dependent, implied that electrostatic interactions played an important role in DNA adsorption on organic clays and montmorillonite. The adsorption of RNA on montmorillonite and kaolinite also seems to protect the biomolecule from degrading agents (Franchi & Gallori, 2005).

Adsorption of glycine and alanine on Na-montmorillonite and on Ca^{2+}- and Mg^{2+}-exchanged forms over a range of pH and temperature have been investigated by UV-spectrophotomety (Kalra *et al.*, 2000). Adsorption of both aminoacids was considerable on all the three adsorbents used. Maximum adsorption was observed at 25°C and neutral pH. Ca^{2+}-montmorillonite exhibited better adsorption than the Mg^{2+}-exchanged form or Na-montmorillonite.

3.2. Oxides, hydroxides and synthetic compounds

The adsorption of serine, leucine and water onto hydrophilic and hydrophobic quartz surfaces was investigated using the radio tracer technique (Alaeddine & Nygren, 1996). The free energy of adsorption of leucine was found to be $\sim$1.15 times greater than that of serine on a hydrophobic surface, and $\sim$1.38 times greater on a hydrophilic surface. Independent of whether serine or leucine is used, the surface concentration is greater on hydrophilic than on hydrophobic surfaces. Furthermore, both aminoacids showed a decrease in the adsorption rate at high bulk concentration except for leucine (on a hydrophobic surface), where the surface density increased linearly with bulk concentration. This linear relation was probably an effect of the medium that screened the electrostatic and hydrophobic side-chain interactions.

Recently, the adsorption of glycine, lysine and other small biomolecules on synthetic silica surfaces was investigated by Stievano *et al.* (2009). They considered the adsorption process from the view-point of the surface, using knowledge, gained previously, of molecular identification of surface functional groups. Different adsorption mechanisms were postulated, corresponding to different spectroscopic signatures of the amino acid/adsorption site complexes. On silica surfaces, this implies cooperative hydrogen bonding in a kind of molecular recognition. The existence of molecular recognition is also proven by experiments on the co-adsorption of different aminoacids, where strong selectivity effects may be observed. Such phenomena may have much practical significance in the fields of biofilm formation and prebiotic chemistry.

Very recently a research group studied the adsorption of lipids on the surface of alumina, titania and Fe oxide films (Nellis *et al.*, 2011). The aim of the work was to create models of a cell membrane (a lipid bilayer) and they discovered that titania is a robust and convenient support to achieve the objective.

In another study, alumina was studied as a substrate for adsorption of carboxylic acids (Megias-Alguacil *et al.*, 2011). Results indicated that the initially hydrophilic nature of the solid substrate changes toward a less hydrophilic character as the bulk concentration and the chain length of the acids increases. The hydrophilic character of the coated alumina decreases with acid concentration upon a certain concentration and beyond that, it increases.

Many research groups are interested in the catalytic effects of magnesium oxide and titania surfaces. The two oxides are currently used in the decontamination of chemical and biological warfare (CBW) agents. Researchers found that aerogel preparations of these catalysts are more reactive toward the CBWs (Sundarrajan *et al.*, 2010).

Magnesium oxide/hydroxide was also studied, using AFM and other experimental techniques (Puriwat *et al.*, 2010), as a possible catalyst for gas-phase isomerization of 1-butene to 2-butene.

Pseudowollastonite (β-$CaSiO_3$) was used as a substrate for the attachment, viability, proliferation and osteogenic differentiation of human mesenchymal stem cells (hMSCs), and provides detailed mechanistic links of surface texture, soluble factors and culture media to cell activities. Soluble factors (Ca and Si concentrations) from the substrate may be more important than small surface roughness variations for osteogenic differentiation in growth medium. Optimum concentration ranges exist for individual soluble factors to balance cell toxicity/growth *vs.* osteogenic differentiation, and soluble factors together have complex, cooperative or opposing, effects on a given cell activity (Zhang *et al.*, 2010).

4. Comparison between experimental data and simulation

Modern studies concerning the interaction of organic molecules with mineral-layer substrates are also conducted by theoretical simulations. It is outside the scope of the present work to deal with the theoretical background of the various simulations. Some previously reported results from minerals are reported.

Computational science was born with the advent of electronic calculators in ~1940. These devices use numerical techniques to solve complex and large-scale problems. Even though quantum mechanic theories were born in the first half of the 20th century, calculators have only become sophisticated enough to be a useful computational tool in the last 30 years. Nowadays, modelling and molecular simulations take an active role in modern mineralogy. One salient characteristic of computational approach is its intrinsic ability to adapt accuracy to the level required by the problem. Switching from atoms, molecules, macro-molecules, *etc.*, there are changes in the size of the system, its phenomena and the appropriate methods to describe them (*i.e.* multiscale approach). The interaction of organic molecules with layer surfaces needs an atomic-scale approach (nanometres) and the computational models used are both *ab initio* (or static) and/or molecular-dynamics methods.

Quantum-chemistry simulations of organic molecules onto individual (silica, alumina, titania), binary (silica/alumina, silica/titania), and ternary (alumina/silica/titania) fumed oxides were conducted to analyse the effects of morphology and surface composition of the materials (Gun'ko *et al.*, 2010). The authors used the software *GAUSSIAN03*, adopting the cluster approach to simulate the adsorption reaction. They achieved a better understanding of adsorbate effects on the materials surfaces and gained information about the hydrogen bonds and energy of adsorption, and reached good agreement with experimental results.

A more complex study is the glycine interaction with a model silica surface terminated by an isolated hydroxyl group (Rimola *et al.*, 2006). The research group adopted a DFT/B3LYP method and CRYSTAL periodic code, which uses a local Gaussian basis set, allowing the users to treat molecules, 1D periodic polymers, 2D periodic surfaces (slabs), and 3D crystals (bulks) with the same level of theory. They evaluated extensively the glycine/silica surface system, from geometry to vibrational features, with both neutral and zwitterionic adsorbate. Whereas glycine is generally adsorbed in its neutral form, two structures show glycine adsorbed as a zwitterion, the surface playing the role of a 'solid solvent' whereas intra-strand hydrogen bonds stabilize the zwitterions cooperatively. These structures are no longer formed at a low glycine coverage, as a simulated enlargement of the unit cell by doubling the value of the *a* cell edge showed the importance of H-bond cooperativity in stabilizing the zwitterionic forms. Furthermore, it has been shown that the amino acid NH_2 group plays only a minor role as a weak hydrogen-bond donor. These results were in good agreement with experimental data.

In a recent review, the state of the art in the study of polymerization of amino acids on mineral surface was reported (Lambert, 2008). The experimental and theoretical results were compared in an attempt to understand the effect of the surface and the mechanism of polymerization. The authors reviewed amino acids in interaction with silicates, oxides and sulphides. It was discovered that amino acid polymerization is favoured by adsorption, at least if the latter is followed by drying at moderately high temperatures. It remains unclear if adsorption without drying could result in significant polymerization, which unfortunately would require many ageing experiments. Basic knowledge of the various possible mechanisms of amino acid adsorption on oxide surfaces at the molecular level is currently being carried out. However, the situation is rather complicated and many different mechanisms may come into play including the specific surface, the specific amino acid, and the experimental conditions (pH, ion strength, *etc.*) under consideration.

In line with the above considerations, molecular dynamics simulations of a simple organic molecule in interaction with α-alumina were performed by Haw & Mosey (2011). The aim of the work was to explore the possible use of aldehydes as functional lubricant. The simulations involved the compression and decompression processes of acetaldehyde molecules between two (0001) surfaces of α-alumina. It was a first-principles molecular dynamics (FPMD) method, adopting PBE functional and a 0.121 fs time step. The authors' results indicated the formation of a protective layer of acetaldehyde molecules on the surface and, after compression, the organic molecules polymerized in polyether. In this case, molecular dynamics is a valuable tool to understand the detailed mechanisms of a process, difficult to obtain experimentally.

Another example of the application is the study of hydronium complexation on Na-bearing montmorillonite (Churakov & Kosakowski, 2010). The evaluation was conducted with an *ab initio* method, using DFT and PBE functionals to take account of electron exchange and correlation. The simulation setup included a 0.17 fs time step. The surfaces analysed were a mono-, bi- and trihydrated Na-montmorillonite. The results are in agreement with experimental data and explain the observed transformation of mono- to di-hydrated Na-montmorillonite upon acid treatment by a strong affinity of hydronium to the water molecules in the interlayer.

Another work by Roark & Feller (2008) regarded the simulations of a phospholipid bilayer interacting with a solid surface of hydroxylated nanoporous amorphous silica, carried out over a range of lipid–solid substrate distances. The porous solid surface allowed the water layer to adjust its thickness dynamically, maintaining equal pressures above and below the membrane bilayer. Qualitative estimates of the force between the surfaces leads to an estimated lipid–silicon distance in very good agreement with the results of neutron scattering experiments. Detailed analysis of the simulation at the separation suggested by experiment shows that for this type of solid support the water layer between surfaces is very narrow, consisting only of bound water hydrating the lipid head groups and the hydrophilic silica surface. The reduced hydration, however, has only minor effects on the head-group hydration, the orientation of water molecules at the interface, and the membrane dipole potential. Whereas these structural properties were not sensitive to the presence of the solid substrate, the calculated diffusion coefficient for translation of the lipid molecules was altered significantly by the silica surface.

In conclusion, theoretical methods provide results in good agreement with experimental data and they are useful tools to help us understand better the interaction of biomolecules with the mineral surface. The choice of the specific computational method depends on the complexity of the system and the level of accuracy required. Both *ab initio* and molecular dynamics methods use quantum mechanic theories and can evaluate fundamental and excited states (reactions, adsorption, *etc.*). The first is adaptable to many type of systems and is very accurate, though with great computational constraints.

Molecular dynamics methods overcome this inconvenience, but with extensive approximations, so are less accurate and need empirical parameters from experimental or *ab initio* data. Despite the lesser accuracy, they can provide 'pictures' of what is happening in the system, as these methods have a timescale.

There are still only a few comparisons at the atomic scale between SPM and computational results. Some very effective results have been reported recently where dedicated experimental set-ups were used and achieved very good agreement with simulations (Fukuma, 2009; Gross *et al.*, 2009).

In conclusion, from the experimental point of view, SPM is a highly effective, flexible and promising technique for the nano-characterization of the surface properties of layer silicates, oxides and hydroxides. Cleavage, morphology, mechanical parameters and surface potential can all be determined quantitatively at the nanometre scale.

The influence of the mineral's surface on nanoconfined liquids or biomolecules and that of the liquid medium on the surface physico-chemical properties can be studied. Furthermore, the SPM has great sensitivity to the interaction forces between the probe and the sample allowing single-molecule investigation of the adsorption of organic matter onto the mineral surface.

References

Alaeddine, S. & Nygren, H. (1996) The adsorption of water and amino acids onto hydrophilic and hydrophobic quartz surfaces. *Colloids and Surfaces B: Biointerfaces*, **6**, 71–79.

Alessandrini, A., Valdrè, G., Valdrè, U. & Muscatello, U. (2007) Defects in ordered aggregates of cardiolipin visualized by atomic force microscopy. *Chemistry and Physics of Lipids*, **146**, 111–124.

Alvarez-Silva, M., Uribe-Salas, A., Mirnezami, M. & Finch, J.A. (2010) The point of zero charge of phyllosilicate minerals using the Mular–Roberts titration technique. *Minerals Engineering*, **23**, 383–389.

Antognozzi, M., Humphris, A.D.L. & Miles, M.J. (2001) Observation of molecular layering in a confined water film and study of the layers viscoelastic properties. *Applied Physics Letters*, **78**, 300–302.

Antognozzi, M., Protti, A., Miles, M.J. & Valdrè, G. (2003) Investigation of nano-confined liquids on muscovite by transverse dynamic force microscopy (TDFM). *GeoActa*, **2**, 101–106.

Antognozzi, M., Wotherspoon, A., Hayes, J.M., Miles, M.J., Szczelkun, M.D. & Valdrè, G. (2006) A chlorite mineral surface actively drives the deposition of DNA molecules in stretched conformations. *Nanotechnology*, **17**, 3897–3902.

Antognozzi, M., Ulcinas, A., Picco, L., Simpson, S.H., Heard, P.J., Szczelkun, M.D., Brenner, B. & Miles, M.J. (2008) A new detection system for extremely small vertically mounted cantilevers. *Nanotechnology*, **19**, 384002.

Baumann, M. & Stark, R.W. (2010) Dual frequency atomic force microscopy on charged surfaces. *Ultramicroscopy*, **110**, 578–581.

Benetoli, L.O.D.B., De Santana, H., Zaia, C.T.B.V. & Zaia, D.A.M. (2008) Adsorption of nucleic acid bases on clays: An investigation using Langmuir and Freundlich isotherms and FT-IR spectroscopy. *Monatshefte für Chemie*, **139**, 753–761.

Cai, P., Huang, Q., Zhang, X. & Chen, H. (2006) Adsorption of DNA on clay minerals and various colloidal particles from an Alfisol. *Soil Biology and Biochemistry*, **38**, 471–476.

Churakov, S.V. & Kosakowski, G. (2010) An *ab initio* molecular dynamics study of hydronium complexation in Na-montmorillonite. *Philosophical Magazine*, **90**, 2459–2474.

Franchi, M. & Gallori, E. (2005) A surface-mediated origin of the RNA world: Biogenic activities of clay-adsorbed RNA molecules. *Gene*, **346**, 205–214.

Fukuma, T. (2009) Subnanometer-resolution frequency modulation atomic force microscopy in liquid for biological applications. *Japanese Journal of Applied Physics*, **48**, 08JA01.

Fukuma, T., Kobayashi, K., Matsushige, K. & Yamada, H. (2005) True atomic resolution in liquid by frequency-modulation atomic force microscopy. *Applied Physics Letters*, **87**, 034101.

Giessibl, F.J. (2003) Advances in atomic force microscopy. *Reviews of Modern Physics*, **75**, 494–983.

Giessibl, F.J., Hembacher, S., Bielefeldt, H. & Mannhart, J. (2000) Subatomic features on the silicon (111)-(737) surface observed by atomic force microscopy. *Science*, **289**, 422–425.

Girard, P. (2001) Electrostatic force microscopy: principles and some applications to semiconductors. *Nanotechnology*, **12**, 485–490.

Gross, L., Mohn, F., Moll, N., Liljeroth, P. & Meyer, G. (2009) The chemical structure of a molecule resolved by atomic force microscopy. *Science*, **325**, 1110–1114.

Gun'ko, V.M., Yurchenko, G.R., Turov, V.V., Goncharuk, E.V., Zarko, V.I., Zabuga, A.G., Matkovsky, A.K., Oranska, O.I., Leboda, R., Skubiszewska-Zie, J., Janusz, W., Phillips, G.J. & Mikhalovsky, S.V. (2010) Adsorption of polar and nonpolar compounds onto complex nanooxides with silica, alumina, and titania. *Journal of Colloids and Interface Science*, **348**, 546–558.

Hanczyc, M.M., Mansy, S.S. & Szostak, J.W. (2006) Mineral surface directed membrane assembly. *Origins of Life and Evolution of the Biosphere*, **37**, 67–82.

Haw, S.M. & Mosey, N.J. (2011) Chemical response of aldehydes to compression between (0001) surfaces of α-alumina. *The Journal of Chemical Physics*, **134**, 014702.

Jurak, M. & Chibowski, E. (2010) Surface free energy and topography of mixed lipid layers on mica. *Colloids and Surfaces B: Biointerfaces*, **75**, 165–174.

Kalra, S., Pant, C.K., Pathak, H.D. & Mehta, M.S. (2000) Adsorption of glycine and alanine on montmorillonite with or without coordinated divalent cations. *Indian Journal of Biochemistry and Biophysics*, **37**, 341–346.

Kendall, T.A., Na, C., Jun, Y.S. & Martin, S.T. (2008) Electrical properties of mineral surfaces for increasing water sorption. *Langmuir*, **24**, 2519–2524.

Krotil, H.-U., Stifter, T., Waschipky, H., Weishaupt, K., Hild, S. & Marti, O. (1999) Pulsed Force mode: a new method for the investigation of surface properties. *Surface and Interface Analysis*, **27**, 336–340.

Kwon, K.D., Vadillo-Rodriguez, V., Logan, B.E. & Kubicki, J.D. (2006) Interactions of biopolymers with silica surfaces: Force measurements and electronic structure calculation studies. *Geochimica et Cosmochimica Acta*, **70**, 3803–3819.

Lambert, J.F. (2008) Adsorption and polymerization of amino acids on mineral surfaces: A review. *Origins of Life and Evolution of the Biosphere*, **38**, 211–242.

Lee, M. & Jhe, W. (2006) General theory of amplitude modulation atomic force microscopy. *Physical Review Letters*, **97**, 036104.

Megias-Alguacil, D., Tervoort, E., Cattin, C. & Gauckler, L.J. (2011) Contact angle and adsorption behavior of carboxylic acids on α-Al_2O_3 surfaces. *Journal of Colloid and Interface Science*, **353**, 512–518.

Montes, S., Atenas, G.M. & Valero, E. (2007) How fine particles on haematite mineral ultimately define the mineral surface charge and the overall floatability behaviour. *Journal of the South African Institute of Mining and Metallurgy*, **107**, 689–695.

Nellis, B.A., Satcher, J.H. & Risbud, S.H. (2011) Phospholipid bilayer formation on a variety of nanoporous oxide and organic xerogel films. *Acta Biomaterialia*, **7**, 380–386.

Nonnenmacher, M., O'Boyle, M.P. & Wickramasinghe, H.K. (1991) Kelvin probe force microscopy. *Applied Physics Letters*, **58**, 2921–2923.

Nyffenegger, R.M., Penner, R.M. & Schierle, R. (1997) Electrostatic force microscopy of silver nanocrystals with nanometer-scale resolution. *Applied Physics Letters*, **71**, 1878–1880.

Pittenger, B., Erina, N. & Su, C. (2010) Quantitative mechanical property mapping at the nanoscale with PeakForce QNM. *Veeco Application Note*, AN128, Rev. A0.

Puriwat, J., Chaitree, W., Suriye, W., Dokjampa, S., Praserthdam, P. & Panpranot, J. (2010) Elucidation of the basicity dependence of 1-butene isomerization on $MgO/Mg(OH)_2$ catalysts. *Catalysis Communications*, **12**, 80–85.

Rimola, A., Sodupe, M., Tosoni, S., Civalleri, B. & Ugliengo, P. (2006) Interaction of glycine with isolated hydroxyl groups at the silica surface: first principles B3LYP periodic simulation. *Langmuir*, **22**, 6593–6604.

Roark, M. & Feller, S.E. (2008) Structure and dynamics of a fluid phase bilayer on a solid support as observed by a molecular dynamics computer simulation. *Langmuir*, **24**, 12469–12473.

Rosa-Zeiser, A., Weilandt, E., Hild, S. & Marti, O. (1997) The simultaneous measurement of elastic, electrostatic and adhesive properties by scanning force microscopy: pulsed-force mode operation. *Measurement Science and Technology*, **8**, 1333–1338.

Stark, R.W., Naujoks, N. & Stemmer, A. (2007) Multifrequency electrostatic force microscopy in the repulsive regime. *Nanotechnology*, **17**, 065502.

Stievano, L., Piao, L.Y., Lopes, I., Meng, M., Costa, D. & Lambert, J.F. (2009) Glycine and lysine adsorption and reactivity on the surface of amorphous silica. *European Journal of Mineralogy*, **19**, 321–331.

Sugimoto, Y., Pou, P., Abe, M., Jelinek, P., Perez, R., Morita, S. & Custance, O. (2007) Chemical identification of individual surface atoms by atomic force microscopy. *Nature*, **446**, 64–67.

Sundarrajan, S., Chandrasekaran, A.R. & Ramakrishna, S. (2010) An update on nanomaterials-based textiles for protection and decontamination. *Journal of the American Ceramic Society*, **93**, 3955–3975.

Valdrè, G. (2007) Natural nanoscale surface potential of clinochlore and its ability to align nucleotides and drive DNA conformational change. *European Journal of Mineralogy*, **19**, 309–319.

Valdrè, G. & Fabbrizioli, S. (2006) The role of layered silicate substrates on immobilization of red blood cells. *Scanning*, **28**(2), 72–73.

Valdrè, G. & Moro, D. (2008a) 3D finite element analysis of electrostatic deflection of commercial and FIB-modified cantilevers for electric and Kelvin force microscopy: I. Triangular shaped cantilevers with symmetric pyramidal tips. *Nanotechnology*, **19**, 405501.

Valdrè, G. & Moro, D. (2008b) 3D finite element analysis of electrostatic deflection and shielding of commercial and FIB-modified cantilevers for electric and Kelvin force microscopy: II. Rectangular shaped cantilevers with asymmetric pyramidal tips. *Nanotechnology*, **19**, 405502.

Valdrè, G., Alessandrini, A., Muscatello, U. & Valdrè, U. (1998) High resolution imaging of n-alkanes crystals by atomic force microscopy. *Philosophical Magazine Letters*, **78**(3), 255–261.

Valdrè, G., Antognozzi, M., Wotherspoon, A. & Miles, M.J. (2004) Influence of properties of layered silicate minerals on adsorbed DNA surface affinity, self-assembly and nanopatterning. *Philosophical Magazine Letters*, **84**(9), 539–545.

Valdrè, G., Malferrari, D. & Brigatti, M.F. (2009) Crystallographic features and cleavage nanomorphology of chlinochlore: specific applications. *Clays and Clay Minerals*, **57**, 183–193.

Valdrè, G., Moro, D. & Ulian, G. (2011) Mineral surface–organic matter interactions and applications. *IOP Conference Series: Materials Science and Engineering, Institute of Physics, London, UK*, in prep.

Yamada, H., Kobayashi, K., Fukuma, T., Hirata, Y., Kajita, T. & Matsushige, K. (2009) Molecular resolution imaging of protein molecules in liquid using frequency modulation atomic force microscopy. *Applied Physics Express*, **2**, 095007.

Yanina, S.V. & Rosso, K.M. (2008) Linked reactivity at mineral-water interfaces through bulk crystal conduction. *Science*, **320**, 218–222.

Zhang, N., Molenda, J.A., Fournelle, J.H., Murphy, W.L. & Sahai, N. (2010) Effects of pseudowollastonite ($CaSiO_3$) bioceramic on in vitro activity of human mesenchymal stem cells. *Biomaterials*, **31**, 7653–7665.

Ziegler, D., Rychen, J., Naujoks, N. & Stemmer, A. (2007) Compensating electrostatic forces by single-scan Kelvin probe force microscopy. *Nanotechnology*, **18**, 225505.

EMU Notes in Mineralogy, Vol. 11 (2011), Chapter 10, 335–370

The surface properties of clay minerals

ROBERT A. SCHOONHEYDT[1] and CLIFF T. JOHNSTON[2]

[1]*Centre for Surface Chemistry and Catalysis, K.U. Leuven,
Kasteelpark Arenberg 23, 3001 Leuven, Belgium
e-mail: robert.schoonheydt@biw.kuleuven.be*
[2]*Department of Crop, Soil and Environmental Sciences, Purdue University,
915 West State Street, West-Lafayette, Indiana 47907-2054, USA
e-mail: cliffjohnston@purdue.edu*

Clay minerals have interlayer surfaces and edge surfaces, the former being the most important, especially in the case of swelling clays or smectites. Water is by far the most important adsorbed molecule in the interlayer space, where it interacts with the exchangeable cations and with the siloxane surface. Transition metal ion complexes are selectively ion-exchanged in the interlayer space of smectites. Poly-amine complexes easily lose their axial ligands to adopt a square planar configuration. The more stable and bulky tris(bipyridyl) and tris(phenanthroline) complexes in the interlayer space give chiral clay mineral composites that can be used in columns for chiral chromatography, in asymmetric catalysis and in non-linear optics. The formation of clay mineral-dye complexes is a two-step process: instantaneous adsorption of the dye molecules, mainly as aggregates, followed by a slower redistribution process over the clay-mineral surface. With careful choice of dye molecules, non-linear optical materials can be prepared which exhibit properties such as second harmonic generation and two-photon absorption. Ion exchange of cationic proteins is a three-step process: (1) instantaneous adsorption at the edges; (2) adsorption in the interlayer space, followed by; (3) weak adsorption in excess of the cation exchange capacity. The extent to which these three processes occur depends on (1) the kind of exchangeable cation in the interlayer; and (2) the molecular weight, shape and charge of the protein molecules.

1. Introduction

Clay minerals are phyllosilicates or layered materials. Each layer consists of either one sheet of SiO_4 tetrahedra joined to one sheet of Al- or Mg-octahedra or one sheet of Al- or Mg-octahedra sandwiched between two sheets of Si-tetrahedra. The former are called 1:1 or TO, the latter 2:1 or TOT clay minerals. Those with an Al-octahedral sheet are called dioctahedral; those with a Mg-octahedral sheet, trioctahedral. Isomorphous substitution of Si^{4+} by mainly Al^{3+} in the tetrahedral sheets; of Al^{3+} by Mg^{2+}, Fe^{2+} or other divalent cations with approximately the same diameter in the Al-octahedral sheets; and of Mg^{2+} by Li^{+} in the Mg-octahedral sheets leads to negatively charged layers. Exchangeable cations, *i.e.* cations which do not belong to the crystalline structure, compensate this negative charge. The classification of clay minerals, shown in Table 1, is based on these principles. More detailed information can be found in textbooks such as those of Bergaya *et al.* (2006) and Nemecz (1981).

DOI: 10.1180/EMU-notes.11.10

Table 1. Classification of phyllosilicates (reproduced from Bergaya et al., 2006, with the permission of Elsevier).

Layer type	Group	Subgroup	Species
1:1	Kaolin-serpentine	Kaolin (dioct.)	Kaolinte, dickite
	$X \approx 0$	Serpentines (trioct.)	Chrysotile, lizardite
2:1	Pyrophillite-talc	Pyrophyllite (dioct.)	Pyrophillite
	$X \approx 0$	Talc (trioct.)	Talc
	Smectite	Dioct. smectite	Montmorillonite
	$X \approx 0.2$–0.6	Trioct. smectite	Saponite, hectorite
	Vermiculite	Dioct. vermiculites	Dioct. vermiculite
	$X \approx 0.6$–0.9	Trioct. vermiculites	Trioct. vermiculite
	Mica	Dioct. micas	Muscovite
	$X \approx 1$	Trioct. micas	Phlogopite, biotite
	Brittle mica	Dioct. brittle mica	Margarite
	$X \approx 2$		
	Chlorite	Dioct. chlorites	Donbassite
	X is variable	Di,trioct. chlorites	Cookeite
		Trioct. chlorites	Clinochlore
2:1 inverted ribbons	Sepiolite-palygorskite X is variable	Sepiolites, palygorskites	Sepiolite, palygorskite

The most important clay minerals, both industrially and scientifically, are the 1:1 clay minerals and the 2:1 clay minerals with moderate charge, also called swelling clays or smectites. This charge is expressed in terms of the number of electrons per half unit cell $[O_{10}(OH)_2]$ (Table 1).

Figure 1 presents the structure of kaolinite, a 1:1 clay mineral and that of a typical 2:1 clay mineral such as smectite. In a 1:1 clay mineral the surface of the Si-sheet consists of oxygen atoms; that of the Al or Mg sheet of hydroxyl groups. In a 2:1 clay mineral the surface consists of oxygen atoms of the Si sheets on both sides of the layers. Both 1:1 and 2:1 clay minerals have edges at which oxygen atoms and cations

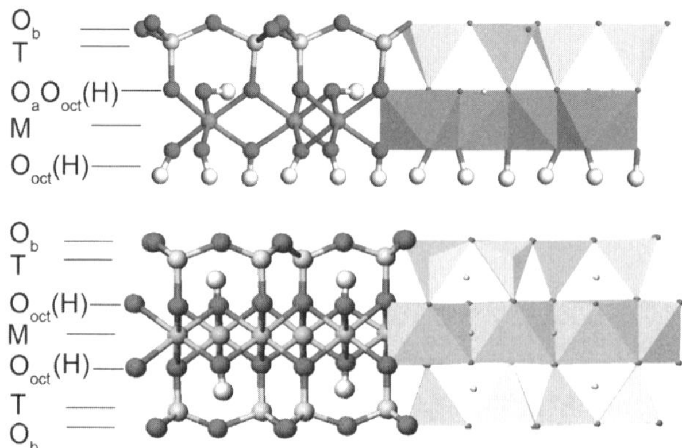

Fig. 1. Arrangement of octahedral and tetrahedral sheets in 1:1 and 2:1 clay minerals. Readers of the paper version of this chapter may wish to download a colour version of this figure from www.minersoc.org/emu-notes/emu-11/11-10-colour.pdf.

(Si^{4+}, Al^{3+} and Mg^{2+}) are exposed. These ions carry charges and react with molecules in the environment, usually water, to maintain charge neutrality and complete their coordination sphere. Thus, the edges consist of $-O^-$, $-OH$ and $-OH_2^+$ groups. Their relative amounts depend on the pH of the aqueous environment in which the clay minerals are residing. The surface chemistry of clay minerals can then be defined as the interaction of adsorbed molecules with the exchangeable cations, with the oxygen atoms, OH and OH_2^+ groups at the surface and with the adsorbed water molecules.

Clay minerals such as smectites occur as aggregates and particles in which the individual layers are more or less organized face-to-face. Interaction of molecules and ions with the planar surfaces of the individual layers is only possible if the molecules and ions are able to diffuse between the individual layers in the aggregate or particle. As molecules occupy this interlayer space, the distance between each layer in the aggregate increases: the clay mineral swells (details on clay structure and classification are reported by Brigatti *et al.*, 2011 and in Guggenheim 2011, this volume). Swelling is a typical phenomenon of smectites or swelling clays. This is due to the hydrated exchangeable cations in the interlayer space. The exchange of these inorganic cations with other cations, especially organic cations such as alkylammonium cations $[(R)_{4-n}NH_n]^+$, changes the interlayer distance, as the size of the cation and the hydration level of the clay mineral are modified. Neutral molecules can also be adsorbed in the interlayer space by replacing some or all of the water molecules and, as a consequence, the interlayer distance changes too. Such adsorption processes, in which the adsorbate penetrates (or intercalates) into the interlayer space, are relatively easy because the interlayer space is pre-opened by the hydrated exchangeable cations, such as Na^+ (see also Christidis, 2011, this volume).

This is not the case for clay minerals without isomorphous substitution such as kaolinite, pyrophyllite and talc (see Guggenheim, 2011, this volume). In these cases adsorption into the interlayer region is difficult, if not impossible, because of the very strong attractive interactions between the layers. There are three types of interaction: H-bonding, dispersive interactions and dipolar interactions. In kaolinite, H-bonding occurs between Al-OH groups of one layer and the Si-O groups of the next layer. In the case of pyrophyllite and talc, dipolar interactions between the Si–O bond dipoles of one layer and the Si–O bond dipoles of the next layer occur. In all cases, dispersive interactions are operative also (for more details see Guggenheim, 2011, this volume).

The surface chemistry of clay minerals can now be defined as the physisorption and chemisorption of molecules and ions in the interlayer space and at the edges of the clay mineral layers. In this chapter, clay mineral–water interactions are discussed in section 2. Section 3 treats transition metal ion (TMI) complexes in the interlayer space, section 4 the properties of adsorbed cationic dye molecules and section 5 the adsorption of proteins. There are four reasons for this choice of subjects: (1) water is the most important adsorbed molecule in clay minerals; (2) TMI complexes and cationic dyes are easy to exchange and an almost quantitative adsorption up to the cation exchange capacity (CEC) is obtained; (3) they can be studied by a variety of spectroscopic techniques such as ultraviolet-visible-near infrared (UV-VIS-NIR) spectroscopy, fluorescence spectroscopy, electron paramagnetic resonance (EPR), Fourier-transform Infrared

Spectroscopy (FTIR) and Raman spectroscopy. An introduction of these techniques and their use in clay mineral studies can be found in Wilson (1994). (4) Hybrid systems, clay mineral-dye molecules and clay mineral-TMI complexes, have been investigated as materials for a range of applications, such as electrodes in electrochemistry, chromatographic columns for chiral separations and clay-based materials in optical devices.

2. Clay mineral–water interactions

For over 80 years, the interaction of water with clay minerals has attracted interest (Bradley *et al.*, 1937; Buswell *et al.*, 1937; Hofmann & Bilke, 1936; von Buzagh, 1929) and has been studied throughout that time (Anderson *et al.*, 2010; Graham, 1964; Henniker, 1949; Johnston, 2010; Low, 1961; Marry *et al.*, 2008; Newman, 1987; Schoonheydt & Johnston, 2007; Sposito & Prost, 1982). From a spatial perspective, the length scales that define clay-water interactions range from short-range H-bonding and ion-dipole interactions (~0.2 nm) to larger-scale interactions that include electrical double layer interactions associated with clay swelling (>10 nm). Because clay particles are small in size and generally have large aspect ratios, these short-range interactions (<10 nm) are manifest at much larger macroscopic scales associated with shrink-swell behaviour in soils, geomorphology and landslides.

2.1 Origin of clay mineral–water interactions

We will limit the scope of our attention here to the group of expandable 2:1 (TOT) clay minerals in the smectite group, as they are among the most interesting and important clay minerals related to clay mineral–water interactions, although there are many interesting aspects of water interactions with other types of clay minerals, such as halloysite (Joussein *et al.*, 2005). To better understand how water interacts with expandable clay minerals, consider the physical picture that has emerged from recent experimental and computational studies. Fundamental particles (Christidis, 2011, this volume) of the clay minerals of the smectite group (referred to as smectites) vary in size with basal dimensions that range from ~20 nm, in the case of Laponite, a synthetic clay, to >1 μm for certain types of montmorillonite. Similar diversity is found in nature with shapes that vary from rhombic to subhedral lamellae, hexagonal lamellae to laths and fibres (Güven, 1988; Ras *et al.*, 2007; Szabo *et al.*, 2009). The thickness of a fundamental particle of smectite is 0.96 nm and multiple layers are generally stacked on top of each other ranging from 1 to 15 layers (Lagaly & Malberg, 1990; Schramm & Kwak, 1982). These particles typically have large aspect ratios and morphologies similar to that of a torn sheet of paper (Cao *et al.*, 2010; Lagaly, 2006; Pizzey *et al.*, 2004; Ras *et al.*, 2003). For some synthetic clays, such as Laponite, the layers are much smaller (~30 nm in diameter) and have correspondingly small aspect ratios. Layers of smectite are not rigid but have some flexibility that is observed experimentally in atomic force microscopy (AFM), scanning electron microscopy (SEM) and transmission electron microscopy (TEM) images and through molecular modelling studies (Cao *et al.*, 2010; Pizzey *et al.*, 2004; Sato *et al.*, 2001; Tamura *et al.*, 1999). In the case of Na^+- and Li^+-exchanged smectites at low ionic strength, the layers can be separated completely from each other (*i.e.* dispersed) into a dispersion of individual 1 nm-thick layers.

The AFM image of a Wyoming montmorillonite is shown in the lower right (2a) portion of Figure 2 showing irregularly shaped layers, that are ~500 nm in diameter and 1 nm thick (Ras *et al.*, 2003). The crystal structure of a representative idealized smectite layer is shown in Figure 2b, comprising 100 unit cells along the *a* axis and 58 unit cells along the *b* axis for a total of 5800 unit cells (214,600 atoms), measuring 51.9 nm × 52.2 nm. The individual rectangular boxes shown in Figure 2b correspond to 'super-cells' consisting of 2 × 2 unit cells measuring 1.039 nm along *a* and 1.8 nm along *b*. The unit-cell composition of a typical Wyoming montmorillonite is $Na_{0.74}(Si_{7.76}Al_{0.24})^{IV}(Al_{3.08}Fe^{3+}_{0.42}Mg_{0.48})O_{20}(OH)_2$ with a molecular mass of 746.2 g/mol (Środoń & McCarty, 2008); the mass of the particle shown in Figure 2b is 7.19×10^{-18} g. Depending on the type of clay mineral, surface-charge density, location of charge and nature of the exchangeable cation, fundamental particles are generally stacked on top of each other with values that range from 1 to 15. A stack of

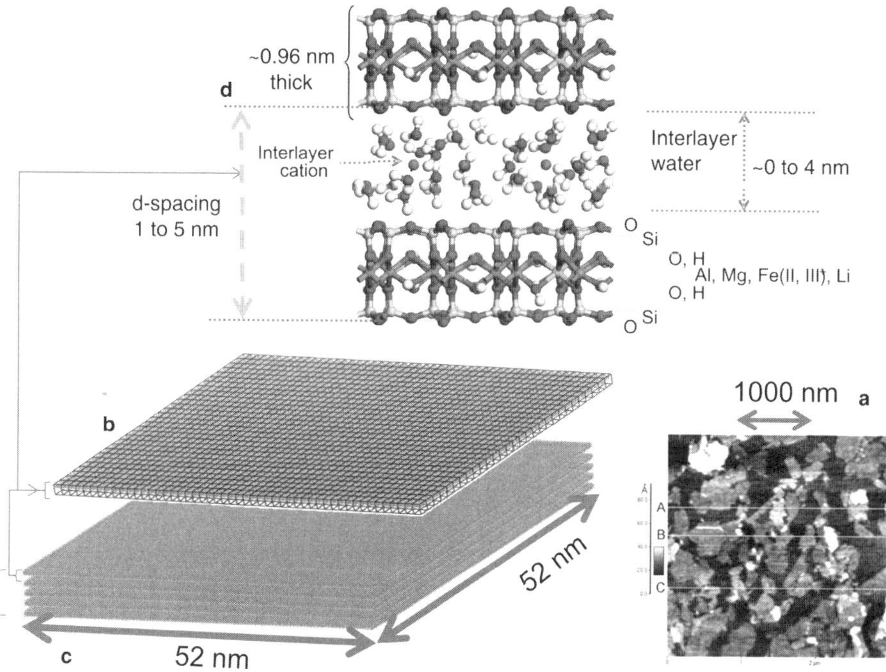

Fig. 2. (***a***) AFM image of Wyoming montmorillonite; (***b***) idealized crystal structure of a smectite particle; (***c***) smectite particle with five layers; (***d***) a three-layer hydrate. Readers of the paper version of this chapter may wish to download a colour version of this figure from www.minersoc.org/emu-notes/emu-11/11-10-colour.pdf.

five fundamental layers separated by three layers of water (~0.75 nm) is shown in Figure 2c. The combined thickness of this five-layer aggregate is 8 nm, excluding the outer layers of hydration. An expanded view of two layers separated by a distance of 0.75 nm of H_2O molecules surrounding interlayer cations is shown in the upper portion of Figure 2d. The perpendicular distance between one 'lattice plane' of atoms in one layer to the same plane in the next layer is referred to as the *d* spacing and can be measured using X-ray diffraction (XRD) or high-resolution transmission electron microscopy (HRTEM) (Fig. 2d). The *d*-spacing values commonly observed for smectites range from 1.25 to 1.75 nm, though fully collapsed values (0.96 nm) and fully expanded (values in excess of 5 nm) are sometimes observed. Taking into account the thickness of the anhydrous structural unit cell (0.96 nm), the typical height of the interlayer space between the layers is commonly 0.29 to 0.84 nm, or roughly one to three layers of water (Fig. 2d).

All smectites have some degree of isomorphous substitution that occurs in either the octahedral or tetrahedral sheets, or more commonly, in a mixture of both, resulting in the development of an overall negative charge on the basal siloxane surface. These negative charges are most commonly compensated in nature by the presence of the exchangeable cations Ca^{2+}, Mg^{2+}, K^+ and Na^+, although many other interesting cationic species exist in nature or in contaminated sites that can effectively compete with these ions in exchange reactions. These include the strongly hydrolytic species (Al^{3+} and Fe^{3+}), redox-sensitive ions (Mn(II/III,IV), $Fe^{2+/3+}$), weakly hydrated ions of importance in nuclear-waste containment (Cs^+ and Sr^{2+}) and a host of organic bases and cationic species including proteins and amino acids. Depending on the extent of isomorphous substitution, the distance between these exchangeable cations ranges from ~0.9 to 1.4 nm (Fig. 2d). This separation is further reduced to values between 0.65 to 0.84 nm in clay interlayers because two opposing clay mineral surfaces are brought into close proximity with each other, which effectively doubles the layer charge density (Johnston & Tombacz, 2002; Schoonheydt & Johnston, 2007). For example, the molar concentration of K^+ ions shown in the interlayer region of the smectite particle shown in Figure 2d is 4.7 M (or 13 H_2O molecules/K^+ ion). Upon further dehydration, the effective molar concentration in the interlayer region of the clay mineral can approach 12 M for a monovalent cation (*e.g.* Na^+) in a 'one layer hydrate' with a *d* spacing of 1.25 nm (4 H_2O molecules surrounding one K^+ ion).

Because the cations Ca^{2+}, Mg^{2+}, K^+ and Na^+ are most common in nature, we will restrict our attention to these ions, which have large negative enthalpies of hydration that range from −350 to −2000 kJ/mol and as a result will be surrounded by varying numbers of water molecules (Burgess, 1978; Cancela *et al.*, 1997; Huheey *et al.*, 1997). The charge–dipole attraction between interlayer water molecules and cations is the origin of clay mineral–water interactions on expandable clay minerals. These hydrated cations in the clay mineral interlayer render the surface hydrophilic, function as molecular props (holding the layers apart allowing the intercalation of additional water molecules and guest molecules), and serve as a precursor to capillary condensation of water and to swelling. Certain divalent cations (*e.g.* Ca^{2+}), hold the layers together so strongly that further expansion is prevented. The interlayer environment is unique in that the clay mineral surface functions as a semi-rigid anion, leaving the interlayer space occupied by cationic species surrounded by varying numbers of water molecules (Fig. 2d). For weakly hydrated cations (*e.g.* K^+ or Cs^+) under strongly desiccating conditions, complete collapse of the clay mineral interlayer can occur, where these cations form inner-surface complexes with the siloxane ditrigonal cavity. At the opposite extreme, clay-swelling resulting in interlayer separations approaching 10 nm can occur for low to medium surface-charge density clay minerals, exchanged with Na^+ or Li^+.

The chemical mechanisms of hydration of interlayer cations are through electrostatic interactions that maximize charge–dipole attraction and minimize water–water repulsion (Feller *et al.*, 1995; Glendening & Feller, 1995; Glendening & Feller, 1996). In the interlayer region of the clay, the negative charge of the clay mineral surface satisfies a portion of the charge of the cation, which reduces the overall hydration energy of the cation. In addition, the 'nanoconfined' region between the clay-mineral interlayers limits the free hydration of interlayer cations and, as mentioned earlier, the basal surface of the clay mineral functions as a planar solvent. Thus, the hydration of interlayer cations is modified by the clay-mineral surface.

The chemical and physical properties of the clay mineral–water–cation complex is critically dependent on the overall water content. From the vapour phase, where the activity of water (a_w) ranges from 0 to 0.98 (Prost *et al.*, 1998), smectites can absorb up to 0.6 g H_2O/g_{clay}, depending on the type of clay and nature of the interlayer cation (Mering, 1946; Mooney *et al.*, 1952a,b; Newman, 1987). The interaction of water with clay minerals at greater water contents ($a_w > 0.98$) can be accessed through pressure-membrane studies (Low, 1980; Prost *et al.*, 1998; Sposito, 1972), where water contents can exceed 10 g of H_2O/g of clay and result in the formation of gels and sols (Abend & Lagaly, 2000). This is illustrated in the combined water-sorption isotherms for Na- and Ca-montmorillonite shown in Figure 3. Pressure-membrane studies are used to study capillary condensation of water that occurs at high water activity values ($a_w > 0.98$) and vapour-phase gravimetric methods are used to study water sorption at lower water contents ($a_w < 0.98$) in the domain of crystalline swelling (Laird, 1996). When these data are plotted as water content *vs.* a_w, the capillary condensation values dominate the isotherms. When restricted to an a_w value of <0.8, the sorption isotherms (Fig. 3c) reveal the domain of crystalline swelling as influenced by Na^+ and Ca^{2+}. Water absorption in the a_w range of 0.1–0.8 is also associated with an increase in *d* spacing and these values are shown in Figure 3c. Because the a_w values in pressure-plate studies are large, water contents are sometimes expressed as log values ($\log(1/a_w)$), as shown in Figure 3a (Prost *et al.*, 1998). The domain of capillary condensation occurs for a_w values > 0.8 and this is represented by the dashed boxes in Figure 3a,b. At large water contents, greater sorption occurs in Na-montmorillonite with values that exceed 10 g of H_2O/g_{clay}.

Having developed a physical picture of the clay mineral–water–cation system, the interaction of water with the clay-mineral surface will be developed from three different perspectives. First, we will consider the H_2O molecule itself as a molecular probe of the clay-mineral surface and briefly review how the properties of H_2O are influenced by their

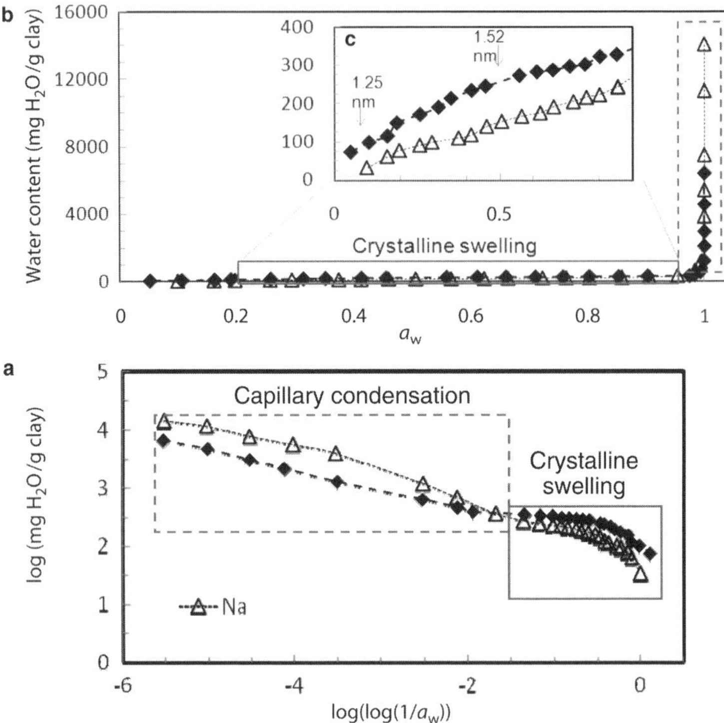

Fig. 3. Water sorption isotherms of Na^+ (Δ) and Ca^{2+} (◆)-montmorillonite. Readers of the paper version of this chapter may wish to download a colour version of this figure from www.minersoc.org/emu-notes/emu-11/11-10-colour.pdf.

close proximity to the interlayer cations and the clay-mineral surface itself. Similarly, the properties of the interlayer cation are influenced by the water content and their interaction with the clay-mineral surface. Third, certain aspects of the clay-mineral structure itself are influenced by water molecules and cations as a function of water content.

2.2. Nanoconfined H_2O molecules in clay-mineral interlayers

In the range of crystalline swelling ($a_w < 0.8$), accessed through vapour-phase sorption studies, the water molecules are strongly polarized by their close proximity to the exchangeable cations in the clay-mineral interlayer. Vibrational spectroscopy (Bishop *et al.*, 1994; Johnston *et al.*, 1992; Poinsignon *et al.*, 1978; Rinnert *et al.*, 2005; Russell & Farmer, 1964; Sposito *et al.*, 1983; Xu *et al.*, 2000) and quasi-inelastic neutron scattering (QENS) (Cebula *et al.*, 1981; Malikova *et al.*, 2005, 2007, 2010; Sobolev *et al.*, 2010; Swenson *et al.*, 2000) are two methods that have provided detailed insight into the nature of 'nanoconfined' water in the clay-mineral interlayer. The combination of gravimetric water adsorption measurements with spectroscopic studies is necessary in order to link the observed molecular behaviour to the overall water content, ratio of water molecules to interlayer ions and basal spacing of the clay mineral (Johnston *et al.*, 1992; Xu *et al.*, 2000). At low water contents (<0.6 g H_2O/g clay), the water molecules are organized around interlayer cations. Vibrational studies have shown that they are less hydrogen bonded to each other and that the water molecules are strongly polarized by the interlayer cation. When correlated with water content, the water molecules in the 1st and 2nd 'spheres' of hydration, extending to ~12 H_2O molecules per interlayer cation, are perturbed relative to bulk water. These studies are in good agreement with quasi-elastic neutron scattering (QENS) studies of water which provides insight into water dynamics in the interlayer region of smectites. In general, QENS studies of water diffusion in clay-mineral interlayers reveals that water diffusion is about an order of magnitude slower in the interlayer space relative to bulk water (Malikova *et al.*, 2007). In addition, QENS can provide information about different populations of water molecules. Nuclear magnetic resonance (NMR) studies of H_2O in clay interlayers accessed through the nuclei (^1H, ^2H, ^{17}O) have also given us some insight into the mobility and dynamics of water (Bowers *et al.*, 2008; Grandjean, 1997; Grandjean & Laszlo, 1989; Grandjean & Laszlo, 1994; Porion *et al.*, 2007). Note that 'macroscopic' methods which measure thermodynamic properties of clay mineral–water–cation interactions continue to provide useful insight. Examples include detailed sorption studies, surface-area analysis (Berend *et al.*, 1996; Cases *et al.*, 1997, 1992), and studies of thermodynamic properties (*e.g.* immersion enthalpies) (Cancela *et al.*, 1997; Rotenberg *et al.*, 2009; Yan *et al.*, 1996).

2.3. Interlayer cations

The molecular environment of interlayer cations and hydrated cations can also be accessed through a wide-range of NMR techniques. Many different cations have NMR nuclei available. These include ^7Li (Alba *et al.*, 1998; Alvero *et al.*, 1994; Porion *et al.*, 2009), ^{23}Na (Gevers & Grandjean, 2001; Hanaya & Harris, 1998; Laperche *et al.*, 1990; Porion *et al.*, 2005), ^{39}K (Bowers *et al.*, 2008), ^{111}Cd and ^{113}Cd (Bank *et al.*, 1989) and ^{133}Cs (Earl & Johnston, 1998; Kim & Kirkpatrick, 1997; Weiss, Jr. *et al.*, 1990). These studies have shown that the chemical shift of the NMR-active nuclei is influenced as a function of hydration. At greater water contents, the NMR signal of the nuclei in the interlayer region is 'solution like' with chemical-shift values similar to that of the ions in aqueous solution and narrow line-widths (Laperche *et al.*, 1990; Prost *et al.*, 1998). Upon reduction of the water content, the chemical-shift values shift to more negative values and the line-widths become broad as the cations are brought into closer proximity with the clay surface.

2.4. Influence of clay mineral–water–cation interactions on clay-mineral structures

For more than sixty years it has been known that cation–water complexes can perturb the structure of clay minerals under desiccating conditions (Hofmann & Klemen, 1950), as observed through the migration of interlayer cations into the clay structure and reduction in layer charge, which occurs at elevated temperatures of >100°C (typically >200°C)

(Calvet & Prost, 1971; Jaynes *et al.*, 1992; Komadel *et al.*, 2005; Sposito *et al.*, 1983). In recent years, more subtle perturbations of the clay structure have been observed with changes in water content at ambient temperature. In kaolinite-intercalation studies, small intercalated organic species have been observed to shift the vibrational frequencies of structural OH groups located between the octahedral and tetrahedral sheets (Costanzo & Giese, 1990; Johnston *et al.*, 2000; Johnston & Stone, 1990), indicating that structural OH groups located within the clay structure were sensitive to interactions in the interlayer space. Sposito *et al.* (1983) observed that the intensity of the structural OH-bending modes of montmorillonite in the 800–950 cm^{-1} region was influenced by the extent of hydration. Later studies quantified the change in intensity and observed that the positions of the structural OH-bending bands shifted with changes in water content (Xu *et al.*, 2000) as shown in Figure 4. The FTIR spectrum of Na-exchanged SWy-1 montmorillonite is characterized by three structural OH-bending vibrations at 918, 886 and 845 cm^{-1} assigned to Al-(OH)-Al, Al-(OH)-Fe^{3+} and Al-(OH)-Mg, respectively (Xu *et al.*, 2000). Upon reduction of the water content to values of <6 H_2O/Na^+ ion, the position of Al-(OH)-Fe^{3+} bands is decreased by ~6 cm^{-1}, whereas the position of the Al-(OH)-Mg band increases by approximately the same

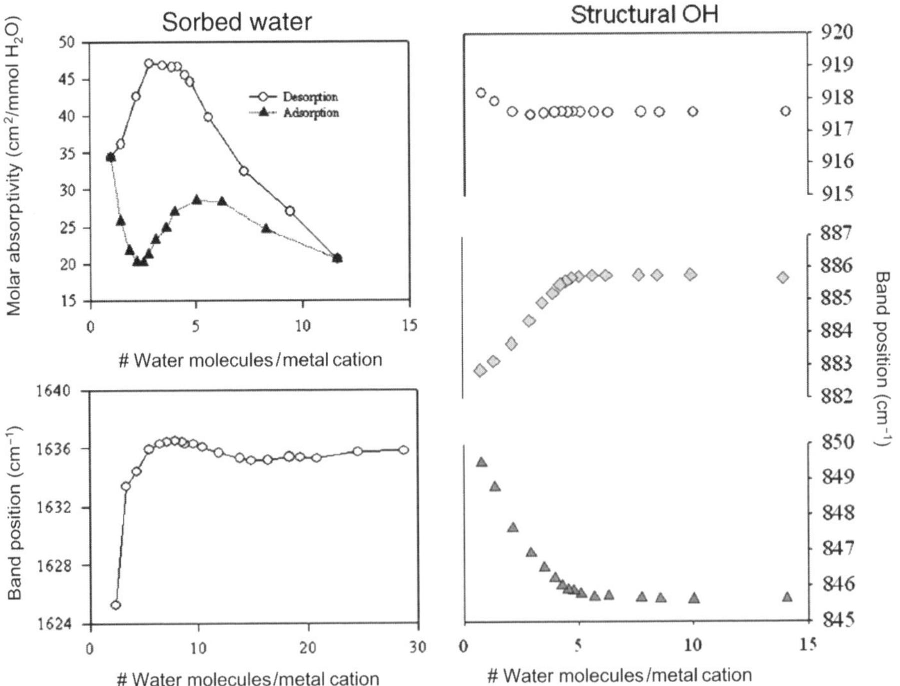

Fig. 4. (left) Molar absorptivity and position of the water deformation band as a function of the hydration level of Na^+-montmorillonite; (right) frequencies of bending modes of structural OH groups: Al-(OH)-Al (upper) Al-(OH)-Fe^{3+} (middle) and Al-(OH)-Mg^{2+} (lower).

amount. In contrast, the position of the OH group in the 'neutral' part of the structure that is not impacted by isomorphous substitution remains unchanged (Fig. 4). Upon reduction of the water content, the Na^+ ions in the interlayer are brought into close proximity with the siloxane ditrigonal cavities and the structural OH groups that reside beneath them. Similarly, introduction of microbial siderophores into the clay-mineral interlayers also resulted in spectral perturbations of the structural OH-bending bands of montmorillonite (Haack *et al.*, 2008).

At small water contents, interlayer H_2O molecules, clustered around interlayer cations, are closely associated with the clay-mineral surface. In a general sense, the water molecules coordinated to metal cations in the clay-mineral interlayers function as 'part' of the clay mineral. Heating the clay mineral or exposing it to a high vacuum can result in the irreversible collapse of the clay mineral, resulting from the Hofmann-Klemen effect (Brindley & Ertem, 1971; Farmer & Mortland, 1966; Hofmann & Klemen, 1950; Jaynes *et al.*, 1992; McBride *et al.*, 1975). Thus, the H_2O molecules clustered around the interlayer cations at small water contents are chemically and physically distinct from those found in bulk water. More importantly, the interaction of any guest molecule introduced into the clay interlayer will be influenced by these water molecules.

3. Transition metal ion complexes

3.1. Planar surfaces like planar complexes

In the 1970s, Cremers and co-workers (Maes *et al.*, 1977, 1978, 1979; Maes & Cremers, 1978, 1979) published a series of papers on the ion-exchange and formation constants of Cu(II)-polyamine complexes on smectites. With ethylenediamine (en) as the key example, the formation reactions can be written as:

$$[Cu(H_2O)_6]^{2+} + en \longleftrightarrow \overline{[Cu(en)(H_2O)_4]}^{2+} + 2\,H_2O; \quad \log \overline{K_{f1}} = 11.6 \qquad (1)$$

$$[Cu(en)(H_2O)_4]^{2+} + en \longleftrightarrow \overline{[Cu(en)_2(H_2O)_2]}^{2+} + 2\,H_2O; \quad \log \overline{K_{f2}} = 11.5 \qquad (2)$$

and overall

$$[Cu(H_2O)_6]^{2+} + 2\,en \longleftrightarrow \overline{[Cu(en)_2(H_2O)_2]}^{2+} + 4\,H_2O; \quad \log \overline{K_f} = 23.1 \qquad (3)$$

The horizontal bar denotes 'on the smectite surface'. Both log $\overline{K_{f1}}$ and log $\overline{K_{f2}}$ are larger than in solution by 1 and 2 units, respectively, thus giving the bis-complex on the clay surface an extra stabilization of three orders of magnitude with respect to aqueous solution. Reactions 1–3 involve exchange of water molecules and one might suspect that the extra stabilization of these complexes has something to do with the change of hydration level of these complexes on the clay-mineral surfaces compared to their hydration in aqueous solution. Figure 5 shows typical UV-VIS-NIR spectra. The d–d band maximum is found in the region 19,600–20,200 cm^{-1}. The exact position of the d–d band maximum depends somewhat on the charge density of the clay and on the exchange level, but it is significantly above the 18,600 cm^{-1} found in aqueous solution (Velghe *et al.*, 1977; Maes *et al.*, 1980). This is indicative of a strengthening of the in-plane ligand field strength and a weakening of the axial ligand field strength. The simplest way to visualize this is to replace the weakly coordinating axial water molecules by the planar clay surfaces as shown in Figure 5. The clay surface can be considered as a weaker ligand than the water molecules, if a ligand at all. It might be better to consider the clay surface as a 'solvent' that partially replaces the water molecules when the complexes are exchanged on the clay-mineral surfaces.

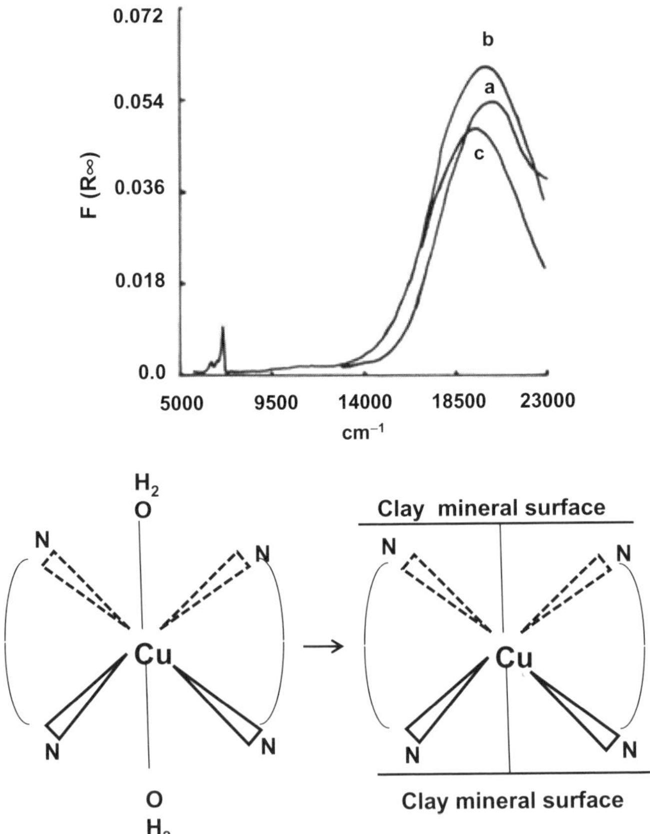

Fig. 5. (upper) Diffuse reflectance spectra of $Cu(en)_2^{2+}$ on Camp Berteau montmorillonite: (a) 0.24 mmol Cu^{2+}/g; (b) 0.36 mmol Cu^{2+}/g; (c) 0.63 mmol Cu^{2+}/g; (lower) aqueous complexes (left) and on the clay mineral surface (right).

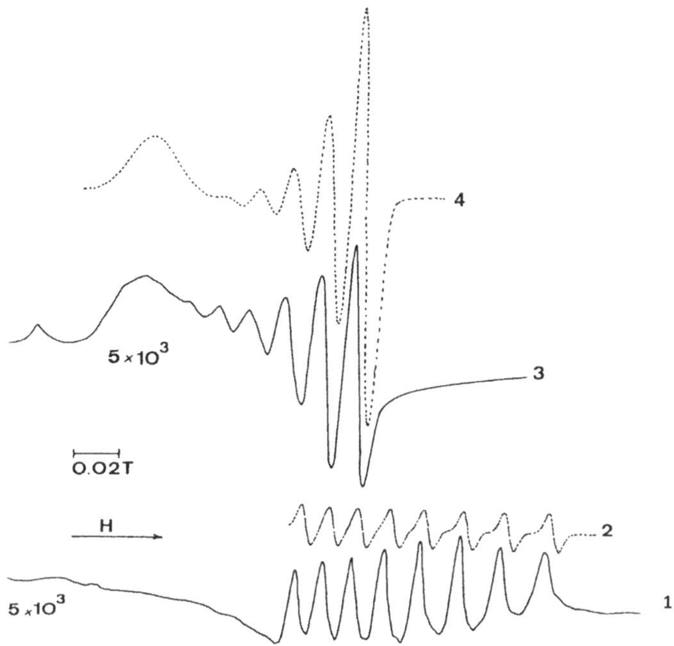

Fig. 6. Electron spin resonance spectra of $Co(en)_2^{2+}$ on hectorite films: experimental (1) and simulated (2) spectrum in the $g_{\parallel}$ region; experimental (3) and simulated spectrum (4) in the $g_{\perp}$ region.

This type of behaviour is not limited to the en complexes of Cu(II), but can also be observed for diethylenetriamine (dien), triethylenetetramine (trien) and tetraethylenepentamine (tetren) complexes and for the en complexes of Ni(II) and Co(II) (Schoonheydt *et al.*, 1979a,b; Schoonheydt & Pelgrims, 1983). Thus, $[Ni(en)_2(H_2O)_2]^{2+}$ is paramagnetic and high spin in solution, but diamagnetic and low spin on the clay-mineral surface (Schoonheydt *et al.*, 1979a). Similarly $[Co(en)_2]^{2+}$, which can be prepared by selective removal of one en ligand from the tris complex on the clay mineral surface, is a low spin complex with a typical axially symmetric EPR spectrum (Fig. 6). The latter is unambiguous proof that the complex is planar with its CoN_4 plane parallel to the planar clay-mineral surface (Schoonheydt & Pelgrims, 1983).

The general conclusion is that the replacement of the two axially coordinating water molecules by the clay mineral surface increases the in-plane ligand field strength and decreases the axial ligand field strength. This leads to an overall stabilization of the complexes. In other words, planar surfaces like planar complexes.

3.2. Chiral clay minerals

$[Ru(bipy)_3]^{2+}$ (bipy = 2,2′-bypridyl) and related complexes such as $[Ru(phen)_3]^{2+}$ (phen = 1,10-phenanthroline) have been studied intensively in various areas of chemistry, including coordination chemistry, electrochemistry, photochemistry, surface chemistry and (photo) catalysis. There are several reasons for this. $[Ru(bipy)_3]^{2+}$ is a kinetically stable complex that is easy to synthesize. It has an intense charge-transfer absorption feature with a maximum at 460 nm and an intense luminescence band at 600 nm. The excited state of the bipyridyl complexes can formally be written as $[Ru^{3+}(bipy)_2(bipy^-]^{2+}$. In addition, it has oxidizing and reducing properties. By transferring an electron from $bipy^-$ to a sacrificial oxidant (*e.g.* $S_2O_8^{2-}$), $[Ru(bipy)_3]^{3+}$ is formed which oxidizes water into O_2 and protons. $Bipy^-$ can also transfer its electron to other organic compounds (*e.g.* methylviologen (MV)) to form MV^-, which, in turn, reduces protons to H_2 over a Pt catalyst. In this way the photocatalytic splitting of water can be realized.

The initial spectroscopic investigations of $[Ru(bipy)_3]^{2+}$ on clay-mineral surfaces were performed with this photocatalytic dissociation of water into H_2 and O_2 as the ultimate goal. In dilute aqueous clay-mineral dispersion and at low loadings (to ensure quantitative adsorption) the maximum of the absorption band is in the range 455–464 nm and that of the emission at 602–620 nm, the exact positions depending somewhat on the type of clay mineral. The 455–620 nm combination is for a synthetic mica-type montmorillonite (Barasym) with $[Ru(bipy)_3]^{2+}$ residing primarily on the external surfaces of the clay particles, while the 464–602 nm combination is for Wyoming bentonite with $[Ru(bipy)_3]^{2+}$ primarily on the planar surfaces (Schoonheydt *et al.*, 1984). In aqueous dispersion, the luminescence of the adsorbed complex can be quenched by an electron acceptor which includes: (1) Fe^{3+} in the clay mineral structure; (2) by a quencher on the clay mineral surface (this must be a cation which can be exchanged, *e.g.* Cu^{2+}); (3) by a quencher in solution such as the anionic quencher $[Fe(CN)_6^{3-}]$ (Vliers *et al.*, 1990). The quenching efficiency follows qualitatively the order Fe^{3+} in the lattice > Cu^{2+} on the clay surface > $Fe(CN)_6]^{3-}$ in solution. However, there is a strong, not yet fully understood dependence on the $[Ru(bipy)_3]^{2+}$ loading, the amount of quencher and the clay mineral type. One of the main difficulties is that the distribution of quenchers and of $[Ru(bipy)_3]^{2+}$ over the clay mineral surfaces is unknown.

$[M(bipy)_3]^{2+}$ and $[M(phen)_3]^{2+}$ (M = transition metal ion) belong to the point group D_3 and are chiral with a Δ and a Λ configuration, as shown schematically in Figure 6. The exchange isotherms prove the strong preference of these complexes for the clay-mineral surfaces. Exchange in excess of the CEC is often observed (Schoonheydt *et al.*, 1978). There is, however, a significant difference in exchange behaviour of the bipyridyl and phenanthroline complexes. At low ionic strength, racemic $[Ru(bipy)_3]^{2+}$ is exchanged up to the CEC and the enantiomers up to 1.5–2.5 times the CEC. At large ionic strength, adsorption in excess of the CEC is obtained both for the racemic and enantiomeric solutions (Villemure, 1990; Yamagishi, 1987, 1993; Sato *et al.* 1992). $[Ru(phen)_3]^{2+}$, on the other hand, adsorbs as a racemic pair, when offered from a racemic solution. It adsorbs as individual molecules from enatiomeric solutions. In the former case, the amount adsorbed reaches twice the CEC; in the case of adsorption of enantiomers, exchange occurs up to the CEC (Yamagishi, 1987, 1993). This difference in exchange behaviour can be traced back to subtle differences in interaction between the complexes on the clay-mineral surfaces (Sato *et al.*, 1992).

The phenanthroline complexes in a racemic solution and on the clay-mineral surfaces form stable (Δ, Λ) pairs with an intermolecular distance of 0.92 nm. The intermolecular distances of the enantiomers, $\Lambda-\Lambda$ and $\Delta-\Delta$, on the smectite surface is 1.34 nm. The bipyridyl complexes do not form (Δ,Λ) pairs and the intermolecular distances, $\Lambda-\Delta$, $\Lambda-\Lambda$ and $\Delta-\Delta$, on the clay-mineral surfaces are almost the same: 0.93 nm for a racemic pair and 0.95 nm for the enantiomers. The corresponding configurations of the bipyridyl and phenanthroline complexes are shown in Figure 7. Each complex covers three hexagonal holes in the T sheets of the clay-mineral surface. In the case of rac-$[M(\text{phen})_3]^{2+}$, all the hexagonal holes are occupied, thus leaving no space for adsorption of molecules between the complexes other than in a second layer on top of the complexes. In the case of Δ- and Λ-$[M(\text{phen})_3]^{2+}$ complexes, only half of the hexagonal holes are occupied. Thus, the surface is chiral and there is room for adsorption of molecules on the clay-mineral surface in between the enantiomers and on top of the monolayer of enantiomers.

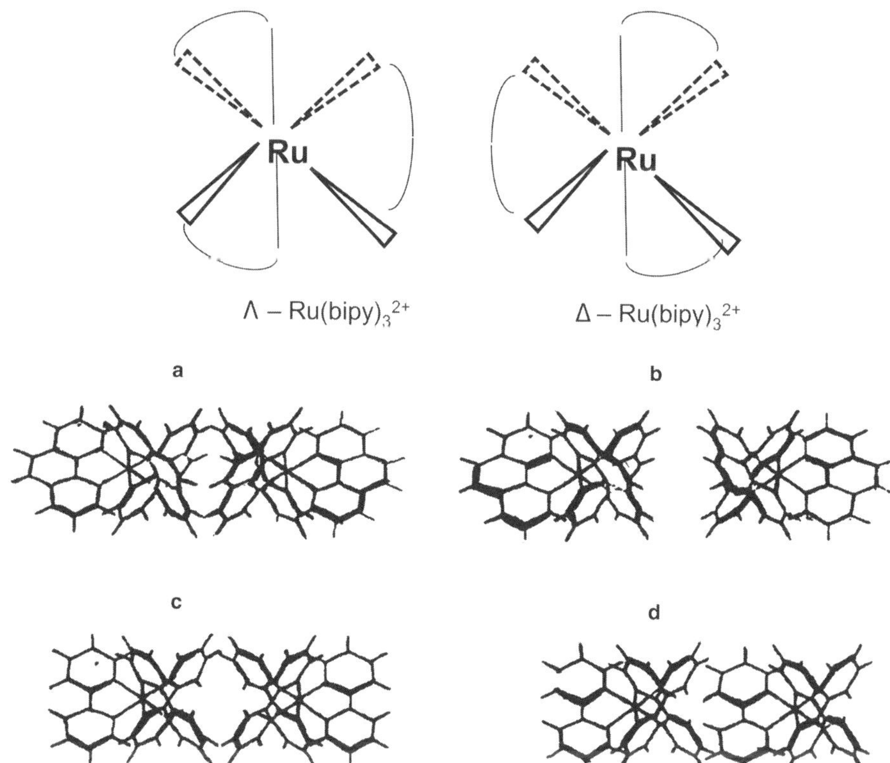

Fig. 7. (upper) Configuration of the chiral tris(bipyridyl)Ru(II) complexes. (lower) Arrangement of racemic pairs (***a***) and enantiomers (***b***) of tris(phenanthroline)Ru(II); (***c***,***d***) the same for tris(bipyridyl)Ru(II) on smectite surfaces. Reproduced from Sato *et al.* (1992) Copyright (1992) the American Chemical Society.

This is exactly what has been observed. With Λ-[*M*(phen)$_3$]$^{2+}$ complexes on the montmorillonite surface, Δ-type enantiomers are preferentially adsorbed and *vice versa* (Yamagishi, 1987, 1993). This leads to three practical applications: (1) improvement of the optical purity of a solution by selective adsorption of one enantiomer; (2) improvement of the optical resolution of a racemic mixture by selective adsorption of one enantiomer; (3) asymmetric synthesis or the conversion of a prochiral molecule into an optically active molecule. The second application means that a smectite loaded with Δ– or Λ–[M(II)(phen)$_3$]$^{2+}$ complexes can be used as packing material in a chromatographic column for separation of two enantiomers in a racemic solution. Examples are the resolution of the Λ and Δ isomers of Co(acac)$_3$ (acac = acetylacetonate) and of the optical isomers of aminoacids, such as glycine, alanine, valine, leucine, proline and serine (Yamagishi, 1987). Factors influencing the resolution are: (1) concentration of the solution; (2) type of solvent; (3) flow rate; and (4) the concentration of the chelate-smectite dispersion. A silica gel column, containing silica beads coated with Λ-[Ru(phen)$_3$]$^{2+}$-montmorillonite, for example, is particularly efficient at separating the enantiomers of organic molecules such as 1,1′-binaphthol, phenylalkylsulphoxides, 1,2-diphenylethane and 2,3-dihydropyrazines (Yamagishi, 1987). The oxidation of alkylphenylsulphides to the corresponding sulphoxides over Δ-[Ni(phen)$_3$]$^{2+}$-montmorillonite in water with selectivities of 90% or more and enantiomeric excesses up to 78% have been reported (Yamagishi, 1985; Nakamura *et al.*, 1988, 1990; Yamagishi, 1982, 1983). The reasons for this chiral discrimination are complex. Molecules interact with solvent, clay-mineral surfaces and the chiral transition metal ion complexes. Separation might occur at the planar surfaces or at the edge surfaces. The molecules to be separated might adsorb between the chiral transition metal ion complexes, or on top of them. This will depend on the number of chiral transition metal ion complexes and thus on the charge density of the clay mineral and the charge of the complex. For example, a divalent complex with a basal surface area of 1.25 nm^2 occupies, on a clay mineral with a CEC of one mmol/g, the full 750 m^2/g, thus leaving no space for adsorption between the complexes.

Organization and compactness of the bipyridyl and phenanthroline complexes can be influenced by putting alkyl chains on one of the bipyridyl and phenanthroline ligands. Thus, for the series [Ru(bipy)$_2$L$_i$] ($i = 1-3$) (L$_1$ = 2,2′-bipyridyl; L$_2$ = 4,4′-diundecyl-2,2′-bipyridyl; L3 = 5,5′-diundecyl-2,2′-bipyridyl), the inclination angle of the C$_3$ axis of the complex with respect to the surface normal is 23°, 30° and 52° respectively (Sato *et al.*, 2005).

Such complexes with two alkyl chains on one of the bipridyl and phenathroline ligands are amphiphilic. A solution of these complexes in chloroform can be spread over the water surface of a dilute clay mineral dispersion in a Langmuir-Blodgett trough. The solvent evaporates and, at the air–water interface, the Ru complexes ion exchange on the clay mineral surfaces. As a consequence, a floating hybrid layer of clay mineral layers and Ru complexes is formed at the air–water interface. This layer can be compressed and transferred to a suitable substrate either by horizontal deposition or vertical deposition. The procedure can be repeated to prepare multilayers. Figure 8 shows that the ordering sequence is dependent on the type of deposition. Horizontal

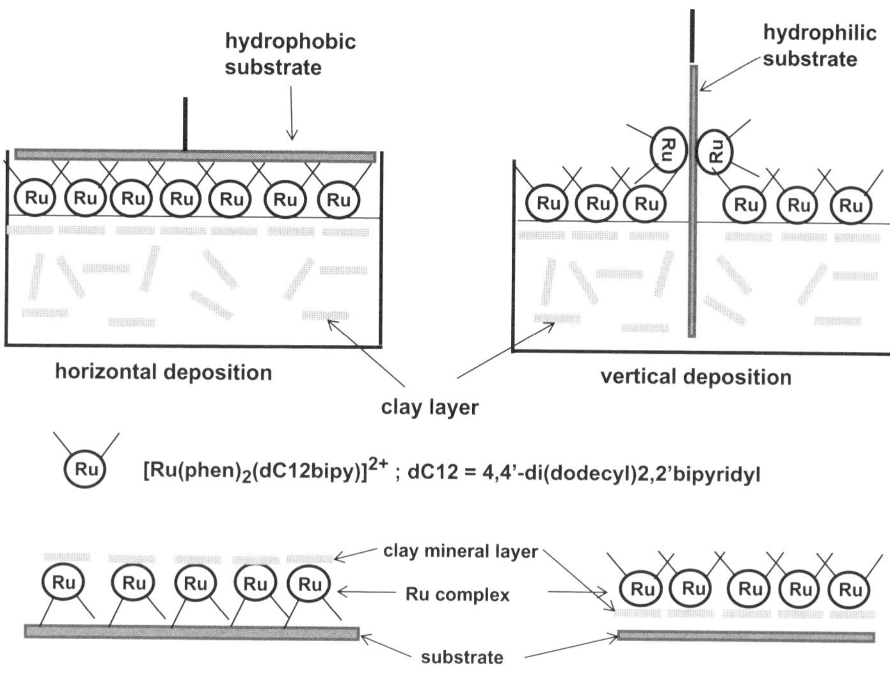

Fig. 8. Ordering sequence of amphiphilic Ru complexes and clay mineral layers, obtained by horizontal and vertical (upstroke) deposition in the Langmuir-Blodgett technique. Readers of the paper version of this chapter may wish to download a colour version of this figure from www.minersoc.org/emu-notes/emu-11/11-10-colour.pdf.

deposition is done with a hydrophobic substrate, which is covered by the amphiphilic Ru complexes. The latter are attached to the clay mineral layers with their hydrophilic heads. The surface of this hybrid monolayer film is composed of clay-mineral layers and is therefore hydrophilic. In the case of vertical deposition, the substrate is hydrophilic. It is covered by clay-mineral layers which carry the Ru complexes with their hydrophilic head turned towards the clay-mineral surfaces (Tamura *et al.*, 1999; Umemura *et al.*, 2002; Umemura, 2002). The surface of the film consists of a layer of Ru complexes with their alkyl groups sticking into the air. The surface is hydrophobic.

The bipyridyl and phenanthroline complexes are chiral. They have a measurable hyperpolarizability and exhibit non-linear optical properties such as second harmonic generation (SHG). Whether SHG can be observed in the films depends on the organization of the Ru complexes. Thus, multilayers of $[Ru(phen)_2(dC12bipy)]^{2+}$ (dC12bipy = didodecyl-2,2′-bipyridyl) deposited by horizontal deposition on a hydrophobic glass plate do not give a SHG signal. However, when the clay-mineral surface is made hydrophobic by exchange of octadecylammonium cations (ODA) prior to deposition of the Ru complxes, a SHG signal is obtained (Fig. 9) (Umemura *et al.*, 2002). Clearly, in the presence of ODA, the $[Ru(phen)_2(dC12bipy)]^{2+}$ complexes are

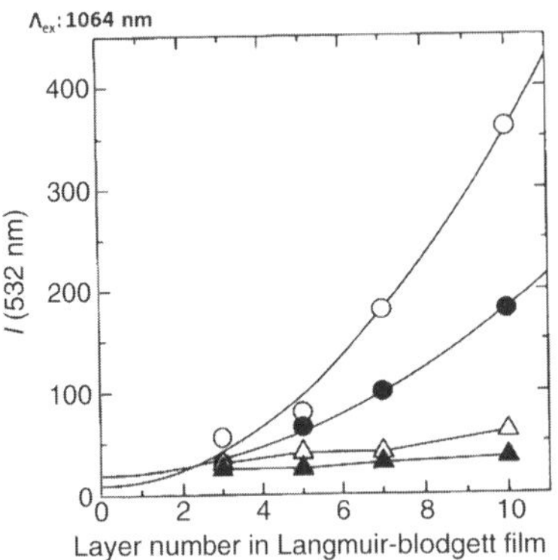

Fig. 9. Intensity of second harmonic signal at 532 nm as a function of the number of hybrid Ru(phen)2(dC12bipy)-saponite layers: (triangles) in the absence of octadecylammonium cations; (circles) in the presence of octadecylammonium cations. Filled and open symbols represent the λ and δ isomers, respectively (reproduced from Umemura *et al.*, 2002). Copyright (2002) the American Chemical Society.

organized in a non-centrosymmetric way. In the absence of ODA, the Ru complexes form islands of aggregated molecules, *i.e.* the chelates prefer to adsorb on neighbouring sites because the hydrophobic/hydrophilic character of the clay-mineral environment does not match that of the $[Ru(phen)_2(dC12bipy)]^{2+}$ complexes.

Hybrid mono- and multi-layers of $[M(phen)_3]^{2+}$ and $[M(bipy)_3]^{2+}$ complexes can be used to study the electron-transfer processes in thin films. This is illustrated in Figure 10. The hydrophobic substrate is covered with a monolayer of $[Ru(acac)_2(acacC12)]$ acting as electron acceptor (acac = acetylacetonate; acaC12 = acetylacetonate derivative with a dodecyl alkyl chain). In the subsequent layers Λ- or Δ-$[Ru(phen)_2(dC18\text{-}bipy)]^{2+}$ molecules are deposited and their luminescence is efficiently quenched by the acac complex. If the $[Ru(acac)_2(acacC12]$ layer is separated from the $[Ru(phen)_2(dC18bipy)]^{2+}$ layers by a hybrid layer of clay, modified with trimethylstearylammonium cations, the luminescence quenching is greatly suppressed (Inukai *et al.*, 2000). Thus, electron transfer through the clay layer is largely prohibited. This observation opens the way: (1) to prepare electrodes modified with mono- and multi-layers of hybrid clay-mineral films carrying amphiphilic transition metal ion complexes; (2) for selective oxidation of

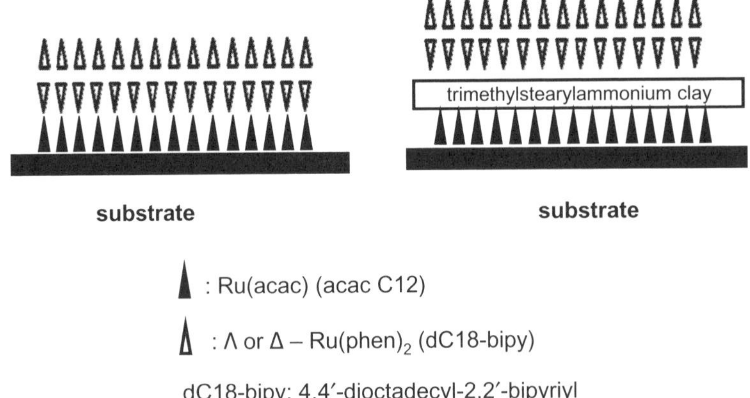

Fig. 10. Arrangement of the amphiphilic acac and phenanthroline complexes of Ru(II) in the absence (left) and presence (right) of an organophilic clay mineral layer.

chiral molecules, (3) for energy-transfer reactions; and (4) for the preparation of artificial photosynthetic systems (Okamoto *et al.*, 2000, 2001; He *et al.*, 2003, 2004).

4. Dyes on clay-mineral surfaces: from spectroscopy to optical materials

The spectra and aggregation tendency of many cationic dyes in aqueous clay-mineral dispersions have been reported. Relevant literature can be found in the textbook by Yariv & Cross (2001) and in review papers (Bujdák, 2006; Ogawa & Kuroda, 1995). Recent research has focused on the study of optical materials. Here, control of the organization of the dye molecules at the clay-mineral surfaces and of the clay-mineral layers themselves are of primary importance. In section 4.1, the spectroscopy of adsorbed dyes is discussed with methylene blue as an example. The ideas generated by the clay mineral-methylene blue system are, to a large extent, also valid for other dyes. In section 4.2, the material aspects of the clay mineral-dye systems are discussed with emphasis on optical materials. Figure 11 shows the structures and acronyms of the dyes discussed in this section.

4.1. Spectroscopy of clay mineral–methylene blue systems

Cationic dyes are strongly adsorbed on smectite surfaces. Below the CEC, the adsorption is quantitative, *i.e.* the amount of dye molecules remaining in solution is below the detection limit. Adsorption easily exceeds the CEC due to co-adsorption of anions. This is due to the efficient screening of the negative clay mineral charge by the dye molecules adsorbed. This charge screening was also invoked as one of the reasons to explain the excess adsorption of bipyridyl and phenanthroline complexes.

Many cationic dyes have typical absorption and fluorescence bands in the visible region with large extinction coefficients. This means that extremely small amounts of adsorbed dyes can be detected easily by either absorption or fluorescence spectrometry. In principle, it is then possible to study spectroscopically the surface chemistry of clay minerals in dilute aqueous dispersion by adsorbing a small amount of dye, typically $<1\%$ of the CEC. At this low loading the dye is adsorbed quantitatively and the colloidal behaviour of the clay mineral dispersion is hardly affected due to the low loadings. This reasoning was the basis of the systematic spectroscopic study of methylene blue (MB^+) adsorbed on smectites in dilute aqueous dispersion (Cenens and Schoonheydt, 1988). Typical spectra are shown in Figure 12.

Four bands are resolved in the 550–750 nm region which are assigned as follows: 760 nm – protonated methylene blue (MBH^{2+}); 725 nm – J dimers; 653 nm and 673 nm – monomers of MB^+ adsorbed in an aqueous environment at the clay-mineral surface and on the planar surface, respectively; 600–610 nm – 0–1 vibronic band of the monomer and H dimer adsorption; and 575 nm (broad) – H aggregates. At low loadings the monomer bands are predominant. As the loading increases, bands of the dimers and higher aggregates grow in intensity. In any case, the relative intensities

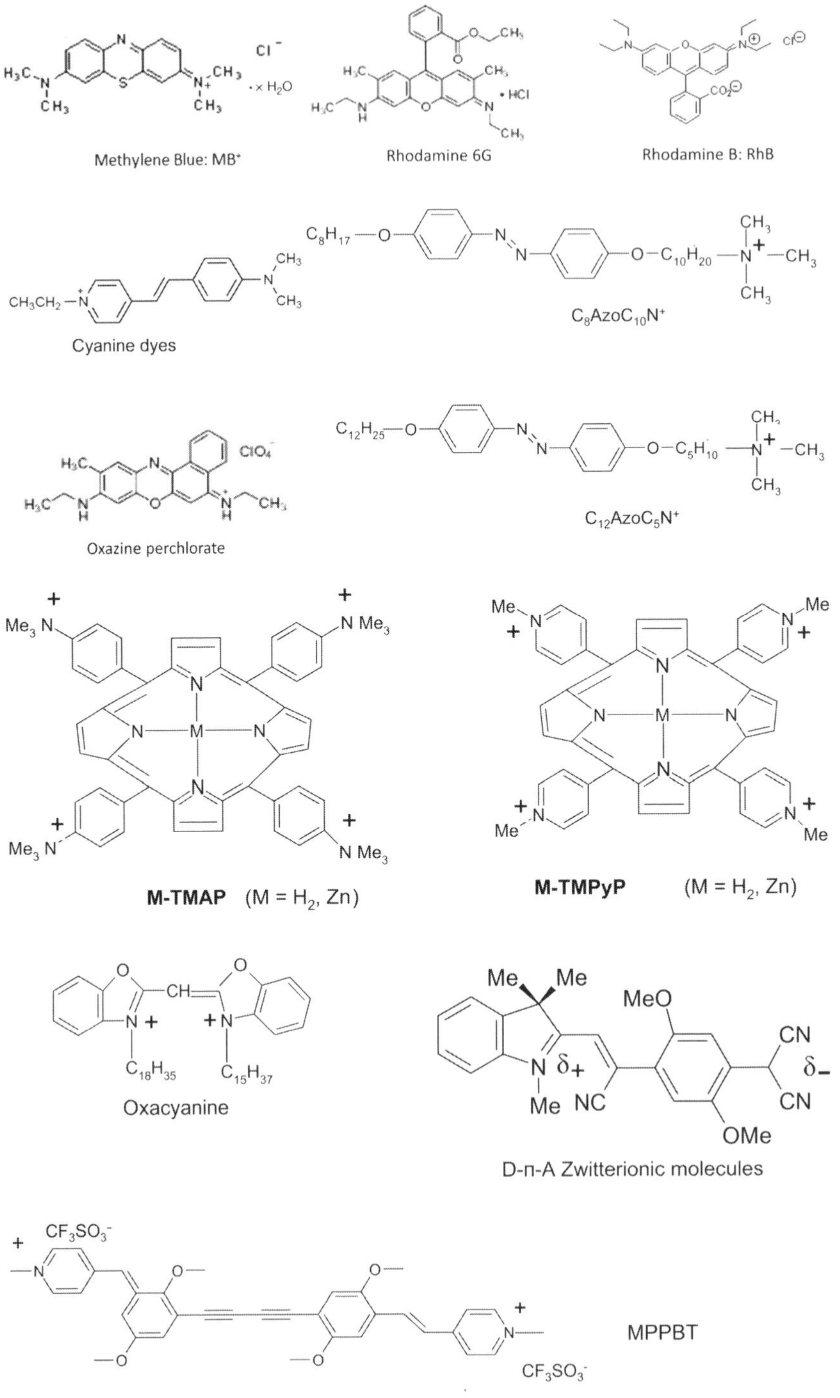

Methylene Blue: MB^+

Rhodamine 6G

Rhodamine B: RhB

Cyanine dyes

$C_8AzoC_{10}N^+$

Oxazine perchlorate

$C_{12}AzoC_5N^+$

M-TMAP ($M = H_2$, Zn)

M-TMPyP ($M = H_2$, Zn)

Oxacyanine

D-π-A Zwitterionic molecules

MPPBT

Fig. 11. Some dye molecules for clay mineral-dye composites.

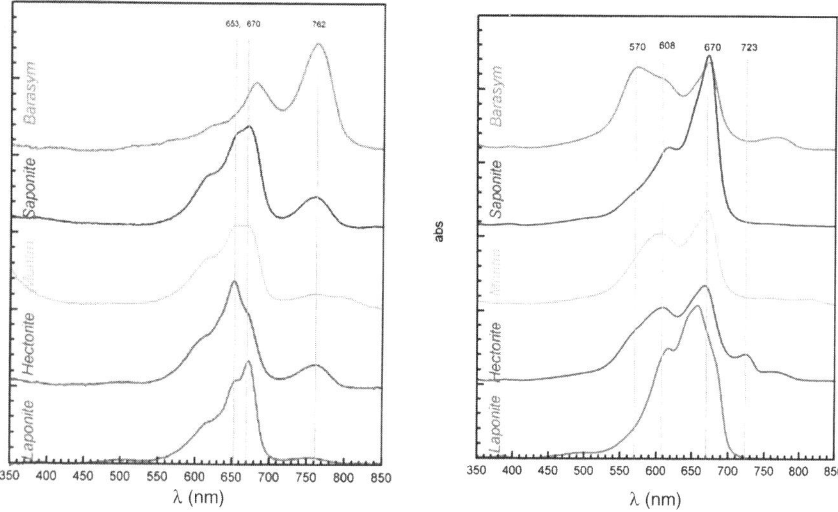

Fig. 12. Absorbance spectra of aqueous methylene blue-smectite dispersions: (left) loading is 0.5% of the CEC; (right) loading is 15% of the CEC. Readers of the paper version of this chapter may wish to download a colour version of this figure from www.minersoc.org/emu-notes/emu-11/11-10-colour.pdf.

of these four bands are different from clay mineral to clay mineral, indicating that the latter offer different environments to the MB^+ molecules. This is due to (1) differences in swelling and in the amount of water in the interlayer region; (2) differences in particle size and thus in the amount of planar or interlayer surface versus amount of edge or external surface; (3) the presence of acidic sites. It is clear from Figure 10, that clay minerals in dilute aqueous dispersion can be identified on the basis of the typical spectra of adsorbed MB^+.

The presence of acidity is particularly evident for Barasym, a synthetic mica-type montmorillonite which also functions as an acid catalyst, with the strong band at 760 nm of MBH^{2+}. Protons may also come from water in the interlayer region, especially at low water content (Mortland *et al.*, 1963). Indeed, MB^+ on Cs^+-saponite is more protonated than MB^+ on Na^+-saponite. Cs^+-saponite is hardly swollen in water and MB^+ molecules in the interlayer space interact with the few strongly polarized water molecules to form MBH^{2+} and presumably OH^-. The presence of H and J dimers with absorption bands of ~600 nm and 575 nm, respectively, and H aggregates with a broad band around 575 nm, indicates the strong tendency of MB^+ to aggregate, even at the low loadings of <1% of the CEC. This means that the MB^+ molecules are not randomly distributed over the surface, but tend to aggregate. Here water plays a crucial role. Two sets of data illustrate this point. Figure 13 shows the spectra of a MB^+-hectorite dispersion and of a MB^+-hectorite film, prepared from the same dispersion. In going from the aqueous dispersion to the film, one observes the disappearance of the bands due to H dimers at 600 nm and J dimers at 725 nm and the

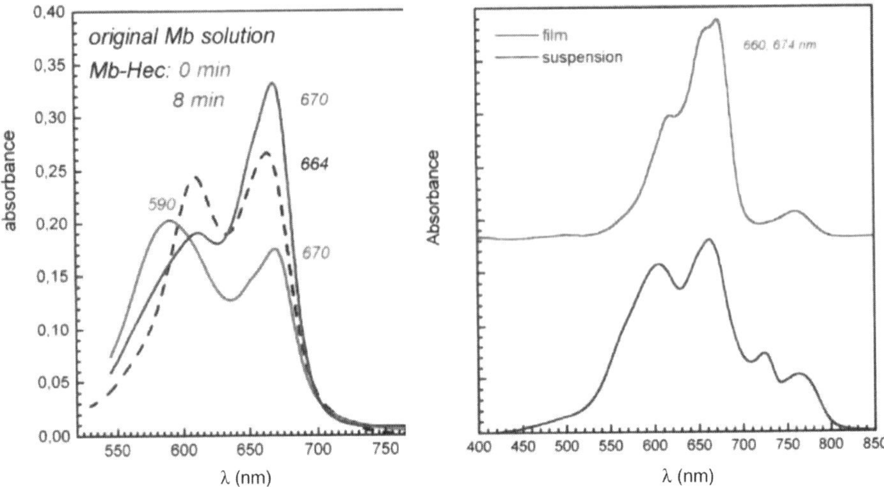

Fig. 13. (right) Absorbance spectra of a methylene blue-hectorite dispersion (blue) and that of the corresponding cast film. (left) Spectrum of a methylene blue solution (dotted line) and of the methylene blue-hectorite dispersions, immediately after preparation (red) and 8 min after preparation (blue). Readers of the paper version of this chapter may wish to download a colour version of this figure from www.minersoc.org/emu-notes/emu-11/11-10-colour.pdf.

appearance of a doublet band with maxima around 660 nm and 674 nm of the adsorbed monomers (Van Duffel *et al.*, 1997). Clearly, upon removal of water molecules, MB^+ molecules interact more strongly with the clay-mineral layers. As a result, dimers and aggregates are broken up into monomers. Some of these monomers are in direct contact with the surface, giving rise to the 670 nm band; some are still in an aqueous environment near the surface, giving the 660 nm band.

Figure 13 also shows spectra of MB^+-hectorite dispersions as a function of time of contact with the clay mineral particles in the dispersion (Jacobs & Schoonheydt, 1999, 2001). The solution spectra consist of monomers (665 nm) and dimers (600 nm). Upon contact with the clay mineral dispersion, this spectrum changes instantaneously with the formation of a broad band in the 600–570 nm range, encompassing H dimers and H aggregates. This is indicative of aggregates of MB^+ molecules at the surface of the clay-mineral particles in the aqueous dispersion. In the next few minutes, the 575 nm band of the H aggregates diminishes in intensity and the monomer band at 670 nm increases in intensity. Thus, the aggregates are broken up and monomers are adsorbed at the clay-mineral surfaces. After ~15 min, a quasi-equilibrium is attained and thereafter, the spectral changes are small.

This mechanism of adsorption-instantaneous adsorption in the aqueous environment of the clay mineral particles, followed by slow monomer adsorption on the clay-mineral surfaces, has also been observed for other dyes such as rhodamine 6G (Bujdák *et al.*, 2002a, 2004). There is, however, a difference between clay minerals with isomorphous substitution in the tetrahedral sheets (saponite) and those with isomorphous substitution in the octahedral sheet (hectorite). In the former case, the negative layer charge is localized and the electrostatic MB^+–surface attraction is strong. As a consequence, the MB^+ molecules can displace the water molecules at the surface and the surface–MB^+ complex so obtained has its characteristic monomer absorption band at 670 nm. In the case of hectorite with isomorphous substitution in the octahedral layer, the negative layer charge is diffuse. The MB^+–clay mineral surface interaction is weak and MB^+ is unable to replace the water molecules at the surface. The typical absorption band of the monomer is then at 653 nm band (Jacobs & Schoonheydt, 1999, 2001).

This effect of layer charge on the spectra of methylene blue in dilute aqueous clay-mineral dispersions has been investigated systematically on reduced-charge montmorillonites (Bujdák & Komadel, 1997; Bujdák *et al.*, 1998, 2001). Thus, on high charge density smectites, which adsorb large amounts of MB^+, the formation of H aggregates is favoured. When the charge density decreases, the relative amount of H aggregates decreases in favour of monomers. At low charge density and low loadings only monomers are found. This observation, mainly H and J aggregates on high-charge smectites and mainly monomers on low-charge smectites, is quite general. It has been observed for MB^+, rhodamines, cyanine dyes and oxazine (see Fig. 11 for their structures) (Bujdák *et al.*, 2002b; Kaneko *et al.*, 2004). Charge reduction by irreversible fixation of Li^+ can be quantified by analysis of the spectra of methylene blue (Czimerovà *et al.*, 2004). Thus, spectroscopy of methylene blue cannot only be applied to the identification of smectites in aqueous dispersion, but also to the determination of their layer charge.

4.2 Organization of cationic dyes on the surface of clay minerals: towards optical materials

Hybrid clay mineral-dye systems can be converted into optical materials if both the organization of the clay mineral layers and of the dye molecules can be controlled. Regular stacking of clay mineral layers on a substrate can be realized in different ways: (1) film casting by slow solvent evaporation from a few drops of the clay mineral dispersion; (2) spin coating of a clay mineral dispersion; (3) layer-by-layer (LbL) assemblage; (4) Langmuir-Schäfer (LS) and Langmuir-Blodgett (LB) techniques, corresponding with horizontal and vertical deposition, respectively (Lopez Arbeloa & Martinez, 2006b; Ulman, 1991; Fendler, 1996; Kotov, 2001). Spectroscopy has revealed the organization of the dye molecules at the clay-mineral surfaces. Monomers and different types of aggregates are found, depending on (1) the charge density of the clay mineral; (2) the particle-size distribution of the clay minerals; (3) the nature of the exchangeable cation; (4) the degree of swelling in water; and (5) the nature of the dye, *i.e.* their specific chemical structure and their orientation at the clay-mineral surfaces in the presence and absence of water molecules.

Molecular orientation can be studied by absorption and fluorescence spectroscopies with polarized light, by X-ray diffraction (XRD) and molecular simulation. For aggregation to occur, the cationic dyes must adopt a tilted orientation with respect to the normal to the clay-mineral surface. Such is the case for methylene blue and rhodamine

6G (Bujdák & Iyi, 2005; Bujdák *et al.*, 2003). In spin-coated Laponite films, rhodamine 6G molecules are also inclined with respect to the normal to the surface. The angle between the transition dipole moment vector of the molecule and the normal to the Laponite surface was determined to be 62° (Martinez Martinez *et al.*, 2006; Lopez Arbeloa & Martinez, 2006a,b).

All these spectroscopic measurements have been performed on hydrated clay mineral-dye systems, but the water molecules influence the orientation of the dye molecules. Molecules are adsorbed in such a configuration that they make maximal contact with the surface. Thus, in the absence of co-adsorbed water molecules, the cationic dyes are adsorbed flat on the surface. This is the case for MB^+, which forms a monolayer on low-charge clay minerals such as Wyoming bentonite and a bilayer on high-charge clay minerals such as Cheto montmorillonite. In the presence of water, a layer of tilted MB^+ molecules is formed on Wyoming bentonite (Klika *et al.*, 2007). Rhodamines, such as rhodamine 6G and rhodamine B, are oriented with their xanthene planes parallel to the surface. In this way they form a bilayer with H dimer arrangement of the individual molecules in the absence of water. In the presence of water, J dimers are formed with tilted xanthene planes of the molecules (Capkova *et al.*, 2004; Klika *et al.*, 2004). Such a transition from horizontally oriented to vertically oriented molecules in the interlayer space can also be realized by increasing the loading (Iwasaki *et al.*, 2000). At low loadings, MB^+ and cyanine dyes are adsorbed in a horizontal fashion. As the loading increases, the dyes adopt a tilted and finally a vertical orientation, in order to accommodate all the molecules at the surface.

Ogawa and Kuroda (1995) have already discussed photochemical intercalation compounds with such functions as photocatalysis, energy storage, photoluminescence, photochromism, photochemical hole burning and nonlinear optics. In the hybrid clay mineral-dye systems the photofuntion is provided by the dye molecules. The clay mineral acts as a host, which provides a two-dimensional organization of the dye molecules. This organization is essential to obtain a measurable photofunction.

The first traces of MB^+, adsorbed on smectites in dilute aqueous dispersion, have a hyperpolarizability, which is 11× greater than that of MB^+ in solution (Boutton *et al.*, 1997). However, the signal quickly fades away, as the loading of MB^+ is increased (Fig. 14). This is attributed to the aggregation of MB^+ molecules at the surface, even at these extremely small loadings, where only monomers are expected.

The cationic porphyrins TMPyP and TMAP (see Fig. 11) with an overall charge of +4 on the N atoms of the four pyridine rings are strictly adsorbed as monomers. This has been explained by the size-matching effect: the average distance between the negative charges in the clay mineral layers (1.25 nm) matches the distance between the positive charges in the porphyrins, which is 1.1 nm for TMPyP and 1.35 nm for TMAP (Takagi *et al.*, 2001, 2002a, 2002b, 2006). TMPyP and TMAP can be co-exchanged or exchanged consecutively on a clay-mineral dispersion. In this way, intra-layer and inter-layer fluorescence resonance energy transfer (FRET) has been realized (Takagi *et al.*, 2002b; 2006). The FRET phenomenon has also been realized in hybrid LB monolayers, containing Laponite, a thiacyanine dye as the donor and rhodamine B as acceptor in a ratio 90:10 (Hussain & Schoonheydt, 2010; Hussain

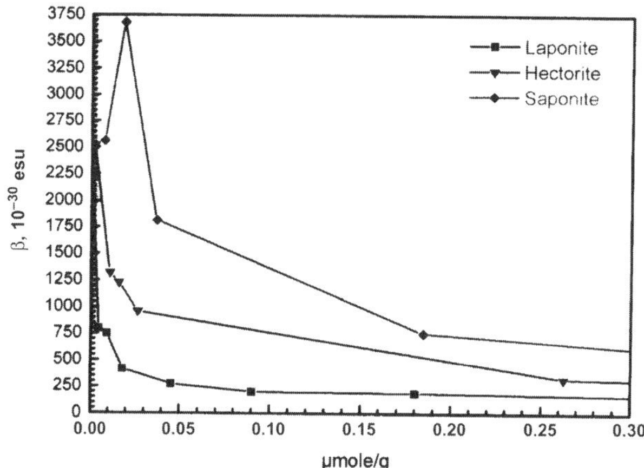

Fig. 14. Hyperpolarizability of methylene blue in methylene blue-smectite dispersions as a function of the loading.

et al., 2010). Thus, upon excitation of the donor, a strong fluorescence of rhodamine is observed.

For nonlinear optics, a non-centrosymmetric arrangement of the molecules at the clay-mineral surface is essential. Such is the case for LB films containing hybrids of clay mineral layers and non-amphiphilic bipyridyl and phenanthroline complexes (Higashi *et al.*, 2006; Kawamata *et al.*, 2006, 2008) and clay mineral layers, hybridized with zwitterionic compounds and cyanine dyes (Ogata *et al.*, 2002, 2003a, 2003b). With MTTPB-clay mineral systems two-photon absorption (TPA) can be realized (Suzuki *et al.*, 2008; Kamada *et al.*, 2007; Kawamata *et al.*, 2010). To obtain a reproducible signal, a low scattering material is necessary. This can be realized using synthetic clay minerals with particle sizes in the range 20–40 nm, significantly below the wavelength of light used in the experiments. Scattering is also reduced by using a mixture of water and dimethylsulfoxide as solvent (Kawamata *et al.*, 2010).

Finally a photofunctionality, useful for sensing, can also be realized by construction of spin-coated Laponite layers, alternating with spin-coated mesoporous TiO_2 layers (Lotsch & Ozin, 2008a, 2008b). The thickness of the layers can be controlled accurately in the 20–500 nm range. As the refractive indices of Laponite and titania are significantly different, the incoming light beam is partially reflected at the interfaces. The reflected light beams interfere constructively when their wavelength, λ, fullfills Bragg's law:

$$\Lambda = 2n_{\text{eff}}D$$

where D is the thickness of the individual layers. The effective refractive index n_{eff} changes with the nature and the amount of adsorbed molecules in the multilayer and so does the interference wavelength. Thus, this wavelength shift of the photonic crystal can be used for chemical sensing.

5. Amino acids and proteins on clay-mineral surfaces

Amino acids are zwitterions and, depending on the pH, they may carry a net positive or a net negative charge. Adsorption of amino acids on clay-mineral surfaces is governed therefore by acid-base chemistry and complexation (Theng, 1974; Lagaly *et al.*, 2006). The positively charged amino-acids are ion exchanged on smectites; the negatively charged amino-acids may be adsorbed at the edges. A second mechanism of adsorption of amino-acids is complexation with polyvalent cations, including transition metal ions (TMI), on clay mineral surfaces. There are two pathways to synthesize amino acid complexes at clay mineral surfaces: (1) adsorption of amino acids from solution on the clay-mineral surfaces, previously loaded with polyvalent cations; and (2) preparation of the complexes of amino acids and transition metal ions in solution, followed by exchange of the complexes on the clay-mineral surfaces. The second route implies that the complexes are positively charged (Theng, 1974; Lagaly *et al.*, 2006; Fu *et al.*, 1996). There is competition between acid-base chemistry and complexation and, as a result, complex adsorption behaviour has been observed (Nagy & Konya, 2004).

Proteins may also carry a net positive charge or a net negative charge, depending on their iso-electric point. There are two mechanisms of adsorption: ion exchange and adsorption by dispersive interactions. Ion exchange implies proteins with a net positive charge. Dispersive interaction is defined here as the interaction between the surface atoms of the clay minerals and the functional groups and atoms on the protein surfaces. In principle, the number of interacting atoms is quite large and, as the dispersive interactions are additive, they can be quite large, whatever the charge of the proteins. Even negatively charged proteins can be adsorbed quite significantly on clay-mineral surfaces, due to dispersive interactions.

The adsorption of positively charged proteins is nicely illustrated with protamine, a protein with a molecular weight of 4500 Da. Figure 15 shows the adsorption isotherms

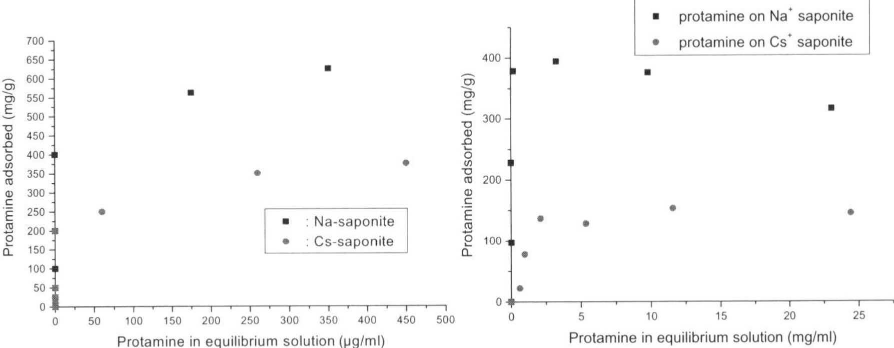

Fig. 15. Adsorption isotherms of protamine on Na^+-saponite (black) and Cs^+-saponite (red). (left) Amount adsorbed, measured in the equilibrium solution. (right) Amount adsorbed, measured after washing and freeze drying. Readers of the paper version of this chapter may wish to download a colour version of this figure from www.minersoc.org/emu-notes/emu-11/11-10-colour.pdf.

for Na^+- and Cs^+-saponite (Szabo *et al.*, 2008). For both Na^+- and Cs^+-saponites, type I isotherms are obtained. However, the amount adsorbed is systematically greater for Na^+-saponite than for Cs^+-saponite and the maximum amount adsorbed is smaller for the washed samples than for samples in the equilibrium solution. As shown in Figure 15, the fractions of protamine above 400 mg/g for Na^+-saponite and above 150 mg/g for Cs^+-saponite are weakly adsorbed and removed by washing the samples with water. What remains on the surface is the ion-exchanged fraction, amounting to 400 mg/g and 150 mg/g for Na^+- and Cs^+-saponite, respectively. The ratio of the amount of Na^+ or Cs^+ released in solution over the amount of protamine adsorbed is 13, suggesting that the average charge per protamine molecule is 13. 400 mg of protamine corresponds to a fully exchanged saponite. The corresponding d_{001} spacing is 1.7–1.8 nm. 150 mg corresponds with a degree of exchange of 36% and XRD reveals that part of the Cs^+-saponite is not intercalated. It is remarkable that such a highly charged molecule as protamine is unable to exchange all the Cs^+ and to provoke complete intercalation.

Extrapolation of these data to proteins with larger molecular weights and smaller positive charges leads to two predictions: (1) difficult intercalation of these bulkier proteins in Na^+-saponite; (2) no intercalation in Cs^+-saponite. Proteins have a strong tendency to adsorb at the edges of the clay mineral layers, however. This was observed in hybrid LB films of clay minerals hybridized with the positively charged lysozyme and the negatively charged bovine serum albumin. In both cases, accumulation of aggregates of proteins at the edges were observed (Miao *et al.*, 2010a, 2010b). It follows that the adsorption of proteins on clay minerals occurs in three steps: (1) adsorption at the edges, followed by (2) intercalation. (3) Finally, a weakly adsorbed fraction is adsorbed on the clay mineral–protein complex obtained in steps 1 and 2.

In the case of adsorbed enzymes, one often observes a shift of the pH of optimal enzymatic activity to higher values with respect to the solution pH (Quiquampoix *et al.*, 1993). Several reasons can be invoked to explain this behaviour: (1) conformational changes of the enzyme in the adsorbed state; (2) pH-dependent orientation of the active site of the enzyme in the adsorbed state; (3) inorganic cations, such as Na^+ and Ca^{2+}, are exchanged by positively charged enzymes. As a consequence, the clay-mineral surface becomes hydrophobic and the enzyme will be adsorbed preferentially with its hydrophobic parts oriented towards the hydrophobic siloxane surface. Finally, the nature and amount of enzymes or proteins adsorbed determines the amount of water, which can be adsorbed by the hybrid clay mineral–protein material. In view of the importance of extra-cellular enzymes in soils and of the development of heterogeneous biocatalysts and hybrid biomaterials, knowledge of all these parameters is important and detailed investigations are needed to understand them (Quiquampoix & Burns, 2007; Ruiz-Hitzky *et al.*, 2008).

6. Conclusions

The surface chemistry of clay minerals, particularly smectites, is rich. This is due to the properties of the clay-mineral layers and of the adsorbed molecules. Layer charge, type of exchangeable cation and the sizes of the clay mineral particles are the most important clay mineral parameters that determine the surface chemistry. For the molecules, the size, shape, charge and hydrophobic/hydrophilic character are the dominant factors. Water plays a crucial role in that it determines: (1) the transition from high-spin to low-spin TMI complexes at the clay-mineral surfaces; (2) the organization of dye molecules at the surface; (3) the intercalation of cationic proteins.

Adsorption of chiral TMI complexes leads to chiral clay minerals, which can be used in chiral chromatographic columns for separation of enantiomers in a racemic mixture. Non-centrosymmetric molecular organization leads to hybrid clay mineral films with nonlinear optical properties, such as enhanced hyperpolarizability of the adsorbed molecules, second harmonic generation and two-photon absorption. Two dyes with suitable absorption and fluorescence spectra, co-adsorbed on the clay mineral surfaces, give more efficient fluorescence resonance energy transfer than in solution.

To realize a device based on clay minerals, good thermal, mechanical and chemical stability of the clay mineral-dye or clay mineral-TMI complex systems in the device are essential. For optical devices, light scattering must be minimized. This can be done with synthetic clay minerals because of their small layer sizes, typically 20–40 nm.

Protein adsorption is a challenge, even for positively charged proteins. This is because these bulky proteins need an expanded clay structure, the 1.25 nm of the monolayer of water in Na^+-saponite being sufficient. In Cs^+-saponite, only 36% of the layers form a monolayer water hydrate and can be intercalated. The remaining 64% is not. Positively and negatively charged proteins adsorb at the edges. Our hypothesis is that the proteins first adsorb at the edges, then diffuse into the interlayer space. The clay mineral–protein hybrid thus obtained can adsorb additional protein molecules weakly, which can be washed off with water. Systematic studies of protein adsorption are required to unravel the effects of size and shape of proteins, of the charge distribution over the surface of the proteins and of the amino acid composition on the adsorption behaviour and on the enzymatic activity of these hybrid clay mineral–protein composites.

Acknowledgments

The authors acknowledge the help of Hugo Leeman in the drawing of the figures. Financial support from the Long-Term Structural Funding-Methusalem Funding of the Flemish Government is gratefully acknowledged.

References

Abend, S. & Lagaly, G. (2000) Sol-gel transitions of sodium montmorillonite dispersions. *Applied Clay Science*, **16**, 201–227.

Alba, M.D., Alvero, R., Becerro, A.I., Castro, M.A. & Trillo, J.M. (1998) Chemical behavior of lithium ions in reexpanded Li-montmorillonites. *Journal Of Physical Chemistry B*, **102**, 2207–2213.

Alvero, R., Alba, M.D., Castro, M.A. & Trillo, J.M. (1994) Reversible migration of lithium in montmorillonites. *Journal of Physical Chemistry*, **98**, 7848–7853.

Anderson, R.L., Ratcliffe, I., Greenwell, H.C., Williams, P.A., Cliffe, S. & Coveney, P.V. (2010) Clay swelling – A challenge in the oilfield. *Earth-Science Reviews*, **98**, 201–216.

Bank, S., Bank, J.F. & Ellis, P.D. (1989) Solid-state ^{113}Cd Nuclear Magnetic Resonance study of exchanged montmorillonites. *Journal of Physical Chemistry*, **93**, 4847–4855.

Berend, I., Cases, J.-M., Francois, M., Uriot, J.P., Michot, L., Masion, A. & Thomas, F. (1996) Mechanism of adsorption and desorption of water vapor by homoionic montmorillonites: 2. The Li^+ Na^+, K^+, Rb^+ and Cs^+-exchanged forms. *Clays and Clay Minerals*, **43**, 324–336.

Bergaya, F., Theng, B.K.G. & Lagaly, G. (editors) (2006) *Handbook of Clay Science*. Developments in Clay Science, Vol. **1**. Elsevier, Amsterdam, 1224 pp.

Bishop, J.L., Pieters, C.M. & Edwards, J.O. (1994) Infrared spectroscopic analyses on the nature of water in montmorillonite. *Clays and Clay Minerals*, **42**, 702–716.

Boutton, C., Kauranen, M., Persoons, A., Keung, M.P., Jacobs, K.Y. & Schoonheydt, R.A. (1997) Enhanced second order optical nonlinearity of dye molecules adsorbed onto laponite particles. *Clays and Clay Minerals*, **45**, 483–485.

Bowers, G.M., Bish, D.L. & Kirkpatrick, R.J. (2008) H_2O and cation structure and dynamics in expandable clays: H-2 and K-39 NMR investigations of hectorite. *Journal of Physical Chemistry C*, **112**, 6430–6438.

Bradley, W.F., Grim, R.E. & Clark, G.L. (1937) A study of the behavior of montmorillonite upon wetting. *Zeitschrift fur Kristallographie*, **97**, 216–222.

Brigatti, M.F., Malferrari, D., Laurora, A. & Elmi, C. (2011) Structure and mineralogy of layer silicates: Recent perspectives and new trends. In: *Layered Mineral Structures and their Application in Advanced Technologies* (M.F. Brigatti & A. Mottana, editors). EMU Notes in Mineralogy, **11**. European Mineralogical Union and the Mineralogical Society of Great Britain & Ireland, pp. 1–71.

Brindley, G.W. & Ertem, G. (1971) Preparation and solvation properties of some variable charge montmorillonites. *Clays and Clay Minerals*, **19**, 399–404.

Bujdák, J. (2006) Effect of the layer charge of clay minerals on optical properties of organic dyes. A review. *Applied Clay Science*, **34**, 58–73.

Bujdák, J. & Iyi, N. (2005) Molecular orientation of rhodamine dyes on surfaces of layered silicates. *Journal of Physical Chemistry B*, **109**, 4608–4615.

Bujdák, J. & Komadel, P. (1997) Interaction of methylene blue with reduced charge montmorillonites. *Journal of Physical Chemistry B*, **101**, 9065–9068.

Bujdák, J., Janek, M., Madejová, J. & Komadel, P. (1998) Influence of layer charge density of smectites on the interaction with methylene blue. *Journal of the Chemical Society, Faraday Transactions*, **94**, 3487–3492.

Bujdák, J., Janek, M., Madejová, J. & Komadel, P. (2001) Methylene blue interaction with reduced charge smectites. *Clays and Clay Minerals*, **49**, 244–254.

Bujdák, J., Iyi, N. & Fujita, T. (2002a) The aggregation of methylene blue in montmorillonite dispersions. *Clay Minerals*, **37**, 121–133.

Bujdák, J., Iyi, N., Hrobáriková, J. & Fujita, T. (2002b) Aggregation and decomposition of a pseudocyanine dye in dispersions of layered silicates. *Journal of Colloid and Interface Science*, **247**, 494–503.

Bujdák, J., Iyi, N., Kaneko, Y. & Sasai, R. (2003) Molecular orientation of methylene blue cations adsorbed on clay surfaces. *Clay Minerals*, **38**, 561–572.

Bujdák, J., Iyi, N. & Sasai, R. (2004) Spectral properties, formation of dye molecular aggregates and reactions in rhodamine 6G/layered silicate dispersions. *Journal of Physical Chemistry B*, **108**, 4470–4474.

Burgess, J. (1978) *Metal Ions in Solution*. Ellis Horwood Limited Publ., Chichester, UK, 481 pp.

Buswell, A.M., Krebs, K. & Rodebush, W.H. (1937) Infrared studies. III. Absorption bands of hydrogels between 2.5 and 3.5 micrometers. *Journal of the American Chemical Society*, **59**, 2603–2605.

Calvet, R. & Prost, R. (1971) Cation migration into empty octahedral sites and surface properties of clays. *Clays and Clay Minerals*, **19**, 175–186.

Cancela, G.D., Huertas, F.J., Taboada, E.R., Sanchez Rasero, F. & Laguna, A.H. (1997) Adsorption of water vapor by homoionic montmorillonites. Heats of adsorption and desorption. *Journal of Colloid and Interface Science*, **185**, 343–354.

Cao, T., Fasulo, P.D. & Rodgers, W.R. (2010) Investigation of the shear stress effect on montmorillonite platelet aspect ratio by atomic force microscopy. *Applied Clay Science*, **49**, 21–28.

Čapková, P., Malý, P., Pospíšil, M., Klika, Z., Weissmannová, H. & Weiss, Z. (2004) Effect of surface and interlayer structure on the fluorescence of rhodamine B – montmorillonite: Modelling and experiment. *Journal Colloid and Interface Science*, **277**, 128–137.

Cases, J.M., Berend, I., Besson, G., Francois, M., Uriot, J.P., Thomas, F. & Poirier, J.E. (1992) Mechanism of adsorption and desorption of water vapor by homoionic montmorillonite. 1. The sodium-exchanged form. *Langmuir*, **8**, 2730–2739.

Cases, J.M., Berend, I., Francois, M., Uriot, J.P., Michot, L.J. & Thomas, F. (1997) Mechanism of adsorption and desorption of water vapor by homoionic montmorillonite .3. The Mg^{2+}, Ca^{2+}, Sr^{2+} and Ba^{2+} exchanged forms. *Clays and Clay Minerals*, **45**, 8–22.

Cebula, D.J., Thomas, R.K. & White, J.W. (1981) Diffusion of water in Li-montmorillonite studied by quasielastic neutron scattering. *Clays and Clay Minerals*, **29**, 241–248.

Cenens, J. & Schoonheydt, R.A. (1988) Visible spectroscopy of methylene blue on hectorite, laponite B and barasym. *Clays and Clay Minerals*, **36**, 214–224.

Christidis, G.E. (2011) The concept of layer charge of smectites and its implications on important smectite-water properties. In: *Layered Mineral Structures and their Application in Advanced Technologies* (M.F. Brigatti & A. Mottana, editors). EMU Notes in Mineralogy, **11**. European Mineralogical Union and the Mineralogical Society of Great Britain & Ireland, pp. 239–260.

Costanzo, P.M. & Giese, R.F. (1990) Ordered and disordered organic intercalates of 8.4-Å synthetically hydrated kaolinte. *Clays and Clay Minerals*, **38**, 160–170.

Czimerová, A., Jankovič, L. & Bujdák, J. (2004) Effect of exchangeable cations on the spectral properties of methylene blue in clay dispersions. *Journal of Colloid and Interface Science*, **274**, 126–132.

Earl, W.L. & Johnston, C.T. (1998) Applications of NMR spectroscopy to the study of the chemistry of environmental interfaces. In: *Structure and Surface Reactions of Soil Particles* (P.M. Huang, N. Senesi & J. Buffle, editors). John Wiley & Sons Ltd., New York, pp. 251–280.

Farmer, V.C. & Mortland, M.M. (1966) Infrared study of the coordination of pyridine and water to exchangeable cations in montmorillonite and saponite. *Journal of the Chemical Society*, **1966A**, 344–351.

Feller, D., Glendening, E.D., Woon, D.E. & Feyereisen, M.W. (1995) An extended basis-set ab-initio study of alkali-metal cation-water clusters. *Journal of Chemical Physics*, **103**, 3526–3542.

Fendler, J.H. (1996) Self-assembled nanostructured materials. *Chemistry of Materials*, **8**, 1616–1624.

Fu, L., Weckhuysen, B.M., Verberckmoes, A.A. & Schoonheydt, R.A. (1996) Clay intercalayed Cu(II) aminoacid complexes: Synthesis, spectroscopy and catalysis. *Clay Minerals*, **31**, 489–498.

Gevers, C. & Grandjean, J. (2001) A multinuclear magnetic resonance study of synthetic clays suspended in water and in dodecyldimethylamine oxide solutions. *Journal of Colloid and Interface Science*, **236**, 290–294.

Glendening, E.D. & Feller, D. (1995) Cation-Water Interactions: The $M^+(H^2O)_n$ clusters for alkali metals, M = Li, Na, K, Rb, and Cs. *Journal of Physical Chemistry*, **99**, 3060–3067.

Glendening, E.D. & Feller, D. (1996) Dication-water interactions: $M(^{2+})(H_2O)_{(n)}$ clusters for alkaline earth metals M = Mg, Ca, Sr, Ba, and Ra. *Journal of Physical Chemistry*, **100**, 4790–4797.

Graham, J. (1964) Adsorbed water on clays. *Reviews of Pure and Applied Chemistry*, **14**, 81–90.

Grandjean, J. (1997) Water sites at a clay interface. *Journal of Colloid and Interface Science*, **185**, 554–556.

Grandjean, J. & Laszlo, P. (1989) Deuterium nuclear magnetic resonance studies of water molecules restrained by their proximity to a clay surface. *Clays and Clay Minerals*, **37**, 403–408.

Grandjean, J. & Laszlo, P. (1994) Deuterium and O-17 Nuclear-Magnetic-Resonance of aqueous clay suspensions. *Magnetic Resonance Imaging*, **12**, 375–377.

Guggenheim, S. (2011) An overview of order/disorder in hydrous phyllosilicates. In: *Layered Mineral Structures and their Application in Advanced Technologies* (M.F. Brigatti & A. Mottana, editors). EMU Notes in Mineralogy, **11**. European Mineralogical Union and the Mineralogical Society of Great Britain & Ireland, pp. 73–121.

Guven, N. (1988) Smectites. In: *Hydrous Phyllosilicates (exclusive of Micas)* (S.W. Bailey, editor). Reviews in Mineralogy, **19**, Mineralogical Society of America, Washington, D.C. pp. 497–559.

Haack, E.A., Johnston, C.T. & Maurice, P.A. (2008) Mechanisms of siderophore sorption to smectite and siderophore-enhanced release of structural Fe^{3+}. *Geochimica et Cosmochimica Acta*, **72**, 3381–3397.

Hanaya, M. & Harris, R.K. (1998) Two-dimensional Na-23 MQ MAS NMR study of layered silicates. *Journal of Materials Chemistry*, **8**, 1073–1079.

He, J.-X., Sato, H., Yang, P. & Yamagishi, A. (2003) Preparation of a novel clay/metal complex hybrid film and its catalytic oxidation to chiral 1,1′-binaphthol. *Journal of Electroanalytical Chemistry*, **560**, 169–174.

He, J.-X., Yamagishi, A., Iwao, M., Abe, Y. & Umemura, Y. (2004) Creation of a novel solid surface as a model photosynthetic system. II. Application of the LB and self-assembly methods to fixation of a light driven polypyridyl Ru(II) complex. *Electrochemistry Communications*, **6**, 61–65.

Henniker, J.C. (1949) The depth of the surface zone of a liquid. *Reviews of Modern Physics* **21**, 322–341.

Higashi, T., Miyazaki, S. Nakamura, S., Seike, R., Tani, S., Hayami, S. & Kawamata, J. (2006) Nonlinear optical properties of Langmuir-Blodgett films consisting of metal complexes. *Colloids and Surfaces A: Physicochemical and Engineering Aspects*, **284–285**, 161–165.

Hofmann, U. & Bilke, W. (1936) Inner crystalline swelling and the ability of montmorillonite to exchange bases. *Kolloid-Zeitschrift*, **77**, 238–251.

Hofmann, U. & Klemen, R. (1950) Verlust der Austauschf'fihigkeit yon Lithiumionen an Bentonit durch Erhitzung. *Zeitschrift für anorganische und allgemeine Chemie*, **262**, 95–99.

Huheey, J.E., Keiter, E.A. & Keiter, R.L. (1997) *Inorganic Chemistry. Principles of Structure and Reactivity*, 4th edition. Prentice Hall, New Jersey, USA, 964 pp.

Hussain, S.A. & Schoonheydt, R.A. (2010) Langmuir-Blodgett monolayers of cationic dyes in the presence and absence of clay mineral layers: N,N'-dioctadecyl thiacyanine, octadecyl rhodamine B and laponite. *Langmuir*, **26**, 11870–11877.

Hussain, S.A., Chakraborty, S., Bhattarcharjee, D. & Schoonheydt, R.A. (2010) Fluorescence resonance energy transfer between organic dyes adsorbed onto nano-clay and Langmuir-Blodgett films. *Spectrochimica Acta Part A: Molecular and Biomolecular Spectroscopy*, **75**, 664–670.

Inukai, K., Jotta, Y., Tomura, S., Takahashi, M. & Yamagishi, A. (2000) Preparation of the Langmuir-Blodgett film of a clay-alkylammonium adduct and its use as a barrier for interlayer photo-induced electron transfer. *Langmuir*, **16**, 7679–7684.

Iwasaki, M., Kita, M., Ito, K., Kohno, A. & Fukunishi, K. (2000) Intercalation characteristics of 1,1′-diethyl-2,2'-cyanine and other cationic dyes in synthetic saponite: orientation in the interlayer. *Clays and Clay Minerals*, **48**, 392–399.

Jacobs, K.Y. & Schoonheydt, R.A. (1999) Spectroscopy of methyelne blue-hectorite suspensions. *Journal Colloid and Interface Science*, **220**, 103–111.

Jacobs, K.Y. & Schoonheydt, R.A. (2001) Time dependence of the spectra of methylene blue-clay minerals suspensions. *Langmuir*, **17**, 5150–5155.

Jaynes, W.F., Traina, S.J., Bigham, J.M. & Johnston, C.T. (1992) Preparation and characterization of reduced-charge hectorites. *Clays and Clay Minerals* **40**, 397–405.

Johnston, C.T. (2010) Probing the nanoscale architecture of clay minerals. *Clay Minerals* **45**, 245–279.

Johnston, C.T. & Stone, D.A. (1990) Influence of hydrazine on the vibrational modes of kaolinite. *Clays and Clay Minerals* **38**, 121–128.

Johnston, C.T. & Tombacz, E. (2002) Surface chemistry of soil minerals. In: *Soil Mineralogy with Environmental Applications* (J.B. Dixon, & D.G. Schulze, editors). Soil Science Society of America, Madison, Wisconsin, USA, pp. 37–67.

Johnston, C.T., Sposito, G. & Erickson, C. (1992) Vibrational probe studies of water interactions with montmorillonite. *Clays and Clay Minerals*, **40**, 722–730.

Johnston, C.T., Bish, D.L., Eckert, J. & Brown, L.A. (2000) Infrared and inelastic neutron scattering study of the 1.03- and 0.95-nm kaolinite-hydrazine intercalation complexes. *Journal of Physical Chemistry B*, **104**, 8080–8088.

Joussein, E., Petit, S., Churchman, J., Theng, B., Righi, D. & Delvaux, B. (2005) Halloysite clay minerals – A review. *Clay Minerals*, **40**, 383–426.

Kamada, K., Tamamura, Y., Ueno, K., Ohta, K. & Misawa (2007) Enhanced two-photon absorption of chromophores confined in two-dimensional nanospace. *Journal of Physical Chemistry C*, **111**, 11193–11198.

Kaneko, Y., Iyi, N., Bujdak, J. Sasai, R. & Fujita, T. (2004) Effect of layer charge density and aggregation of a cationic laser dye incorporated in the interlayer space of montmorillonite. *Journal of Colloid and Interface Science*, **269**, 22–25.

Kawamata, J., Seike, R., Higashi, T., Inada, Y., Sasaki, J., Ogata, Y., Tani, S. & Yamagishi, A. (2006) Clay templating Langmuir-Blodgett films of a non-amphiphilic ruthenium(II) complex. *Colloids and Surfaces A: Physicochemical and Engineering Aspects*, **284–285**, 135–139.

Kawamata, J., Yamaki, H., Ohshiye, R., Seike, R., Tani, S., Ogata, Y. & Yamagishi, A., (2008) Fabrication of hybrid LB films consisting of a smectite clay and a nonamphiphilic chiral ruthenium complex. *Colloids and Surfaces A: Physicochemical and Engineering Aspects*, **321**, 65–69.

Kawamata, J., Suzuki, Y. & Tenma, Y. (2010) Fabrication of clay mineral-dye composites as nonlinear optical materials. *Philosophical Magazine*, **90**, 2519–1527.

Kim, Y. & Kirkpatrick, R.J. (1997) Na-23 and Cs-133 NMR study of cation adsorption on mineral surfaces: Local environments, dynamics, and effects of mixed cations. *Geochimica et Cosmochimica Acta*, **61**, 5199–5208.

Klika, Z., Weissmannová, H., Čapková, P. & Pospíšil, M. (2004) The rhodamine B intercalation of montmorillonite. *Journal of Colloid and Interface Science*, **275**, 243–250.

Klika, Z., Čapková, P., Horáková, P., Valášková, M., Malý, P., Machán, R. & Pospíšil, M. (2007) Composition, structure and luminescence of montmorillonites saturated with different aggregates of methylene blue. *Journal of Colloid and Interface Science*, **311**, 14–23.

Komadel, P., Madejová, J. & Bujdak, J. (2005) Preparation and properties of reduced-charge smectites – A review. *Clays and Clay Minerals*, **53**, 313–334.

Kotov, N.A. (2001) Ordered layered assemblies of nanoparticles. *Materials Research Society Bulletin*, **26**, 992–997.

Lagaly, G. (2006) Colloid Clay Science. In: *Handbook of Clay Science* (F. Bergaya & B.K.G. Theng, editors). Developments in Clay Science, **1**, Elsevier, Amsterdam, pp. 141–245.

Lagaly, G. & Malberg, R. (1990) Disaggregation of alkylammonium montmorillonites in organic-solvents. *Colloids and Surfaces*, **49**, 11–27.

Lagaly, G., Ogawa, M. & Dekany, I. (2006) Clay mineral organic interactions. In: *Handbook of Clay Science* (F. Bergaya, B.K.G. Theng & G. Lagaly, editors). Elsevier, Amsterdam, pp. 309–378.

Laird, D.A. (1996) Model for crystalline swelling of 2:1 phyllosilicates. *Clays and Clay Minerals* **44**, 553–559.

Laperche, V., Lambert, J.F., Prost, R. & Fripiat, J.J. (1990) High-resolution solid-state NMR of exchangeable cations in the interlayer surface of a swelling mica: ^{23}Na, ^{111}Cd, and ^{133}Cs vermiculites. *Journal of Physical Chemistry*, **94**, 8821–8831.

López Arbeloa, F. & Martinez Martinez, V. (2006a) New fluorescent polarization method to evaluate the orientation of adsorbed molecules in uniaxial 2D layered materials. *Journal of Photochemistry and Photobiology A: Chemistry*, **181**, 44–49.

López Arbeloa, F. & Martinez Martinez, V. (2006b) Orientation of adsorbed dyes in the interlayer spaces of clays. 2. Fluorescence polarization of rhodamine 6G in laponite films. *Chemistry of Materials*, **18**, 1407–1416.

Lotsch, B.V. & Ozin, G.A. (2008a) Clay Bragg Stack Optical Sensors. *Advanced Materials*, **20**, 4079–4084.

Lotsch, B.V. & Ozin, G.A. (2008b) Photonic clays: a new family of functional 1D photonic crystals. *ACS Nano*, **2**, 2065–2074.

Low, P.F. (1961) Physical chemistry of clay-water interactions. *Advances in Agronomy* **13**, 269–327.

Low, P.F. (1980) The swelling of clay: II. Montmorillonites. *Soil Science Society of America Journal*, **44**, 667–676.

Maes, A. & Cremers, A. (1978) Stability of metal uncharged ligand complexes in ion exchangers. Part 3. Complex ion selectivity and stepwise stability constants. *Journal of the Chemical Society, Faraday Transactions 1*, **74**, 2470–2480.

Maes, A. & Cremers, A. (1979) Stability of metal uncharged ligand complexes in ion exchangers. Part 4. Hydration effects and stability changes of copper-ethylenediamine complexes in montmorillonite. *Journal of the Chemical Society, Faraday Transactions 1*, **75**, 513–524.

Maes, A., Marynen, P. & Cremers, A. (1977) Stability of metal uncharged ligand complexes in ion exchangers. Part 1. Quantitative characterization and thermodynamic basis. *Journal of the Chemical Society, Faraday Transactions 1*, **73**, 1297–1301.

Maes, A., Peigneur, P. & Cremers, A. (1978) Stability of metal uncharged ligand complexes in ion exchangers. Part 2. The copper + ethylenediamine complex in montmorillonite and sulphonic acid resin. *Journal of the Chemical Society, Faraday Transactions 1*, **74**, 182–189.

Maes, A., Schoonheydt, R.A., Cremers, A. & Uytterhoeven, J.B. (1980) Spectroscopy of $Cu(en)_2^{2+}$ on clay surfaces. Surface and charge density effects. *Journal of Physical Chemistry*, **84**, 2795–2799.

Malikova, N., Cadene, A., Marry, V., Dubois, E., Turq, P., Zanotti, J.M. & Longeville, S. (2005) Diffusion of water in clays – microscopic simulation and neutron scattering. *Chemical Physics*, **317**, 226–235.

Malikova, N., Cadene, A., Dubois, E., Marry, V., Durand-Vidal, S., Turq, P., Breu, J., Longeville, S. & Zanotti, J.M. (2007) Water diffusion in a synthetic hectorite clay studied by quasi-elastic neutron scattering. *Journal of Physical Chemistry C*, **111**, 17603–17611.

Malikova, N., Dubois, E., Marry, V., Rotenberg, B. & Turq, P. (2010) Dynamics in clays – combining neutron scattering and microscopic simulation. *Zeitschrift fur Physikalische Chemie – International Journal of Research in Physical Chemistry & Chemical Physics*, **224**, 153–181.

Marry, V., Rotenberg, B. & Turq, P. (2008) Structure and dynamics of water at a clay surface from molecular dynamics simulation. *Physical Chemistry Chemical Physics*, **10**, 4802–4813.

Martinez Martinez, V., Salleres, S., Bañuelos Prieto, J. & López Arbeloa, F. (2006) Application of fluorescence with polarized light to evaluate the orientation of dyes adsorbed in layered materials. *Journal of Fluorescence*, **16**, 233–240.

McBride, M.B., Pinnavia, T.J. & Mortland, M.M. (1975) Electron spin resonance studies of cation orientation in restricted water layers on phyllosilicate (smectite) surfaces. *Journal of Physical Chemistry*, **79**, 2430–2435.

Mering, J. (1946) On the hydration of montmorillonite. *Transactions of the Faraday Society*, **42B**, 205–219.

Miao, S., Leeman, H., De Feyter, S. & Schoonheydt, R.A. (2010a) Facile preparation of Langmuir-Blodgett films of water-soluble proteins and hydrated protein-clay films. *Journal of Materials Chemistry*, **20**, 698–705.

Miao, S.D., Leeman, H., De Feyter, S. & Schoonheydt, R.A. (2010b) Three-component Langmuir-Blodgett films consisting of surfactant, clay mineral and lysozyme: construction and characterization. *Chemistry, a European Journal*, **16**, 1–10.

Mooney, R.W., Keenan, A.G. & Wood, L.A. (1952a) Adsorption of water vapor by montmorillonite. I. Heat of desorption and application of BET theory. *Journal of the American Chemical Society*, **74**, 1367–1374.

Mooney, R.W., Keenan, A.G. & Wood, L.A. (1952b) Adsorption of water vapor by montmorillonite. II. Effect of exchangeable ions and lattice swelling as measured by X-ray diffraction. *Journal of the American Chemical Society*, **74**, 1371–1374.

Mortland, M.M., Fripiat, J.J., Chaussidon, J. & Uytterhoeven, J.B. (1963) Interaction between ammonia and the expanding lattices of montmorillonite and vermiculite. *Journal of Physical Chemistry*, **67**, 248–258.

Nagy, N.M. & Konya, J. (2004) The adsorption of valine on cation-exchanged montmorillonites. *Applied Clay Science*, **25**, 57–69.

Nakamura, Y., Yamagishi, A., Iwamoto, T. & Koga, M. (1988) Adsorption properties of montmorillonite and synthetic saponite as probing materials in liquid column chromatography. *Clays and Clay Minerals*, **36**, 530–536.

Nakamura, Y., Yamagishi, A. & Iwamoto, T. (1990) Clay column chromatography for optical resolution: partial resolution of 1,1'-binaphthol on optically active $[Co(phen)_{3-x}]^{n+}$-montomorillonite columns. *Clay Science*, **8**, 17–23.

Nemecz, E. (1981) *Clay Minerals*. Akadémiai Kiado, Budapest, 547 pp.

Newman, A.C.D., editor (1987) The interaction of water with clay mineral surfaces. In: *Chemistry of Clays and Clay Minerals*, Monograph **6**. Mineralogical Society, London, 480 pp.

Ogata, Y., Kawamata, J., Chong, C.-H., Mahikari, M., Yamagishi, A. & Saito, G. (2002) Optical second harmonic generation of zwitterionic molecules aligned on clays. *Molecular Crystals, Molecular Liquids*, **376**, 245–250.

Ogata, Y., Kawamata, J., Chong, C.-C., Yamagishi, A. & Saito, G. (2003a) Strcutural features of a clay film hybridized with a zwitterionic molecule as analyzed by second-harmonic generation behavior. *Clays and Clay Minerals*, **51**, 181–185.

Ogata, Y., Kawamata, J., Yamagishi, A., Chong, C.-H. & Saito, G. (2003b) A novel film with a non-centrosymmetric molecular alignment of D-π-A zwitterionic molecules fabricated at an air-water interface. *Synthetic Metals*, **133–134**, 671–672.

Ogawa, M. & Kuroda, K. (1995) Photofunctions of intercalation compounds. *Chemical Reviews*, **95**, 399–438.

Okamoto, K., Tamura, K., Takahashi, M. & Yamagishi, A. (2000) Preparation of a clay-complex hybrid film by the Langmuir-Blodgett method and its applications as an electrode modifier. *Colloids and Surfaces A: Physicochemical and Engineering Aspects*, **169**, 241–249.

Okamoto, K., Taniguchi, M., Takahashi, M. & Yamagishi, A. (2001) Studies on energy transfer from chiral, polypyridyl Ru(II) to Os(II) complexes in cast and Langmuir-Blodgett films. *Langmuir*, **17**, 195–201.

Pizzey, C., Klein, S., Leach, E., van Duijneveldt, J.S. & Richardson, R.M. (2004) Suspensions of colloidal plates in a nematic liquid crystal: a small angle X-ray scattering study. *Journal of Physics – Condensed Matter*, **16**, 2479–2495.

Poinsignon, C., Cases, J.M. & Fripiat, J.J. (1978) Electrical-polarization of water molecules adsorbed by smectites. An infrared study. *Journal of Physical Chemistry*, **82**, 1855–1860.

Porion, P., Faugere, A.M., and Delville, A. (2009) Long-time scale ionic dynamics in dense clay sediments measured by the frequency variation of the Li-7 multiple-quantum NMR relaxation rates in relation with a multiscale modeling. *Journal of Physical Chemistry C*, **113**, 10580–10597.

Porion, P., Faugere, A.M. & Delville, A. (2005) Analysis of the degree of nematic ordering within dense aqueous dispersions of charged anisotropic colloids by Na-23 NMR spectroscopy. *Journal Of Physical Chemistry B*, **109**, 20145–20154.

Porion, P., Michot, L.J., Faugere, A.M. & Delville, A. (2007) Structural and dynamical properties of the water molecules confined in dense clay sediments: A study combining H-2 NMR spectroscopy and multiscale numerical modeling. *Journal of Physical Chemistry C*, **111**, 5441–5453.

Prost, R., Koutit, T., Benchara, A. & Huard, E. (1998) State and location of water adsorbed on clay minerals: Consequences of the hydration and swelling-shrinkage phenomena. *Clays and Clay Minerals*, **46**, 117–131.

Quiquampoix, H. & Burns, R.G. (2007) Interactions between proteins and soil mineral surfaces: environmental and health consequences. *Elements*, **3**, 401–406.

Quiquampoix, H., Staunton, S., Baron, M.H. & Ratcliffe, R.G. (1993) Interpretation of the pH dependence of protein adsorption on clay mineral surfaces and its relevance to the understanding of extra-cellular enzyme activity in soils. *Colloids and Surfaces A: Physicochemical and Engineering Aspects*, **75**, 85–93.

Ras, R.H.A., Johnston, C.T., Franses, E.I., Ramaekers, R., Maes, G., Foubert, P., De Schryver, F.C. & Schoonheydt, R.A. (2003) Polarized infrared study of hybrid Langmuir-Blodgett monolayers containing clay mineral nanoparticles. *Langmuir*, **19**, 4295–4302.

Ras, R.H.A., Umemura, Y., Johnston, C.T., Yamagishi, A. & Schoonheydt, R.A. (2007) Ultrathin hybrid films of clay minerals. *Physical Chemistry Chemical Physics*, **9**, 918–932.

Rinnert, E., Carteret, C., Humbert, B., Fragneto-Cusani, G., Ramsay, J.D.F., Delville, A., Robert, J.L., Bihannic, I., Pelletier, M. & Michot, L.J. (2005) Hydration of a synthetic clay with tetrahedral charges: A multidisciplinary experimental and numerical study. *Journal Of Physical Chemistry B*, **109**, 23745–23759.

Rotenberg, B., Morel, J.P., Marry, V., Turq, P. & Morel-Desrosiers, N. (2009) On the driving force of cation exchange in clays: Insights from combined microcalorimetry experiments and molecular simulation. *Geochimica et Cosmochimica Acta*, **73**, 4034–4044.

Russell, J.D. & Farmer, V.C. (1964) Infra-red spectroscopic study of the dehydration of montmorillonite and saponite. *Clay Minerals Bulletin*, **5**, 443–464.

Ruiz-Hitzky, E., Ariga, K. & Lvov, Y. (editors) (2008) *Bio-Inorganic Hybrid Nanomaterials: Strategies, Synthesis, Characterization and Applications*. Wiley, New York, 503 pp.

Sato, H., Yamagishi, A. & Kato, S. (1992) Monte Carlo simulations for the interactions of metal complexes with silicate sheets of clay. Comparison of binding sites between tris(1, 10-phenanthroline)metal(II) and tris(2,2'-bipyridyl)metal(II) chelates. *Journal of the American Chemical Society*, **114**, 10933–10940.

Sato, H., Yamagishi, A. & Kawamura, K. (2001) Molecular simulation for flexibility of a single clay layer. *Journal of Physical Chemistry B*, **105**, 7990–7997.

Sato, H., Hizoe, Y., Tamura, K. & Yamagishi, A. (2005) Orientational tuning of a polypyridyl Ru(II) complex immobilized on a clay surface towards chiral discrimination. *Journal Physical Chemistry B*, **109**, 18935–18941.

Schoonheydt, R.A. & Johnston, C.T. (2007) Surface and interface chemistry of clay minerals. In: *Handbook of Clay Science* (F. Bergaya & B.K.G. Theng, editors). Developments in Clay Science, **1**, Elsevier, Amsterdam, pp. 87–112.

Schoonheydt, R.A. & Pelgrims, J. (1983) Preparation, spectroscopy and reaction with O_2 of $Co(en)_2^{2+}$ on the surface of hectorite. *Journal of the Chemical Society, Faraday Transactions 2*, **79**, 1169–1180.

Schoonheydt, R.A., Pelgrims, J., Heroes, Y. & Uytterhoeven, J.B. (1978) Characterization of tris(2,2'-bipyridyl) ruthenium (II) on hectorite. *Clay Minerals*, **13**, 435–438.

Schoonheydt, R.A., Velghe, F. & Uytterhoeven, J.B. (1979a) Characterization of $[Ni(en)_x]^{2+}$ (x = 1, 2, 3; en = ethylenediamine) on the surface of montmorillonite. *Inorganic Chemistry*, **18**, 1842–1847.

Schoonheydt, R.A., Velghe, F., Baerts, R. & Uytterhoeven, J.B. (1979b) Complexes of diethylenetriamine (dien) and tetraethylenepentamine (tetren) with Cu(II) and Ni(II) on hectorite. *Clays and Clay Minerals*, **27**, 269–278.

Schoonheydt, R.A., De Pauw, P., Vliers, D. & De Schryver, F.C. (1984) Luminescence of tris(2,2'-bipyridyl)-ruthenium(II) in aqueous clay mineral dispersions. *Journal Physical Chemistry*, **88**, 5113–5118.

Schramm, L.L. & Kwak, J.C.T. (1982) Influence of exchangeable cation composition on the size and shape of montmorillonite particles in dilute suspension. *Clays and Clay Minerals*, **30**, 40–48.

Sobolev, O., Buivin, F.F., Kemner, E., Russina, M., Beuneu, B., Cuello, G.J. & Charlet, L. (2010) Water-clay surface interaction: A neutron scattering study. *Chemical Physics*, **374**, 55–61.

Sposito, G. (1972) Thermodynamics of swelling clay-water systems. *Soil Science*, **114**, 243–249.

Sposito, G. & Prost, R. (1982) Structure of water adsorbed on smectites. *Chemical Reviews*, **82**, 553–573.

Sposito, G., Prost, R. & Gaultier, J.P. (1983) Infrared spectroscopic study of adsorbed water on reduced-charge Na/Li montrmorillonites. *Clays and Clay Minerals*, **31**, 9–16.

Środoń, J. & McCarty, D.K. (2008) Surface area and layer charge of smectite from CEC and EGME/H_2O-retention measurements. *Clays and Clay Minerals*, **56**, 155–174.

Suzuki, Y., Hizakawa, S., Sakamoto, Y., Kawamata, J., Kamada, K. & Ohta, K. (2008) Hybrid films consisting of a clay and a diacetylenic two-photon absorption dye. *Clays and Clay Minerals*, **56**, 487–493.

Swenson, J., Bergman, R. & Howells, W.S. (2000) Quasielastic neutron scattering of two-dimensional water in a vermiculite clay. *Journal of Chemical Physics*, **113**, 2873–2879.

Szabo, T., Mitea, R., Leeman, H., Premachandra, G.S., Johnston, C.T., Szekeres, M., Dekany, I. & Schoonheydt, R.A. (2008) Adsorption of protamine and papain proteins on saponite. *Clays and Clay Minerals*, **56**, 494–504.

Szabo, T., Wang, J., Volodin, A., van Haesendonck, C., Dekany, I. & Schoonheydt, R.A. (2009) AFM study of smectites in hybrid Langmuir-Blodgett films: Saponite, Wyoming bentonite, hectorite, and Laponite. *Clays and Clay Minerals*, **57**, 706–714.

Takagi, S., Shimada, T., Yui, T. & Inoue, H. (2001) High density adsorption of porphyrins into clay layers without aggregation: characterization of smectite-cationic porphyrin complex. *Chemistry Letters*, **2**, 1218–129.

Takagi, S., Shimada, T., Eguchi, M., Yui, T., Yoshida, H., Tryk, D.A. & Inoue, H. (2002a) High-density adsorption of porphyrins on clay layer surfaces without aggregation: the size-matching effect. *Langmuir*, **18**, 2265–2272.

Takagi, S., Tryk, D.A. & Inoue, H. (2002b) Photochemical energy transfer of cationic porphyrin complexes on clay surfaces. *Journal of Physical Chemistry B*, **106**, 5455–5460.

Takagi, S., Eguchi, M., Tryk, D.A. & Inoue, H. (2006) Porphyrin photochemistry in inorganic/organic hybrid materials: clays, semicondcutors, nanotubes and mesoporous materials. *Journal Photochemistry and Photobiology C: Photochemistry Reviews*, **7**, 104–126.

Tamura, K., Setsuda, H., Taniguchi, M. & Yamagishi, A. (1999) Application of the Langmuir-Blodgett technique to prepare a clay-metal complex hybrid film. *Langmuir*, **15**, 6915–6920.

Theng, B.K.G. (1974) *Clay-Organic Reactions*. A. Hilger, London.

Ulman, A. (1991) *An Introduction to Ultrathin Organic Films: from Langmuir-Blodgett to Self-Assembly*. Academic Press, New York.

Umemura, Y. (2002) Hybrid films of a clay mineral and an iron (II) complex cation prepared by a combined method of the Langmuir-Blodgett and self-assembly techniques. *Journal Physical Chemistry B*, **106**, 11168–11171.

Umemura, Y., Yamagishi, A., Schoonheydt, R., Persoons, A. & De Schryver, F.C. (2002) Langmuir-Blodgett films of a clay mineral and ruthenium (II) complexes with a noncentrosymmetric structure. *Journal of the American Chemical Society*, **124**, 992–997.

Van Duffel, B., Jacobs, K.Y. & Schoonheydt, R.A. (1997) Methylene blue-hectorite complexes: from suspensions to films. *Proceedings of the 11th International Clay Conference* (H. Kodama, editor), pp. 475–481.

Velghe, F., Schoonheydt, R.A., Uytterhoeven, J.B., Peigneur, P. & Lunsford, J.H. (1977) Spectroscopic characterization and thermal stability of copper (II) ethylenediamine complexes on solid surfaces. 2. Montmorillonite. *Journal of Physical Chemistry*, **81**, 1187–1194.

Villemure, G. (1990) Effect of negative surface-charge densities of smectite clays on the adsoprtion isotherms of racemic and enantiomeric tris(2,2'-bipyridyl)ruthenium(II) chloride. *Clays and Clay Minerals*, **38**, 622–630.

Vliers, D.P., Van De Vliet, B., Schoonheydt, R.A. & De Schryver, F.C. (1990) Luminescence quenching of $Ru(bipy)_3^{2+}$ on clays with Cu^{2+} and $Fe(CN)_6^{3-}$. Proceedings of the 9th International Clay Conference (V.C. Farmer and Y. Tardy, editors). *Sciences Géologiques Mémoires*, **86**, 51–58.

von Buzagh, A. (1929) On current birefringence and thixotropics of bentonite suspensions. *Kolloid-Zeitschrift*, **47**, 223–229.

Weiss, C.A., Jr., Kirkpatrick, R.J. & Altaner, S.P. (1990) The structural environments of cations adsorbed onto clays: Cesium-133 variable-temperature MAS NMR spectroscopic study of hectorite. *Geochimica et Cosmochimica Acta*, **54**, 1655–1659.

Wilson, M.J. (editor) (1994) *Clay Mineralogy: Spectroscopic and Chemical Determinative Methods*. Chapman & Hall, London, 367 pp.

Xu, W., Johnston, C.T., Parker, P. & Agnew, S.F. (2000) Infrared study of water sorption on Na-, Li-, Ca- and Mg-exchanged (SWy-1 and SAz-1) montmorillonite. *Clays and Clay Minerals*, **48**, 120–131.

Yamagishi, A. (1982) Racemic adsorption of dicyanobis(1,10-phenanthroline) iron (II) on colloidally dispersed sodium montmorillonite. *Inorganic Chemistry*, **21**, 1778–1782.

Yamagishi, A. (1983) Chirality recognition of a clay surface by an optically active metal chelate. *Journal of the Chemical Society, Dalton Transactions*, **4**, 679–681.

Yamagishi, A. (1985) Chromatographic resolution of enantiomers having aromatic groups by an optically active clay-chelate adduct. *Journal of the American Chemical Society*, **107**, 732–734.

Yamagishi, A. (1987) Optical resolution and asymmetric syntheses by use of adsorption on clay minerals. *Journal of Coordination Chemistry*, **16**, 131–211.

Yamagishi, A. (1993) Chirality recognition by a clay surface modified with an optically active metal chelate. In: *Dynamic Processes on Solid Surface* (K. Tamaru, editor). Plenum Press, New York, pp. 307–347.

Yan, L.B., Low, P.F. & Roth, C.B. (1996) Enthalpy changes accompanying the collapse of montmorillonite layers and the penetration of electrolyte into interlayer space. *Journal of Colloid and Interface Science*, **182**, 417–424.

Yariv, S. & Cross, H. (2001) *Organo-Clay Complexes and Interactions*. Marcel Dekker, New York, 566 pp.

EMU Notes in Mineralogy, Vol. 11 (2011), Index, 371–376

Subject and Author Index

Numbers refer to the page where a definition or explanation of, and/or a *figure* or a **table** for, a given subject is found. Author names are given with the number of the first page of the paper in which they are involved.

1:1 layer structure, 14
2:1 layer structures, 31

A
AAM, alkylammonium method, 245
Absorbance spectra of, aqueous methylene blue-smectite dispersions, *355*, hectorite dispersions, *356*
Adsorption, and intercalation, 272, phenomena on phyllosilicate surfaces, 229
AEM-TEM, analytical electron microscopy-transmission electron microscopy, 243
AFM, atomic force microscopy, 314, *316*, *317*, *327*, 338, *339*
Aggregates, 251
AIPEA, 261
Aldega, Luca, 285
Alkylammonium ions in the interlayer space, 245
Allophane, 46
Almeida Paz, Filipe A., 123
Amesite, 28
Amino acids on clay-mineral surfaces, 360
Antigorite, 26, *27*
Aparacio-Galán-Ferrell index, 10
Árkai index, 9
Armbrusterite, 29
Astrophyllite, 139
Atomistic methods, applied to phyllosilicates, 203, 215
AXANES, 285

B
Bafertisite, 132, *138*
Bementite, 29, *30*
Bergaya, Faïza, 259
Berthierine, 29
Brigatti, Maria Franca, 1
Brindleyite, 31
Brucite, *252*
BWA, Bertaut-Warren-Averbach method, 285, 290

C
Carlosturanite, 28
Caryopilite, 31
Catalysis, applications of intercalation/deintercalation processes, 278
Cationic dyes, 357
CEC, 237
CFF91, 208
Charge of fundamental particles, 253
Chiral clay minerals, 347, *349*
Chlorite, 40, 177, 86, 97
Christidis, George E., 237
Chrysotile, 26
Cis-vacant octahedral site, 11
Cis-vacant polymorphism in dioctahedral phyllosilicates, 223
Classic mechanics, 206
Classification, of clay minerals' layer charge, **263**, of phyllosilicates, 336, of planar hydrous phyllosilicates, **82**, scheme for natural smectite, **242**, scheme of dioctahedral smectites, **241**, of non-planar hydrous phyllosilicates, **108**
Clay Minerals Society, 261
Clay minerals, spectroscopic properties of, 225, surface properties of, 335
Clay, structure of the interlayer, 304
CLAYFF, 175, 208
Clay-polymer nanocomposites, 259, 275
Clinochlore, structural interpretation of, *317*, AFM images, *317*
Co-intercalation of two compounds, 271
Commensurate polytype fragments, 183
Complexation, 270
Computational atomistic methods applied to phyllosilicates, 203
Computational mineralogy, 204
Cookeite, 44
CPN, clay-polymer nanocomposites, 259, 275
Cronstedtite, 28, 177
Crystal thickness measured by XRD, 288
CVFF, 208
Cylindrical structures, 107

D
Data management for analysis experiments, 205
Dehydroxylation-rehydroxylation of phyllosilicates, 227, *228*
Delindeite, 134, *135*
Density functional theory, 210
Dickite, octahedral sites in, *15*, 19
DIFFaX, 10, 160
Diffraction, order and disorder effects on, 102

DOI: 10.1180/EMU-notes.11.index

Diffuse reflectance spectra, of $Cu(en)_2^{2+}$ on Camp Berteau montmorillonite, *346*
Discover, 208
DL_POLY, 208
DMOL, 211
DMSO, dimethyl sulphoxide, 264
DTA, differential thermal analysis, 128
Dye on clay-mineral surfaces, 353

E
ED, *see* electron diffraction
EFM, electric force microscopy, 313
Electron diffraction, 11, Co asbolane, *186*
Electron paramagnetic resonance, EPR, 337
Electron spin resonance, ESR spectra of $Co(en)_2^{2+}$ on Camp Berteau hectorite films, *347*
Electron transfer and redox reactions, 270
Elmi, Chiara, 1
Electron probe microanalysis, 243
EPMA, *see* electron probe microanalysis
EPR, *see* electron paramagnetic resonance
ESR, *see* electron spin resonance
EXAFS, *see* Extended X-ray Absorption Fine Structure
Extended X-ray Absorption Fine Structure, EXAFS, 216, 294, 303
Exchange in swelling clays, 102
Exchange selectivity between inorganic cations, 268
Exfoliated structures, 277
Experimental data compared with simulation, 330
Exsolution, 81
EXSY, 11

F
Ferripyrophyillite, 32
FM-AFM, *see* frequency modulation AFM
FMS, *see* full multiple scattering
Forcite, 208
Fourier transform infrared spectroscopy, FTIR, 337, 344
Fraipontite, 31
Frequency modulation atomic force microscopy, 315, *316*
FTIR, *see* Fourier transform infrared spectroscopy
Full multiple scattering, FMS, 301
Fullerene, 306
Fundamental particles, 251, *252*

G
GALOPER, 10
GAMESS, 211
GAUSSIAN, 211
GAUSSIAN03, 330
Geometry, layered clay minerals, 261
GIXRD, *see* Grazing Incidence X-ray Diffraction
Grafting, 271
Graphene and its derivative structure, 306, 308
Grazing Incidence X-ray Diffraction, GIXRD, 285, 292, *293*
Greenalite, 29
Guggenheim, Stephen, 73
Guidottiite, 29
GULP, 208

H
Halloysite, 21
Hectorite, 241, 38
Heterophyllosilicates, 136
High-Resolution Transmission Electron Microscopy, HRTEM, 307, images, kaolinite, *156*, pyrophyllite, *179*
Hisingerite, 23
HONDO, 211
HRTEM see High-Resolution Transmission Electron Microscopy
Hydrous phyllosilicates, 73
Hydroxides, interaction with organic molecules, 313
HyperCHEM, 211

I
Idealized shapes of smectite particles, *238*
Illite, structure of the interlayer, 304
Illite-smectite, *156*
Imogolite, 46, nanotube, *6*
Infrared, IR, 216, *217*
Interatomic potentials, **207**
Intercalation, 100, /deintercalation processes, 278, of CPN from solvents, 276, processes of layered minerals, 259, 264, 265, processes, 285
Interlayer cations, 343
Interstratification, 100, 154
Interstratified phyllosilicates, **100**
Ion exchange 268, influence of layer charge on, 247
IR, *see* Infrared

J
Johnston, Cliff, 335
Jonesite, 131
Junction probability diagram, *163*

K
K-saturation method, KSM, 246
Kalifersite, 46
Kanemite, 259
Kelvin probe force microscopy, KPFM, *320*
Kaolin, 14, 83, 96
Kaolinite, HRTEM image, *156*, non-swelling minerals, 264, octahedral sites in, *15*
Kellyite, 29
Kenyaite, 259

KPFM, *see* Kelvin probe force microscopy
KSM, *see* K-saturation method
Kübler index, 9

L
Lagaly, Gerhard, 259
Lamellar structure, defective, 153
Lamphrophyllite, 132, *133*
LAMMPS, 208
Langmuir-Blodgett technique, 350, *351*
Lanson, Bruno, 151
Laurora, Angela, 1
Layer charge, 12, 80, of smectite, 237, 243
Layer periodicity in halloysite, *21*
Layer silicates, 1
Layer stacking in a 1*H* polytype, *158*
Layer-charge distribution in montmorillonite, *88*
LayerCharge program, 246
Layered Double Hydroxides, LDH, 180, 259
Layered minerals, intercalation processes of, 259
Layered oxides, 187
Layered titanosilicates, 123, *125, 126*
LDH, *see* Layered Double Hydroxides
Liètard index, 10
Lin, Zhi, 123
Lintisite-type structures, 128, *129*
Lizardite, *24*

M
Magadiite, 259
Magnesioastrophyllite, *140*
Malferrari, Daniele, 1
Maximum possible degree of disordering, 159
Medicine, applications of intercalation/deintercalation processes, 278
Melt intercalation, 276
Mering principle, *106*
Methylene blue-clay systems, 353, *354*
Metropolis-Monte Carlo method, 205
Mica, 33, 84, 91, 266, structure of the interlayer, 304
Minimization of geometry, 212
Minnesotaite diffraction pattern, *113*
MINTEQA2, 205
Modulated phyllosilicates, 107
MODXRSD, 10
Molecular cluster models, 211
Molecular dynamics, first-principles, 331, simulations, 213
Monte Carlo simulations, 205, 213, *219, 220*
Montmorillonite, 267, 88
Moro, Daniele, 313
Mössbauer, 216
Mottana, Annibale, 285
MUDMASTER, 10, 285
Murmanite, *139*

N
Nacrite, 20
Nafertisite, *141*
Nanocharacterization techniques, 313, 314
Nanoconfined H_2O molecules in clay mineral interlayers, 342
Nanomechanical properties, 321
Nano-size layer silicates, 1
Narsarsukite, 128
Natisite, 130
Nepouite, 31
Near-edge Extended X-ray Absorption Fine Structure, NEXAFS, 285, 294, *307*
NEWCHEM, 211
NEWMOD, 10, 165
NEXAFS, *see* Near-edge Extended X-ray Absorption Fine Structure
NMR, *see* Nuclear Magnetic Resonance
Non-swelling minerals, 264
Nuclear Magnetic Resonance, NMR, 206, 295, 343

O
Octahedral and tetrahedral sheets in 1:1 and 2:1 clay minerals, *336*
Octahedral *cis*- and *trans*-configurations in smectite, 242
Octahedral sheet, 2, *3*
Octahedral sites for kaolin minerals, *96*
Odinite subgroup, 23
OH orientations on the octahedral surface of kaolinite, *17*
Order-disorder, 74, in layer stacking, *8*
Ordering in cation substitutions in phyllosilicates, 215
Organic molecules, interaction with layer silicates, 313, 325
Organic-inorganic composite membranes, 124
Organo-clay minerals, 260, 268, *269*
Oxides, interaction with organic molecules, 313

P
Palygorskite, *4*, 46, -sepiolite minerals, *111*
Peakforce Quantitative Nanomechanical Mapping, 322, *324*
Periodical models for crystalline solids, 212
PE-XAFS, *see* Pre-edge X-ray Absorption Fine Structure
P-EXAFX, *see* Polarized Extended X-ray Absorption Fine Structure
PHREEQE, 205
Phyllosilicate, 75, *77*, 100, classification of, 336, computational atomisitic methods applied to, 203, structure of the interlayer, 304
PILC, pillared clays, 259, 273
Pillared clays, 259, 273
Planar trioctahedral 1:1 layers, 94
Plastic viscosity, **249**

Polarized Extended X-ray Absorption Fine Structure, P-EXAFS, 285, 294
Polysomatic structures, 107
Polysomes, 113
Polytypes of chlorite, *98*, of mica, *92*
Porosity, layered clay minerals, 261
Pre-edge X-ray Absorption Fine Structure, PE-XAFS, 298
Proteins on clay-mineral surfaces, 360
Protonation, 269
Pyrophyllite, 31, 83, 99, crystal structure of, *228*
Pyrosmalite group, 30

Q
QENS, *see* Quasi-Elastic Neutron Scattering
Quantum mechanical methods, 208
Quasicrystals, 251
Quasi-Elastic Neutron Scattering, QENS, 342

R
Raman spectroscopy, 338
Random stacking faults in XRD patterns, *155*
Range and Weiss index, 10
Rectorite, *252*
Reichweite, 9, 101, 161
Rietveld code, *BGMN*, 160
Rocha, João, 123
Rolled, tubular and cylindrical structures, 112
Rolling mechanism in halloysite, *22*

S
SAD, *see* Selected Area Diffraction
Sainz-Díaz, C. Ignacio, 203
Saponite, 37, *360*
Sauconite, 38
Scanning Electron Microscopy, SEM, 338
Scanning Probe Microscopy, SPM, 313, *322*
Schoonheydt, Robert, 335
Second harmonic generation, SHG, 351, *352*
Selected area diffraction, SAD, 10
SEM, *see* Scanning Electron Microscopy
Sepiolite, *4*
Serpentine, 23, 83, *95*
SFM, *see* Structural Formula Method
SHG, second harmonic generation, 351, *352*
Smectite, 35, 86, 99, dioctahedral, classification, **241**, natural, classification, **242**, layer charge, 237
Solid solution, 81
Solvation in swelling clays, 102
Spartan, 211
Spectroscopic properties of clay minerals, 225
SPM, *see* Scanning Probe Microscopy
Stacking defects, 153
Stacking, 91, *95*
Stevensite, 39, 242
Stoch index, 9
Structural characterization of mixed layers, 169
Structural distortions, 78, *79*
Structural features of smectite, *238*, 239
Structural Formula Method, SFM, 243, 244
Structural overview, layer silicates, 2
Structure defects, 153
Structure model describing a twinning fault in a 1*H* polytype, *159*
Sudoite, 45
Surface potential, 318
Surface properties of clay minerals, 335
Surfaces, layered clay minerals, 261
Swelling clay minerals, intercalation in, 267
Swinefordite, 39
SXLSQA, 205

T
Talc, 31, 83, 99
TDFM, *see* Transverse Dynamic Force Microscopy
TEM, 11, 97, 338, image of illite-smectite, *156*, image of smectite, *253*, image of Wyoming montmorillonite, *262*, image of yofortierite, *114*
Tetrahedral sheet, 2, *3*
Thermal disorder, 74
Titanosilicates, 123
TOT minerals, 264
Transition metal ion complexes, 345
Translation vection fluctuations, *154*
Trans-vacant octahedral site, 11
Trans-vacant polymorphism in dioctahedral phyllosilicates, 223
Transverse dynamic force microscopy, TDFM, 323
Trioctahedral chlorite, 43
Tuperssuatsiaite, 47
Turbostratic effects, smectite, 99

U
Ulian, Gianfranco, 313
UV-VIS-NIR, UV-VIS near infrared spectroscopy, 337

V
Valdré, Giovanni, 313
Vermiculite, 39, 86, 98
Vibration spectra, 214
Viscosity, influence of layer charge on, 248
Vuonnemite, crystal packing, *137*

W
Water molecules adsorbed on kaolinite, *18*
Water–clay mineral interactions, 338
Water-smectite properties, 247
Water-sorption isotherms of montmorillonite, 342
Willemsite, 33

X

XAFS, *see* Absorption Fine Structure
XANES, *see* Absorption Near Edge Spectroscopy
XAS, *see* X-ray Absorption Spectroscopy
X-ray Absorption Fine Structure, XAFS, 294
X-ray Absorption Near Edge Spectroscopy, XANES, 300, *298*
X-ray Absorption Spectroscopy, XAS, 285, 294
X-ray diffraction, XRD, 9, 126, 152, 160, *164, 167, 170, 172, 174, 175, 185, 188, 189*, 206, 243, 287, 340, patterns showing disorder effects, *77, 103, 105*, profiles, modelling, 151
XRD *see* X-ray diffraction

Y

Yofortierite, TEM image of, *114*
Young's modulus measurement, *324*

Z

Zeolites, 123